High Latitude Limnology

Developments in Hydrobiology 49

Series editor
H. J. Dumont

High Latitude Limnology

Edited by
W. F. Vincent and J. C. Ellis-Evans

Reprinted from Hydrobiologia, vol. 172 (1989)

Kluwer Academic Publishers

Dordrecht / Boston / London

Library of Congress Cataloging in Publication Data

High latitude limnology / editors W.F. Vincent, J.C. Ellis-Evans.
p. cm. -- (Developments in hydrobiology ; 49)
Includes index.
Proceedings of the first High Latitude Limnology Symposium, held Feb. 8-14, 1987, during the 23rd Congress of the Societas Internationalis Limnologiae in Hamilton, New Zealand.
Reprinted from Hydrobiologia, 1988.
ISBN 9061936608
1. Limnology--Polar regions--Congresses. I. Vincent, Warwick F. II. Ellis-Evans, J. C. (J. Cynan), 1951- III. High Latitude Limnology Symposium (1st : 1987 : Hamilton, New Zealand)
IV. Series.
QH84.1.H54 1988
574.5'26322'0911--dc19
DNLM/DLC
for Library of Congress 88-13003
CIP

ISBN 90-6193-660-8

Kluwer Academic Publishers incorporates the publishing programmes of Dr W. Junk Publishers, MTP Press, Martinus Nijhoff Publishers, and D. Reidel Publishing Company.

Distributors

for the United States and Canada: Kluwer Academic Publishers, 101 Philip Drive, Norwell, MA 02061, U.S.A.
for all other countries: Kluwer Academic Publishers Group, P.O. Box 322, 3200 AH Dordrecht, The Netherlands

PRINTED IN BELGIUM

Contents

Preface

This volume is derived from a symposium on High Latitude Limnology held during the 23rd Congress of the Societas Internationalis Limnologiae in Hamilton, New Zealand. The symposium stemmed from our belief that an exchange of views between limnologists working in the north and south polar zones would be timely and productive. Over the last decade there has been a major increase in the limnological research effort in Antarctica with the expansion of science programmes there by many nations from both the northern and southern hemispheres. Freshwater research has also continued to develop in the Arctic, stimulated by the need for basic information to assess environmental impacts of the oil industry, mining, urbanisation and other human activities. By bringing together aquatic investigators from both poles we hope to draw attention to the distinctive features that high latitude systems hold in common, and to the marked contrasts between and within each zone.

The dominant impression from the assemblage of papers presented here is one of great limnological diversity. The studies include clear, turbid and brown water rivers in the sub Arctic (LaPerriere, Van Nieuwenhuyse & Anderson); chlorophyte dominated streams in the maritime Antarctic (Hawes); streams on the antarctic continent lined with thick cyanobacterial mats (Howard-Williams & Vincent); meromictic waters in the Arctic (Ouellet, Dickman, Bisson & Pagé) and Antarctic (e.g. Gallagher, Burton & Calf); tundra ponds on the Yukon Delta (Maciolek); hypersaline pools in Antarctica (Torii, Nakaya, Matsubaya, Matsumoto, Masuda, Kawano & Murayama) and a range of freshwater lakes in both regions. Even at one geographical location, Broady's studies on Ross Island, in McMurdo Sound, Antarctica, underscore the variety of aquatic and semi-aquatic habitats that act as major foci for biological activity: shallow freshwater or saline lakes and pools; hypersaline rookery ponds; melting snowbanks enriched by nearby penguin colonies; shoreline ponds derived from sea ice; cryoconite pools in glacier ice; and water-flushed soil, rock and stone environments, as well as persistent flowing waters in well defined perennial stream channels.

Despite this diversity a common theme throughout this volume is the pervasive influence of the extreme solar radiation and temperature cycles. The abrupt seasonality of light and temperature at polar latitudes is clearly illustrated by a continuous annual record obtained by Wharton, McKay & Simmons from the Dry Valleys of Antarctica: at this location (77° 38′ S) there are c. 4 months of continuous sunlight and c. 4 months of continuous darkness with winter temperatures falling to −40 °C. This strong seasonal forcing restricts many high latitude ecosystems to a brief period of activity each year: at Barrow Point, Alaska, the tundra ponds only thaw out for the months July to September (Alexander, Whalen & Klingensmith); on Signy Island the period of streamflow is restricted to November-March (Hawes) whilst the lakes there are only ice-free from January-April (Ellis-Evans & Lemon).

In other polar environments liquid water persists throughout the year beneath a thick ice-cap, but the ice (plus snow) has a wide-ranging influence on ecosystem processes. The increased hydraulic stability in these permanently ice-covered lakes favours a highly stratified water column with major depth variations in many properties, illustrated here, for example, by microbial and organic carbon distributions

(Matsumoto), zooplankton community structure (Bayly & Eslake), inorganic nutrient and metal concentrations (Green, Gardner, Ferdelman, Angle, Varner & Nixon) and phytoplankton distribution and physiology (Priscu).

The attenuating effect of the snow and ice cover on energy supply further influences the timing and intensity of biological processes (e.g. Ellis-Evans & Lemon; Priscu). Low summer temperatures particularly in the Antarctic, additionally affect the activity, species composition and population size of the biota. In Dry Valley streams, for example, the near-zero water temperatures depress autotrophic growth processes, but may favour the high standing stock of cyanobacterial biomass by more strongly inhibiting loss processes such as respiration and bacterial decomposition (Vincent & Howard-Williams).

Although the seasonal cycles of population growth and metabolism are likely to be much more pronounced in the polar zones than at temperate latitudes, full annual data sets are still very rare. The contributions in this volume show a strong bias towards mid-summer data collection and reflect the extreme logistic difficulties of sampling these harsh and often remote environments at other times of year: only three papers present mid-winter data, one using automated instruments.

If seasonal records are sparse then long term records for high latitude lakes and streams are virtually non-existent, yet there are reasons to believe that polar aquatic ecosystems may be more prone to large interannual fluctuations than in the temperate zone. Air temperatures around the margins of Antarctica hover near 0 °C throughout the summer growing season and small changes in local climate are likely to have a major impact on the timing and extent of meltwater discharge through streams and into lakes and pools. Apart from altering the availability of habitat for colonisation these variations may generate important shifts in other environmental variables e.g. changes in the various nitrogen constituents of streamwaters (Howard-Williams, Priscu & Vincent). The potential influence of such effects is illustrated by the Kuparuk River, Alaska, where the stream epilithon was apparently limited by phosphorus availability in one year but not another, with evidence of a switch to nitrogen limitation (Hullar & Vestal). These findings from the Arctic contrast completely with data from a continental antarctic stream where application of the same continuous flow bioassay technique indicated that the cyanobacterial mats were limited neither by phosphorus nor nitrogen supply (Howard-Williams & Vincent).

The extreme conditions of the polar winter effectively strip the biota from some environments which are then newly colonized each summer. Hawes discusses the importance of these 'resetting events' for algal growth in the streams of the maritime Antarctic. From an arctic and subarctic stream Miller & Stout present evidence that winter resetting or 'intermediate disturbance' results in a stochastic pattern of macroinvertebrate colonization, generating large year-to-year variations in species composition and dominance.

The popular idea that high latitude communities are highly simplified assemblages of plant and animal species is not well supported by many of the papers in this volume. In the tundra ponds of the Yukon delta the macroinvertebrate species diversity included more than 20 species of caddisflies in 10 genera and 6 families (Maciolek). In a comprehensive review of benthic invertebrate records for interior Alaskan streams and rivers Oswood shows that many taxa decrease in proportional contribution from south to north, with a marked shift towards community dominance by chironomid species. The overall species diversity is still high and the food chain relatively long, especially by comparison with the antarctic zone where macro-invertebrates are entirely absent (e.g. more than one hundred macroinvertebrates were recorded in arctic Imnavait Creek, Miller & Stout). Even in Antarctica, however, there may be aquatic food webs of unknown complexity particularly in lakes where activity can be maintained year round. These interactions involve microherbivores such as rotifers, nematodes, tardigrades, enchytraeids and protozoa, with complex interactions between the plankton and benthos (e.g. Clarke, Ellis-Evans, Sanders & Holmes). As yet the ecology of these animal populations has received little scientific attention.

High latitude waters also offer exciting opportunities for geochemical studies. Gallagher *et al.* trace the evolution of hypersaline meromictic lakes in the Vestfold Hills, Antarctica. From geochemical studies on nearby Ellis Fjord they suggest that meromixis in these lakes may have developed prior to their isolation from the sea by isostatic uplift of the land some 8–10 thousand years ago. Isostatic trapped seawater is also invoked as the ultimate origin of salts in hypersaline Garrow Lake in the Canadian Arctic, but in their hypothesised evolutionary sequence for this meromictic water body Ouellet *et al.* suggest that underground freeze-concentration processes and brine migration beneath the permafrost have resulted in additional salt inputs during the development of the lake.

Extreme environments such as polar lakes and rivers have been invoked as ideal experimental systems in which certain chemical, physical or biological properties are exaggerated and therefore more readily studied. The findings from such investigations could potentially offer insights into the structure and functioning of less extreme systems elsewhere. Antarctic lakes, for example, here provide a useful testing ground for the development of lipid markers of distinct microbial groups (Matsumoto, Watanuki & Torii); the formulation of new nutrient-light models for planktonic algae (Priscu); for evaluating the significance of geomicrobiological/geochemical cycling (Ellis-Evans & Lemon); and for understanding the lipid physiology of freshwater zooplankton in temporally variable food regimes (Clarke *et al.*). Wharton *et al.* extend this type of argument to apply their antarctic observations beyond planet Earth. These authors suggest that ice-covered lakes of the Dry Valleys, Antarctica, might provide instructive sedimentation analogues of the deep lakes that once existed in the Valles Marineris canyon system on Mars, as well as insights into the coldwater environments that may have supported the earliest forms of life on Earth.

Entire subject areas still await investigation in the polar regions. One of the most significant must surely be that of catchment/lake interactions. The work of Green *et al.* reported here begins to consider this important subject, but problems such as groundwater hydrology over permafrost are yet to be addressed in detail. Biological problems such as colonization processes, life cycle strategies and community development, particularly in the Antarctic, also remain as yet poorly understood. Further there is clearly considerable scope for useful bipolar comparisons, particularly when considering factors such as isolation as an ecological parameter.

The papers presented at the first High Latitude Limnology Symposium will hopefully give some indication of the breadth of research opportunities available to polar limnologists. High latitude regions are now very much in the public eye and there is a growing concern for the fragile ecosystems of these regions as political and economic considerations increasingly come to bear. Given this greater public awareness, it is to be hoped that more limnologists will seize the opportunity to study high latitude systems.

We greatly appreciate the co-operation and assistance we have had from many individuals and several institutions in bringing this volume through to completion. We especially thank our referees: W. P. Adams, V. Alexander, R. Axler, I. A. E. Bayly, B. Biggs, P. A. Broady, H. Burton, P. A. Cranwell, M. T. Downes, G. E. Fogg, D. J. Forsyth, J. B. Gallagher, I. Hawes, C. Howard-Williams, D. McKnight, J. C. Maciolek, M. C. Miller, P. Mulholland, M. W. Oswood, J. G. Paren, P. Pagé, R. A. Ragotzkie, J. R. Sargent, G. M. Simmons, J. Skei, J. R. Vestal, A. B. Viner. We also thank the SIL steering and organizing committees for their support, particularly Carolyn Burns who helped stimulate this symposium; and Wil Peters and Henri Dumont for their advice, assistance and encouragement throughout the preparation of this volume.

Warwick F. Vincent
Taupo Research Laboratory, Division of Marine & Freshwater Science,
DSIR, Taupo, New Zealand

J. Cynan Ellis-Evans
British Antarctic Survey, Natural Environment Research Council,
Cambridge, United Kingdom

Hydrobiologia 172: 1–18, 1989
W. F. Vincent and J. C. Ellis-Evans (eds) High Latitude Limnology

Filamentous green algae in freshwater streams on Signy Island, Antarctica

Ian Hawes
British Antarctic Survey, Natural Environment Research Council, High Cross Site, Madingley Road, Cambridge CB3 OET, United Kingdom

Key words: Antarctica, Chlorophyta, distribution, filamentous algae, *Klebsormidium*, *Mougeotia*, periphyton, streams, *Zygnema*

Abstract

The streams of Signy Island are varied and extremely seasonal environments. Water flows from November/December to March/April; streams are frozen for the rest of the year. Streams usually flow through small, barren catchments and are nutrient poor, though they may be enriched by dense summer populations of seabirds and seals. Temperatures are consistently low. Stream depth is maximal during the spring melt period, declining over the course of the summer. Vegetation is exclusively algal, and filamentous chlorophytes from a particularly conspicuous component. Small numbers of vegetative cells survive the long frozen period in situ. A steady increase in standing crop results in a maximum 2 to 3 months after flow begins. Sloughing is the major loss mechanism and grazers are effectively absent. Three taxa of filamentous algae are common in Signy streams, species of *Zygnema*, *Mougeotia* and *Klebsormidium*. The distributions of these algae are described and related to physical and chemical features of their environment.

Introduction

There is a wide variety of freshwater habitats on the Antarctic continent and its off-lying islands, ranging from small meltwater streams to large permanently ice-covered lakes. Recent reviews of Antarctic freshwater biology have not presented the full extent of this range and instead have concentrated on the relatively well understood lake ecosystems with scant, if any, reference to streams (Heywood, 1977a, 1984; Priddle, 1985). This omission reflects more the lack of information on Antarctic streams than their abundance and probable significance as centres of biological activity (Heywood, 1977b; Broady, 1982; Howard-Williams *et al.*, 1986). It is only in the last few years that these ephemeral ecosystems have begun to be studied.

Antarctica has been divided into 2 climatic zones, the continental and the maritime Antarctica (Holdgate, 1964). The latter comprises the western side of the Antarctic Peninsula and its off-lying islands and is characterised by mean monthly winter temperatures which rarely fall below −20 °C but where at least one month each summer has a mean temperature above freezing. The former has mean monthly temperatures which do not rise above freezing in summer and regularly fall below −20 °C in winter.

Antarctic stream studies have until now been confined to continental Antarctica (Hirano, 1979; Broady, 1982; Howard-Williams *et al.*, 1986).

Here, streams are often dominated by perennial, slow growing cyanobacteria, though filamentous chlorophytes and xanthophyceae have also been recorded. There is little information available on maritime Antarctic streams despite their widespread occurence. Heywood (1977b) and Priddle and Belcher (1982) reported Zygnematales (which have not been noted from continental Antarctica – Hirano, 1965) as the dominant stream vegetation at Alexander Island (at the southern limit of the maritime Antarctic) and Signy Island (towards the northern limit of the maritime Antarctic) respectively. The purpose of this study was to investigate the distribution and seasonality of filamentous chlorophytes in streams at Signy Island in relation to environmental variables.

Study sites

Signy Island (Fig. 1) is a small (8 × 5 km), roughly triangular island forming part of the South Orkney Islands. It comprises a mostly ice-capped central system of ridges which isolates areas of relatively flat, low-lying coastal plain. The ice cap covers a third of the island and descends to the sea on the south and east coasts as the McLeod and Orwell Glaciers respectively. Geologically, the island comprises heavily folded metamorphic rocks of pre-Cambrian age. Garnetiferous quartz-mica-schists predominate (Mathews & Maling, 1967). Most deglaciated areas comprise frost-shattered rock or scree or are covered in a heterogenous layer of unconsolidated glacial till.

The island accumulates a considerable depth of snow (largely as drifts) during winter, most of which melts during summer. Mean monthly temperatures typically rise above zero for two months each year, and large numbers of meltstreams are found during this period.

The length and discharge of streams at Signy Island is limited by the size of the island. The maximum observed length was 535 m and the maximum drainage basin area 99.2 ha. All samples described in this paper were collected from shallow, freshwater streams. Only streams which were flowing for most or all of the summer period were sampled. Sites within these streams were delimited in the field by clear differences in substratum, current velocity, nature of the terrain through which they flowed or after the confluence of two smaller streams. Samples were taken from locations considered to be typical of each site. A total of 36 sites were chosen. All of these sites were sampled at least twice (mid January and mid February 1985) and some on more frequent occasions.

In general, the streams fell into three categories,
i. Meltwater runnels
ii. Larger streams and lake outflows
iii. Glacial streams

Meltwater runnels were found mainly at the heads of catchments and were fed by melting snowbanks. They were shallow and diffuse, apparently lacking sufficient volume or duration of flow to excavate clearly defined channels. Substrata were often unstable, comprising sand and silt as well as gravel and small stones. Meltwater runnels fed into larger streams or more usually drained into one of the island's sixteen lakes. The larger streams, which formed lower down catchments after the confluence of a number of meltwater runnels, and outflows from the lakes could have substantial discharges. They frequently occupied well-defined channels which had been swept clear of sedimentary material to leave a substratum of stable rocks and boulders. Glacial streams drained the island's glaciers, which are in retreat. Substrates comprised well-rounded boulders and stones in the faster flowing regions, interspersed with unstable areas of gravel and silt. They were fast flowing streams, with a high load of suspended silt giving a milky appearance.

Stream vegetation also fell into three categories. First, a sparse flora of epilithic diatoms was the only vegetation in the most torrential streams, mostly *Achnanthes austricia* and *Synedra rumpens*. Secondly, a patchy covering of epilithic, epiphytic and epipsammic diatoms and cyanophytes – mostly *A. austricia*, *S. rumpens*, *Gomphonema angustatum*, *Cymbella* spp., *Phormidium fragile*, *P. antarcticum* and *P. frigidum* – was found in the

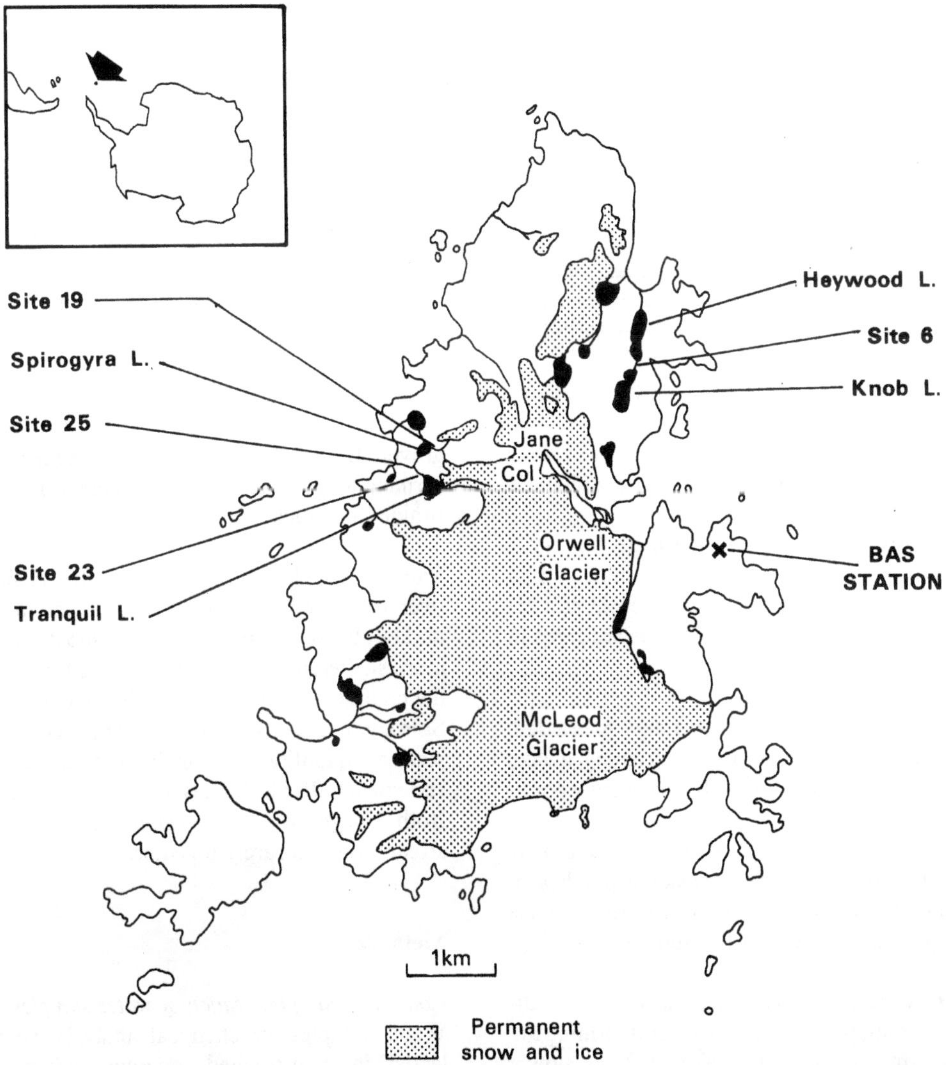

Fig. 1. Map of Signy Island, showing lakes and major streams (black) and areas of permanent ice cover. Inset shows the location of Signy Island.

slower flowing streams, particularly those enriched with nutrients. This community was poorly developed in streams with unstable substrata and/or relatively high current velocities. Splash zones often had encrusting colonies of *Gloeocapsa magma* growing on them. Filamentous chlorophytes were the third major component of stream vegetation and were found in conspicuous amounts in 31 of the 36 sites examined. Three genera were found to predominate, *Mougeotia*, *Zygnema* and *Klebsormidium*. The last of these, *Klebsormidium* was identified as such on the basis of chloroplast morphology. This is a variable feature and this alga may have belonged to the

Table 1. Physical characteristics of main study streams at Signy Island.

Stream Number	Max Obs Depth[1] (cm)	Min Max Depth[2] (cm)	Width[3] (m)	Length (m)	Gradient	Catchment area (ha)	% Snow/ ice in catchment
6	11.0	3.0	3.5	115	1 : 30	31.5	7
19	3.5	1.0	7.0	140	1 : 12	12.5	39
23	21.0	7.5	6.0	55	1.10	74.0	58
25	7.5	1.5	7.0	100	1.5	99.2	48

[1] Maximum depth at sites used for estimating cover over the entire sampling period.
[2] Maximum depth at sampling site at time of minimum water level.
[3] Width of transect site at maximum water level.

similar genus *Ulothrix*. Filaments were epilithic in the larger and glacial streams and formed loosely attached mats on softer substrata.

Four stream sites were selected for more detailed study of the seasonal dynamics of filamentous algae. These were sites 6, 19, 23 and 25. Physical characteristics of these sites are given in Table 1.

Site 6. The outflow from Knob Lake. A braided, moderately fast flowing stream draining a lake which is surrounded by moss banks heavily populated by fur seals during summer. The stream bed comprises stones and rock which are mostly swept clear but there are occasional patches of sand and gravel. *Klebsormidium* formed a dense cover particularly early in the summer.

Site 19. A broad, shallow meltwater runnel indirectly draining part of the ice-cap and forming the main inflow to Spirogyra Lake. It flows slowly through an extensive area of glacial till with little or no terrestrial vegetation. The stream is poorly defined with no confining channel, the substrate mostly comprising small stones and silt. *Zygnema* forms a near complete cover.

Site 23. Situated on the Tranquil Lake outflow approximately 50 m downstream of the lake. The stream is approximately 5 m wide and 5–15 cm deep at this point and current velocity is moderate to high. The stream bed again comprises boulders and large stones but is swept free of sand and gravel. *Mougeotia* is the most abundant filament forming long (up to 90 cm) trailing strands in the braided channel.

Site 25. A high velocity, braided stream flowing over a steep, raised beach of large, rounded stones and boulders with patches of soft substrata in the more sheltered areas. This stream is formed after the confluence of the outflows from Tranquil and Spirogyra Lakes and in addition, receives substantial quantities of meltwater direct from snowfields. It discharges directly into the sea. Both *Mougeotia* and *Zygnema* are found in this stream, though rarely together.

Methods

Collection and preparation of water samples

Water samples for chemical analysis were collected in acid-washed, opaque, 500 ml polypropylene bottles. Each bottle was rinsed three times and filled, expelling all air. Separate samples were taken in 125 ml polyethylene bottles for determination of pH and alkalinity. Sample bottles were returned to the laboratory in an insulated box to prevent freezing. The pH and alkalinity samples were then equilibrated to 5 °C in a cooled water bath. Water for other analyses was filtered through a GF/C filter and then used for analysis of ammonia, nitrate, DRP (dissolved reactive phosphorus), TDP (total dissolved phosphorus) and DRSi (dissolved reactive silicate). At each collection site, the water depth and temperature were measured and current velocity was estimated on a scale of 1–5 where:

1 = no flow
2 = slight
3 = moderate
4 = fast
5 = very fast

Analysis of water samples

pH was measured using a Phillips PW9409 pH meter, and alkalinity of the same sample by Gran titration (Talling, 1973). Nitrate was determined according to Morries (1971). DRP was determined using the method of Murphy and Riley (1962 – as given by Mackereth *et al.*, 1978). Preparation of samples for TDP involved digestion with acid persulphate followed by determination of DRP (Mackereth *et al.*, 1978). Ammonia was determined according to Chaney & Marbach (1962) and DRSi according to Mullin & Riley (1955).

Preparation and analysis of algal samples

Algal samples were removed from streams with forceps and placed in acid washed polythene bottles or 'whirl-pack' bags with a small volume of water; these were returned to the laboratory in an insulated box. Samples were rinsed three times in filtered stream water and sub-samples set aside for identification. Further sub-samples were dried at 105 °C for 24 h then frozen at −40 °C for return to the UK. There they were analysed for C, N, P and ash content. C and N were estimated using a Carlo-Erba 1106 elemental analyser. P was estimated as FRP (Bartlett, 1959) after digestion of a weighed sample containing approx 0.1 μmol P with 800 μl of 72% perchloric acid at 180 °C for 30 min. Ash content was taken as that remaining after combustion at 500 °C for 12 h.

Standing crop estimation

Two measures of algal standing crop were made. The cover of filamentous green algae was estimated using a point quadrat system. A line marked at 5 cm intervals was stretched between 2 stakes permanently fixed at each side of the stream. The presence or absence of filamentous algae was noted at each mark (marks were approximately 3 mm wide). Three traverses were made at each site and cover expressed as a% of the number of sites occupied at the time of maximum observed cover. This method enabled a rapid assessment of the development of algal communities.

The standing crop was also estimated in terms of chlorophyll *a*. At 2–3 weekly intervals 5 stones with maximum linear dimensions between 5 and 15 cm were taken from the stream bed. Algal cover was removed by first scraping and then scrubbing with a stiff nylon brush. The algal material was collected by centrifugation, then resuspended in 95% methanol and stored at 4 °C for 24 h for extraction of pigments. Absorbance of the extracts at 665 and 750 mm was measured with a Pye Unicam SP6-550 spectrophotometer. Absorbance measurements were repeated after acidification with dilute hydrochloric acid and neutralisation with ethanolamine to permit estimation of phaeopigments (Marker *et al.*, 1980). The equations given in Marker *et al.* (1980) were used for calculations of pigment concentrations in extracts. Chlorophyll *a* was then expressed on an areal basis after estimating the surface area of each stone by dividing its shape into component triangles and measuring each side with calipers. The tendency of the quartz-mica-schist bedrock to split along the flat planes of schistosity facilitated this approach.

Taxonomy

Filamentous green algae could be classified to genus level only. All samples examined were sterile, with no species specific morphological characteristics.

Results

Distribution of filamentous algae

Three genera were found to predominate in Signy Island streams, *Mougeotia*, *Zygnema* and *Klebsormidium*. There was no indication from cell morphology that more than one species of each genus

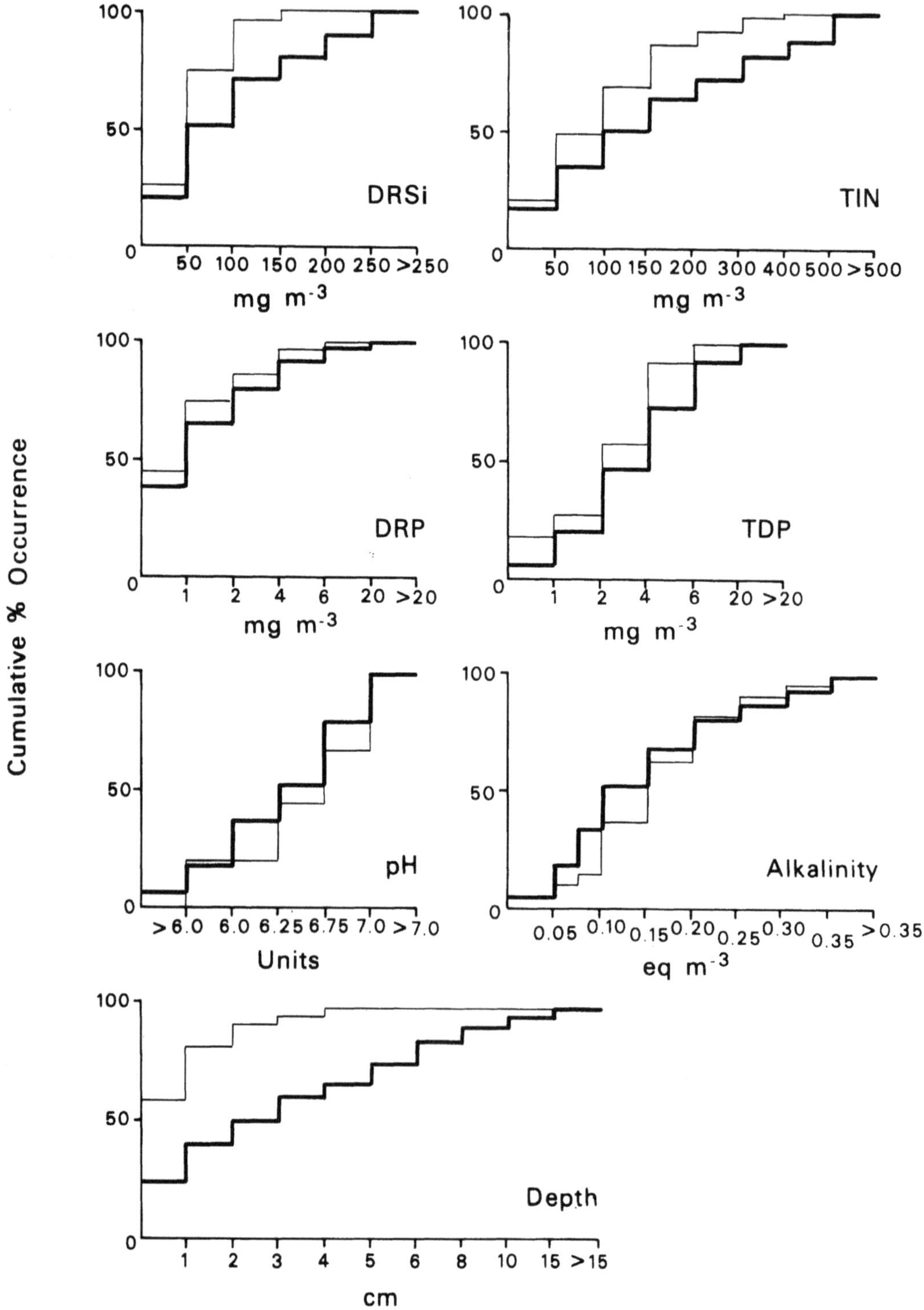

Fig. 2. Kolmogorov curves for *Zygnema* (see text). The cumulative distribution of a given variable at all streams sampled is shown by the bold line (the reference curve), that of the streams where *Zygnema* occurred is shown by the narrower line.

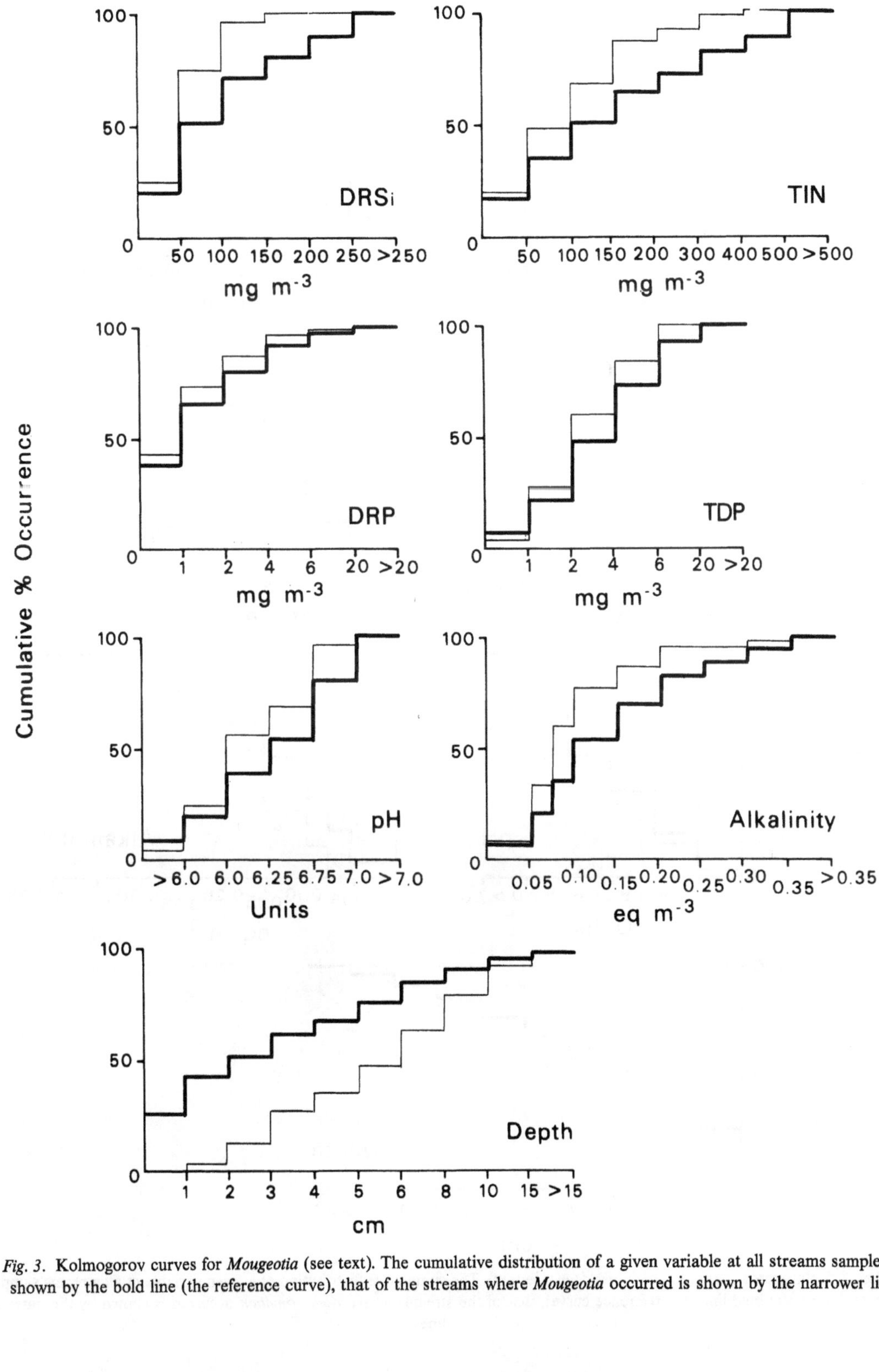

Fig. 3. Kolmogorov curves for *Mougeotia* (see text). The cumulative distribution of a given variable at all streams sampled is shown by the bold line (the reference curve), that of the streams where *Mougeotia* occurred is shown by the narrower line.

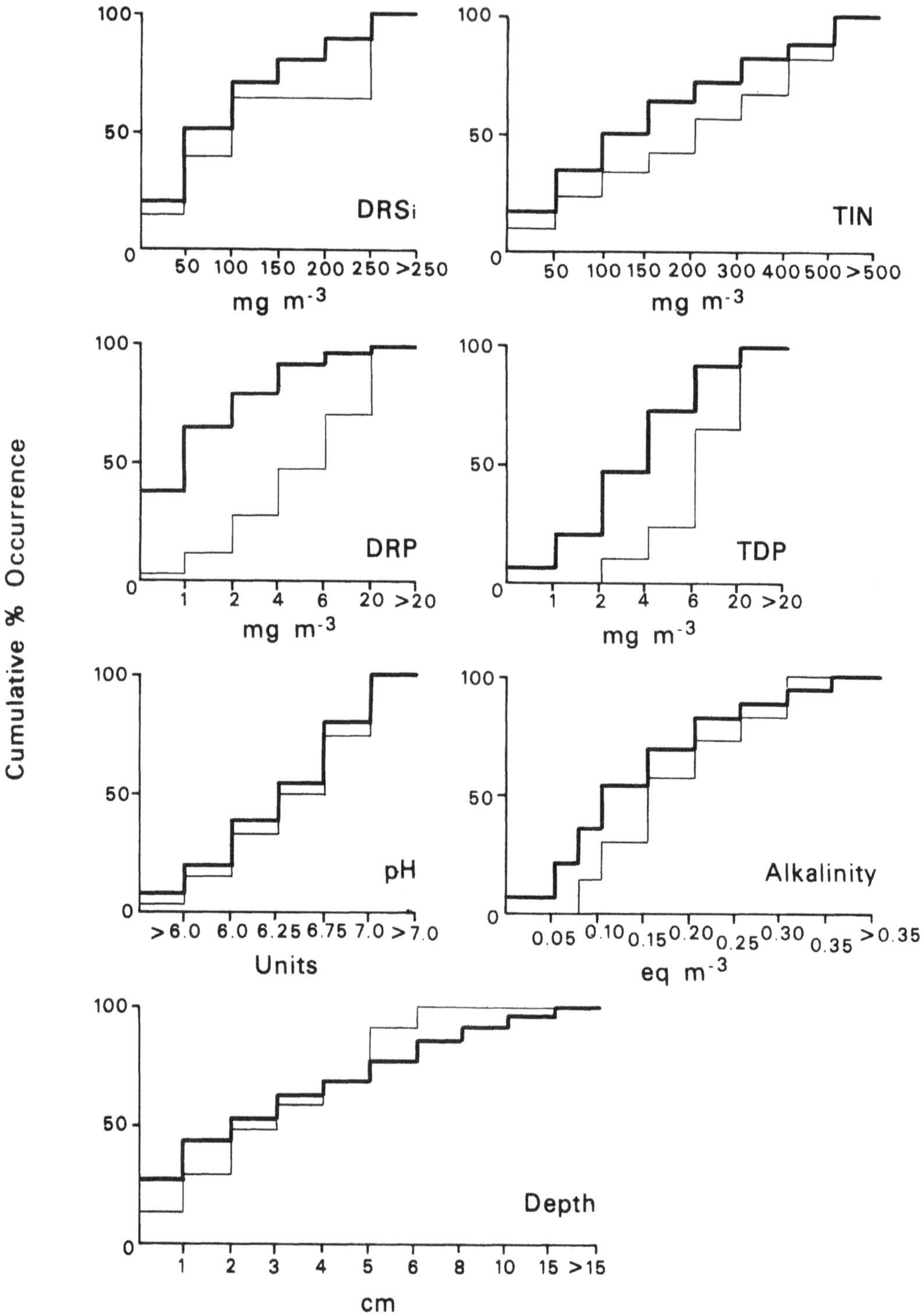

Fig. 4. Kolmogorov curves for *Klebsormidium* (see text). The cumulative distribution of a given variable at all streams sampled is shown by the bold line (the reference curve), that of the streams where *Klebsormidium* occurred is shown by the narrower line.

Table 2. Correlation matrix for determinands included in Kolmogorov analysis. TIN, Total inorganic nitrogen; DRP, Dissolved reactive phosphorus; TDP, Total dissolved phosphorus; SRSi, Dissolved reactive silicon; ALK, Alkalinity. The significance of the correlations and their signs are indicated: + + + or − − − $p < 0.001$; + + or − − $0.001 < p < 0.01$; + or − $0.01 < p < 0.05$; ns $p > 0.05$.

	TIN	DRP	TDP	DRSi	ALK	pH	DEPTH
TIN	X						
DRP	ns	X					
TDP	ns	+ + +	X				
DRSi	+ + +	ns	ns	X			
ALK	+ + +	ns	ns	+ + +	X		
pH	− −	ns	ns	+ +	+ +	X	
DEPTH	− −	ns	ns	−	−	ns	X

was present. A Kolmogorov analysis (Conover, 1971) was used to attempt to relate the distribution of these genera at the 36 sites to measured environmental variables. The Kolmogorov approach is designed to compare the cumulative distribution curves of an environmental variable (e.g. depth) at all sites sampled (the reference curve), with the cumulative distribution of that variable at sites supporting one particular taxon. Cumulative distribution curves for seven variables and the three genera are shown in Figs. 2–4. Confidence intervals (not shown in the figures) were applied after Kolmogorov and used to assess the significance of deviations from the reference curves (Conover, 1971: 299). Care had to be taken in interpreting these data as there were strong correlations between some variables (Table 2).

Significant deviations from the reference curves were found for all three genera (Table 3). *Klebsormidium* was most frequently encountered in P-rich streams. This was the only genus which may have had a nutrient response. The most significant deviations from the reference curves for both *Mougeotia* and *Zygnema* related to depth; the former being most abundant in deep streams, the latter in shallow ones. Significant relationships between low nitrate ($p < 0.10$) and silicate ($p < 0.05$) concentrations and *Mougeotia* abundance may be due to the negative relationships between these variables. Quantitative measures of current velocity were not made. Visual estimates suggested that the shallow streams which tended to be occupied by *Zygnema* were slow flowing, while in faster streams *Zygnema* was rare and *Mougeotia* or *Klebsormidium* predominated. *Mougeotia* was also found in slow flowing and even static water but *Klebsormidium* was not. Filamentous algae were very rarely encountered in very fast (category 5) streams.

Table 3. Significance of deviations from the reference curve in the Kolmogorov analyses for the three main genera and selected determinands. 109 stream sites were included in the analysis. Abbreviations and symbols as in Table 2.

Variable	*Zygnema*	*Mougeotia*	*Klebsormidium*
TIN	NS	−	NS
DRP	NS	NS	+ + +
TDP	NS	NS	+ + +
DRSi	NS	− −	NS
ALK	NS	NS	+
pH	NS	NS	NS
DEPTH	− − −	+ + +	NS

The reference curves are useful indicators of the distribution of variables in the streams. Most streams can be described as shallow (65% less than 5 cm deep), slightly acidic (80% pH less than 7) with low concentrations of DRP (70% less than 2 mg m^{-3}) but relatively high TIN concentrations (50% above 150 mg m^{-3}).

Main study sites

Year-round observations were made at the four main study sites. These were frozen from late April to November. During this winter period, samples were excavated from the streams which,

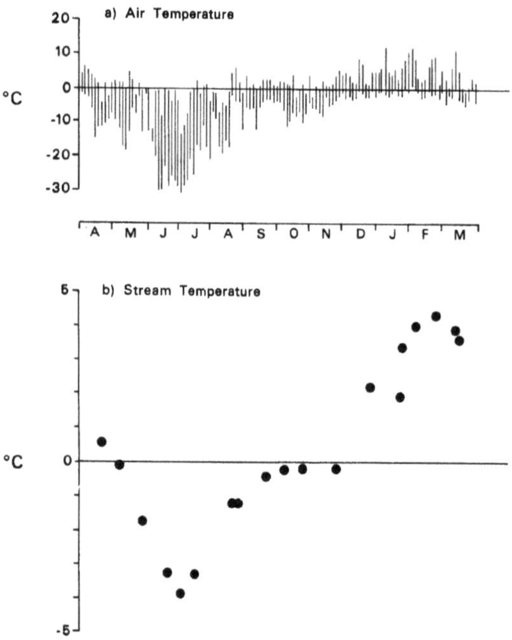

Fig. 5. Seasonal changes in a) air temperature and b) water temperature at stream site 23. Bars in a) represent the range of air temperatures over consecutive three day periods.

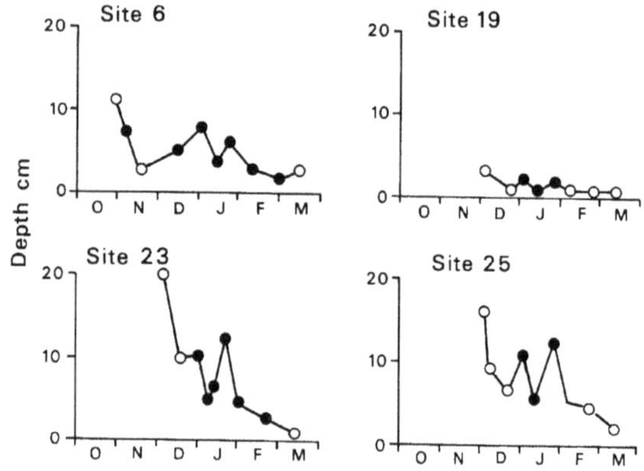

Fig. 6. Changes in stream depth at controlled locations in the four main study sites over the 1984/85 austral summer. Open symbols indicate the presence of surface ice on the transect at the time of sampling, closed symbols indicate no ice present.

on transfer to sterile culture medium, yielded viable filaments of all three genera. A thermistor buried in the thick ice (30–50 cm) at site 23 showed a minimum temperature of only −4 °C (Fig. 5), while air temperatures fell to −25 °C or below over the same period. This minimum was reached shortly after stream freezing but by August, when 30 cm of snow had accumulated on the stream surface the temperature at the stream bed had risen to just below freezing point. Accumulation of snow and ice clearly offers very effective insulation.

The opening of the streams coincided with the melting of the winter snow accumulation in their catchments. Maximum stream depths were recorded at this time and flow velocities were high (categorised as 3–5) in all streams. Almost all of the thin, patchy cover of filamentous algae revealed during the melt was rapidly sloughed away by these high discharges. The subsequent pattern of stream depth (which is taken here to reflect discharge) was similar at all sites. Depth tended to decrease gradually over the course of the summer, with peaks in early and late January (Fig. 6) associated with periods of high insolation. Liquid water was found on all sampling occasions, though surface ice was frequently encountered (Fig. 6).

Stream chemistry showed some temporal variation. Streams mainly comprising meltwater (e.g. site 19) were initially very dilute with nutrient concentrations, particularly TIN, tending to increase as the summer progressed (Fig. 7). At site 19, TIN rose from 58 to 616 mg m^{-3} from 3/1/85, to 14/3/85, while TDP rose from 0.5 to 2.5 mg m^{-3}. Streams initially comprising displaced lake water (e.g. sites 6 and 23) had relatively high nutrient concentrations which declined to lower, relatively stable, summer levels. Summer TDP concentrations in unenriched streams were relatively constant at around 1–2 mg m^{-3}, with enriched site 6 considerably higher at 20 mg m^{-3}.

Increase in algal standing crop after the initial meltwater pulse was rapid. Both cover of filamentous algae and chlorophyll *a* showed similar patterns of increase, with maxima reached 2–2.5 months later (Figs. 8, 9). This maximum was towards the end of the summer at all but site 6, the outflow from nutrient-enriched Knob Lake, which began to flow in October and attained a

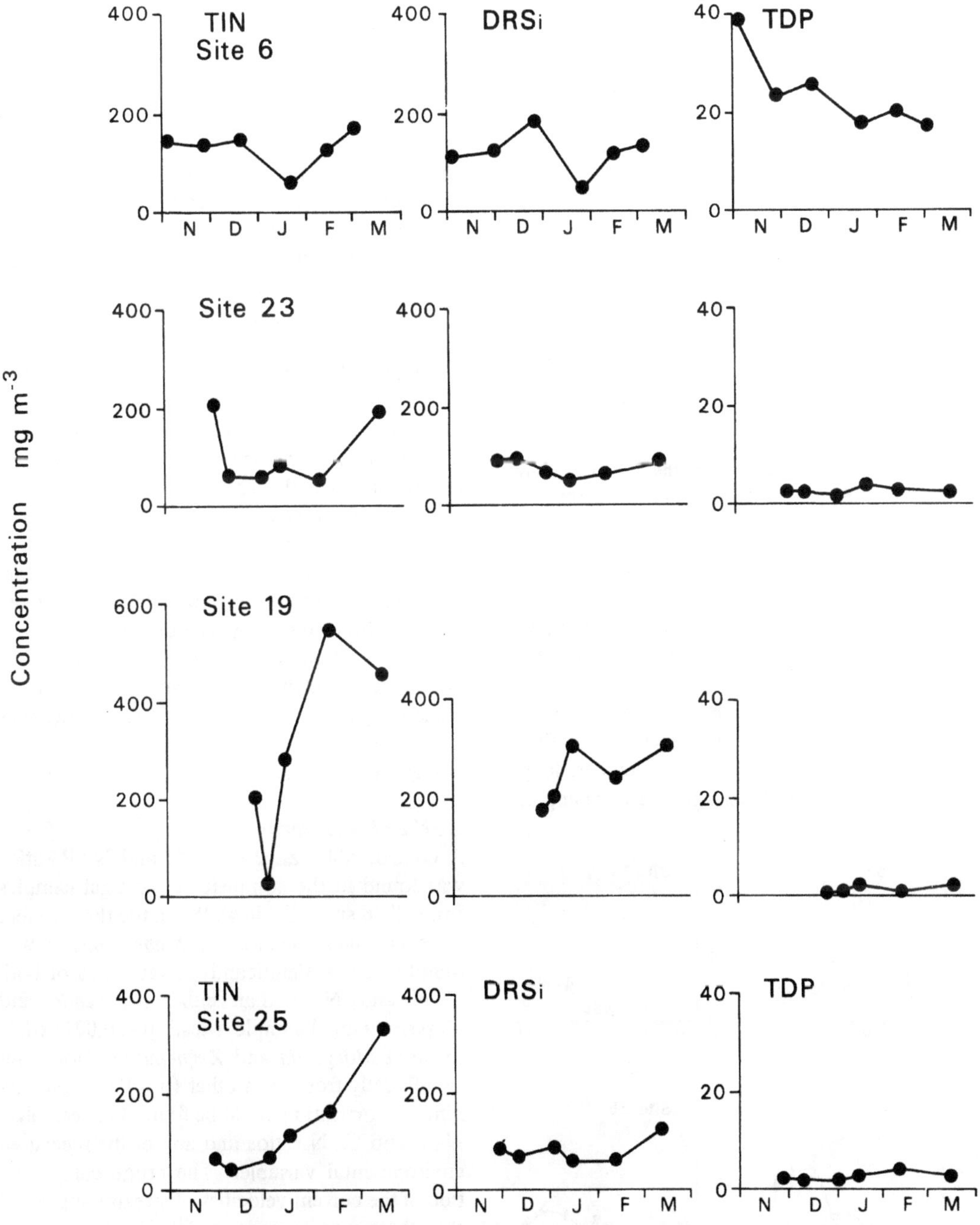

Fig. 7. Concentrations of total inorganic nitrogen (TIN), dissolved reactive silicate (DRSi) and total dissolved phosphorus (TDP) at main study sites during the 1984/85 austral summer. Each point is the mean of three determinations.

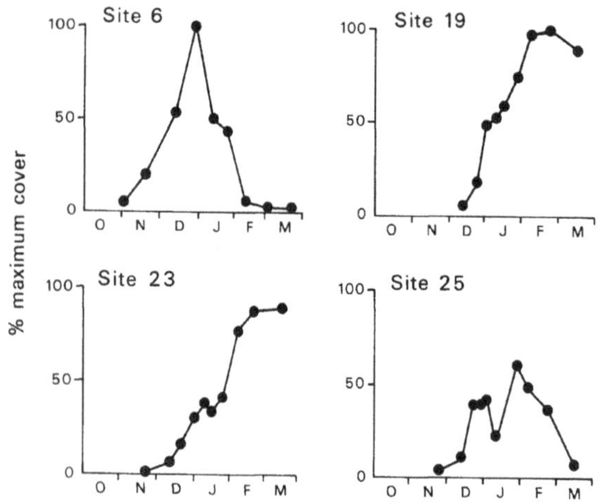

Fig. 8. Cover of filamentous algae at the four main study sites over the 1984/85 austral summer. Results are expressed as a percentage of the maximum observed cover at each site during the study period.

maximum algal standing crop, expressed both as cover of *Klebsormidium* and chlorophyll *a*, by late December. Early growth comprised *Klebsormidium* alone, but later, increasing amounts of *Phormidium* spp. began to overgrow the filamentous cover. By the end of December, long (> 30 cm), thick strands of algae were trailing in

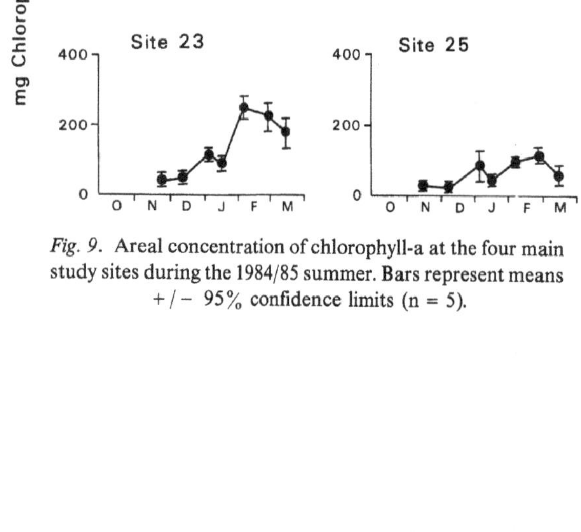

Fig. 9. Areal concentration of chlorophyll-a at the four main study sites during the 1984/85 summer. Bars represent means +/− 95% confidence limits (n = 5).

the stream flow. The dramatic decline in algal cover in January (Fig. 8) was due to sloughing. It coincided with the loss of ice cover from the lake itself, and one of the periods of high discharge mentioned above. Wind induced turbulence in shallow Knob Lake (mean depth 1.2 m) resulted in increased suspensoids. This was incorporated into the *Klebsormidium*/*Phormidium* mat and large areas were seen to slough away. Little new development occurred after this loss.

At sites 19, 23 and 25, cover of filamentous algae increased at similar rates (Fig. 8), until most of the available substrate was occupied (Fig. 10). *Zygnema* was not found in the deeper, faster flowing runnels of site 19, and *Mougeotia* did not occur in areas of site 23 which were periodically exposed to the air (Fig. 10). This is consistent with the depth distributions identified by the Kolmogorov analysis. Chlorophyll *a* cover in the *Mougeotia* and *Zygnema* dominated streams did not attain such high levels as at site 6, but filamentous green algae in these streams were not subject to overgrowth by *Phormidium*. Sloughing losses were nonetheless evident in these streams, with clear checks to the steady increase of cover or declines in existing cover occurring during the two discharge peaks in early and late January (Fig. 6).

C : N and N : P ratios

A considerable range of C : N and N : P ratios was found in the complete set of algal samples from all 36 sites (Table 4). When the three genera were examined separately, *Klebsormidium* was found to show significantly lower ratios of both C : N and N : P than either *Mougeotia* and *Zygnema* (small sample 't-test', $p < 0.001$) while those of *Mougeotia* and *Zygnema* did not differ significantly from each other ($p > 0.05$). No significant correlation could be found between algal N : P and C : N ratios and any of the measured environmental variables. The crude categorisation of the current velocities of streams suggested that there was a tendency for C : N and N : P ratios to be lower in faster-flowing water than in slow (Table 5). This applied both when the complete set of samples was considered together and when the genera were examined separately.

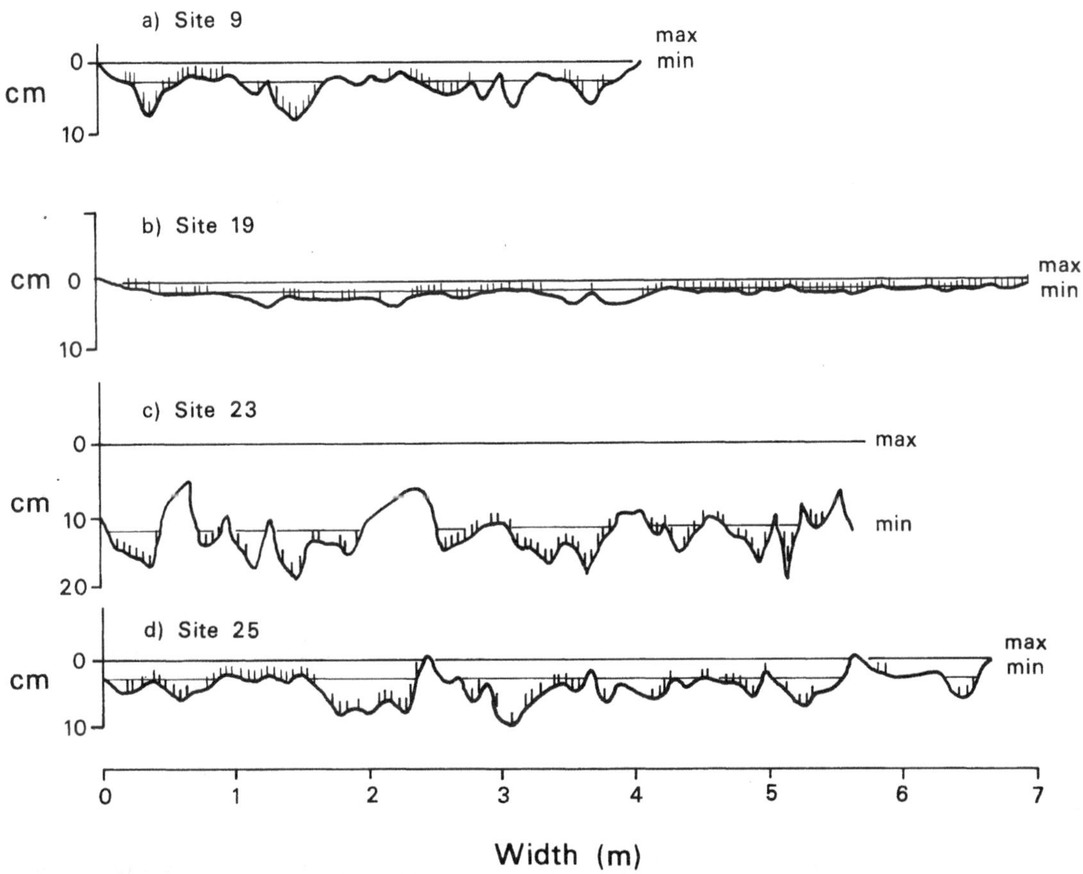

Fig. 10. Transects across the four main study sites to show minimum and maximum water levels. Vertical bars indicate the maximum extent of filamentous algal coverage. Depths were measured at 10 cm intervals and the presence or absence of filaments every 5 cm.

Table 4. C : N and N : P ratios of stream algal samples.

	C:N			N:P		
	Mean	Range	n	Mean	Range	n
All samples	16.3	5.6–20.6	55	12.3	4.4–22.1	42
Zygnema	12.4	8.9–20.6	22	12.7	6.0–22.1	14
Mougeotia	13.3	10.1–19.5	24	14.7	6.6–20.9	19
Klebsormidium	7.6	5.6–11.9	9	6.5	4.4– 8.9	9

Table 5. C : N and N : P ratios in algal samples taken from streams with different current velocities. Velocity was estimated visually on a scale of 1–5.

Current Velocity	C:N			N:P		
	Mean	Range	n	Mean	Range	n
1	20.6	–	1	15.8	–	0
2	13.4	10.1–19.5	21	14.2	7.6–22.1	18
3	11.3	5.6–15.7	23	10.9	5.4–17.9	15
4	8.9	6.5–13.2	10	9.9	5.6–14.7	10
5	–	–	0	–	–	0

Seasonal as well as spatial differences in elemental composition were seen. Those at site 23 are shown in Table 6. C : N and N : P ratios increased in late summer as stream depth and discharge rates declined, despite increasing nutrient concentrations.

Discussion

Freshwater lakes at Signy Island are assumed to be typical of the maritime Antarctic. This assumption is based on considerations of climate and the scant information available on other maritime Antarctic freshwater ecosystems (see review by Heywood, 1978). If Signy Island streams are also typical of the maritime Antarctic climatic zone, it seems there are major physical and biological differences between these and the streams of continental Antarctica for which relatively comprehensive information is now available (Howard-Williams *et al.*, 1986; Vincent & Howard-Williams, 1986).

Table 6. C : N and C : P ratios in samples of *Mougeotia* from site 23 over the course of the 1984/85 austral summer. Each point represents the mean of 3 replicates.

Date	C:N	N:P	C:P
11/12/84	10.7	24.4	261
3/01/85	12.3	15.9	196
16/01/85	15.1	15.3	231
8/02/85	12.5	16.5	206
28/01/85	16.9	19.6	331
2/03/85	17.2	25.0	430

Physical differences are consistent with the climatic differences between the two areas. Higher air temperatures allow maritime Antarctic streams to flow for longer than continental ones, typically 4 to 5 compared to 1 to 2 months, although water temperatures are similar. Relatively high precipitation (Signy Island received 230 mm water equivalent in 1984 – D. Limbert pers. comm.) permits extensive winter snow accumulation in maritime Antarctica, which has two important effects. Firstly, it insulates streams from the low winter air temperatures and secondly, it permits very high discharges of meltwater to accumulate during the first flows of spring (see also Hawes, 1983). This contrasts with continental Antarctica, where there is little snow accumulation and frozen streams often ablate over winter, leaving a freeze-dried periphyton exposed to air temperatures down to − 55 °C (Howard-Williams *et al.*, 1986).

The pattern of discharge seen in the four main study streams, where a spring peak declines over the course of summer, is essentially similar to that of Arctic streams (Harper, 1981). These spring flushes can be very abrasive (Stockner & Hynes, 1976) and at Signy Island clear much of the existing benthic flora, particularly the filamentous chlorophytes, from the stream bed. Catastrophic effects of high flow rates on periphyton are frequently recorded (e.g. Wehr, 1981). Filamentous algae appear to be particularly susceptible (McIntire, 1966; Antoine & Benson-Evans, 1982). In the absence of grazers, sloughing would appear to be the only important source of loss of periphyton from Signy Island streams. The compounding effect of heavy overgrowth with epiphytes was clearly seen in stream 9, which drained Knob Lake. This is consistent with one of the ideas from the river continuum concept (Vannote *et al.*, 1980), which proposes that stream ecosystems are in a continual state of change along temporal and spatial axes, never attaining the sort of climax communities seen in terrestrial ecosystems.

The seasonality of growth of filamentous algae in Signy Island streams stems from this 're-setting' of the environment in spring each year. Such re-setting events favour the dominance of a periphyton which is capable of rapidly attaining high cover at low temperatures, a role which filamentous chlorophytes appear to fill at Signy Island. Similar communities have been found in cold, fast-flowing streams in northern Scandinavia (Round, 1981). Moore (1974a, b) noted a similar seasonality in periphyton growth in a detailed study of three streams in the Canadian Arctic, with maximum standing crop developing towards the end of the period of flow. Although he found diatoms and cyanobacteria to be the dominant groups, he also reported *Zygnema* and *Mougeotia* as locally abundant.

This essentially annual pattern of development contrasts with the situation in continental Antarctic streams where there is extensive overwinter survival and little development of new biomass over the summer period (Howard-Williams *et al.*, 1986). It should be noted that these last authors have assumed, in calculating survival rates, that all chlorophyll *a* extracted from stream bed mats is viable. Such an assumption is not necessarily valid considering the low production:biomass ratios they report, and the near ideal conditions for preservation of dead material which exist in freeze dried continental streams during winter. While lush growths of trailing, filamentous algae can occur in continental streams, the biomass is usually dominated by cyanophytes (Hirano, 1965; Broady, 1982, 1984, 1985; Howard-Williams *et al.*, 1986; Howard-Williams & Vincent, 1987). The persistence of these perennial cyanophyte populations, in contrast to the fast growing annuals of Signy Island, may rest on the absence of a scouring spring re-setting event. There is little winter snow accumulation in the catchments of continental streams and discharges usually, but not always, build up gradually to a mid-summer peak (Chinn, 1981).

Standing crops of periphyton (as chlorophyll *a*) in Signy Island streams are similar to those in the continental Antarctic streams (2–40 $\mu g\ cm^{-2}$; Howard-Williams *et al.*, 1986), with the proviso noted above that all chlorophyll-*a* may not be viable in the latter. Chlorophyll data is not available for the Arctic streams studied by Moore (1974a, b), but, if a dry weight to chlorophyll-a conversion factor of 50:1 is assumed (Hawes, unpublished data), similar standing crops were present. These biomasses are similar to those attained in temperate streams (e.g. Wehr, 1981; Round, 1981) so over the course of the summer temperature does not appear to adversely affect the total amount of biomass produced. However, Antarctic algae have been found to be consistently operating below their optimum temperatures (Seaburg *et al.*, 1981; Vincent and Vincent, 1982, Tilzer *et al.*, 1986) and it is likely that temperature will impose an upper limit to the rate at which this biomass is attained.

Nutrients, particularly P, have been implicated in limiting the productivity of the phytoplankton in Signy Island lakes (Hawes, 1983). Nutrient availability may be similarly important to filamentous algae, particularly in the nutrient-poor meltwater streams. Nutrient limitation of periphyton in streams is more difficult to demonstrate as it is complicated by the need to consider discharge as well as nutrient concentration (Whitford & Schumacher, 1964; Lock & John, 1979). Identification of nutrient limitation in lentic algae has usually involved fabricated experimental channels (Wurhman & Eichenberger, 1975; Krewer & Holm, 1982) or extensive perturbations of natural communities (Petersen *et al.*, 1983). In this study, the intracellular concentrations of C, N and P have been used to identify changes in the nutritional status of the algae. This approach was made possible by the tendency of the filamentous algae to occur in dense, mono-generic stands. Significant inter-generic differences were apparent. Increasing ratios of C : N and N : P were taken as evidence of increasing deficiency of N and P respectively with C : N and N : P ratios of around 10–15 frequently thought to indicate balanced nutrient supply (Droop, 1973; Rhee, 1978; Senft, 1978; Goldman *et al.*, 1979; Tett *et al.*, 1985).

Ratios recorded in algae from Signy Island streams suggest that, for the most part, neither N nor P is particularly limiting. The low ratio of N : P in *Klebsormidium*, growing in P-rich streams, suggests that, if any nutrient were to be limiting, it is more likely to be N than P. P enrichment at Signy Island is associated with the activity of seals and sea-birds and is confined to low-lying coastal areas (Hawes, 1983). Low concentrations of TDP in many upland meltwater streams (e.g. site 19) led to relatively high N : P ratios in stream water and would be consistent with potential P limitation as indicated by high N : P ratios in some *Mougeotia* and *Zygnema* samples from such streams. Unlike some recent reports (e.g. Gibson & Whitton, 1987), no statistically significant relationships were found between ambient nutrient concentrations and algal N : P ratios. High current velocity was statistically more important in maintaining a low N : P ratio. It is likely that declining current velocity was responsible for the increase in algal N : P ratio at site 22 as the summer season progressed as dissolved nutrient concentrations did not show any marked decrease.

Care must be taken in interpreting statistical relationships. The techniques used here for sampling and analysis did not permit causative relationships to be established. Samples of algae and water were taken at the same, discrete time, from environments known to be temporally variable. This variation could mask real relationships where they develop over long time scales. Intercorrelation of variables which were not measured with those that were could also lead to erroneous identification of ecologically important variables. This criticism applies both to intra- and interstream comparisons. While it appeared that *Klebsormidium* was absent from the streams with lowest TDP concentrations, both this and the apparent relationship between the distribution of *Mougeotia* and *Zygnema* and depth potentially represents the effects of a number of other variables. In the latter case, current speed, substrate type, winter temperatures and frequency of summer freezing or drying may be particularly important.

Conclusions

The variables which have been identified as important in determining the spatial and temporal distributions of the three genera of filamentous chlorophytes are essentially the same as those important in streams in any other geographical location. However, maritime Antarctic streams can be characterised by their particular combination of these variables:

1. Barren catchments, giving low concentrations of dissolved nutrients, particularly phosphorus, except where enriched by seal or seabird activity.
2. Consistently low temperatures during the short (3 to 5 month) period of flow.
3. Variable discharge with periodic freezing and drying during the period of flow.
4. Prolonged winter freezing under sometimes thick snow cover.
5. Spring discharge maxima associated with snow-melt.
6. Absence of invertebrate grazers.
7. Highly seasonal vegetation, predominantly algal, with filamentous chlorophytes often abundant, and a single standing crop maximum in mid-summer.

Acknowledgements

Thanks are due to the British Antarctic Survey for providing facilities for this work and to all members of the base at Signy Island for assistance with field work. Many people commented on various drafts of this paper, particularly Drs. J. C. Ellis-Evans, B. A. Whitton, C. Howard-Williams and P. A. Broady.

References

Antoine, S. E. & K. Benson-Evans, 1982. The effect of current velocity on the rate of growth of benthic algal communities. Int. Revue ges. Hydrobiol. Hydrogr. 67: 575–583.

Bartlett, G. R., 1959. Phosphorus assay in column chromatography. J. Biochem. 234: 466–468.

Broady, P. A., 1982. Taxonomy and ecology of algae in a freshwater stream in Taylor Valley, Victoria Land, Antarctica. Archiv fur Hydrobiol. Suppl. 63: 3431–339.

---------, 1984. The vegetation of Cape Bird, Ross Island, Antarctica. Melbourne University Programme in Antarctic Studies No 62.

---------, 1985. A preliminary report of phycological studies in northern Victoria Land and on Ross Island during 1984–85. Melbourne University Programme in Antarctic Studies No 66.

Chaney, A. L. & E. P. Marbach, 1962. Modified reagents for the determination of urea and ammonia. Clin. Chem. 8: 130–132.

Chinn, T. J. H., 1981. Hydrology and climate in the Ross Sea area. J. Roy. Soc. N.Z. 11: 373–386.

Conover, W. J., 1971. Practical non parametric statistics. Wiley, New York. 462 pp.

Droop, M. R., 1973. Some thoughts on nutrient limitation in algae J. Phycol. 9: 264–277.

Gibson, M. & B. A. Whitton, 1987. Relationship between water chemistry, algal morphology and phosphatase activity in freshwater Chaetophorales. Br. J. Phycol. 22: 11–22.

Goldman, J. C., J. J. McCarthy & D. G. Peavey, 1979. Growth rate influence on the chemical composition of phytoplankton in oceanic waters. Nature 279: 210–215.

Harper, P. P., 1981. Ecology of streams at high latitudes. In M. A. Lock & D. D. Williams, (eds) Perspectives in running water ecology. Plenum Press, New York: 313–337.

Hawes, I., 1983. Nutrients and their effects on phytoplankton populations in lakes on Signy Island, Antarctica. Polar Biol. 2: 115–126.

Heywood, R. B., 1977a. Antarctic freshwater ecosystems – a review and synthesis. In G. A. Llano (ed) Adaptations within Antarctic Ecosystems. Third SCAR symposium on Antarctic Biology, 1974. Gulf Publishing, Texas: 801–828.

Heywood, R. B., 1977b. A limnological survey of the Ablation Point area, Alexander Island, Antarctica, Phil. Trans. R. Soc. Lond. Ser. B. 279: 39–54.

---------, 1978. Maritime Antarctic lakes. Verh. int. Ver. Limnol. 20: 1210–1215.

---------, 1984. Antarctic inland waters. In R. M. Laws (ed), Antarctic Ecology Vol. 1. Academic Press, Lond.: 279–343.

Hirano, M., 1965. Freshwater algae in the Antarctic regions. In J. Van Miegham, P. Van Oye & J. Schell (eds), Biogeography and Ecology in Antarctica. Junk, The Hague: 127–193.

---------, 1979. Freshwater algae from Yukidori Zawa near Syowa Station, Antarctica. National Institute of Polar Research Memoirs, Tokyo. No. 11: Special Issue. pp 1–25.

Holdgate, M. W., 1964. Terrestrial ecology in the Maritime Antarctic. In. R. Carrick, M. W. Holdgate & J. Prevost (eds), Biologie Antarctique. Hermann, Paris: 181–194.

Howard-Williams, C. & Vincent, W. F., 1987. Microbial communities in southern Victoria Land streams. I. Photosynthesis. This Volume.

---------, C. L. Vincent, P. A. Broady & W. F. Vincent, 1986. Antarctic stream ecosystems: Variability in environmental properties and algal community structure. Int. Revue ges. Hydrobiol. 71: 511–544.

Krewer, J. A. & H. W. Holm, 1982. The phosphorus-chlorophyll-a relationship in periphytic communities in a controlled ecosystem. Hydrobiologia 94: 173–176.

Lock, M. & P. John, 1979. The effect of flow patterns on uptake of phosphorus by river periphyton. Limnol. Oceanogr. 24: 376–383.

Mackereth, F. J. H., J. Heron & J. F. Talling, 1978. Water Analysis: some revised methods for limnologists. Freshwater Biological Association Scientific Publications No. 36. 120 pp.

Marker, A. F. H., E. A. Nusch, H. Rai & B. Rieman, 1980. The measurement of photosynthetic pigments in freshwaters and standardisation of methods: Conclusions and recommendations. Arch. Hydrobiol. 14: 91–106.

Mathews, D. H. & D. H. Maling, 1967. The geology at the South Orkney Islands: I. Signy Island. F. I. D. Surv. Sci. Rep. No. 25. 32 pp.

McIntyre, C. D., 1966. Some effects of current velocity on periphyton communities in laboratory streams. Hydrobiologia 27: 559–570.

Moore, J. W., 1974a. Benthic algae of southern Baffin Island. I. Epipelic communities in rivers. J. Phycol. 10: 50–57.

---------, 1974b. Benthic algae of southern Baffin Island. III. Epilithic and epiphytic communities. J. Phycol. 10: 456–462.

Morries, P., 1971. A note on the ultra-violet spectrophotometric method for the determination of nitrate in water. Wat. Treat. Exam. 20: 132–137.

Mullin, J. B. & J. P. Riley, 1955. The colourimetric determination of silicate with special reference to sea and natural waters. Analytica chim. Acta, 12: 162–176.

Murphy, J. and J. P. Riley, 1962. A modified single solution method for the determination of phosphate in natural waters. Analytica Chim. Acta, 27: 31–36.

Petersen, B. J., J. E. Hobbie & T. L. Corliss, 1983. A continuous-flow periphyton bioassay: Tests of nutient limitation in a tundra stream. Limnol. Oceanogr. 28–583–591.

Priddle, J., 1985. Terrestrial habitats – Inland waters. In. W. N. Bonner and D. W. H. Walton (eds), Key Environments Antarctica. Pergamon Press, Oxford: 118–132.

Priddle, J. & J. H. Belcher, 1982. An annotated list of benthic algae (excluding diatoms) from freshwater lakes on Signy Island. Br. Ant. Surv. Bull. 57: 41–53.

Rhee, G. -Y., 1978. Effects of N:P atomic ratios and nitrate

limitation on algal growth, cell composition and nitrate uptake. Limnol. Oceanogr. 23: 10–25.
Round, F. E., 1981. The ecology of algae. Cambridge University Press, Cambridge. 633 pp.
Seaburg, K. C., B. C. Parker, R. A. Wharton & G. M. Simmons, 1981. Temperature growth responses of algal isolates from Antarctic oases. J. Phycol. 17: 353–360.
Senft, W. H., 1978. Dependence of light-saturated rates of algal photosynthesis on intracellular concentrations of phosphorus. Limnol. Oceanogr. 23: 709–718.
Stockner, Z. S. J. & H. B. N. Hynes, 1976. Studies on the tributaries of Char Lake, Cornwallis Island, Canada. Hydrobiologia 49: 97–102.
Talling, J. F., 1973. The application of some electrochemical methods to the measurement of photosynthesis and respiration in fresh water. Freshwat. Biol. 3: 335–362.
Tett, P., S. I. Heaney & M. R. Droop, 1985. The Redfield ratio and phytoplankton growth rate. J. Mar. Biol. Ass. UK 65: 487–504.
Tilzer, M. M., M. Elbrachter, W. W. Geiskes & B. Beese, 1986. Light-temperature interactions in the control of photosynthesis in Antarctic phytoplankton. Polar Biol. 5: 105–111.
Vannote, R. L., G. W. Minshall, K. W. Cummins, J. R. Sedell & C. E. Cushing, 1980. The river continuum concept. Can. J. Fish. Aqu. Sci 37: 130–137.
Vincent, W. F. & C. L. Vincent, 1982. Factors controlling phytoplankton production in Lake Vanda (77° S). Can. J. Fish. aquat. Sci. 39: 1602–1609.
Vincent, W. F. & C. Howard-Williams, 1986. Antarctic stream ecosystems: physiological ecology of a blue-green algal epilithon. Freshwat. Biol. 16: 219–233.
Wehr, J. D., 1981. Analysis of seasonal succession of attached algae in a mountain stream, the North Alouette River, British Columbia. Can. J. Bot. 59: 1465–1474.
Whitford, L. A. & G. J. Schumacher, 1964. Effect of a current on respiration and mineral uptake in *Spirogyra* and *Oedogonium*. Ecology 45: 168–170.
Wuhrmann, K. & E. Eichenberger, 1975. Experiments on the effects of inorganic enrichments of rivers on periphyton primary production. Verh. int. Ver. Limnol. 19: 2028–2034.

Hydrobiologia 172: 19–26, 1989
W. F. Vincent and J. C. Ellis-Evans (eds) High Latitude Limnology

The effects of nutrient limitation and stream discharge on the epilithic microbial community in an oligotrophic Arctic stream

Meredith A. Hullar & J. Robie Vestal
Department of Biological Sciences, University of Cincinnati, Cincinnati, Ohio 45221, USA

Key words: Arctic, epilithic microbiota, microbial activity, microbial biomass, nitrogen limitation, oligotrophic water, phosphorus limitation

Abstract

Nutrient limitation of epilithic microbial activity is modified by stream discharge and drainage from the tundra surrounding the Kuparuk River, Alaska, USA. During 1984, after three weeks of whole stream enrichment with phosphorus, autotrophic activity per unit biomass had increased in the enriched section of the stream suggesting that phosphorus availability was limiting productivity. In contrast, after three weeks of phosphorus enrichment during 1985, heterotrophic and autotrophic activity was similar in the control and enriched sections of the stream. However, when ammonia or nitrate and phosphorus were added to an *in situ* bioassay chamber for two weeks, higher community biomass and heterotrophic activity resulted. Ten days later biomass significantly dropped in the unenriched section. Nitrate levels over this period increased four fold concomitantly with decreased stream discharge. Apparently during 1985, nitrogen was limiting epilithic microbial community in the phosphorus enriched section of the Kuparuk River. The significant negative relationship between nitrate concentration and stream discharge observed during 1984 supported the trends seen in 1985. These data suggest that nutrient concentrations which limit epilithic microbial activity and biomass are regulated by the stream discharge and drainage from the surrounding tundra.

Introduction

Nutrient availability is one factor which controls productivity in aquatic systems. In temperate lotic systems, the extent to which nitrogen and phosphorus concentrations mediate algal and bacterial productivity varies according to high natural concentrations in groundwater (Wuhrman & Eichenburger, 1975), elevated levels due to a watershed disturbance (Meyer & Johnson, 1983) or agricultural inputs (Kilkus *et al.*, 1975; Moore, 1977).

An ideal setting for studying the effects of nutrient limitation on productivity in lotic systems can be found at high latitudes. The Arctic provides systems which are simplified, exposed to continuous light during the summer with no shading due to riparian vegetation and free of impact due to anthropogenic sources (Peterson *et al.*, 1985). This study investigated the response of the epilithic microbiota to lotic enrichment with phosphorus and nitrogen. Nutrient manipulations included whole stream enrichment with phosphorus and an *in situ* bioassay chamber enriched with phosphorus and nitrate or ammonia. Epilithic microbial community biomass, alkaline phosphatase activity, heterotrophic and autotrophic activity were measured to determine the effect of nutrient enrichment on the microbial community in an oligotrophic arctic stream.

Study site

The Kuparuk River originates in the foothills of the Brooks Range of northern Alaska, USA. It then flows North-Northeast draining an area of 8 107 km^2 and empties into the Arctic Ocean. The study reach was located at approximately 68°35'N, 149°20'W. It consisted of a phosphorus enriched section and an unenriched, control section. The point source of nutrient addition was approximately 1.5 km above the point where the Dalton Highway crosses the river. The river consists of a single meandering channel from 3 to 6 meters wide, bordered by sedges, dwarf willow and tussock tundra underlaid by continuous permafrost. A complete description of the study is available (Peterson *et al.*, 1985).

Methods

Whole stream enrichment

Water velocity was measured with a Gurly pigmy current meter. The velocity was measured 30 cm above the stream bed and 30 cm below the water's surface across a river transect. The stream stage was measured at the site where the Dalton Highway crosses the Kuparuk River and the discharge was calculated after the discharge-height relationship was determined (Peterson *et al.*, 1986).

Phosphoric acid (P) was added to the Kuparuk River from one half inch (ID) tubing connected to a 20 l carboy, as a Mariotte bottle. During 1984, the rate of P addition was adjusted to 10 $\mu g\,l^{-1}$ above ambient concentrations based on the estimated river discharge. Actual concentrations were 7.7 $\mu g\,l^{-1} \pm 6.0$ greater than ambient. The whole stream enrichment lasted from 28 June to 12 August, 1984. During 1985, phosphoric acid was added to the Kuparuk River at constant drip rate using a peristaltic pump. The mean addition concentration was 55.5 $\mu g\,l^{-1} \pm 34.39$ greater than ambient. The experiment lasted from 30 June to 12 August, 1985.

During 1984, clear glass beads (4 mm diameter) were strung on monofilament line and placed on plexiglass rafts which were submersed approximately 20 cm beneath the water's surface. The beads were incubated for 21 days along a river transect, one station above and four stations below (100 m, 340 m, 1 000, and 3 000 m) the point of the phosphorus addition. After incubation, the epilithic community which developed on the beads was assayed for biomass and autotrophic activity.

Two sampling methods were used during 1985 to test the effect of the whole stream enrichment on epilithic microbiota. Clear glass beads (4 mm diameter) were incubated on the bottom of the Kuparuk River in nylon mesh bags (mesh size = 1.0 mm^2). After 21 or 35 days of incubation, the beads were sampled to measure epilithic biomass and activity. Also, after − 1 and 14 days of incubation in the river, epilithic microbiota were removed from a known surface area of indigenous stream rock using a wire brush. A composite sample was collected from nine rocks found in each of the river sections. A slurry was made by adding the epilithon to a known volume of filter sterilized (0.22 μm pore size) Kuparuk River water. Aliquots of the slurry were used for measurements of epilithic community biomass and activity.

In situ bioassay chamber

During 1985, an *in situ* flow-through microcosm (Peterson *et al.*, 1983) was used to test whether nitrogen was co-limiting the activity and biomass of the epilithic community in the phosphorus enriched section of the river. The microcosm, Fig. 1, was placed 1 000 m from the phosphorus dripper site. Glass beads strung on monofilament line were incubated in Lucite tubes in the light from 15 July to 25 July, 1985. Nutrient enrichment consisted of 100 $\mu g\,l^{-1}$ as NH_4Cl-N or 100 $\mu g\,l^{-1}$ as NO_3-N and the ambient levels of phosphorus in the phosphorus enriched section of the river (Table 2). The epilithic community which developed on the glass beads was sampled for various microbiological assays.

Fig. 1. In situ bioassay chamber used for the nitrogen enriched treatments in the phosphorus enriched section of the river during 1985.

Measurements of biological response to treatments

Lipid phosphate (White *et al.*, 1979) was used as a measure of total epilithic microbial biomass. Depending upon the sample type, a sample series consisted of three, one meter bead strands per treatment, or 18 ml of epilithic slurry composed of 2 ml from each of the nine rocks sampled per site.

Microbial heterotrophic activity and autotrophic activity potentials were measured by ^{14}C-acetate or ^{14}C-bicarbonate incorporation into microbial lipids (McKinley *et al.*, 1982). A sample consisted of 5 cm of glass beads which had been incubated under various nutrient regimes. To measure heterotrophic activity, 100 μl of ^{14}C-acetate (2.18 μCi/ml; specific activity = 58.6 mCi $mmol^{-1}$) were pipetted into scintillation vials containing samples and 1.9 ml of filter sterilized water. Vials were capped and incubated in the dark for 2 h at 10 °C. Autotrophic activity was measured using 100 μl of ^{14}C-bicarbonate (5.27 μCi ml^{-1}; specific activity = 57.1 mCi $mmol^{-1}$, Amersham). Clear and darkened vials were capped and incubated in a lighted incubator (85 μmol photons m^{-2} sec^{-1}) at 10 °C for 2 hours. After incubation, all samples were extracted with 2.5 ml of chloroform plus 5.0 ml of methanol as previously described (McKinley *et al.*, 1982). Production rates were calculated following the method of Wetzel & Likens (1979).

Alkaline phosphatase was measured on 5 cm of colonized glass beads. Two ml of 5 mM p-nitrophenyl (PNP)-phosphate was added to a glass vial containing a sample and incubated for 2–4 hours at 10 °C. The reaction was stopped by placing the samples on ice. After the incubation periods, 3 ml of 50 mM $NaHCO_3$ (pH = 9) were added to enhance color. Activity was quantified as a change in absorbance at 410 nm due to the hydrolysis of p-nitrophenyl phosphate to p-nitrophenol by alkaline phosphatase (Federle & Vestal, 1980).

Nutrient analysis

Water samples were retrieved from the Kuparuk River once a week along the stream transect. Nutrient analyses for NO_3-N, NH_4Cl-N and PO_4 were determined colorimetrically (Wetzel & Likens, 1979) on fresh water samples by Dr. M. C. Miller.

Statistical analysis

ANOVA was used to assess the effects of nutrient enrichment on the epilithic community for the *in situ* bioassay chamber. T-tests were used to assess the effect of whole stream phosphorus enrichment on the epilithon which developed later in the season. Regression analysis was used to assess the relationship between stream discharge and nitrate concentration in the water column.

Results

The epilithic microbial community in the Kuparuk River responded differently to whole stream nutrient enrichment during the two seasons of study in 1984 and 1985. During 1984, after twenty one days of incubation, autotrophic activity (Fig. 2) was about two times greater in the phosphorus enriched than the unenriched section of the river. However, during 1985, little difference between the unenriched and enriched reach was seen for autotrophic and heterotrophic activity after twenty one days of phosphorus enrichment (Figs. 3 and 4).

In 1985, after twenty one days of incubation in the stream, alkaline phosphatase activity (Fig. 5) was lower for epilithon associated with the phosphorus enriched sections of the river. Although heterotrophic activity and autotrophic activity associated with whole stream phosphorus enrich-

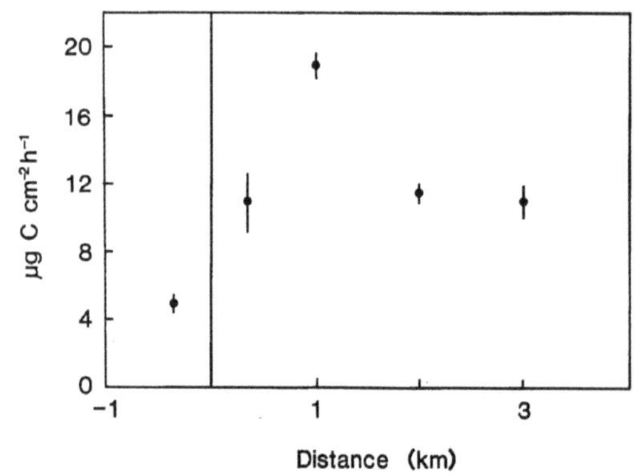

Fig. 2. The effect of whole stream enrichment with phosphorus on the autotrophic activity (μgC cm^{-2} hr^{-1}) of epilithic microbiota on glass beads in the Kuparuk River, AK after twenty-one days of incubation during 1984. The data are presented as the mean and standard deviation of three replicates. Distances are in reference to the point of nutrient addition indicated by the vertical line (positive distances are downstream).

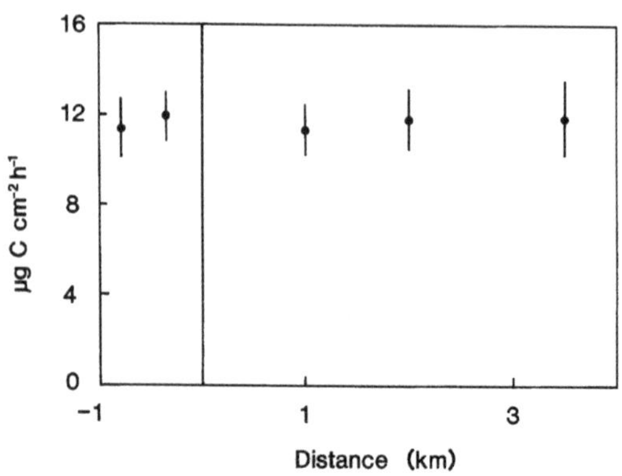

Fig. 3. The effect of whole stream enrichment with phosphorus on the autotrophic activity (μgC cm^{-2} hr^{-1}) associated with the epilithic microbial community which developed on glass beads after twenty one days of incubation during 1985. The data are presented as the mean and standard deviation of three replicates. Distances as in Fig. 2.

ment were not significantly different (Table 1) earlier in the season, heterotrophic activity in the phosphorus enriched section increased significantly (Fig. 6; $p < 0.05$) when ammonia or nitrate was added to an *in situ* bioassay chamber. Nitrate and to a greater extent ammonia, also resulted in a significant increase in microbial biomass in this P-enriched section (Fig. 7). Later in the season, when nitrate levels in the Kuparuk river increased as discharged decreased (Days 28–35; Table 2), microbial biomass and heterotrophic activity significantly dropped in the unenriched section of the stream, but increased in the phosphorus enriched section of the stream (Table 1).

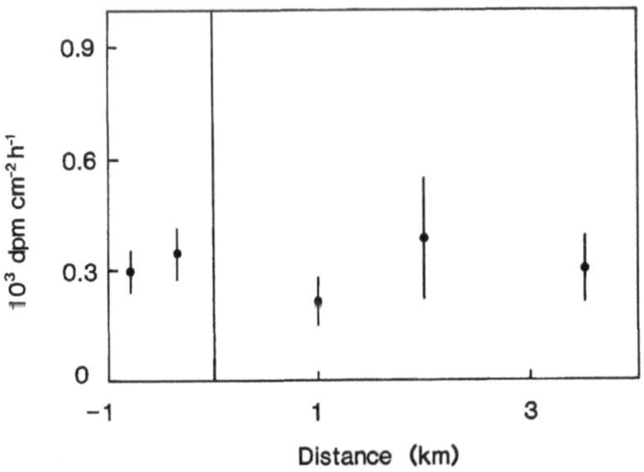

Fig. 4. The effect of whole stream enrichment with phosphorus on the heterotrophic activity (dpm $cm^{-2} hr^{-1}$) associated with the epilithic microbial community which developed on glass beads in the Kuparuk River, AK after twenty one days of incubation during 1985. The data are presented as the mean and standard deviation of three replicates. Distances as in Fig. 2.

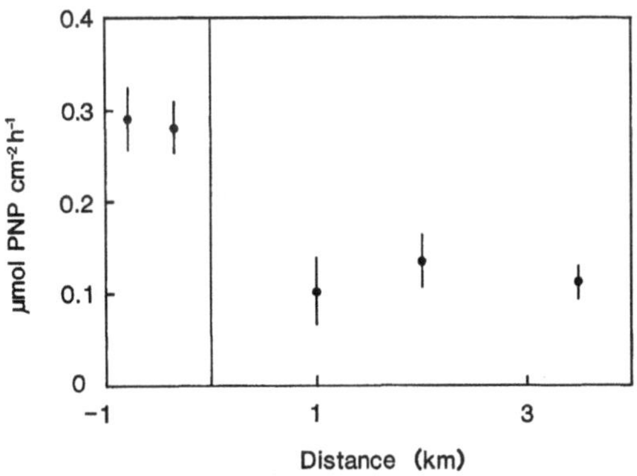

Fig. 5. The effect of whole stream enrichment with phosphorus on alkaline phosphatase activity (umol PNP released $cm^{-2} hr^{-1}$) associated with the epilithic community which developed on glass beads after twenty one days of incubation during 1985. The data are presented as the mean and standard deviation of three replicates. Distances as in Fig. 2.

Table 1. The effect of nutrient enrichment on the activity and community biomass of the epilithic microbiota of the Kuparuk River, Alaska, U.S.A. The epilithon was sampled from glass beads or directly from the river rock surface. Heterotrophic activity was measured as the ^{14}C-acetate incorporation into microbial lipids. Autotrophic activity was measured as the ^{14}C-bicarbonate incorporation into microbial lipids. Samples were retrieved at indicated intervals from 30 June, 1985 to 4 August, 1985. Data are represented as the mean plus or minus the standard deviation of three replicate samples. T-tests were used to assess the effect of the treatments on the epilithic microbial activity and biomass.

Day	Method	Treatment	Heterotrophic activity dpm $cm^{-2} hr^{-1}$	Autotrophic activity μgC $cm^{-2} hr^{-1}$	Microbial biomass μmol PO_4 cm^{-2}
−1	Rocks	−P	1149 ± 234	15 ± 3	
		+P	1076 ± 431	14 ± 8	
14	Rocks	−P	1244 ± 569	22 ± 6	0.532 ± 0.002
		+P	2694 ± 1727	37 ± 13	0.511 ± 0.001
21	Beads	−P	1400 ± 200	11 ± 2	0.173 ± 0.016
		+P	1000 ± 50	11 ± 1.5	0.183 ± 0.057
35	Beads	−P	500 ± 200		0.08 ± 0.01
		+P	1800 ± 100**		0.23 ± 0.03**

** $p < 0.05$ for the comparison of means between the control (−P) and enriched (+P) section of the river.

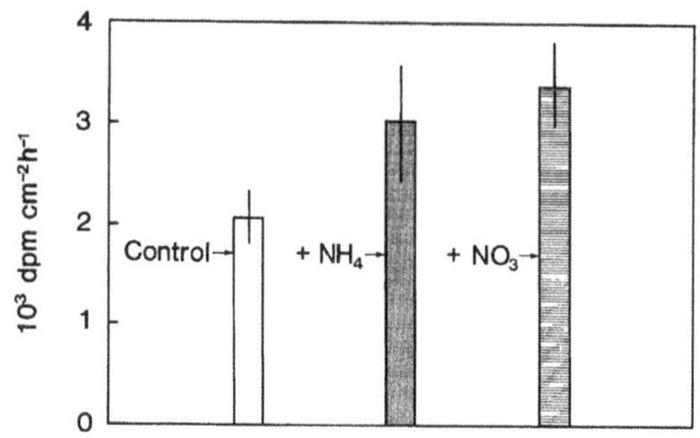

Fig. 6. The heterotrophic activity (DPM cm^{-2} hr^{-1}) associated with microbiota on strands of glass beads in an *in situ* bioassay chamber enriched with nitrogen as NH_4-N or NO_3-N. The chamber was located in the phosphorus enriched section of the river. Control was ambient phosphate enriched river water. The data are presented as the mean and standard deviation of three replicates.

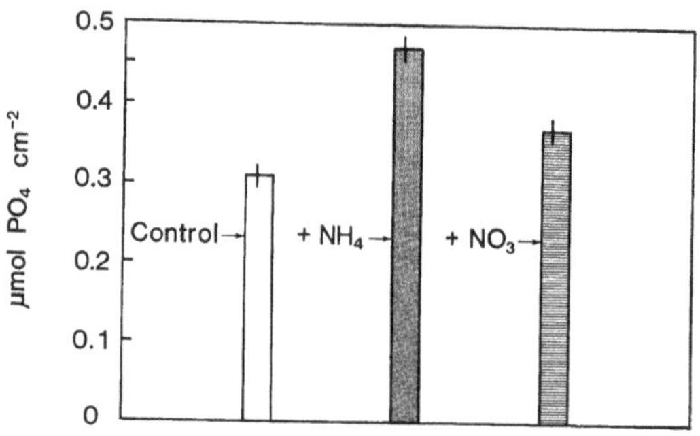

Fig. 7. The microbial biomass (μmol PO_4 cm^{-2} hr^{-1}) associated with microbiota on strands of glass beads in an *in situ* bioassay chamber enriched with nitrogen as NH_4-N or NO_3-N. The chamber was located in the phosphorus enriched section of the river. Control was ambient phosphate enriched river water. The data are presented as the mean and standard deviation of three replicates.

Discussion

The concentration of nitrate in the water column was a function of river discharge. A significant negative relationship between nitrate concentration and stream flow was noted during the summer of 1984 ($r^2 = 0.74$; $p < 0.01$, slope = -0.18, $n = 7$, Hullar, 1986). During 1985, the amount of nitrate increased as discharge decreased (Table 2), suggesting a constant amount of nitrate was seeping from the surrounding tundra, possibly as the soil thawed, and became more or less concentrated with a change in river discharge. Other possibilities could include increased nitrification rates in tundra soils as the soil warmed. Since nitrate is not retained within the humic matrix of the Arctic soils (Gersper *et al.*, 1980), it would run over the top of the permafrost and into the river. Howard-Williams *et al.* (1986) also found an inverse relationship between inorganic nutrients and stream flow in several antarctic streams.

Table 2. Discharge (m^3 s^{-1}) and mean seasonal nutrient concentration (μmol l^{-1}) above and below the phosphorus input in the Kuparuk River, AK., 30 June to 9 August, 1985.

Days	Discharge	NO_3	NH_4	DRP	N : P
− 1	2.3				
above		0.23	0.28	0.06	7.97
below		0.04	0.22	0.07	4.31
1	1.94				
above		0.21	0.17	0.03	14.08
below		0.10	0.36	0.13	3.44
14	0.82				
above		0.81	0.64	0.02	92.18
below		0.59	0.58	0.45	2.69
21	0.42				
above		1.68	0.15	0.03	88.43
below		1.31	0.56	0.61	3.29
28	0.30				
above		4.19	0.31	0.04	158.30
below		2.31	0.15	0.80	8.57
35	0.24				
above		5.20	0.44	0.12	48.52
below		4.00	0.47	1.21	3.76
39	0.20				
above		4.93		0.03	211.70
below		3.62		1.24	3.18

The microbiota of the Kuparuk River appeared to be phosphorus limited as are many arctic ecosystems (Miller *et al.*, 1986; Peterson *et al.*, 1986). During 1984, autotrophic activity increased with phosphorus enrichment (Fig. 2). High levels of alkaline phosphatase activity (Fig. 5) were also associated with the epilithon in the control section of the Kuparuk River suggesting phosphorus limited conditions during the 1985 season. Algal and bacterial production of alkaline phosphatase has been associated with phosphorus limitation in other aquatic systems (Sayler *et al.*, 1979; Smith & Kalff, 1981; Federle & Vestal, 1980; Jones, 1972). However, during this year the epilithic microbial response to in-stream phosphorus enrichment was only evident towards the end of the sampling season.

The enrichment tube experiments in the Kuparuk River suggested that nitrogen may also play a role in controlling epilithic biomass and activity, possibly acting as a co-limiting nutrient in some years. There was no response in the control to the P-enriched river water alone, but a significant response to added nitrate or ammonium (Figs. 6 and 7). During the early part of the year when nitrogen levels were lower (Table 2), the epilithic microbial biomass did not increase when phosphorus was added to the stream (Table 1), as it had in previous years (Fig. 2, Peterson *et al.*, 1986). The low N : P ratios (Table 2) and reduced alkaline phosphatase activity (Fig. 5) in this P-enriched section are consistant with the response seen in the nitrogen enriched tubes. It remains possible, however, that nitrogen acted as a primary limiting nutrient in the upstream, unenriched section and that the alkaline phosphatase activity was indicative of only a mild phosphorus shortage, or a dependence on organic phosphorus compounds. Later in the season, although ammonia concentrations remained similar (Days 28–38; Table 2), nitrate levels increased two fold as discharge decreased. Microbial biomass and activity decreased in the upstream section, but increased in the downstream, P-enriched section suggesting a shift back towards phosphorus as the single limiting nutrient in the natural, unenriched epilithon.

Acknowledgements

We thank Mr. D. Feldhake for field assistance during 1984, and Ms. N. K. Coleman and Ms. M. J. Kaufman for assistance in the field during 1985. Many thanks to Dr. M. C. Miller for nutrient analysis data; Dr. B. J. Peterson and Mr. J. Helfrich for the Kuparuk River discharge data; Dr. M. A. Lock and Ms. S. Ditner for maintenance of the nitrogen bioassay chamber during 1985. This work was make possible through funds received from Division of Polar Programs Grand 8320544 from the USA National Science Foundation.

References

Elwood, J. W., J. D. Newbold, A. F. Trimble & R. W. Stark, 1981. The limiting role of phosphorus in a woodland stream eco-system: Effects of P enrichment on leaf decomposition and primary producers. Ecology 62: 146–158.

Federle, T. W. & J. R. Vestal, 1980. Microbial colonization and decomposition of *Carex* litter in an arctic lake. Can. J. Microbiol. 28: 686–695.

Gersper, P. L., V. Alexander, S. A. Barkley, R. J. Barsdate & P. S. Flint, 1980. The soils and their nutrients. In J. Brown, P. S. Miller, L. L. Tiezen and F. L. Bunnell (eds.), An Arctic Ecosystem. Dowden, Hutchinson & Ross, Inc., Stroudsburg, PA 219–254.

Howard-Williams, C., C. L. Vincent, P. A. Broady & W. F. Vincent, 1986. Antarctic stream ecosystems: Variability in environmental properties and algal community structure. Int. Revue ges. Hydrobiol. 71: 511–544.

Hullar, M. A., 1986. The effects of nutrient enrichment and light regimes on the epilithic microbiota of an oligotrophic arctic river. M. Sc. Thesis, University of Cincinnati, Cincinnati, OH, USA. 100 pp.

Kilkus, S. P., J. D. Laperriere & R. W. Bachman, 1975. Nutrients and algae in some central Iowa streams. J. Wat. Pollut. Cont. Fed. 47: 1870–1879.

Jones, J. G., 1972. Studies on freshwater microorganisms: Phosphatase activity in lakes of differing degrees of eutrophication. J. Ecol. 60: 777–791.

McKinley, V. L., T. W. Federle & J. R. Vestal, 1982. Effects of petroleum hydrocarbons on plant litter microbiota in an arctic lake. Appl. Environ. Microbiol. 43: 129–135.

Miller, M. C., G. R. Hater, P. Spatt, P. Westlake & D. Yeakel, 1986. Primary production and its control in Toolik Lake, Alaska. Arch. Hydrobiol. 1: 97–131.

Moore, J. W., 1977. Some factors affecting algal densities in an eutrophic farm stream. Oecologia 29: 257–267.

Peterson, B. J., J. E. Hobbie & T. L. Corliss, 1986. Carbon flow in a tundra stream ecosystem. Can. J. Fish. Aquat. Sci. 43: 1259–1270.

Peterson, B. J., J. E. Hobbie, A. E. Hershey, M. A. Lock, T. E. Ford, J. R. Vestal, V. L. McKinley, M. A. J. Hullar, M. C. Miller, R. M. Ventullo & G. S. Volk, 1985. Transformation of a tundra river from heterotrophy to autotrophy by addition of phosphorus. Science 229: 1383–1386.

Peterson, B. J., J. E. Hobbie, T. L. Corliss & K. Kriet, 1983. A continuous-flow periphyton bioassay: Tests of nutrient limitation in a tundra stream. Limnol. Oceanogr. 28: 583–591.

Rhee, G. Y. & I. J. Gotham, 1980. Optimum N:P ratios and co-existance of planktonic algae. J. Phycol. 16: 486–489.

Sayler, G. S., M. Puziss & M. Silver, 1979. Alkaline phosphatase assay for freshwater sediments: Application to perturbed sediment systems. Appl. Environ. Microbiol. 38: 922–927.

Smith, R. E. & J. Kalff, 1981. The effect of phosphorus limitation on algal growth rates: Evidence from alkaline phosphatase. Can. J. Fish. Aquat. Sci. 38: 1421–1427.

Terry, K. L., 1980. Nitrogen and phosphorus requirements of *Pavlova literi* in continuous culture. Bot. Mar. 23: 757.

Wetzel, R. G. & G. E. Likens, 1979. Limnological analyses. W. B. Saunders, Philadelphia, PA. 357 pp.

White, D. C., W. M. Davis, J. S. Nickels, J. D. King & R. J. Bobbie, 1979. Determination of the sedimentary microbial biomass by extractable lipid phosphate. Oecologia. 40: 51–62.

Wuhrmann, K. & E. Eichenberger, 1975. Experiments on the effects of inorganic enrichment of rivers on periphyton primary production. Verh. Int. Ver. Limnol. 19: 2028–2034.

Hydrobiologia 172: 27–38, 1989
W. F. Vincent and J. C. Ellis-Evans (eds) High Latitude Limnology

Microbial communities in southern Victoria Land streams (Antarctica) I. Photosynthesis

Clive Howard-Williams & Warwick F. Vincent
Taupo Research Laboratory, DSIR, P.O. Box 415, Taupo, New Zealand

Key words: Antarctica, biofilms, cyanobacteria, photosynthesis, chlorophyll, pigments, nitrogen, phosphorus, epilithic algae

Abstract

The glacier-fed ephemeral streams of southern Victoria Land (*ca.* 78° S, 164° E) are colonised by an epilithon dominated by cyanobacterial mats and films. Biomass levels are often high (> 15 μg Chl *a* · cm^{-2}). The mat structure, pigment and photosynthetic characteristics of these communities have been investigated on site. The mats in high light environments have a layered structure with high levels of light shielding accessory pigments in the upper layers and elevated chlorophyll *a* and phycocyanin concentrations in the lower layers. Photosynthetic rates per unit area (0.4–3.5 μg C · cm^{-2} · hr^{-1}) fall within the range reported for temperate communities. P *vs.* I curves were used to separate high, intermediate and low light communities. I_k values for high light communities were at or lower than PAR recorded at midnight in the polar midsummer (*ca.* 100 μE m^{-2} · s^{-1}). We did not detect photoinhibitory responses at the midday light intensities. *In situ* continuous nutrient enrichment experiments failed to demonstrate N or P limitation to pigment content or photosynthetic rates. We suggest that the growth of these communities is controlled by factors other than light and nutrients.

Introduction

Rivers and streams are a common feature of the coastal ice-free areas of Victoria Land in Antarctica. These southern polar streams are derived from melting glaciers and typically flow for several weeks during mid-summer (Vincent & Howard-Williams, 1986). Where slopes are not too steep and channel beds are relatively stable, a dense epilithic community can develop. This is dominated by cyanobacteria with associated heterotrophic bacteria, fungi and a few microherbivores. A large proportion (> 50%) of the biomass of the summer epilithic communities remains in a frozen but viable state over the 10 months or so of the year when the streams do not flow and the channels are dry. This period includes the Antarctic winter when temperatures in this area fall to − 55 °C.

With the onset of glacier melt, and the presence of flowing water in the stream channels the dry frozen epilithic communities respond to rewetting rapidly (Vincent & Howard-Williams, 1986, 1987) and within 30 minutes $H^{14}CO_3$ label enters a wide range of photosynthetic end products including low molecular weight compounds, protein and polysaccharides. However, although the cyanobacteria can re-initiate metabolism rapidly on rehydration, full recovery of the populations to streamflow conditions is much slower. A time frame of days to weeks is required.

A previous study (Vincent & Howard-

Williams, 1986) demonstrated rates of gross photosynthesis for the midsummer epilithon of *ca.* 1–2 $\mu g\ C\ cm^{-2}\ hr^{-1}$ of which a variable but generally high proportion was consumed in respiration. On the basis of colonisation experiments on artificial substrates in several of the streams of different nutrient status we hypothesised that nutrients and light were unlikely to be controlling factors in the productivity of the epilithon although Goldman *et al.* (1972) had suggested that the marked photoinhibition they detected in pond algal mats in the area could repress daily photosynthetic rates.

The aims of this study were to examine in more detail the pigment composition and photosynthetic characteristics of the stream communities and attempt to better quantify the primary control on photosynthesis. To this end we report here on the specific effects of variable light intensity and nutrient regimes on pigment composition and photosynthetic rates.

Study sites

Experiments and field sampling were carried out at three sites in the McMurdo Sound area (78° S, 164° E). These were the Adams and Fryxell streams and the Onyx river. All these typically flow from mid December to late January or early February. Some typical environmental features are listed in Table 1. Flows are highly variable both diurnally, with the streams often reduced to a trickle between 2400 and 0600 hrs, and over longer time scales when days of warm weather can cause large volumes of water to discharge from the glaciers. Conversely cold weather can stop flow completely. Like discharge, nutrient concentrations are very variable with highest levels occurring at lowest discharge during freeze periods. The Fryxell stream was found to contain among the lowest nutrient concentrations in a wide range of streams we have examined (Howard-Williams *et al.*, 1986b). The algal communities of the streams typically consisted of;

- *Nostoc* dominated mats and films in seepage areas near glacier faces and on moist ground where water flows very slowly.
- Dark 'crusts' of *Gloeocapsa* and *Schizothrix* occur on upper surfaces of rocks and are particularly abundant at or near the water line.
- Mat communities in stream channels dominated by *Phormidium, Oscillatoria, Lyngbya, Microcoleus* and *Schizothrix* (Broady, 1982). These mats are the most common type of epilithic community and occur on and between rocks, across pebble beds and gravels. Typically the mats are brown pink or orange.
- Green or grey green or yellow green filaments and mats of *Prasiola calophylla, Tribonema* sp., *Binuclearia tectorum* and *Oscillatoria* spp. occur in crevices between large boulders or beneath

Table 1. Selected environmental features of three south Victoria Land streams (Data from Howard-Williams *et al.*, 1986a, 1986b). ND = not detectable.

		Adams	Fryxell	Onyx	Alph
Length (km)		2.7	2	30	12
Midseason discharge* $m^3\ s^{-1}$		0.15	0.06	1.1	1.0
Suspended sediments $g\ m^{-3}$		8	2	27–131	5
Nutrients midseason					
Upstream	DRP $mg\ m^{-3}$	28	6.5	0.3	12.1
	NH_4-N $mg\ m^{-3}$	35	1.5	2	3.4
	NO_3-N $mg\ m^{-3}$	46	17.6	13	5.1
Downstream	DRPmg m^3	5.4	0.2	1.9	12.3
	NH_4-N	8.5	ND	8.0	2.8
	NO_3-N	7.5	2.0	4.0	1.3

* Daily maximum
# DRP = Dissolved Reactive Phosphorus

boulders in fast flowing water where sediments cannot accumulate.

The biomass dominants were *Nostoc*, *Phormidium*, *Binuclearia* and *Prasiola* and our experiments concentrated upon these communities.

Methods

Environmental measurements

Stream temperatures were measured with a YSI combined O_2-temperature probe and solar radiation as photon flux density in the 400–700 nm waveband with a LiCor Instruments sensor LI 190 and meter LI 185. Stream discharges were obtained from velocity profiles obtained with a Teledyne Gurley pygmy current meter. Water for nutrient analyses was filtered on site through pre-rinsed Whatman GF/C filter papers stored frozen and later analysed by Technicon Auto-Analyser (Howard-Williams *et al.*, 1986b).

Biomass, pigments and tissue analyses

Samples of epilithon were collected by small plastic syringe corers and kept frozen in storage in the dark until return to New Zealand. Dry weight, ash free dry weight and% carbon were analysed as in Vincent & Howard-Williams (1986). Tissue N and P analyses were done by Technicon Auto-Analyser following acid digestions.

Extracts of the core samples for chlorophyll *a* were made with dimethylsulphoxide at room temperature (15–20 °C) for 24 hrs and then assayed by the method of Shoaf & Lium (1976). Pigment scans were run from 350–750 nm on a Cary dual beam recording spectrophotometer blanked against dimethylsulphoxide/acetone mix. Relative carotenoid concentrations were expressed as the ratio of the extinction at 480 nm (Strickland & Parsons, 1968) to that at 663 nm. Phycocyanin extracts were made on an aqueous extract of cell material following maceration in a tissue grinder and then cell rupture with glycerol/distilled water (Wyman & Fay, 1986). Concentrations were determined on an Aminco scanning spectrofluorometer against prepared C-phycocyanin standards.

Carbon dioxide exchange

Cores of the epilithon were placed in 30 ml Universal vials with a covering of 2–3 ml of stream water and sealed with a rubber septum and screw cap. The vials were connected with a Binos portable Infra-Red Gas Analyser (IRGA) set up on site at the stream (Vincent & Howard-Williams, 1986). The IRGA was set up in a differential closed circuit configuration. Field calibration was done with two CO_2/air standards prepared by the laboratory of New Zealand Industrial Gases Ltd.

Field experiments

(a) *Low light intensity and pigment composition.* A 0.25 m^2 area of 95% shade cloth was placed over intact *Phormidium* mats in the Fryxell stream, and at 0, 2, and 4 days, triplicate core samples were collected from the sunlit and shaded portions of the stream for analysis.

(b) *Photosynthesis vs. Irradiance* (P *vs.* I). Experiments were conducted on several mat communities. Sample and reference vials of the IRGA set up were placed in a series of shade cloths in stream water (temperatures 5–8 °C). The light series was 100% (full sun), 60%, 36%, 27%, 16%, 10%, 6%, 2%, 1% and 0% (darkness). The IRGA system was flushed out after each shade incubation to ensure an adequate CO_2 supply in high light and to prevent a CO_2 buildup in low light. Ambient photon flux density and temperature were monitored throughout the experiments.

(c) *The effects of nutrient supply.* The experiment was a 9 day incubation of several epilithic communities in flow through enrichment tubes placed in the middle of the bed of Fryxell stream. The design closely followed that of Peterson *et al.* (1983) except that in the shallow Fryxell stream we could place the clear plexiglass tubes on the stream bed (Plate 1). Continuous enrichment with dissolved inorganic N (final concentration 200 mg N m^{-3}) or dissolved reactive phosphorus (final concentration 20 mg DRP m^{-3}) or both together was maintained with 3 litre reservoirs coated in black, which discharged through a 4 mm tube to a tip of fine calibrated auto-analyser

Plate 1. Fryxell stream during early morning low flow conditions showing the nutrient enrichment tubes. The stream water source, the Canada glacier, lies in the background.

tube which delivered a constant flow of 35 ml hr^{-1} at a 50 cm head. Variability in flow was influenced only by small changes in the reservoir hydrostatic heads as they drained. Reservoirs were topped up every two days and the incubation continued for nine days. Some freezing problems were encountered on days 7, 8 and 9 when the tubes froze briefly in the early hours of the morning. A control tube had stream water only in the reservoir. During tube construction, placing of the mixing baffles was modified relative to Peterson *et al.* (1983) with the aid of dye tracers to ensure an homogeneous mix of added nutrient and stream water by the time it reached 25 cm from the upstream end of the tube. All study samples were placed at this distance in the tubes. Stream discharge and temperatures were monitored several times each day and velocities through the tubes were regularly checked to ensure equal flow through each tube. Communities of *Nostoc*, *Binuclearia* and two *Phormidium* types were subject to the experimental treatments. Initially and at the end of the incubation pigment analyses, N and P tissue concentrations and photosynthetic rates were measured on 3 to 6 replicate samples of each community.

Results

Epilithic biomass

In midsummer, maximum biomass was frequently measured in excess of 15 μg Chl *a* cm^{-2} (Table 2). In the low light zone on the underside of the rocks mean maximum values were always high (30–40 μg Chl *a* cm^{-2}) but values approaching this could also be found on the exposed upper surfaces of the stream bed in the mats dominated by *Nostoc* and *Phormidium*. A major proportion (20–100%) of this chlorophyll survives the winter in a dry frozen state (Vincent & Howard-Williams, 1986). Some of this may, however, not be photosynthetically active, or even chlorophyll *a* (Vincent & Howard-Williams, 1988).

Standing stocks for three epilithic communities in terms of C, N and P are shown in Table 3. The relatively high inorganic component (silts, sand grains etc) which makes up 60–80% of the dry weight in the *Nostoc* and *Phormidium* mats is reflected by low% C values and hence tissue N and P (Table 3). Carbon as% of dry weight of the mat communities was 5.2 (± SE 0.51) for *Nostoc* and only 1.4 (± SE 0.14) for *Phormidium*. The high inorganic component of the mats probably contributes to considerable light scattering and attenuation within them. The C/N ratio is similar

Table 2. Biomass (μg Chl *a* cm^{-2}) of epilithon in three south Victoria Land streams. NF = not found, ND = no data. Values as means (± SD). n ranged from 3–13.

Community	Stream		
	Adams	Fryxell	Onyx
High light			
Nostoc mats and films	27.3 (9.5)	28.0 (9.5)	NF
Gloeocapsa, Schizothrix crusts	11.7		12.0 (18.5)
Phormidium mats			
Pink mats in channels	30.1(17.6)	17.2(10.5)	
Brown furry mats on stones	11.7 (2.6)	1.8 (1.4)	3.0 (2.5)
Crevices			
Binuclearia tectorum tufts	ND	32.6 (7.3)	NF
Low light			
Prasiola calophylla	32.0 (3.5)	29.6(12.1)	NF

Table 3. Stocks of carbon, nitrogen and phosphorus and N/P ratios in epilithon from Adams and Fryxell streams. Data as mg cm^{-2}, means ($\pm$ SD) (n > 6). – = no data.

Community	N	P	C	N/P	C/N
Nostoc	1.8(0.45)	0.13(0.02)	28.1(13.2)	28.1(3.95)	15.6
Phormidium	0.4(0.14)	0.09(0.04)	7.9 (3.2)	4.8 (1.01)	17.5
Binuclearia	1.2(0.28)	0.11(0.04)	–	10.9(0.99)	–

to that reported for antarctic stream chlorophytes (Hawes, 1988) and is high for cyanophyte communities. Perhaps this is because a proportion of the mat is comprised of a polysaccharide matrix (Vincent & Howard-Williams, 1987) and because dead cells are incorporated in this matrix too. The high N/P ratio of *Nostoc* may reflect the N-fixing ability of this community.

Pigment composition and mat structure

The cyanobacterial mats of the high light zone are a complex structure of layers of trichomes embedded in a mucilage (matrix). Associated bacteria (rods and cocci) and a few fungal hyphae have been identified from the matrix (Vincent & Howard-Williams, 1987), and scanning electron micrographs also show fine sediment particles trapped here. Mats in the high light environment produce polysaccharide as the dominant end product of photosynthesis, presumably in part for incorporation into the polysaccharide mat matrix. In the low light, non-mat forming *Prasiola calophylla* community, low molecular weight compounds were produced as the dominant end product (Vincent & Howard-Williams, 1988).

Light is rapidly attenuated by the mat such that cells in the lower layers of the 0.5–3 mm thick mats are in a low light environment. Those in the upper layers are exposed to bright 24 hr insolation. Photosynthetically active radiation (PAR) is reduced to *ca.* 1% of above surface levels half way through a 9 mm thick *Nostoc* mat, and to 0.01% at the base of the mat. The cells in the lower layers appear to be shade adapted in terms of the amount and composition of their photosynthetic pigments (Table 4). This layer has twice as much chlorophyll *a* on a unit area basis, and when normalised to carbon, a fourfold increase is seen in the lower layers. The phycobilin pigment C-phycocyanin also increased in the bottom layers, but carotenoid accessory pigments were twice as high in the upper portion of the mat. Spectral scans of acetone extracts of the layers (Fig. 1a) clearly show the carotenoid shielding below wavelengths of 500 nm in the mat upper layers and the high chlorophyll peak at 663 nm in the lower layers.

Table 4. PAR and pigment characteristics of the upper and lower layers of an epilithic *Nostoc* mat community, Fryxell Stream. Total mat thickness was 9 mm.

Parameter	Layer		Bottom/Top
	Top	Bottom	
% of mat	63	37	
PAR at base (% of mat surface value)	0.6	0.01	
Chlorophyll *a* µg cm^{-2}	35.4 ± 11.4	74.6 ± 7.3	2.1
Chlorophyll *a* µg mg C^{-1}	4.1 ± 0.7	17.5 ± 2.3	4.3
Carotenoid ratio	4.9 ± 0.2	2.4 ± 0.2	0.5
C-Phycocyanin *a* µg cm^{-2}	5.2	15.0	2.9

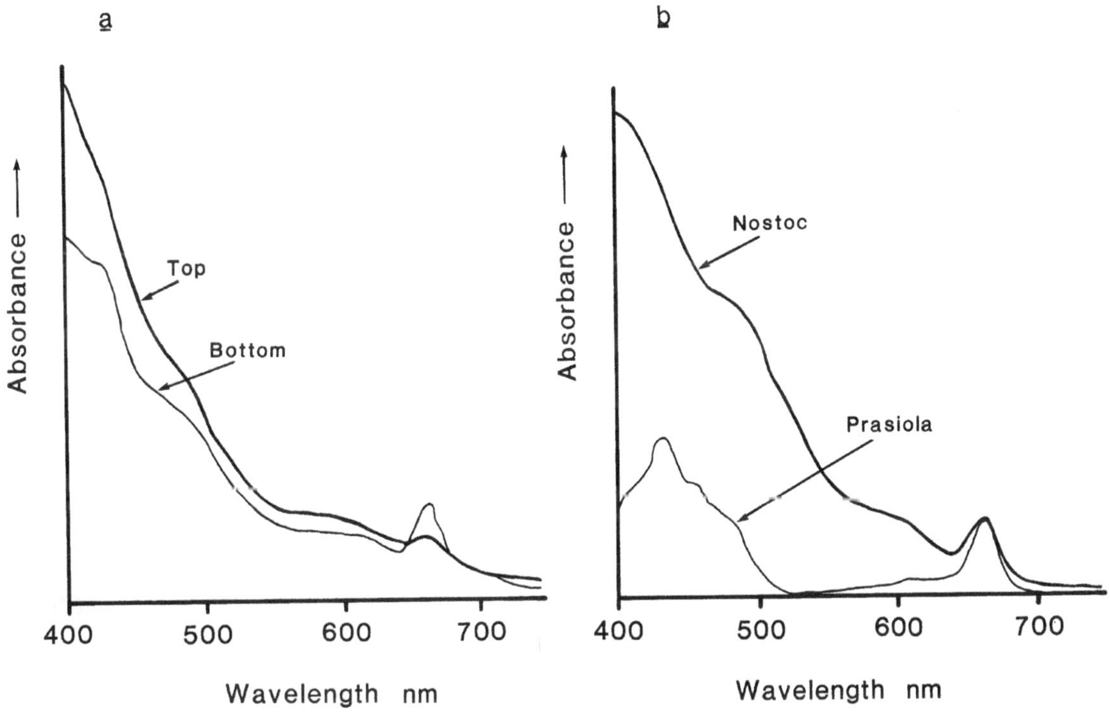

Fig. 1. Absorbance scans of DMSO/acetone extracts of some epilithic communities. (a) Top and bottom sections of a *Nostoc* mat from Fryxell stream. (b) Comparison of a sun (*Nostoc*) and shade (*Prasiola*) community from Adams stream. Data normalised to absorbance at 663 nm.

Distinct differences in pigment composition are seen in the low light habitat community *Prasiola calophylla* when compared with a high light *Nostoc* mat (Fig. 1b). While the spectral scan of the *P. calophylla* community is similar to that for chlorophyll, the scan of *Nostoc* shows a high accessory pigment shielding below 500 nm.

The shade adaptation experiment was run to examine the short term pigment responses to changing light intensity. Within two days of covering a section of the Fryxell stream bed with 95% shade cloth a major increase in chlorophyll *a* was recorded relative to adjacent unshaded control sites (Table 5).

The effect of light intensity on photosynthesis

P *vs.* I curves for five algal communities comprising four dominant epilithic genera show distinct differences (Figs. 2, 3). Site to site variability in P *vs.* I for a given community appears minimal (Fig. 2), as the same response was found for *Nostoc* mats of similar thickness in two very different streams.

However, variability in P *vs.* I curves between communities is considerable, and follows an expected pattern based on the light regime in the natural habitat (Fig. 3). The shade community,

Table 5. Chlorophyll *a* analyses (μg Chla cm^{-2}) on *Phormidium* mats in the sun and shaded with 95% shade cloth over an 8-day period. Data as mean ± SE for 5 replicate core samples from the mats in Fryxell Stream 6–10 January 1986.

Day	Sun	Shade
0	15.9 ± 3.1	15.9 ± 3.1
2	18.1 ± 3.6	31.8 ± 5.1
4	17.9 ± 3.6	28.6 ± 2.2

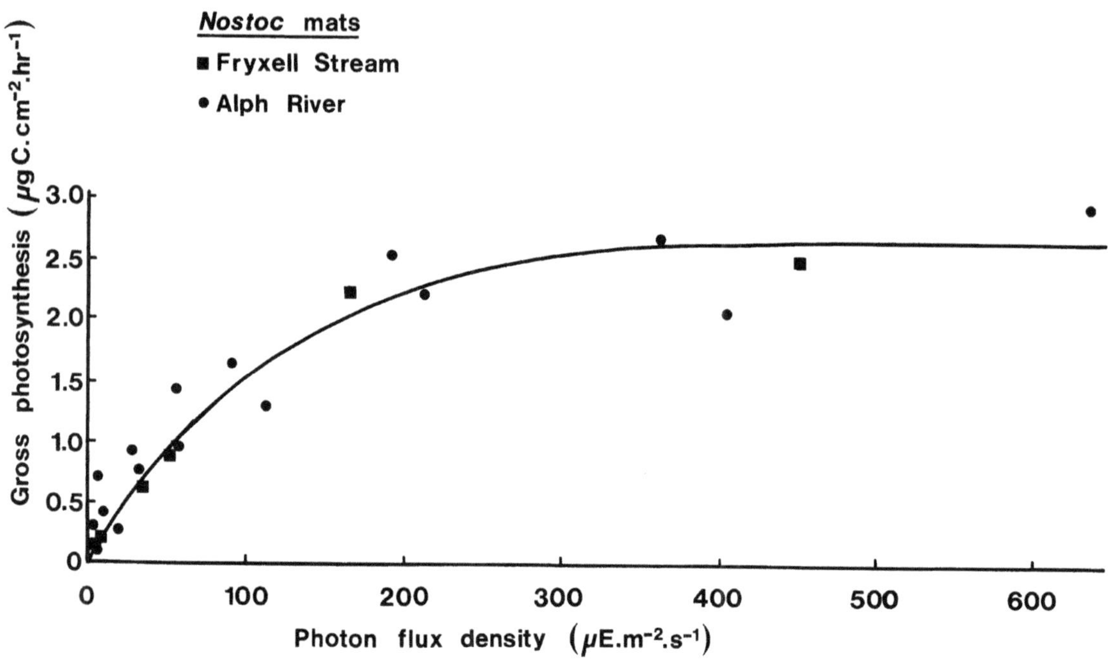

Fig. 2. P *vs* I relationship for *Nostoc* mats of similar thickness from two streams.

Prasiola, has an I_k value of *ca.* 20 $\mu E\, m^{-2}\, s^{-1}$, the intermediate light community of rock crevices (*Binuclearia*) and I_k of *ca.* 70 $\mu E\, m^{-2}\, s^{-1}$, and that for *Phormidium* is *ca.* 105 $\mu E\, m^{-2}\, s^{-1}$. The highest I_k was found for *Nostoc* and probably results from the high light shielding effect of the dark accessory pigments of this community.

Absolute rates of photosynthesis on a unit area basis increase from sun to shade (Fig. 3) but assimilation numbers (μg C fixed (μg Chl *a*)$^{-1}$ hr^{-1}) are higher (0.17) in the sun communities than in the shade (0.09 for *Prasiola*). These assimilation numbers lie well below those recorded for temperate communities. This discrepancy may be due to a proportion of the chlorophyll being in a preserved, non-viable state in these cold conditions. Photoinhibition was not observed in these short-term incubations and we did not detect hysteresis effects when incubations were taken down from bright light to shade.

The effect of nutrients

Nine days of continuous enrichment in the tube experiment (Plate 1) had surprisingly little effect on three test epilithic communities (Table 6). Fryxell stream was chosen for this study as the ambient nutrient levels were lower than those recorded in other Victoria Land streams (Table 1) (Howard-Williams *et al.*, 1986b). During the incubation period ambient stream inorganic nitrogen levels ranged from 3.6 to 6.1 mg m^{-3} of NO_3-N and 1.9 to 7.0 mg m^{-3} NH_4-N. DRP varied between 1.3 and 2.0 mg m^{-3}. Enrichment of the tubes was calculated to increase the NO_3-N to 200 mg m^{-3} and the DRP to 20 mg m^{-3}. Spot checks were carried out twice during the incubation by sampling the outflow ends of the tubes. The mean DRP value in the +P treatment was 23.6 mg m^{-3}, the NO_3-N in the +N treatment was 186 mg m^{-3} and in the N + P treatment DRP and NO_3-N were 36.0 and 262 mg m^{-3}.

A positive response in terms of tissue nitrogen

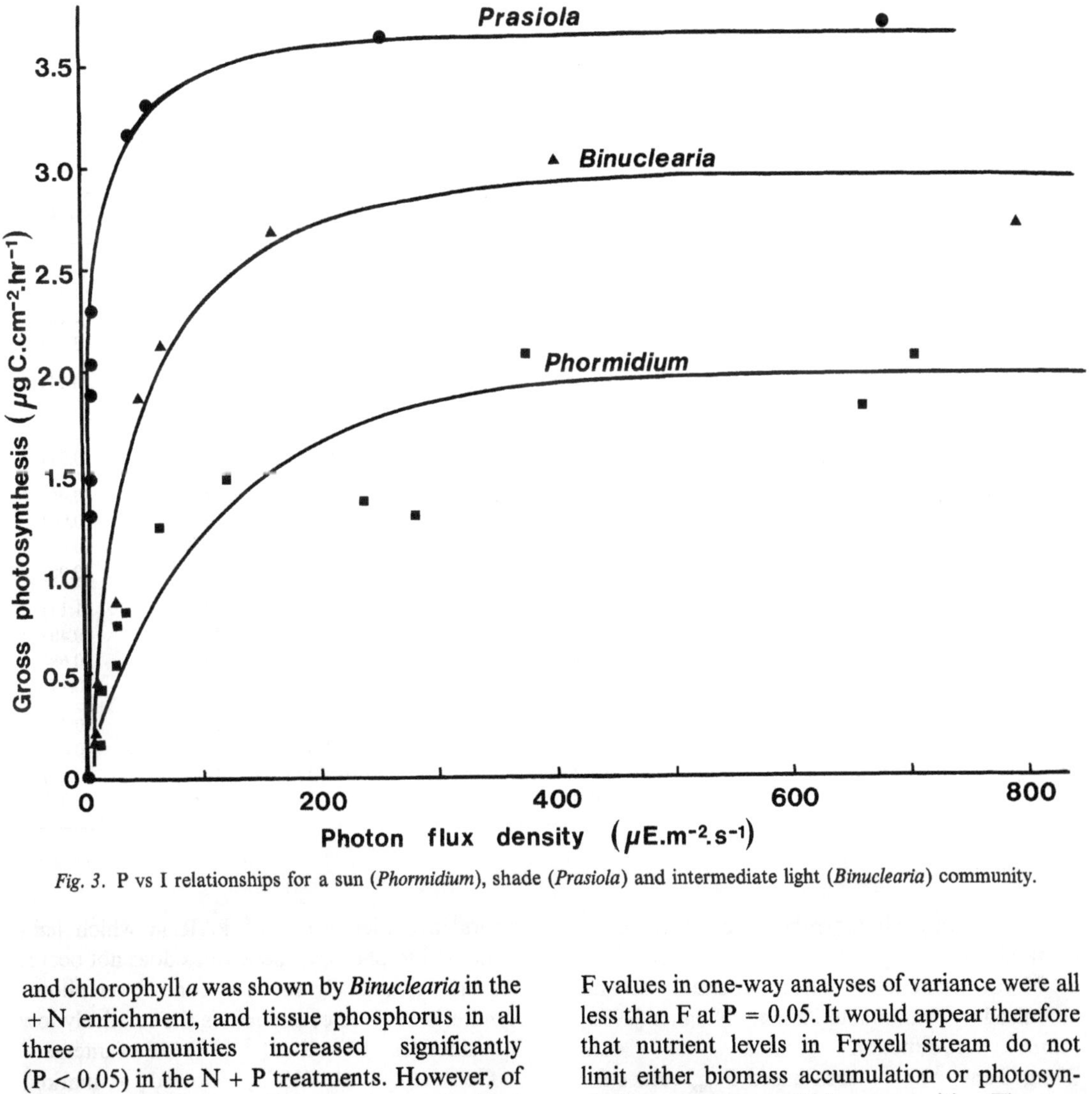

Fig. 3. P vs I relationships for a sun (*Phormidium*), shade (*Prasiola*) and intermediate light (*Binuclearia*) community.

and chlorophyll *a* was shown by *Binuclearia* in the +N enrichment, and tissue phosphorus in all three communities increased significantly ($P < 0.05$) in the N + P treatments. However, of note is that the N/P ratio in the tissues declined significantly ($P < 0.01$) in *Nostoc* in the +P and N + P treatments (Table 6). Neither *Nostoc* nor *Phormidium* mats showed any biomass response to enrichment. However, as we were examining changes in an established community, comparison with other studies (e.g. Peterson *et al.*, 1983; Hullar & Vestal, 1988) which measured colonisation of new substrates, is difficult.

There was no response to enrichment in terms of community photosynthetic rates (Table 6) and F values in one-way analyses of variance were all less than F at $P = 0.05$. It would appear therefore that nutrient levels in Fryxell stream do not limit either biomass accumulation or photosynthetic rates of the epilithic communities. The continuous supply of nutrients by the stream water with a mean velocity of 0.26 m^{-3} s^{-1} through the tubes was adequate for growth.

Discussion

Photosynthetic rates of epilithon for a variety of streams are described in the review by Lock (1981). A range of mean values (from a mixture of

Table 6. Response of three communities to 9 days of continuous enrichment with nitrogen (+N), phosphorus (+P) and N + P combined. Controls (C) had no enrichment. Data as means ± SE. Significant ($P < 0.05$) responses above (or in the case of N/P ratios, below) control values are indicated with an asterisk.

Measurement		Species					
		Binuclearia		Phormidium		Nostoc	
		x	(± SE)	x	(± SE)	x	(± SE)
	C	27.4	(3.02)	17.2	(1.96)	28.0	(1.71)
Chlorophyll *a*	+P	21.9	(2.76)	9.1	(0.71)	11.8	(1.14)
$\mu g\ cm^{-2}$	+N	39.7*	(3.62)	9.2	(0.69)	28.3	(4.18)
	+NP	26.2	(2.75)	12.2	(2.37)	25.7	(3.98)
	C	2.64	(0.28)	1.58	(0.47)	1.10	(0.64)
Gross Photosynthesis	+P	1.56	(0.30)	1.94	(0.32)	1.64	(0.43)
$\mu g\ C\ cm^{-2}\ hr^{-1}$	+N	1.42	(0.12)	1.94	(0.17)	2.16	(0.21)
	+NP	2.26	(0.37)	1.60	(0.09)	1.25	(0.07)
	C	1.72	(0.01)	0.74	(0.01)	1.01	(0.06)
Microbial Phosphorus	+P	2.30	(0.17)	0.77	(0.01)	1.15	(0.04)
$\mu g\ g^{-1}$	+N	2.52	(0.07)	0.81	(0.03)	0.99	(0.04)
	+NP	4.10*	(0.53)	1.15*	(0.03)	1.56*	(0.06)
	C	18.9	(0.17)	3.2	(0.24)	12.4	(1.13)
Microbial Nitrogen	+P	21.2	(1.13)	3.8	(0.05)	11.1	(0.40)
$\mu g\ g^{-1}$	+N	27.8*	(0.06)	3.5	(0.30)	12.1	(0.68)
	+NP	21.4	(0.29)	3.8	(0.14)	10.4	(0.14)
	C	10.98	(0.16)	4.32	(0.38)	12.16	(0.39)
N/P ratio	+P	9.21	(1.18)	4.94	(0.01)	9.65*	(0.04)
	+N	11.03	(0.36)	4.32	(0.23)	12.22	(0.22)
	+NP	5.21*	(0.65)	3.30	(0.02)	6.66*	(0.31)

net and gross photosynthesis measurements) from 0.8 to 11.3 $\mu g\ C\ cm^{-2}\ hr^{-1}$ is reported. In a North American desert thermal stream colonised by cyanobacterial mats Naiman (1976) found production rates of ca. 29 $\mu g\ C\ cm^{-2}\ hr^{-1}$. Area based P_{max} values for the Victoria Land stream communities reported here (Figs. 3, 4) vary from 2.0–3.6 $\mu g\ C\ cm^{-2}\ hr^{-1}$ are within, but are at the lower end of, the range for temperate communities. However, assimilation numbers of less than 0.2 μg C (μg Chl*a*)$^{-1}$ hr^{-1} are much lower than those of temperate communities. During stream flow in the short summer of this region PAR varies between 1 000 $\mu E\ m^{-2}\ s^{-1}$ at midday and 100 $\mu E\ m^{-2}\ s^{-1}$ at midnight. Reference to Figs. 2 and 3 shows that the sun communities operate for 24 hours in a diel regime of PAR in which light limitation to photosynthetic rates does not occur. As light is unlikely to be a control on photosynthesis in these communities the relatively low I_k values (*ca.* 100 $\mu E\ m^{-2}\ s^{-1}$) are of interest.

Goldman *et al.* (1972) showed a marked midday depression in photosynthetic rates of benthic algal mats in ponds at Cape Evans, Ross Island at the same latitude as Fryxell stream. They attributed this to marked photoinhibition during this time of day with maximum photosynthetic rates occurring close to midnight on bright days. The light shielding effect of the accessory pigments in the upper layers of the mats of the high light stream communities (Fig. 1, Table 4) thus appears to have the effect of reducing the I_k of the community as the bulk of the chlorophyll is found in the lower layers. High carotenoid/chlorophyll *a* ratios in antarctic lake benthic algal felts has been suggested to be due to chlorophyll degradation in the high light rather than a carotenoid buildup (Goldman *et al.*, 1963). Layering and high carotenoid pigment levels were not observed in the low light community (*Prasiola*) which exhibited a very low I_k value (Fig. 3) and even in this community we did not detect photoinhibition. Given the saturation values and the absence of photoinhibiting effect we would expect saturated photosynthesis to operate throughout the 24 hour cycle. It is puzzling that the shade adapted *Prasiola* community did not show photoinhibition and it is of interest that another antarctic low light and temperature community (*Trachyneis aspersa*) also has this property (Palmisano *et al.*, 1985).

Peterson *et al.* (1983) point out that information on the response of stream periphyton to nutrient enrichment is scarce as streams are difficult to treat experimentally. However, studies on temperate and arctic streams have indicated that in non-shaded conditions, the supply of N and/or P to algal mats could limit photosynthesis (Peterson *et al.*, 1983; Triska *et al.*, 1983; Hullar & Vestal, 1988).

Lock & John (1979) suggest that at low concentrations nutrient uptake by stream periphyton may be limited by diffusion because of laminar boundary layers over the biofilm. This effect is likely to be enhanced by the presence of the polysaccharide matrix surrounding the cells (Vincent & Howard-Williams, 1987). Where diffusion limitation occurs the concentration of nutrients in the stream water would be the critical factor, even though the supply is continuous. Hence the findings of Wuhrmann & Eichenberger (1975) that enrichment of an unshaded Swiss stream resulted in no increase in periphyton growth may be explained by relatively high concentrations of DRP (78 mg m^{-3}) and NO_3-N (> 500 mg m^{-3}) in the ambient stream water. In deep shade, where light is a primary control, nutrients were found to be unimportant even where nitrate levels were undetectable in summer months (Triska *et al.*, 1983). The benthic algal felts of the ponds of Cape Evans, however, showed no enhanced photosynthesis on the addition of inorganic N, P or trace elements (Goldman *et al.*, 1972) in high light conditions. Although these authors concluded that nutrients did not limit benthic algal photosynthesis in the ponds, the incubations were of short duration (4 hrs) and the ponds were nutrient rich due to Antarctic skua (*Catharacta maccormicki*) bathing activities here. Vincent & Vincent (1982) also recorded no nutrient limitation of phytoplankton of ponds along the edge of the Ross Sea, and similarly attributed this lack of response to the high natural nutrient levels here. Hawes (1988) has shown that in maritime Antarctic Signy Island, the chlorophyte *Klebsormidium* sp. was significantly associated with streams of high dissolved phosphorus content. He was careful to emphasise, however, that other, particularly physical factors may also have been of importance in determining the distribution of this species.

Our data from low nutrient Fryxell stream consistently showed no response in photosynthetic rates or chlorophyll *a* to nutrient enrichment after 9 days. Even in *Binuclearia* and *Nostoc* where a response in terms of increased tissue phosphorus occurred which significantly reduced the N/P ratio (Table 6), this enhanced nutrient status was not translated to growth. We assume that the phosphorus uptake was a luxury response. A similar effect was reported by Wuhrmann & Eichenberger (1975) where, in their stream treatments with P, the tissue N/P ratio decreased due to luxury uptake of P but no response to algal growth was recorded.

While we have identified the primary controls of epilithic biomass in the Victoria Land streams as bed stability and sediment load (Vincent & Howard-Williams, 1986; Howard-Williams *et al.*, 1986b), the explanation of the low photosynthetic rates clearly does not lie with light or nutrients. In this way these streams differ from their temperate and even arctic counterparts where light and/or nutrients (Triska *et al.*, 1983; Peterson *et al.*, 1983; Elwood *et al.*, 1981; Hullar & Vestal, 1988) and even space (Matson & Klotz, 1983) control lotic primary production. The absence of light or nutrient limitations in south Victoria Land streams further emphasises the need to consider other potential controls on plant growth here.

Acknowledgements

This work was performed within the New Zealand Antarctic Research Programme. We thank Antarctic Division DSIR, the US Navy and the Royal New Zealand Air Force for logistic support, Dr John Priscu for helpful discussions, Geoff Spearpoint and Gillian Wratt for skilled field assistance and Lynell May and Virginia Reid for laboratory help. The manuscript was reviewed by Prof. G. E. Fogg, Dr Eddie White and Dr Robie Vestal. We thank Janet Simmiss for typing the many drafts.

References

Broady, P. A., 1982. Taxonomy and ecology of algae in a freshwater stream in Taylor Valley, Victoria Land, Antarctica. Arch. Hydrobiol. Supplement 63: 331–339.

Elwood, J. W., J. D. Newbold, A. F. Trimble & R. W. Stark, 1981. The limiting role of phosphorus in a woodland stream ecosystem: effects of P enrichment of leaf decomposition and primary producers. Ecology 62: 146–158.

Goldman, C. R., D. T. Mason & B. J. B. Wood, 1963. Light injury and inhibition in antarctic freshwater phytoplankton. Limnol. Oceanogr. 8: 313–322.

Goldman, C. R., D. T. Mason & B. J. B. Wood, 1972. Comparative study of the limnology of two small lakes on Ross Island, Antarctica. In G. A. Llano (ed.), Antarctic terrestrial biology. Antarctic Research Series, American Geophysical Union 20: 1–48.

Hawes, I., 1988. Filamentous green algae in freshwater streams on Signy Island, Antarctica. Hydrobiologia (this volume).

Howard-Williams, C., W. F. Vincent & G. S. Wratt, 1986(a). The Alph River ecosystem: A major freshwater environment in southern Victorialand. N.Z. Ant. Rec. 7: 21–33.

Howard-Williams, C., C. L. Vincent, P. A. Broady & W. F. Vincent, 1986(b). Antarctic stream ecosystems: variability in environmental properties and algal community structure. Int. Revue ges. Hydrobiol. 71: 511–544.

Hullar, M. L. & J. R. Vestal, 1988. The effects of nutrient limitation and stream discharge on the epilithic microbial communities in an oligotrophic arctic stream. Hydrobiologia (this volume).

Lock, M. A., 1981. River epilithon – a light and organic energy transducer. In M. A. Lock & D. D. Williams (eds.), Perspectives in Running Water Ecology. Plenum Press NY: 3–40.

Lock, M. A. & P. H. John, 1979. The effect of flow patterns on uptake of phosphorus by river periphyton. Limnol. Oceanogr. 24: 376–383.

Matson, E. A. & R. L. Klotz, 1983. Organic carbon supply and demand in the Shetucket River of eastern Connecticut. In T. D. Fontaine & S. M. Bartell (eds.), Dynamics of Lotic Ecosystems. Ann Arbor Science, Michigan: 247–169.

Naiman, R. J., 1976. Primary production, standing stock and export of organic matter in a Mohave Desert thermal stream. Limnol. Oceanogr. 21: 60–73.

Palmisano, A. C., J. B. Soo Hoo, D. C. White, G. A. Smith, G. R. Stanton & L. H. Burckle, 1985. Shade adapted benthic diatoms beneath antarctic sea ice. J. Phycol. 21: 664–667.

Peterson, B. J., J. E. Hobbie & T. L. Corliss, 1983. A continuous periphyton assay: tests of nutrient limitation in a tundra stream. Limnol. Oceanogr. 28: 583–591.

Shoaf, W. T. & B. W. Lium, 1976. Improved extraction of chlorophyll *a* and *b* from algae using dimethylsulfoxide. Limnol. Oceanogr. 12: 926–928.

Strickland, J. D. H. & T. R. Parsons, 1968. A practical handbook of seawater analysis. Fisheries Research Board of Canada Bulletin 167: 1–311.

Triska, F. J., V. C. Kennedy, R. J. Avanzino & B. N. Reilly, 1983. Effect of simulated canopy cover on regulation of nitrate uptake and primary production by natural periphyton assemblages. In T. D. Fontaine & S. M. Bartell (eds.), Dynamics of Lotic Ecosystems. Ann Arbor Science, Michigan: 129–159.

Vincent, W. F. & C. Howard-Williams, 1986. Antarctic stream ecosystems: physiological ecology of a blue-green algal epilithon. Freshwat. Biol. 16: 219–233.

Vincent, W. F. & C. Howard-Williams, 1987. Microbial ecology of antarctic streams. In F. Megusar (ed.), Proceedings of the Fourth International Symposium on Microbial Ecology. Ljubljana: 201–6.

Vincent, W. F. & C. Howard-Williams, 1988. Microbial communities in Southern Victoria Land streams (Antarctica) II. The effects of low temperature. Hydrobiologia (this volume).

Vincent, W. F. & C. L. Vincent, 1982. Response to nutrient enrichment by the plankton of Antarctic coastal lakes and the inshore Ross Sea. Polar Biol. 1: 159–165.

Wuhrman, K. and E. Eichenberger, 1975. Experiments on the effects of inorganic enrichment of rivers on periphyton primary production. Verh. int. Ver. Limnol. 19: 2028–2034.

Wyman, M. & P. Fay, 1986. Underwater light climate and the growth and pigmentation of planktonic blue-green algae (cyanobacteria) II. The influence of light quality. Proc. Royal Soc. Lond. ser.B 227: 381–393.

Hydrobiologia 172: 39–49, 1989
W. F. Vincent and J. C. Ellis-Evans (eds) High Latitude Limnology

Microbial communities in southern Victoria Land streams (Antarctica) II. The effects of low temperature

Warwick F. Vincent & Clive Howard-Williams
Taupo Research Laboratory, Division of Marine and Freshwater Science, DSIR, Box 415, Taupo, New Zealand

Key words: Antarctica, bacteria, biofilms, cyanobacteria, microbial mats, *Nostoc*, *Phormidium*, photosynthesis, respiration, stream algae, temperature

Abstract

Water temperatures in southern Victoria Land streams fluctuated over the range 0 to 10 °C but generally lay close to freezing. In a series of controlled assays at Fryxell Stream, Taylor Valley, the benthic microbial mats showed strongly positive metabolic responses to increases in temperature well above ambient. Rates of polysaccharide and lipid biosynthesis increased with temperature over the range 0 to 25 °C. Between 0 and 10 °C, Q_{10} values for the cyanobacterial mats were 1.7 to 3.2 for gross photosynthesis, 2.5 to 5.7 for respiration, 2.2 to 2.5 for acetate incorporation into lipid, 1.9 to 3.8 for glucose catabolism, and 1.9 to 2.8 for thymidine incorporation. Respiration accounted for a high percentage of gross photosynthesis, and a net respiratory loss of carbon from three communities was either induced or worsened by an increase in temperature from 0 to 10 °C. The chlorophyll *a* content of *Nostoc* discs incubated for one month in darkness decreased by 27% at 5 °C, but by 99% at 25 °C. This set of assays suggests that the cyanobacterial mats contained large amounts of chlorophyll *a* and carbon associated with inactive or senescing cells. This unusual standing stock could probably not persist under warmer conditions, which would promote both increased respiratory losses and faster rates of bacterial decomposition.

Introduction

The cyanobacterial mat and film communities of southern Victoria Land streams inhabit an unusual environment of highly variable discharge, near-zero water temperatures and periodic freezing and desiccation. Large aquatic invertebrates are absent from this region of Antarctica and without the influence of grazing arthropods, the epilithic communities have been able to accumulate to a high biomass that persists from year to year. Despite these large standing stocks, however, the photosynthetic rates per unit chlorophyll *a* or organic carbon are low (Vincent & Howard-Williams, 1986, 1987). These normalised rates do not appear to respond to nutrient enrichment, nor do they seem to be a photoinhibitory response to high photon flux densities during the 24 hour irradiance cycle (Howard-Williams & Vincent, 1988).

These southern high latitude streams are fed by glacier meltwater and flow for only a few weeks each summer. Even at this time of year, however, the air temperatures lie close to zero and streamwater temperatures are correspondingly low. Cold temperatures are an obvious feature of most polar environments and are known to have a restrictive effect on various microbial processes in Antarctica including pelagic photosynthesis and bacterial heterotrophy in the Southern Ocean

(Tilzer *et al.*, 1986; Morita *et al.*, 1977), sea ice algal production in McMurdo Sound (Palmisano *et al.*, 1986), and CO_2 exchange by cryptoendolithic lichens that live beneath rock surfaces in the Dry Valleys (Kappen & Friedmann, 1983). In this paper we examine the influence of water temperatures on the epilithon of Fryxell Stream in the McMurdo Dry Valley region. We measured the temperature responsiveness of community metabolism over the ambient range 0 to 10 °C and compared the influence of temperature on production processes (photosynthesis, synthesis of protein and other photosynthate fractions) and on metabolic loss processes (respiration, bacterial decomposition). Our experiments suggest that the low normalised photosynthetic rates are not simply an inhibitory response to low temperature, but rather that the thick epilithic mats contain a large proportion of non-active organic material (including chlorophyll *a*) that under warmer conditions would rapidly decompose.

Methods

Water temperatures at each of the stream sites were measured either with a Yellow Springs Instrument combined oxygen and temperature probe calibrated against a mercury thermometer, or directly with an unshielded mercury thermometer. Irradiance was measured either with a Li-Cor pyranometer (Vanda Station) or a Li-Cor photosynthetically available radiation sensor (all other measurements).

With the exception of the decomposition experiment (see below) all of the microbial assays were conducted at Fryxell Stream during January 1986. Four types of streambed communities were recognised on the basis of their biomass dominants: thick mucilaginous mats of cyanobacteria dominated by Oscillatoriaceae ('*Phormidium* mat') or *Nostoc commune* with a much smaller population of Oscillatoriaceae ('*Nostoc* mat'); tufts and streamers dominated by *Binuclearia tectorum*, and straps of *Prasiola calophylla*. Further details of these communities are given in Broady (1982).

For several of the assays (end products of photosynthesis, lipid biosynthesis, nucleoprotein synthesis, glucose catabolism) the cyanobacterial communities were sampled with a 3 ml plastic syringe that instead of a needle had its tip sharpened into a 2 mm diameter cutting edge. The mats were cut and then expressed from the tip by gently applying pressure with the syringe plunger. For each assay three of these 2 mm diameter 'microcores' from different samples of the same community were expressed into 30 ml Universal vials filled with freshly collected water from Fryxell Stream. This method was not possible for the *Binuclearia* and *Prasiola* communities, and therefore samples of these algae from a known area were suspended in streamwater, and then subsamples were transferred to the Universal vials.

All of the temperature experiments were conducted in water bath incubators under ambient light conditions near the stream. The incubators consisted of 5 litre polystyrene boxes that were filled to their brim with water and further insulated with clear bubble plastic. The water temperatures were frequently checked and adjusted manually. The assay samples were also gently agitated at these time intervals. The large mass of water gave a good level of control and the temperatures varied over less than a ± 1.5 °C range throughout the experiments.

For the end products of photosynthesis assay the microcore vials were injected with $^{14}C\text{-}HCO_3$ (final activity of 0.6 μCi per ml). The cores were then filtered either immediately (control) or after 2 h in the controlled temperature incubators onto 25 mm diameter GF/C filters. The filters were washed with streamwater, injected with 0.2 ml of 5% glutaraldehyde and stored frozen until analysis. The filters were later ground in 2 ml of methanol (1.4 ml), and water (0.6 ml) with a Teflon tissue grinder. Four major classes of photosynthate (low molecular weight compounds, lipids, protein and polysaccharide) were then coarsely separated using the method of Li *et al.* (1980), and the ^{14}C-label in each fraction was counted by liquid scintillation spectrometry. Dissolved inorganic carbon levels in the incubation streamwater (with ^{14}C added) were measured by infra-red gas analysis.

For the glucose catabolism assays the microcore vials were injected with (U-^{14}C)D-glucose, final activity and concentration of 0.1 μCi per ml, 0.4 μM. At t_o and after 4 h incubation the samples were filtered through 25 mm GF/F filters and the filtrate was preserved with NaOH to pH 10. These filtrates were stored cool and unfrozen and later acidified and bubbled with a stream of oxygen through a CO_2 trapping solution (β-phenethylamine). A subsample of this solution was then counted by liquid scintillation.

The microbial lipid assay was a modification of the method described by White *et al.* (1977). Microcore vials were injected with 2-^{14}C-acetate (final activity and concentration of 0.1 μCi, 2.2 μM). The samples were incubated for 90 mins and then uptake terminated by filtration and glutaraldehyde as above. The filters were ground in a single phase mixture of methanol (2.6 ml), chloroform (1.3 ml), water (1.0 ml) and concentrated HCl (0.02 ml) and then sonicated for 5 mins. The phases were then broken by shaking the extract with 5 ml of chloroform plus 5 ml of water. A subsample of the chloroform phase was transferred to a vial, evaporated to dryness at 55 °C and counted by liquid scintillation.

The incorporation of thymidine into DNA was measured using a modified version of the Fuhrman & Azam (1980) assay. Microcore vials were injected with methyl-^3H-thymidine at a final concentration of 6 nM. At t_o and after 2 h of incubation the microcores were filtered on to 25 mm GF/F filters, washed with ice-cold 10% trichloracetic acid (TCA) and stored frozen. The filters were later ground in 3 ml of ice-cold TCA with a Teflon tissue grinder. A subsample of this material was then filtered onto another GF/F filter, washed again with TCA, transferred to a vial containing Brays cocktail, and then counted by liquid scintillation. Overnight dialysis of the TCA extracts in running water produced no major shifts in counts. The efficiency of nucleic acid recovery with this assay (70%) was established using ^{14}C-labelled DNA. The TCA-precipitated material was not washed with a lipid solvent as recently recommended by Robarts *et al.* (1986), and therefore the measurements may be biased towards uptake rather than incorporation.

The CO_2 exchange characteristics of the microbial mat communities were measured by infra-red gas analysis (IRGA). The assays were run with 18 mm diameter cores of the epilithon (details in Howard-Williams & Vincent, 1988).

A decomposition experiment was run on *Nostoc* mats incubated in darkness or under 350 $\mu E\ m^{-2}\ s^{-1}$ cool white fluorescent lighting. Dry frozen *Nostoc* was collected from the lower reaches of the Fryxell Stream bed in early November prior to streamflow. These mats were then stored in the dark and frozen (−20 °C) in sterile plastic containers for 8 months prior to the experiment. The dry frozen mat was then cut into 4 mm diameter cores with a cork borer and these sections were each transferred to 15 ml of inorganic nutrient media sterilised in 25 ml scintillation vials. The only inoculum of heterotrophic bacteria was therefore the natural assemblage of cells that survived in the dry *Nostoc* mat. The cores were allowed to hydrate for 2 days under the various treatments and then the first set of samples was removed for chlorophyll and phaeophytin analysis. The remaining cores were incubated for 1 month (31 days) and then harvested for pigment analysis.

Two methods were adopted for chlorophyll *a* determinations depending on the size of the mat community sample. The large discs used for the IRGA measurements were stored frozen and then extracted in dimethyl sulfoxide overnight. This extract was diluted 1 : 1 with 90% acetone and the pigment content then assayed spectrophotometrically (Strickland & Parsons, 1968). In the microcore and decomposition experiments the microbial cells were initially filtered on to 25 mm diameter glass fibre filters (Whatman grade GF/C) which were stored frozen and then later cut up and transferred to 90% acetone. This material was then ground with a Teflon tissue grinder and the extracts cleared by centrifugation and assayed by spectrofluorometry (Strickland & Parsons, 1968).

Results

Temperature

Throughout our 3 year study of the streams in the McMurdo Sound region water temperatures rarely exceeded 8 °C and were typically in the range 0 to 5 °C (Table 1). For example, for 40 measurements in the mid afternoon (the warmest time of day) at a downstream site on the Onyx River (Vanda Weir) the mean temperature over the December–January season of streamflow was 4.4 °C with 95% of the measurements ($\bar{x} \pm 2SD$) within the range 0.2 to 8.6 °C. Stream water temperatures were generally close to freezing for several hours over each 24 h cycle or during days of persistent cloud cover or high wind velocities. The warmest water temperature measured was 13 °C in a shallow portion of the *Phormidium* mat area in the Onyx River 3 km upstream from Vanda Weir.

In almost all of the streams there were large diel fluctuations in water temperature that more or less followed the diel radiation cycle at a lag (Fig. 1). The amplitude of these 24 h variations was typically about 5 °C (Table 1). An exception to these observations was the Alph River where temperatures and also discharge remained constant, probably reflecting the influence of water storage in Trough Lake which feeds this system (Howard-Williams *et al.*, 1986). In most of the streams the 24 h radiation cycle also generated large variations in discharge (Fig. 2), and as a result the algal communities at the edge of the stream were periodically exposed to sub-zero temperatures.

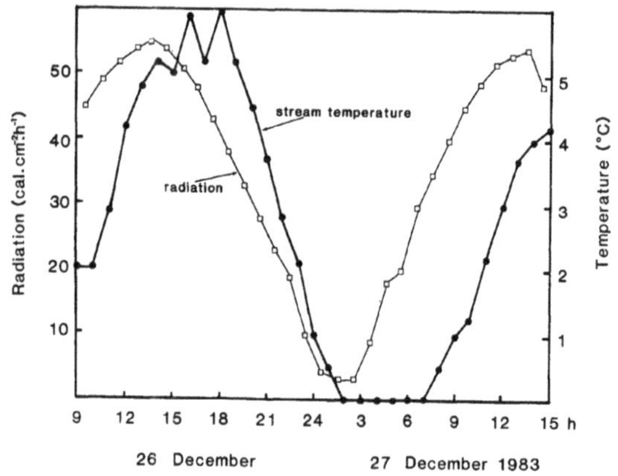

Fig. 1. Diel variations in water temperature and incoming radiation for the Onyx River at Vanda Weir, Wright Valley.

Measurements of the stream water temperature next to the mats has never revealed any streambed warming relative to the flowing water above. However, a 10–15 °C warming of the black *Nostoc* thalli in a shallow-water flush area draining into Fryxell Stream has previously been recorded (K. Thompson, University of Waikato Antarctic Research Unit, unpublished data).

Photosynthesis

Photosynthesis as measured by $^{14}C\text{-}CO_2$ incorporation was strongly responsive to temperature over the range 0 to 10 °C for all of the commu-

Table 1. Stream water temperatures over the 24 h cycle during mid to late summer in the McMurdo Sound region. The mean values are from measurements or interpolated estimates for each hour over the 24 h cycle.

Stream	Date	Stream temperature (°C)		
		mean	minimum	maximum
Onyx River	26/27 Dec 1983	2.6	0	6.0
Onyx River	2/3 Jan 1986	3.3	0	10.5
Onyx River	17/18 Dec 1983	3.4	0	6.2
Alph River	17/18 Jan 1986	0.7	0.5	1.0
Bird Stream	18/19 Jan 1984	1.4	0	3.8
Fryxell Stream	14/15 Jan 1985	6.1	2.5	9.0
Adams Stream	12/13 Jan 1984	2.1	0	5.2

nities tested (Table 2). Q_{10} values varied from 1.8 to 2.7, but there were no major differences in the pattern of labelling of photosynthate fractions between the two temperatures (Table 2). Polysaccharide was the dominant end product of photosynthesis for the *Phormidium*, *Nostoc* and *Binuclearia* communities at both 0 °C and 10 °C, but in the *Prasiola* incubation the low molecular weight fraction (LMW) was more heavily labelled than polysaccharide or protein. The lipid fraction contained the lowest percentage (1–6%) of total label in all of the assays.

The effect of temperature on the end products of photosynthesis was further examined in a series of incubations of *Phormidium* micro-cores over the wider range 0 °C to 25 °C. The biosynthesis of all four photosynthate fractions responded to increasing temperatures, but with the exception of the lipid component the temperature response was much greater over the range 0 to 15 °C than from 15 to 25 °C (Fig. 3). For example, the labelling of polysaccharide (the dominant fraction) had a Q_{10} value of 2.7 between 5 and 15 °C, but dropped to 1.3 between 15 °C and 25 °C. The Q_{10} for protein biosynthesis similarly dropped from 3.1 (5–15 °C) to 1.6 (15–25 °C).

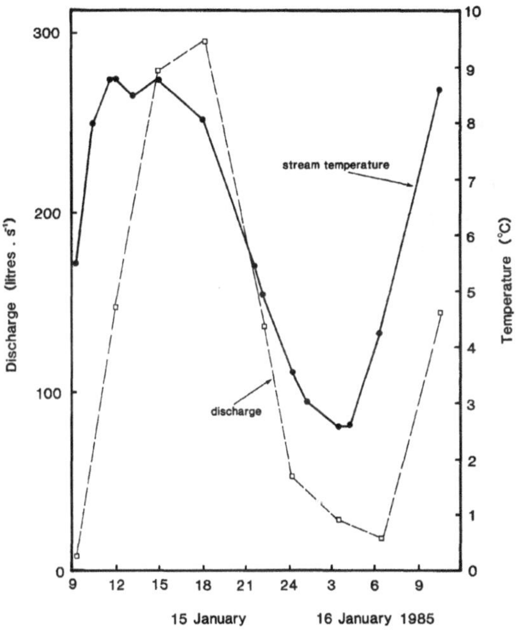

Fig. 2. Diel variation in water temperature and discharge in Fryxell Stream, Taylor Valley.

Table 2. Photosynthesis and the percent labelling of photosynthate fractions at two temperatures for the various mat communities in Fryxell Stream.

	Total photosynthesis ($\mu g\,C \cdot cm^{-2} \cdot h^{-1}$)	% Photosynthate labelling			
		LMW	Lipid	Protein	Polysaccharide
Phormidium					
0 °C	1.2	9	2	36	50
10 °C	3.1	11	1	39	49
Q_{10}	2.6				
Nostoc					
0 °C	9.7	13	3	22	62
10 °C	23.7	7	3	23	66
Q_{10}	2.4				
Binuclearia					
0 °C	7.6	6	6	17	71
10 °C	14.0	4	5	24	68
Q_{10}	1.8				
Prasiola					
0 °C	2.0	43	2	27	27
10 °C	5.4	36	4	29	32
Q_{10}	2.7				

Community CO_2 exchange
Gross photosynthetic rates were measured by IRGA for each of the stream communities as the light minus dark CO_2 exchange (final 3 columns in Table 3). These rates were low per unit chlorophyll *a*, less than 1 μg C (μg chl *a*)$^{-1}$ h^{-1} at 0 °C, but were stimulated by a factor ranging between 1.6 and 3.2 by a 10 °C rise in temperature. The Q_{10} values for respiration (dark CO_2 production, Table 3) were comparable or higher. Respiration as a percentage of gross photosynthesis varied greatly between stream communities; e.g. at 0 °C from 23% for *Binuclearia* to 171% for the thick *Nostoc* mat. This percentage respiratory loss increased for all of the communities tested except the *Nostoc* mats with the 10 °C increase in temperature.

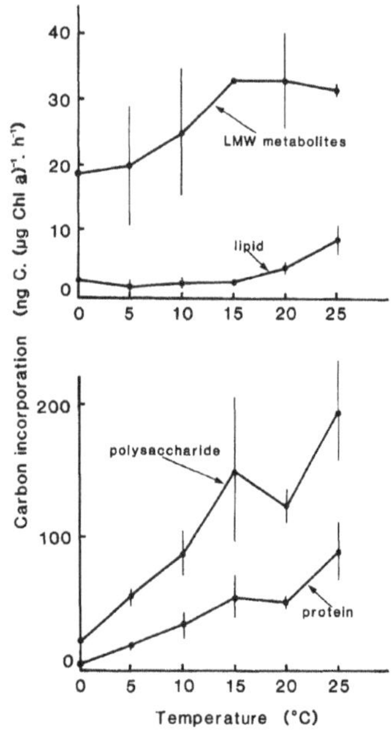

Fig. 3. The influence of temperature on the incorporation of ^{14}C-CO_2 into end products of photosynthesis, for *Phormidium* mats at Fryxell Stream.

The combined effect of temperature on photosynthesis and respiration is expressed by the CO_2 exchange rate in the light (first two columns in Table 3) which may be considered the community net productivity. The *Binuclearia* community showed a simple, almost twofold increase in net photosynthesis between 0 and 10 °C. The *Prasiola* community showed no significant response to temperature because the increase in gross photosynthesis was offset by a greater increase in respiration. For the cyanobacterial mats the net community photosynthesis at 0 °C was below zero (i.e. more CO_2 was respired than was fixed by photosynthesis) or only slightly above zero. The net carbon loss from the community was either induced (*Phormidium* b) or further stimulated (*Phormidium* a, *Nostoc*) by the 10 °C rise in temperature.

Heterotrophic activity
Three measures of microbial heterotrophy suggested that the stream mat communities, particularly the thick *Nostoc* and *Phormidium* mats,

Table 3. CO_2 exchange by the Fryxell Stream mat communities, January 23/24, 1986. + = CO_2 production; − = CO_2 uptake. The production or uptake values are expressed as μg C per μg chlorophyll *a* per hour. Incubations were under full ambient sunlight (320–900 μE m^{-2} s^{-1}).

Community dominant	Light		Dark			Light minus dark		
	0 °C	10 °C	0 °C	10 °C	Q_{10}	0 °C	10 °C	Q_{10}
Phormidium a*	+0.22	+1.01	+0.75	+1.92	2.6	−0.53	−0.91	1.7
Phormidium b**	−0.14	+2.06	+0.80	+4.56	5.7	−0.94	−2.5	2.7
Nostoc	+0.12	+0.16	+0.29	+0.71	2.5	−0.17	−0.55	3.2
Binuclearia	−0.61	−1.18	+0.18	+0.69	3.9	−0.79	−1.87	2.4
Prasiola	−0.45	−0.42	+0.08	+0.42	5.2	−0.53	−0.84	1.6

* Thick pink mat
** Orange tufted mat

contained active populations of heterotrophic bacteria. The cyanobacterial as well as *Binuclearia* communities incorporated ^{14}C-acetate into lipid, and these rates were stimulated more than two-fold by a 10 °C rise in temperature (Table 4). Comparable or higher Q_{10} values were recorded for the catabolism of ^{14}C-glucose to CO_2 by the cyanobacterial mats (Table 5). No incorporation of thymidine into microbial nucleoprotein was detected with *Binuclearia* samples, but the three other communities showed significant uptake and incorporation (Table 6) suggesting bacterial DNA synthesis and growth within the mats. The thymidine incorporation responded positively to the temperature increase from 0 to 10 °C, consistent with the acetate and glucose assays.

As a further measure of the effects of temperature on bacterial metabolism and growth, microcores of *Nostoc* mat were incubated for 48 h at 0 °C or 10 °C, and then assayed with ^3H-thymidine for 2 h at the two temperatures. In all of the assays the rate of thymidine incorporation was stimulated by the 10 °C increase in temperature, but the greatest effect was seen in response to the 48 h pre-incubation (Table 7). If the incorporation values can be considered a rough measure of the population size of the heterotrophically active bacteria then their growth rates over 48 hours may be approximated as 0.5 ln [(48 h at T_1)/(0 h at T_1)] where T_1 is the temperature. This gives growth estimates of 0.13 and 0.21 d^{-1} at 0 °C, but 0.51 and 0.55 d^{-1} at 10 °C.

Decomposition

The dark incubations of *Nostoc* mat showed a strong response to temperature (Table 8). The samples incubated at 5 °C averaged a relatively small loss in chlorophyll *a* indicating that large quantities of chlorophyll can persist inactive but undecomposed in the algal mats at low stream temperatures. At 25 °C, however, the chlorophyll *a* was almost completely decomposed. In both dark treatments the amount of phaeophytin increased as a percentage of total pigment, but at 25 °C even the phaeophytin concentrations were reduced to almost zero after the month long period of decomposition.

Table 4. Incorporation of 2-^{14}C-acetate into lipid in the stream mat communities. Each value is the mean ± 2SE for triplicates.

	Lipid biosynthesis ($ng\ ^{14}C\ cm^{-2} \cdot h^{-1}$)		
	0 °C	10 °C	Q_{10}
Nostoc	2.6 ± 0.8	5.7 ± 0.7	2.2
Phormidium	3.3 ± 0.8	8.2 ± 2.6	2.5
Binuclearia	2.0 ± 0.3	5.1 ± 1.4	2.5

Table 5. Catabolism of U-^{14}C-glucose by Fryxell Stream mat communities at two temperatures. Each value is the mean ± range for duplicates.

Community dominant	CO_2 production ($ng\ ^{14}C\ cm^{-2} \cdot h^{-1}$)		
	0 °C	10 °C	Q_{10}
Phormidium a*	0.29 ± 0.11	0.85 ± 0.21	2.9
Phormidium b**	0.38 ± 0.04	0.73 ± 0.05	1.9
Nostoc	0.23 ± 0.03	0.88	3.8

* Thick pink mat
** Orange tufted mat

Table 6. Incorporation of ^3H-thymidine by the Fryxell Stream mat communities. Each value is the mean ± 2SE for triplicates.

Community dominant	Thymidine incorporation ($pmol \cdot cm^{-2} \cdot h^{-1}$)		
	0 °C	10 °C	Q_{10}
Phormidium	0.18 ± 0.04	0.53 ± 0.19	2.8
Nostoc	0.45 ± 0.22	0.85 ± 0.23	1.9
Prasiola	0.31 ± 0.11	0.45 ± 0.11	1.5

Table 7. Effect of preincubation temperature on the incorporation of ^3H-thymidine by *Nostoc* dominated mats from Fryxell Stream. Each value is the mean ± 2SE for triplicates.

Preincubation	Thymidine incorporation ($pmol \cdot cm^{-2} \cdot h^{-1}$)		
	0 °C	10 °C	Q_{10}
0 h	0.45 ± 0.22	0.85 ± 0.23	1.9
48 h at 0 °C	0.68 ± 0.08	1.11 ± 0.44	1.6
48 h at 10 °C	1.36 ± 0.08	2.34 ± 0.33	1.7

Table 8. Changes in the chlorophyll *a* content of *Nostoc* discs incubated in nutrient media in darkness or 350 $\mu E\ m^{-2}\ s^{-1}$ at two temperatures. Day 2 values are means for triplicates ± 2SE. Day 33 values are means for duplicates ± range.

Treatment	μg Chl $a \cdot cm^{-2}$		
	Day 2	Day 33	% change
5 °C dark	26 ± 0.6	19 ± 5	-27
25 °C dark	27 ± 3	0.16 ± 0.04	-99
5 °C light	18 ± 3	28 ± 8	+56
25 °C light	27 ± 10	14 ± 0.5	-46

Similar effects are suggested by the light incubation data (Table 8), but the results are not conclusive because of the high variability between certain replicates. The chlorophyll *a* levels increased in the 5 °C treatment (but $t = 1.8$, $p > 0.05$) and decreased in the 25 °C treatment (but $t = 1.2$, $p > 0.05$).

Discussion

All of the microbial processes measured in this study responded positively to an increase in temperature from 0 to 10 °C. However, although Arrhenius-like effects operated over this range there was no abrupt suppression of metabolism at 0 °C, indicating that the communities are able to function reasonably well at the near-zero stream temperatures.

The microbial phototrophs and heterotrophs on the Fryxell streambed appear to be cold-tolerant but they do not seem to be closely adapted to the low temperatures of this environment. Ambient water temperatures typically lay in the range 0 to 5 °C yet the temperature optima for protein, polysaccharide, and lipid biosynthesis from carbon fixed by the *Phormidium* mats lay at or above 25 °C (Fig. 3). These responses contrast with various antarctic marine communities that have optima below 10 °C e.g. 6 °C for sea ice algal photosynthesis (Palmisano *et al.*, 1987), 3–10 °C for growth of phytoplankton isolates from the Southern Ocean (Jacques, 1983), 7 °C for bacterial growth rates in the Scotia Sea (Azam *et al.*, 1981). Much higher optima, however, have been reported from freshwater environments in Antarctica e.g. 14–19 °C for the growth of bacterial isolates from lakes in the maritime zone (Ellis-Evans & Wynn-Williams, 1985); 12.5–18 °C for growth of phytoflagellate isolates from Dry Valley lakes (Seaburg *et al.*, 1981); > 14 °C for benthic algal mats from a pond on Ross Island (Goldman *et al.*, 1963). The Q_{10} values for photosynthesis by the Fryxell epilithon typically lay in the range 2–3, and these also were comparable with other freshwater measurements in the region (e.g. Q_{10} of 2.3 for benthic algal photosynthesis, Goldman *et al.*, 1963) but well below the values derived from various marine studies e.g. 4.6 for phytoplankton photosynthesis in the Scotia Sea, 5.9 for congelation ice algal photosynthesis in McMurdo Sound (Palmisano *et al.*, 1986). These lower temperature optima and the relatively high responsiveness to temperature change by the marine phototrophs is presumably a reflection of the persistently low but highly stable temperature condition in the sea. Micro-algae typically have their temperature optima for photosynthesis several degrees higher than the temperature at which they are growing (Li, 1980). A similar generalisation would seem to apply to heterotrophs as well as phototrophs for a wide range of microbial communities in Antarctica with the temperature optimum for metabolic activity above the upper range of habitat temperatures. This effect is especially pronounced for the epilithic microbiota of Fryxell Stream.

The highly unstable environmental conditions that characterise the Dry Valley streams probably select for broad tolerances to chemical and physical fluctuations, including temperature. Our data are consistent with the observations of Seaburg *et al.* (1981) who measured the temperature minimum and maximum for growth in culture of a wide range of algal isolates (including cyanobacteria) from the Dry Valleys. Most of their meltwater stream isolates grew within the temperature range 5 to 25 °C, although almost one third of them ceased growth at 2 °C. Although there were some differences in the upper and lower temperature limits between strains of the same species there was no evidence that any of the isolates grew especially fast at low temperature, or had an optimum for growth near 0 °C.

Our IRGA analyses of the Fryxell Stream mat communities (Table 3) show that there was a fine dynamic balance between photosynthetic carbon fixation and respiratory carbon loss. Respiration accounted for 90% or more of gross photosynthesis in the cyanobacterial mats at 0 °C, and a 10 °C increase induced or worsened the net loss of carbon from the mats sampled at this time of year. This type of effect has been similarly reported from the cryptoendolithic lichen communities living within the rocks of the Dry Valleys. Under full sunlight net carbon uptake by the community was measured over the range −4 to +8 °C but higher temperatures continued to stimulate respiration and not photosynthesis and the community then experienced a net carbon loss (Kappen & Friedmann, 1983). Like the rock microbial assemblage the cyanobacterial mats from Fryxell Stream would seem to have a temperature compensation point (where net photosynthesis equals zero) at the top of the temperature range typically experienced in the natural habitat. This effect underscores the difficulty of applying optima measured with isolates in culture or from assays of a single metabolic parameter to the performance of a consortium of heterotrophs and phototrophs in a microbial mat community.

In the cryptoendolithic lichen communities assayed by Kappen & Friedmann (1983) the high, temperature-responsive respiratory losses were presumably dictated in part if not dominated by the high biomass of symbiotic fungi and associated bacteria. In the streambed mat communities the primary sources of respiratory carbon are less clearly identifiable. Fungal hyphae were relatively rare in virtually all of the microbial mats we have examined from this region and bacterial rods and cocci, although numerically common, accounted for a small percentage of total biomass. These mat communities are composed mostly of cyanobacterial trichomes and associated mucilage binding together silt particles, and it is probably the *Nostoc* and Oscillatoriacean cells that dominate the respiratory CO_2 production. Light is strongly attenuated within the mat profile by the surface layer of carotenoid-rich cells and the high concentrations of light-scattering silt particles (Howard-Williams & Vincent, 1988). The low saturation irradiance value for the community suggests that at least some of the phototrophs are shade-adapted, and such organisms are known to have a low P_{max} relative to respiration (Raven & Glidewell, 1975). It is also possible that a high percentage of cells in the mat are senescing or have been damaged by the seasonal (or more frequent) freeze-thaw cycle.

Although the heterotrophic bacteria in the cyanobacterial mats were present in a relatively low biomass their metabolic activities were strongly stimulated by a 10 °C rise in temperature with Q_{10} values in the range 2 to 4. The implications of this temperature effect were most clearly illustrated by the *Nostoc* decomposition experiment. The 25 °C treatment resulted in almost the complete decomposition of the thick mat within one month, whereas the 5 °C incubation experienced little change under the same conditions of darkness. This experiment implies that high levels of chlorophyll *a* can remain inactive and undecomposed at low temperatures but not at high temperatures, and these effects were also suggested by the light incubations. Thus the high standing stock of biomass and organic carbon in the Dry Valley streambeds may be a function not only of the absence of large grazing animals, but also due to the cold temperature-depressed growth rates of bacterial decomposers. In this environment slow rates of consumption by secondary producers, both animal and bacterial, as well as slowed respiratory losses may be the main factors allowing large standing stocks of microbial mat to accumulate.

Low temperatures operating directly on photosynthesis do not seem to completely account for the low normalised photosynthetic rates. Photosynthesis was not abruptly suppressed by near-zero stream temperatures and even assuming a Q_{10} of 4 most of the assimilation numbers reported in Howard-Williams & Vincent (1988) would still lie at or below 1 μg C per μg chlorophyll *a* per hour at 10 °C. A more likely explanation is that a large proportion of organic carbon and chlorophyll *a* persists in the microbial mat associated with metabolically inactive or senescing cells, and that respiratory losses and bacterial degradation of this material are inhibited by the low ambient stream temperatures. It is possible that some of the measured pigment is the degradation product chlorophyllide which would be analysed as apparent chlorophyll *a* in the assay procedure adopted here (T. Jacobsen, personal communication).

Like other aquatic environments in which microbial mats are found (Cohen *et al.*, 1984) the Dry Valley streambeds allow a gradual net accumulation of biologically-derived organic material, but the rate of accumulation and ultimate standing stock is controlled by an unusual combination of slow growth rates, large respiratory losses, slow secondary consumption rates and the influence of low temperatures on all community processes.

Acknowledgements

This research was conducted within the 1983/84, 1984/85 and 1985/86 seasons of the New Zealand Antarctic Research Programme. We thank Antarctic Division (DSIR), the US Navy and the Royal New Zealand Airforce for logistic support, Ms Gillian Wratt for field assistance, Ms Lynell May for laboratory assistance, Professor G. E. Fogg, Dr J. R. Vestal and Dr E. White for reviewing the manuscript and Ms Janet Simmiss for typing each draft.

References

Azam, F., J. W. Ammerman & N. Cooper, 1981. Bacterioplankton distribution patterns and metabolic activities in the Scotia Sea. Ant. J. US. 16: 164–165.

Broady, P. A., 1982. Taxonomy and ecology of algae in a freshwater stream in Taylor Valley, Victoria Land, Antarctica. Arch. Hydrobiol. Suppl. 63: 331–339.

Cohen, U., R. W. Castenholz & H. O. Haldorson, 1984. Microbial mats (stromatolites). Allan R. Liss Inc., NY. 498 pp.

Ellis-Evans, J. C. & D. Wynn-Williams, 1985. The interaction of soil and lake microflora at Signy Island. In: W. R. Siegfried, P. R. Condy & R. M. Laws (eds), Antarctic Nutrient Cycles and Food Webs. Springer-Verlag, Berlin: 200–222.

Fuhrman, J. A. & F. Azam, 1980. Bacterioplankton secondary production estimates for coastal waters of British Columbia, Antarctica and California. Appl. Envir. Microbiol. 39: 1085–1095.

Goldman, C. R., Mason, D. T. & B. J. B. Wood, 1963. Light injury and inhibition in antarctic freshwater phytoplankton. Limnol. Oceanogr. 8: 313–322.

Howard-Williams, C. & W. F. Vincent, 1988. Microbial communities in southern Victoria Land streams (Antarctica). I. Photosynthesis. Hydrobiologia, this volume.

Howard-Williams, C., W. F. Vincent & G. S. Wratt, 1986. The Alph River ecosystem: a major freshwater environment in southern Victoria Land. Ant. Rec. (NZ) 7: 21–33.

Jacques, G., 1983. Some ecophysiological aspects of the antarctic phytoplankton. Polar Biol. 2: 27–33.

Kappen, L. & E. I. Friedmann, 1983. Ecophysiology of lichens in the Dry Valleys of southern Victoria Land, Antarctica. II. CO_2 gas exchange in cryptoendolithic lichens. Polar Biol. 1: 227–232.

Li, W. K. W., 1980. Temperature adaptation in phytoplankton: cellular and photosynthetic characteristics. In P. G. Falkowski (ed.) Primary Productivity in the Sea. Plenum Press, NY: 259–279.

Li, W. K. W., H. E. Glover & I. Morris, 1980. Physiology of carbon photoassimilation by *Oscillatoria thiebautii* in the Caribbean Sea. Limnol. Oceanogr. 25: 447–456.

Morita, R. Y., R. P. Griffiths & S. S. Hayasaka, 1977. Heterotrophic activity of micro-organisms in Antarctic waters. In G. R. Llano (ed), Adaptations within Antarctic Ecosystems. Gulf Publishing Co., Texas: 99–113.

Palmisano, A. C., J. B. Soohoo & C. W. Sullivan, 1987. Effects of four environmental variables on photosynthesis-irradiance relationships in antarctic sea ice microalgae. Mar. Biol. 94: 299–306.

Raven, J. A. & S. M. Glidewell, 1975. Photosynthesis, respiration and growth in the shade alga *Hydrodictyon africanum*. Photosynthetica 9: 361–367.

Robarts, R. D., R. J. Wicks & L. M. Sephton, 1986. Spatial and temporal variations in bacterial macromolecule labeling with methyl-^3H thymidine in a hypertrophic lake. Appl. Envir. Microbiol. 52: 1368–1373.

Seaburg, K. G., B. C. Parker, R. A. Wharton & G. M. Simmons, 1981. Temperature-growth responses of algal isolates from antarctic oases. J. Phycol. 17: 353–360.

Shoaf, W. T. & B. W. Lium, 1976. Improved extraction of chlorophyll *a* and *b* from algae using dimethylsulphoxide. Limnol. Oceanogr. 12: 926–928.

Strickland, J. D. H. & T. R. Parsons, 1968. A practical handbook of seawater analysis. Bull. Fish. Res. Bd. Can. No. 167, 311 pp.

Tilzer, M. M., M. Elbrachter, W. W. Gieskes & B. Beese, 1986. Light-temperature interactions in the control of photosynthesis in antarctic phytoplankton. Polar Biol. 5: 105–111.

Vincent, W. F. & C. Howard-Williams, 1986. Antarctic stream ecosystems: physiological ecology of a blue-green algal epilithon. Freshwat. Biol. 16: 219–233.
Vincent, W. F. & C. Howard-Williams, 1987. Microbial ecology of antarctic streams. In F. Megusar (ed.) Proceedings of the Fourth International Symposium on Microbial Ecology, Ljubljana. 201–6.
White, D. C., R. J. Bobbie, S. J. Morrison, D. K. Oosterhof, C. W. Taylor & D. A. Meeter, 1977. Determination of microbial activity of estuarine detritus by relative rates of lipid biosynthesis. Limnol. Oceanogr. 22: 1089–1099.

Hydrobiologia 172: 51–61, 1989
W. F. Vincent and J. C. Ellis-Evans (eds) High Latitude Limnology

Nitrogen dynamics in two antarctic streams

Clive Howard-Williams[1], John C. Priscu[2] & Warwick F. Vincent[1]
[1] *Taupo Research Laboratory, Division of Marine and Freshwater Science, DSIR, Box 415, Taupo, New Zealand*; [2] *Department of Biology, Montana State University, Bozeman, Montana, USA*

Key words: ammonium, Antarctica, blue-green algae, cyanobacteria, nitrogen, nitrogen-fixation, organic nitrogen

Abstract

The many glacier meltwater streams of southern Victoria Land flow through catchments where life forms are almost entirely microbial. Allochthonous inputs of nitrogen from two study streams near McMurdo Sound were derived mostly from the melting glaciers (ca. 100–200 mg N m^{-3}) with some originating from N_2-fixation by heterocystous cyanobacteria (max. 939 mg N m^{-2} $year^{-1}$). Thirty to fifty per cent of the glacier derived N was dissolved organic N and a major proportion of this was identified as urea N which was utilised by the rich algal and cyanobacterial mats in the streams. A nutrient budget for Fryxell Stream was estimated, quantifying uptake of urea-N and dissolved inorganic N and the release of dissolved organic (non urea) and particulate N by the stream communities. An index of in-stream nitrogen processing, the Net Uptake Length Constant in these streams was compared with that from temperate climates and was found to be similar. Despite the influence of low temperatures on microbial activity (mean daily water temperature = 5 °C) nutrient removal rates from these antarctic streams are high because of the large standing stock of microbial biomass there.

Introduction

In headwaters and first order streams, catchment derived allochthonous inputs play an integral role in the river continuum and in nutrient spiralling (Vannote *et al.*, 1980; Newbold *et al.*, 1981). By contrast streams at high latitude and altitudes are likely to show a reduced terrestrial dependence, and the extremes of this may be seen in the world's most southerly documented group of streams in the McMurdo Sound region of southern Victoria Land, Antarctica at 78°S (see Vincent & Howard-Williams, 1986; Howard-Williams *et al.*, 1986).

This group of streams share the following characteristics. The streams are glacier-fed, rather than snowfed and all water originates from a well defined point source (the glacier). The catchments are essentially unvegetated; allochthonous inputs should be minimal and shading non-existent. Catchment biota consist of a few small isolated patches of mosses and lichens and a sparse soil microflora. Invertebrate grazing is minimal, and a limited trophic structure exists; arthropod herbivores are non-existent. Micro-herbivores consist of ciliates, nematodes, rotifers and tardigrades, but biomass levels are low. Most of the streams contain microbial mats dominated by cyanobacteria. In some streams these develop to a degree comparable with nutrient-rich waters in the temperate zone, and uptake by this mixed autotrophic heterotrophic community results in downstream nutrient gradients. Low temperatures dictate low turnover rates of the autotrophs in spite of their apparently high biomass.

This stream ecosystem type is especially well developed in the McMurdo Sound region, but is likely to be found at many coastal sites along the margins of Antarctica. Recent studies at Signy Island in the maritime antarctic region show that streams there also have barren catchments, an absence of invertebrate grazers and a highly seasonal algal vegetation, but predominantly of chlorophytes (Hawes, 1988).

Some of the common cyanobacteria in the streams of southern Victoria Land are nitrogen fixers (Allnutt *et al.*, 1981) and organic nitrogen compounds have been recorded in the stream waters at levels which are similar to those in temperate situations in vegetated catchments (Downes *et al.*, 1986). In some antarctic streams, dissolved inorganic-N decreased with downstream distance, a feature that was attributed to uptake by the epilithon (Vincent & Vincent, 1982; Howard-Williams *et al.*, 1986).

The present study was undertaken to examine in more detail these in-stream processes. By adopting partly an experimental, and partly a nutrient budget approach, we have measured the rates of nitrogen-fixation, inorganic nitrogen uptake and the downstream fluxes of nitrogenous compounds. From these data we have attempted to explain the source and production of dissolved organic and particulate nitrogen in these streams.

Study sites

The study sites were the Adams stream, Miers Valley (78° 08′ S, 163° 53′ E) and the Fryxell Stream, Taylor Valley (77° 37′ S, 163° 02′ E). Descriptions of these streams are given in Howard-Williams *et al.* (1986), and some basic stream characteristics are given in Table 1. Sampling and experiments for this study were conducted in December and January of 1984-85 and 1985-86, during maximal streamflow conditions.

Table 1. Characteristics of the two study streams (see also Howard-Williams & Vincent, 1987). Mean ± SD. DON* excludes urea. ND = not detectable. – = no data. Chlorophyll *a* and biomass nitrogen data are for regions of continuous cyanobacterial mat.

	Stream	
	Fryxell	Adams
Length (km)	2	2.7
Mid season discharge $m^3\ s^{-1}$	0.06	0.15
Glacier source	Canada	Adams
Glacier ice melt:		
NO_3-N mg m^{-3}	95.8	31.3
NH_4-N mg m^{-3}	17.3	14.5
DON mg m^{-3}	77.5*	55.0
Urea-N mg m^{-3}	15.5	–
Particulate N	ND	ND
Mat communities:		
Chl*a* (μg cm^{-2})		
Nostoc	28 (±9.5)	27.3 (±9.5)
Phormidium	17.2 (±10.5)	30.1 (±17.6)
Nitrogen (mg cm^{-2})		
Nostoc	1.80 (±0.45)	1.91 (±0.20)
Phormidium	0.45 (±0.14)	1.21 (±0.01)

Methods

Sampling procedure and nitrogen analyses:

Water samples were collected in acid-washed polythene bottles and filtered through pre-rinsed Whatman GF/C filter paper, and then frozen for transport to New Zealand for analysis.

The fine suspended particulate fraction in the water was collected on acid washed Whatman GF/C filters after initial sieving through 1 mm nylon mesh, whilst the coarse particulate fraction was that trapped by the 1 mm mesh and included inorganic and organic material; the latter was composed mostly of intact pieces of cyanobacterial mat.

Cyanobacterial mats were also sampled from the substrate for biomass and nitrogen analyses. Samples were collected with corers made from 10 mm diameter cut off syringes. Particulate samples were frozen intact for transport to New Zealand.

Water analysis for NH_4-N, NO_2-N, NO_3-N, total dissolved N (TDN) and urea was conducted on filtered samples on a Technicon model II Auto-Analyzer. NH_4-N was analysed by the Crooke & Simpson (1971) colorimetric method. Nitrate was determined by the method of Downes (1978), and nitrite by the same method but excluding the reduction step. Total dissolved nitrogen (TDN) was measured by pumping the sample (mixed with 10% sodium perchlorate and buffered to pH 9.0) around a 1 kW ultraviolet lamp in a 3 mm (internal diameter) silica coil for 25 min. All forms of nitrogen, thus converted to nitrate and nitrite, were analysed together as above. Dissolved inorganic nitrogen (DIN) was calculated as: DIN = NO_3-N + NO_2-N + NH_4 -N.

Dissolved organic nitrogen (DON) was calculated as: DON = TDN − DIN. The urea-N component of the DON was estimated by the diacetyl monoxime colorimetric technique of Demanche *et al.* (1973). Particulate nitrogen was estimated by Kjeldahl digestion and subsequent NH_4-N analysis on the Auto-Analyzer as above.

Nitrogen uptake kinetics: $^{15}NO_3$ and $^{15}NH_4$ tracers were added to solutions with samples of *Nostoc* and *Phormidium* mats from Adams Stream. The mats were dispersed in the incubation chambers, left to settle during the incubation, and finally dispersed again prior to filtering. The experimental material contained *ca.* 40 μg N of *Phormidium* mat or *ca.* 100 μg N of *Nostoc*. The material was added to 19 ml of stream water containing 17 and 40 mg m^{-3} of NH_4 and NO_3 respectively. The added nitrogen concentrations were as follows; 8, 12, 16, 25, 35, 50, 75, 100, 150 and 200 mg m^{-3} of either $^{15}NH_4$ or $^{15}NO_3$. Chambers were incubated at the edge of the Adams stream for 6 hrs (1400–2000 hrs) at a temperature of 6 °C under ambient light. At the end of the incubation, mat material was collected on GF/C filters and stored frozen for one month before analysis for ^{15}N by optical emission spectrometry (Timperley & Priscu, 1986).

Nitrogen fixation experiments: N_2-fixation rates were estimated from acetylene reduction assays (Flett *et al.*, 1975). Experiments were performed on triplicate light and duplicate dark samples of three epilithic communities in Adams Stream and two communities in Fryxell Stream. Incubations were carried out in sealed 30 cm^3 Universal vials on 2.5 cm^2 cores of *Nostoc* and *Phormidium* communities in stream water. The headspace/water volume ratio was 0.25 and acetylene was injected through the seal to give an initial concentration of 15% in the water phase. Controls included vials with no acetylene, and vials in which the cores had been killed with formalin (final concentration = 4%).

Experiments included time course assays up to 7 hours and assays at different temperatures. The time course assays were carried out at ambient stream temperatures which varied between 3 and 8 °C. The effect of temperature was studied by incubations at various light intensities at two temperatures 0.5 ± 0.5 °C, 10 ± 1 °C. Incubations for the controlled temperature experiments were conducted in polystyrene incubation boxes with clear insulated plastic lids with interchangeable light screens and temperatures were kept constant by additions of small amounts of warm water or ice to the incubation boxes. At the end of each experiment (up to 5 hrs) 0.2 ml of headspace gas was removed by syringe and injected into pre-evacuated 4 ml tubes with sealed rubber septa for transportation. Analysis for ethylene was carried out later in New Zealand on a Hewlett Packard 5890A gas chromatograph.

Nutrient mass flows: Detailed mass flow estimates of N species in Fryxell stream were made from data collected every 3 hours over a 36 hour period in mid-summer by multiplying the concentrations of the dissolved and the fine and coarse particulate N fractions by the stream discharge at a point in the stream 500 m from Lake Fryxell (Site F2 of Howard-Williams *et al.*, 1986). Stream velocity was measured with a Teledyne-Gurly pygmy current meter at a series of segments across the stream and when the stream became too low in the early morning, at a single mid point. Discharge was obtained from the sum of velocity times cross sectional area for each segment across stream. Estimates of particulate nitrogen in algal mat material caught in drift nets were made by multiplying the concentration of nitrogen in the mats (Howard-Williams & Vincent, 1988) by the dry weight of material collected in the mats at each time interval.

Stream mapping: For calculations of the short-term nutrient budget in Fryxell Stream the total stream standing stock of epilithic biomass was mapped. This was done by estimating the percent cover of epilithon over 42 transects laid across the stream. The transects were placed at 25 m intervals upstream from the sampling point and on each the proportion occupied by *Nostoc* and *Phormidium* was estimated as was the proportion which was wet. The total area covered was then obtained by summing over the total mapped stream length the areas for each 25 m length (see Fig. 1).

Results

The extent of the epilithic cover of the stream can be gauged by reference to Fig. 1 which shows that the total area occupied by *Nostoc* and *Phormidium* in the stream area was a function of stream wettable area. *Nostoc*, the N-fixing cyanobacterium, was centred on the zone 300–700 m up the stream from the sampling point. In this region the stream widths were maximal, and depth and velocity minimal so that the community occupied an extensive flat spongy area. At this point also the stream was adjacent to and flowed away from the glacier. Total area covered by each major cyanobacterial community in the mapped section was 2 977 and 9 569 m^2 for *Nostoc* and *Phormidium* respectively.

The primary source of nitrogen for the streams was the surface of the melting glaciers. Only small quantities of nitrogen can be derived from the stream bed soils (Howard-Williams *et al.*, 1986). High concentrations of nitrogenous compounds were found associated with the glacier surfaces (Table 2) and particularly with the first meltwaters early in the summer season and with low flow conditions later in the season. In an ice pond on

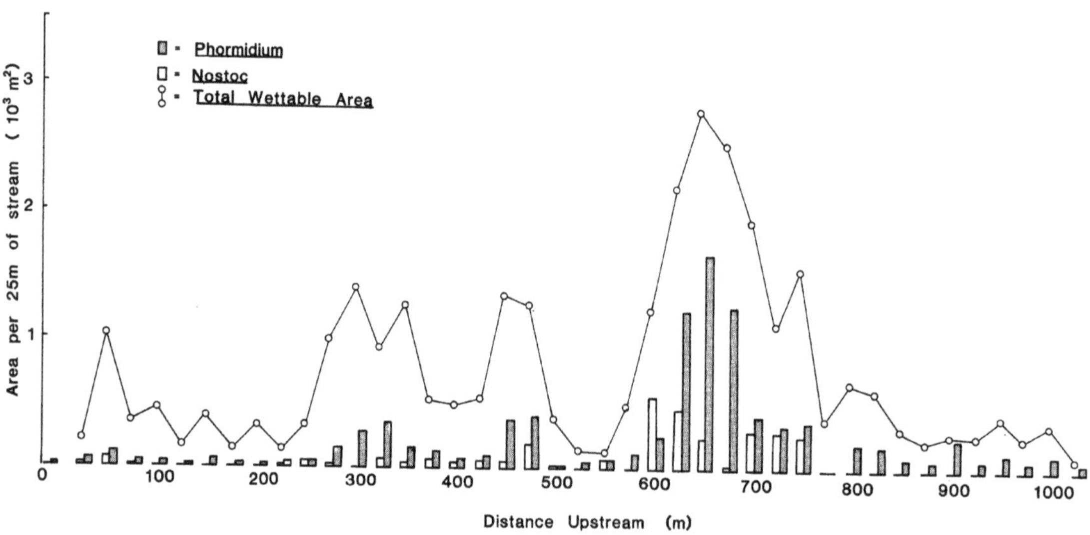

Fig. 1. Areas of *Nostoc* and *Phormidium* and the moist and water covered area of the central section of Fryxell Stream in January 1985. The area mapped can be seen in relation to total stream length on Fig. 2.

Table 2. Nitrogenous compounds on glacier surfaces and initial meltwaters on the Adams and Canada glaciers. Data, as mg m^{-3}, are means of two replicates. – = no data.

Glacier	Sample	NH_4-N	NO_3-N	DON	Urea-N
Adams	Snow on surface	11.6	9.5	63	–
	Yellow snow layer beneath surface snow	16.1	27.7	150	–
	Stream on glacier surface	7.0	26.0	63	–
	Ice bed of glacier stream	8.8	20.5	113	–
	Ponded water on glacier surface	35.5	150	308	222
Canada	Surface ice	9.9	4.9	50	–
	Early season dripping icicles (November)	17.3	96	112	–
	Late season dripping icecles (January)	10.8	8.6	56	45
	Stream on glacier surface at low flow	8.5	137	97	–
	Stream on glacier surface at high flow	2.8	26.7	23	–

top of the Adams Glacier, and in the early season melt from the Canada glacier (Table 2), both organic and inorganic forms of N were much higher than in the ice and snow a few cm down below the upper melt surface (Table 2). Seventy to eighty per cent of the dissolved organic nitrogen component in these snow and ice samples was composed of Urea-N.

A downstream plot of several dissolved nitrogen species for the Fryxell Stream is given in Fig. 2. Dissolved inorganic nitrogen levels were depleted with increasing distance downstream although changes in ammonium concentration were relatively small. The total dissolved organic nitrogen increased after the first km downstream but the urea component (a labile organic fraction) decreased in absolute terms and as a proportion of DON (Fig. 2). It would appear that inorganic and labile organic N compounds were removed by the stream biota and other dissolved organic N compounds generated within the stream.

During low flow conditions caused by freezing, dissolved nutrients became concentrated, and values were often much higher then than during rapid melt and high flow conditions (Table 3). This was particularly marked in the upper reaches of the streams. Fig. 3 shows the diurnal changes in particulate and dissolved N compounds. Particulate-N tends to bear a direct relationship to flow, while dissolved-N tends to be inversely related to flow as a result of freeze concentration.

In situ nitrogen uptake rates for Fryxell Stream were calculated from the concentration differences in Fig. 2 times the discharge, and then normalised to unit biomass from the total stream biomass (obtained from the stream mapping, Fig. 1) times the concentration of N in each species (Table 1), so that:

$$\text{N uptake } (\mu\text{g N } (\mu\text{g mat N})^{-1}\,\text{hr}^{-1} = 10^3 \times Q(C_1 - C_2)/(B_N PN_N + B_P PN_P)$$

Table 3. The effects of discharge on dissolved inorganic nitrogen concentrations in Fryxell Stream.

Distance from glacier (km)	Low Flow ($0.01\ m^3\,s^{-1}$)		High Flow ($0.14\ m^3\,s^{-1}$)	
	NO_3-N	NH_4-N	NO_3-N	NH_4-N
0	135	8	26	3
0.9	24	5	10	2
1.5	2	5	<1	2
1.8	4	3	<1	4
2.0	5	4	<1	<2

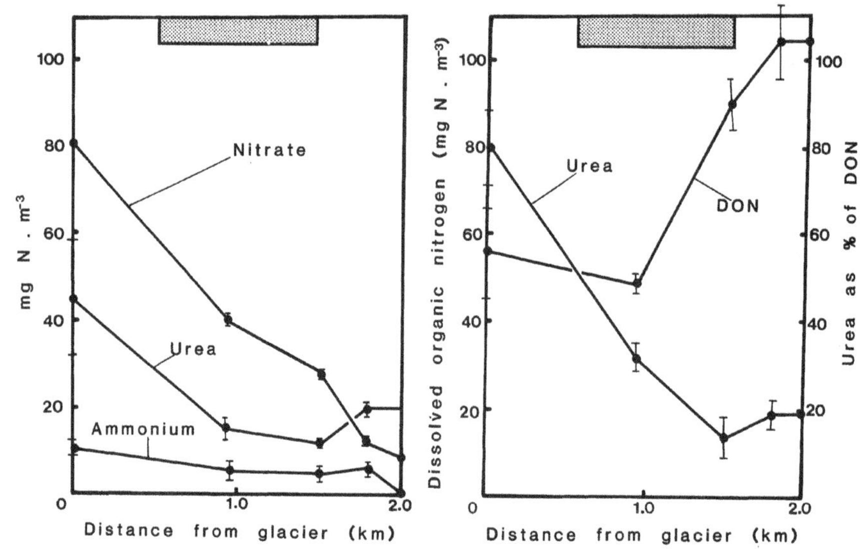

Fig. 2. Downstream changes in dissolved nitrogen species in Fryxell Stream (December, 1984). Urea-N expressed as concentration and as a% of Dissolved Organic Nitrogen (DON). The shaded length at the top of the figure shows the distance covered by the stream mapping in Fig. 1. Data as means (±range), n = 2.

where: Q = discharge ($m^3\ hr^{-1}$), C_1 and C_2 are DIN concentrations in the water at points 1 and 2 ($mg\ m^{-3}$). B_N and B_P are the total biomasses (g dry wt.) of *Nostoc* and *Phormidium* and PN_N and PN_P are the mat concentrations of nitrogen for *Nostoc* and *Phormidium* ($\mu g\ N\ g^{-1}$ dry wt) and 10^3 corrects the concentrations units from mg N to μg N.

In situ loss rates were calculated for DON in the same way as uptake rates (above) and for coarse and fine particulate N we obtained a mean mass flow in the stream over the diel cycle from data in Fig. 3. All the particulate N was assumed to come from the upstream biological community as particulate N was not detectable in the glacier ice melt samples (Table 1) in spite of particulate P values of up to 9 $mg\ m^{-3}$. The gains and losses are shown in Table 4. The data suggest a measurable balance between organic nitrogen formation and loss in this community, with gains slightly exceeding losses during this mid-season period.

A further gain to the community was expected to arise from N_2-fixation by heterocystous cyanobacteria in the mats. Acetylene reduction assays showed that the *Phormidium* mats, which contained a small population of heterocystous species, were either not able to fix atmospheric N or were not efficient at this process (Table 5). The highest rates were found in *Nostoc* communities in Adams Stream where dissolved reactive phosphorus was some 5 times higher than in Fryxell Stream. These fixation rates (71 ng N (mg mat N)$^{-1}$ h^{-1} were some 6.5 times lower than nitrogen uptake (Table 4) and maximal fixation rates for Fryxell Stream of 45.6 ng N (mg mat N)$^{-1}$ h^{-1} were only 10% of total N gains derived from the nutrient uptake measurements. The values reported in Table 5 show that C_2H_2 reduction proceeded at very low rates, the values being in the same range as those reported by Allnutt *et al.* (1981) for the benthic cyanobacterial mat communities of Lakes Fryxell, Hoare and Chad. Table 6 shows N-fixation rates as a function of light intensity and temperature. At 10 °C N_2-fixation in the light was significantly higher than in the dark only in the highest light intensities (>310 $\mu E\ m^{-2}\ s^{-1}$). At 0 °C the

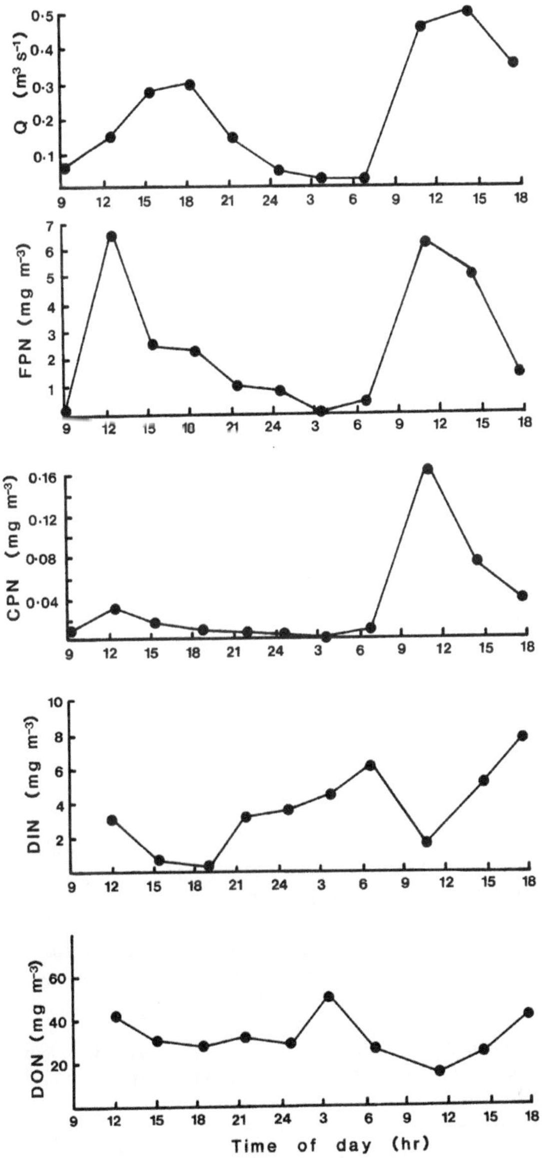

Fig. 3. Diel changes in discharge (Q) and concentration of fine and coarse particulate nitrogen (FPN, CPN) and dissolved inorganic and organic nitrogen (DIN, DON).

highest fixation rates were found at 175–310 $\mu E\,m^{-2}\,s^{-1}$ (34–60% of full light on the day of sampling). The Q_{10} values for N_2 -fixation above 175 $\mu E\,m^{-2}\,s^{-1}$ were 2.3–2.7.

Table 4. A nitrogen budget for the cyanobacterial community of Fryxell Stream derived from the particulate and dissolved nitrogen mass flow data.

		$\mu g\,N\,(mg\ mat\ N)^{-1}\,h^{-1}$
GAINS		
Nitrate uptake		0.370
Ammonium uptake		0.021
Urea uptake		0.074
	Total	0.465
LOSSES		
Coarse particulate organic-N		0.001
Fine particulate organic-N		0.031
Dissolved organic-N (non urea)		0.280
	Total	0.312

Dissolved inorganic nitrogen uptake *versus* concentration was studied for both *Phormidium* and *Nostoc* mats with ^{15}N labelled NH_4 and NO_3 separately. In the first experiment we tested trace and high concentration additions of $^{15}NH_4$-N and $^{15}NO_3$-N on 2.5 cm intact cores of both communities. V_{max} rates determined experimentally in this way (Table 7) were similar to *in situ* uptake calculated from Table 4. Net uptake rates were higher in the light than in the dark for both NO_3-N and NH_4-N.

In the next experiment we tested a wide concentration range of ^{15}N-labelled NH_4 and NO_3. As, initially, we were unsure of the variability associated with intact mats we broke up the mats and took homogeneous subsamples from slurries of mat for the incubations. During the incubations the mats resettled as a layer at the bottom of the chambers. We recognise that V_{max} may have been overestimated by this approach as some of the diffusion barriers present in intact mats may no longer exist. However, later inspection of the data showed that the rates for dispersed (Table 8) and intact mats were very similar. The half saturation constants, K_s, were low and in the case of *Nostoc* were below our minimum tested concentrations of 20 mg NH_4-N m^{-3} and 15 mg NO_3-N m^{-3}. NH_4-N was preferred source of nitrogen for both *Nostoc* and *Phormidium* (Table 7) suggesting that any regenerated N would be retained within the mats.

Table 5. Nitrogenase activity in cyanobacterial mats from the two streams. Conversion to nitrogen fixed is based on 3 : 1 ratio of C_2H_2 reduced to N_2 fixed. Data as means of 3 replicates ± SD.

Stream	Community	Treatment	nmol C_2H_2 (g dry wt)$^{-1}$ hr^{-1}	ng N (mg mat N)$^{-1}$ hr^{-1}
Adams	*Nostoc* mat	Light	8.5 ± 3.24	71.0 ± 0.3
		Dark	1.8 ± 0.22	14.7 ± 1.8
	Phormidium mat	Light	<0.1	<0.1
	(pink)	Dark	<0.1	<0.1
	Phormidium mat	Light	0.4 ± 0.33	1.7 ± 1.3
	(orange tufted)	Dark	0.3 ± 0.02	1.1 ± 0.1
Fryxell	*Nostoc* mat	Light	5.7 ± 1.14	45.6 ± 9.1
		Dark	2.8 ± 0.21	22.4
	Phormidium mat	Light	0.7 ± 0.01	0.6 ± 0.1
	(pink)	Dark	0.02	0.2

Discussion

Our estimates of nitrogen transformation rates and of the nitrogen budget are based on experimental and field evidence collected over two summer seasons. We are aware that such calculations in temperate streams are often made on year(s) long data sets with intensive sampling. Our estimates are therefore, at best, at order of magnitude level and it may be fortuitous that we have achieved a good balance of gains and losses (e.g. Table 4). However, the study streams were

Table 6. Effects of temperature and light intensity on nitrogenase activity in *Nostoc* mats at Fryxell Stream. Data as nmol C_2H_2 reduced (g dry wt)$^{-1}$ hr^{-1}. Data as means of 3 replicates ± SE except * = 2 replicates. ** = significantly different from dark value ($P < 0.05$).

Light intensity μE m^{-2} s^{-1}	Temperature °C	
	0.5 ± 0.5 °C	10 ± 1 °C
510	3.1 ± 0.14	8.5 ± 0.50**
310	4.1 ± 0.34**	9.4 ± 2.72**
175	5.7 ± 1.14**	2.4 ± 0.01*
100	3.0 ± 0.58*	3.0 ± 0.72
45	2.5 ± 0.48	2.2 ± 0.20
25	1.5 ± 0.12	3.4 ± 0.50
15	2.4 ± 0.11	3.7 ± 0.58
Dark	2.8 ± 0.21	2.3 ± 0.62

Table 7. Ammonium and nitrate uptake as μg N (mg mat N)$^{-1}$ h^{-1} in light and dark from 2.5 m^2 cores of *Nostoc* and *Phormidium* mats from Adams Stream. Data as means ± range from duplicate samples. The assays were conducted at two concentrations of each substrate: Low = 3.2 mg m^{-3} NH_4-N or 6.0 mg m^{-3} NO_3-N; High = 160 mg m^{-3} NH_4-N or 160 mg m^{-3} NO_3-N.

N form	Treatment		Species	
			Nostoc	*Phormidium*
NH_4-N	Low	Light	1.96 ± 0.76	0.78
	High	Light	0.69 ± 0.03	0.48 ± 0.03
	High	Dark	0.50 ± 0.16	0.40 ± 0.09
NO_3-N	Low	Light	0.76 ± 0.04	0.40
	High	Light	0.51 ± 0.05	0.44 ± 0.01
	High	Dark	0.30 ± 0.02	0.31 ± 0.08

Table 8. Ammonium and nitrate uptake parameters for dispersed mats from Adams Stream determined from ^{15}N-labelled substrates.

	V_{max} μg N (mg mat-N)$^{-1}$ hr^{-1}	K_s mg N m^{-3}
Phormidium mat		
NH_4	2.0 ± 0.1	34 ± 7
NO_3	0.4 ± 0.04	12 ± 10
Nostoc mat		
NH_4	0.5 ± 0.06	*
NO_3	0.1 ± 0.04	*

* V_{max} already attained at the minimum concentration tested (20 mg NH_4-N m^{-3} or 15 mg NO_3-N m^{-3})

simple first order streams with a strong single point source (their respective glaciers) for the water and nutrient supply. The biotic communities were also relatively simple consisting only of microbial mats strongly dominated by one type of organism (cyanobacteria) and with a negligible herbivore activity. The streams flowed for only a short period each year and diel periodicity was the major temporal scale causing variance in stream properties. Our calculations and data sets have covered this time scale so we believe the results are valid as first order estimates.

Although initial impressions of the essentially non-vegetated catchments of the southern Victoria Land streams suggest minimal allochthonous inputs of organic material, dissolved organic forms of nitrogen were supplied by the glaciers (Fig. 2, Table 2). A high proportion of this was urea-N and as such was available for uptake by the stream biota (Fig. 2, Table 4). The studies of Timperley *et al.* (1985) show a high utilisation of the low molecular weight fractions of dissolved organic nitrogen originating in both Northern and Southern Hemisphere precipitation. Downes *et al.* (1986) suggested that old marine and lacustrine sediments in the dry valleys of southern Victoria Land, as well as marine-derived aerosols, could be potential sources of organic material windblown on to the glaciers. They found dissolved organic carbon levels on ice and snow in the area were not much lower than in New Zealand freshwaters (*c.* 1–3 mg of DOC m^{-3}). While the urea component of the DON from the glaciers was apparently utilised in the streams, total DON increased downstream (Fig. 2 and see Downes *et al.*, 1986) presumably due to activities of the cyanobacterial films. Seventy two percent of the inorganic nitrogen taken up by these communities was regenerated to the stream water as non-labile (non-urea) DON, and approximately 7% was recovered as particulate N (Table 4). We suspect, however, that there may be a major loss of particulate N relative to DON in the first flows of the season.

Transformations of nitrogen during passage down the Fryxell stream were considerable. The effectiveness of a stream community in reducing the nutrient load of a stream is a function of discharge, uptake rate and area of the biotic community, and for many streams can be summarised by the Net Uptake Length Constant K_w with units of m^{-1} (Howard-Williams, 1985) where:

$$K_w = [\ln(C_z/C_o)]/z$$

and C_o is the dissolved nutrient concentration at the upstream end of a study reach and C_z is the concentration at z metres downstream. An exponential decline in concentration downstream would be expected in oligotrophic waters such as Fryxell stream where nutrient levels are near or below V_{max} for the communities (cf. Table 8). Rutherford *et al.* (1987) show a strong inverse relation of K_w with stream discharge (Q) for a wide range of temperate rivers and streams ($Q = 0.0001\ m^3\ s^{-1}$ to $Q = 30\ m^3\ s^{-1}$). K_w values for Fryxell stream derived from Fig. 2 and Table 3 fit the Rutherford *et al.* (1987) regression of Kw vs Q very closely, demonstrating that nutrient uptake rates here were not abnormally slow. For instance, for flows of 0.01 and 0.06 $m^3\ s^{-1}$ K_w values for dissolved inorganic-N were 0.004 and 0.007 m^{-1} in Fryxell stream. K_w values derived for temperate streams for these flow rates are 0.002 and 0.0007 m^{-1} respectively (Rutherford *et al.*, 1987).

Nitrogenase activity was a relatively small component of the N-dynamics, accounting for only 10% of total N gains to the mats. Rates of activity were of the same order as those reported for south Victoria Land lake edge cyanobacterial communities (Alnutt *et al.*, 1981). They were slightly higher than for the *Nostoc muscorum* containing moss turfs and carpets of Signy Island (Christie, 1987) but one to two orders of magnitude lower than rates measured on cultures of *N. muscorum* from those turfs and carpets where values of 119 and 200 nmol C_2H_2 (g dry wt)$^{-1}$ h^{-1} were recorded (Christie, 1987). Conversion of the acetylene reduction data (Table 5) to nitrogen fixed gave a range from 9–939 mg N m^{-2} $year^{-1}$ for the two south Victoria Land streams. This spans the range of N_2-fixation rates reported for the terrestrial communities of the Vestfold Hills of 52–119 mg N m^{-2} $year^{-1}$ (Davey & Marchant, 1983) and Signy Island of 46–192 (mg N m^{-2} $year^{-1}$ (Christie, 1987). The higher rates found in our study may be due to the aquatic nature of the N_2-fixing cyanobacteria here, and both the studies from Signy Island and Vestfold Hills indicated that weather conditions in these terrestrial communities resulted in higher N_2-fixation rates. However, in terms of a stream N budget, the maximal fixation rate for Fryxell stream of 630 mg N m^{-2} $year^{-1}$ was only 10% of total N gained by the mat which derived virtually all its nitrogen from dissolved forms.

Temperature exerts a major metabolic control on southern Victoria Land stream communities (Vincent & Howard-Williams, 1988). The effects of low temperature on nitrogenase activity were seen only in the high light intensities ($>175\ \mu E\ m^{-2}\ s^{-1}$). The I_k value for light saturation of photosynthesis by the *Nostoc* community was found to be $130\ \mu E\ m^{-2}\ s^{-1}$ (Howard-Williams & Vincent, 1988). Reduced nitrogenase activity and the lack of a temperature response below this light level indicates a shift towards light rather than temperature limitation of N_2-fixation. However, through most of the polar summer 24 hour cycle the I_k is likely to be exceeded and temperature would be a controlling factor.

Dissolved inorganic N is often in plentiful supply in both maritime (Hawes, 1988) and continental antarctic streams and, also in at least one high arctic stream (Hullar & Vestal, 1988). Evidence for N limitation in polar streams, has not been conclusively established. Even in low nitrogen waters of Fryxell Stream an enrichment assay failed to detect evidence of N-limitation (Howard-Williams & Vincent, 1988). Net uptake rates of inorganic N from these streamwaters are not abnormally low when compared with temperate streams in spite of a distinct low temperature control on growth per unit biomass in the antarctic (Vincent & Howard-Williams, 1988). Area based production rates are comparable with temperate streams because of the high epilithic biomass in these antarctic streams (Howard-Williams & Vincent, 1988). This large standing stock of microbial biomass also accounts for the observed inorganic N uptake rates. Our evidence shows that a major loss of N from these microbial communities is as DON, but the dynamics organic as well as inorganic nutrient regeneration from these communities needs further study.

Acknowledgements

This research was carried out within the 1984/85 and 1985/86 seasons of the New Zealand Antarctic Research Programme. We thank Antarctic Division (DSIR), the US Navy and the Royal New Zealand Air Force for logistic support, Geoff Spearpoint and Gillian Wratt for field assistance and L. May and M. Gibbs for laboratory assistance. The manuscript was reviewed (and greatly improved) by Dr Tony Viner and Dr John Ferris. Thanks to Jan Simmiss for typing.

References

Allnutt, F. C. T., Parker, B. C., Seaburg, K. G. & G. M. Simmons, 1981. *In situ* nitrogen (C_2H_2) fixation in lakes of southern Victorialand, Antarctica. Hydrobiol. Bull. 15: 99–109.

Christie, P., 1987. Nitrogen in two contrasting antarctic bryophyte communities. J. Ecol. 75: 73–93.

Crooke, W. M. & W. E. Simpson, 1971. Determination of ammonium of Kjeldahl digests of crops by an automated proceedure. J. Sci. Food Agric. 22: 9–10.

Davey, A. & H. Marchant, 1983. Seasonal variation in nitrogen fixation by *Nostoc commune* Vaucher at the Vestfold Hills, Antarctica. Phycologia 22: 377–385.

Demanche, J. M., Curl H. & D. D. Coughenower, 1973. An automated analysis for urea in seawater. Limnol. Oceanogr. 18: 686–689.

Downes, M. T., 1978. An improved hydrazine reduction method for the determination of low nitrate levels in freshwater. Wat. Res. 12: 673–675.

Downes, M. T., Howard-Williams, C. & W. F. Vincent, 1986. Sources of organic nitrogen, phosphorus and carbon in antarctic streams. Hydrobiologia 134: 215–225.

Flett, R. J., Hamilton, R. D. and N. E. R. Campbell, 1975. Aquatic acetylene-reduction techniques: solutions to several problems. Can. J. Microbiol. 22: 43–51.

Hawes, I., 1988. Filamentous green algae in freshwater streams on Signy Island, Antarctica. Hydrobiolgia (this volume).

Howard-Williams, C., 1985. Cycling and retention of nitrogen and phosphorus in wetlands: a theoretical and applied perspective. Freshwat. Biol. 15: 391–431.

Howard-Williams, C., Vincent, C. L., Broady, P. A. & W. F. Vincent, 1986. Antarctic stream ecosystems: variability in environmental properties and algal community structure. Int. Revue ges. Hydrobiol. 71: 511–544.

Howard-Williams, C. & W. F. Vincent, 1988. Microbial communities in southern Victoria Land streams (Antarctica) I: Photosynthesis. Hydrobiologia (this volume).

Hullar, M. A. & J. R. Vestal, 1988. The effects of nutrient limitation and stream discharge on the epilithic microbial community in an oligotrophic Arctic stream. Hydrobiologia (this volume).

Newbold, J. D., Elwood, J. W., O'Neill, R. V. & W. van Winkle, 1981. Measuring nutrient spiralling in streams. Can. J. Fish. aquat. Sci. 38: 860–863.

Rutherford, J. C., Williamson, R. B. & A. B. Cooper, 1987. Nitrogen, phosphorus and oxygen dynamics in rivers. In: A. B. Viner (ed), Inland Waters of New Zealand. Science Information Publishing Centre, Wellington, New Zealand: 139–165.

Timperley, M. H. & J. C. Priscu, 1986. Determination of nitrogen-15 by optical emission spectrometry using an atomic absorption spectrometer. Analyst 111: 23–28.

Timperley, M. H., Vigor-Brown, R. J., Kawashima, M. & M. Ishigami, 1985. Organic nitrogen compounds in atmospheric precipitation. Their chemistry and availability to phytoplankton. Can. J. Fish. aquat. Sci. 42: 1171–1177.

Vannote, R. L., Minshall, G. W., Cummins, K. W., Sedell, J. R. and C. E. Cushing, 1980. The river continuum concept. Can. J. Fish. aquat. Sci. 37: 130–137.

Vincent, W. F. and C. Howard-Williams, 1986. Antarctic stream ecosystems: physiological ecology of a blue-green algal epilithon. Freshwat. Biol. 16: 219–233.

Vincent, W. F. and C. Howard-Williams, 1988. Microbial communities in southern Victoria Land streams II. The effect of low temperature. Hydrobiologia (this volume).

Vincent, W. F. and C. L. Vincent, 1982. Factors controlling phytoplankton production in Lake Vanda. Can. J. Fish. aquat. Sci. 39: 1602–1609.

Hydrobiologia 172: 63–75, 1989
W. F. Vincent and J. C. Ellis-Evans (eds) High Latitude Limnology

Benthic algal biomass and productivity in high subarctic streams, Alaska

Jacqueline D. LaPerriere[1], Erwin E. Van Nieuwenhuyse[2] & Patricia R. Anderson[3]
Alaska Cooperative Fishery Research Unit, University of Alaska, Fairbanks, AK 99775-0110; [1] U.S. Fish and Wildlife Service; [2] *Present address: School of Forestry, Fisheries and Wildlife, University of Missouri, Columbia, MO 65201*; [3] *Present address: California Department of Fish and Game, Monterey, CA 93940*

Key words: alkalinity, benthic algae, phosphorus, primary productivity, sestonic chlorophyll *a*, standing crop

Abstract

Year-round measurements of the standing crop of epilithic algae (as chlorophyll *a* concentration) in two streams – one second and one fourth order (map scale 1:63 360) – in interior Alaska (64°-65° N) were only about one tenth that reported from streams of temperate North America. Cell densities in these streams, however, were similar to those in comparable temperate streams. Year-round domination of the benthic flora by very tiny diatoms (*Achnanthes* spp.) may explain the apparent disparity between low chlorophyll *a* content and nearly average cell densities. Chlorophyll *a* standing crop in a more alkaline groundwater-fed stream, however, was higher and within the range of similarly sized temperate streams. Maximum chlorophyll *a* standing crop varied positively with alkalinity in 5 clear-water streams where standing crop was measured on natural or artificial substrates. Seasonal mean concentrations of sestonic chlorophyll *a* (used as estimates of benthic algal chlorophyll *a* standing crop) varied directly and significantly with alkalinity among ten clear-water streams; and, with total phosphorus among 8 of 10 clear-water and 5 brown-water streams studied. During the summer, when there is little darkness, gross primary productivity (as estimated by the diurnal dissolved-oxygen method) was similar to that of northern temperate streams. Gross primary productivity was also seen to vary directly with alkalinity in 5 clear-water streams of this region.

Introduction

Essentially three types of streams occur in the high subarctic of Alaska: clear water, brown water, and glacial. Clear-water streams can be subdivided into two sub-types; one dominated by surface water and one by groundwater. Brown-water streams are colored by tannins and lignins from slowly decomposing vegetation and peat deposits lying over frozen or wet ground. Glacial streams are fed in part by glacial meltwaters, and transport large amounts of fine suspended sediment which generate high turbidity.

We measured standing crop and primary productivity of benthic algae in various clear-and brown-water streams in interior Alaska, in order to examine two hypotheses: that in high subarctic Alaska streams benthic algal standing crop (as measured directly and estimated as sestonic chlorophyll *a*) and benthic algal productivity (1) varied directly with alkalinity, and (2) varied directly with total phosphorus. Primary productivity of the benthic algae of these streams was compared to that of streams of other regions as reported in the literature.

Study sites

The entire study area lies in the high subarctic and, being continental, is subject to wide temperature extremes. At Fairbanks the mean maximum July temperature is about 21 °C and the mean minimum January temperature is − 27 °C. The mean annual temperature is − 4 °C (Hartman & Johnson, 1978); therefore, the area is underlain by discontinuous permafrost. Ice-rich stream valleys contain extensive muskeg, while valleys that are more ice-free have boreal vegetation. Most of the sites of the placer mining study were near the tree line and included muskeg. At lower altitude sites, streams had a narrow band of riparian vegetation.

Most of the precipitation (only about 51 cm per year) occurs in summer, although snowpack runoff is usually the largest single flow event. Nevertheless, reduced infiltration due to the occurrence of ice-rich permafrost, as well as low evaporation and transpiration in the area (Hartman & Johnson, 1978), allow stream runoff to be significant throughout the ice-free season. The low flow in winter is provided only by groundwater.

The area is covered with recent alluvium through the valleys, with some granitic intrusions (common along the upper Chena River where they can be seen as tors on the ridgetops near Angel Creek, and where quarrying has been conducted near the highway). The bedrock is mainly Birch Creek schist, and the soil is primarily micaceous loess (Péwé, 1955).

Thirteen stations on tributaries of the Tanana River (Fig. 1) were sampled five times at about

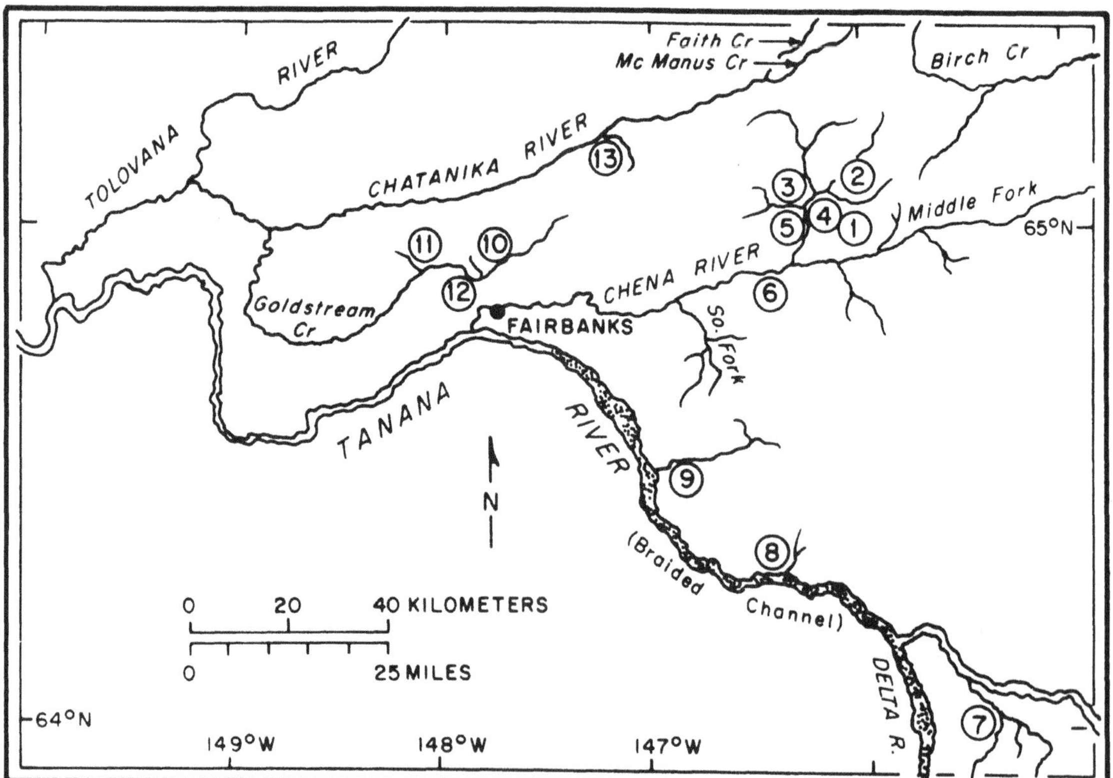

Fig. 1. Study sites 1–13 on tributaries of the Tanana River. Stations as described in Table 1.

monthly intervals from May 15 to September 12, 1979 (LaPerriere, 1983). One of these streams, Delta Clearwater Creek, (Station 7, Fig. 1) was sampled at least once a month from February 1978 until June 1979, in a baseline water quality study.

Additional work, done to describe the seasonal changes in primary producers of the upper Chena River at Monument Creek and the West Fork (Stations 1 and 3, Fig. 1) was conducted from May 1982 to May 1983 (Anderson, 1984), and provided detailed information on the fluctuations in abundance of certain taxa at these sites, in addition to the information we present here.

The Tanana River is a major feature of the Intermontane Plateau of interior Alaska, which lies between the Brooks Range (Rocky Mountains) to the north, and the Alaska Range (Coastal Mountains) to the south. Ultimately the Tanana flows into the Yukon River which flows to the sea. Delta Clearwater Creek is fed by a massive aquifer of undelineated size. Because of the great size of this aquifer the flow of the Clearwater is very steady, varying only 10% in the three years it was gauged by the United States Geological Survey (Dorothy Wilcox, Fairbanks, Alaska, 1979, personal communication). Its surface tributaries are intermittent with visible flow only occurring at spring runoff, and during some summer storm events. All of the remaining study streams have smaller aquifers, and surface runoff during summer storm events causes their flow to be more variable.

The Tanana River Valley stations were selected to represent the morphometric, hydrologic and chemical characteristics of streams of the region (Table 1). They were sited near highway bridges (for easy access), but in reaches unaltered by the structures.

Some human disturbance was noted on Banner, O'Connor, Spinach and Goldstream creeks (Stations 8, 10, 11 & 12, Fig. 1). In the early 1900's gold was dredged in the headwaters of Goldstream Creek and along the valley of Banner Creek. Placer gold mining was underway with four placer operations reported active in the Banner Creek valley in summer 1979. Construction activities – particularly road building, pipeline crossing, and gravel extraction – were recent in upper Goldstream Creek. A gold claim was still

Table 1. Selected descriptive characteristics of the study stations during summer 1979.

Station No.	Station Name	Location	Watershed area upstream (km²)	Mean width (m)	Local Slope[b]	Watershed vegetation	Disturbance observed
1	Moment Creek	65° 04′ N, 146° 05′ W	85	6.48	0.012	boreal	none
2	North Fork	65° 04′ N, 146° 05′ W	229	12.00	0.015	boreal	none
3	West Fork	65° 04′ N, 146° 10′ W	433	20.73	0.004	boreal	none
4	Big Bend	65° 03′ N, 146° 12′ W	779	35.26	0.005	boreal	none
5	Angel Creek	65° 02′ N, 146° 12′ W	99	10.73	0.004	boreal	none
6	Gauging Station	64° 54′ N, 146° 25′ W	2437	39.62	0.003	boreal	none
7	Delta Clearwater	64° 05′ N, 145° 35′ W	911[c]	48.77	0.002	boreal	none
8	Banner Creek	64° 17′ N, 146° 21′ W	52	4.04	0.009	boreal	placer mining
9	Little Salcha	64° 30′ N, 146° 59′ W	173	8.47	0.001	muskeg	beaver dam[d]
10	O'Connor Creek	64° 55′ N, 147° 53′ W	32	1.77	0.013	muskeg	none
11	Spinach Creek	64° 55′ N, 148° 10′ W	26	3.52	0.006	muskeg	none
12	Goldstream Creek	64° 54′ N, 147° 51′ W	258	6.30	0.014	muskeg	gold dredged
13	Kokomo Creek	65° 10′ N, 147° 18′ W	105	6.16	0.004	muskeg	beaver dam[e]

[a] See Fig. 1 for location of stations.
[b] Slope within 305 m of the station.
[c] The tributaries of this stream are influent, but are included in measurement of the watershed area.
[d] Dam located approximately 305 m upstream of station.
[e] Dam located approximately 183 m upstream of station.

maintained in the headwaters of O'Connor Creek, but there was no evidence of recent activity. Along Spinach Creek, prospecting had occurred but mining had not yet begun.

Five streams – two at the headwaters of the Chatanika River and three at the headwaters of Birch Creek (Fig. 2) (all of which are located north of the streams of Fig. 1, and east of 147° W) were sampled intensively during the open-water seasons of 1982 and 1983 in a study of the effects of placer gold mining on primary productivity in streams (Van Nieuwenhuyse & LaPerriere, 1986). The main study sites were at two paired watersheds, each pair having a mined and an unmined stream. One pair, at the headwaters of the Chatanika River, consisted of a moderately mined stream, Faith Creek, and the unmined control, McManus Creek. The other pair was east of Twelvemile Summit, at the headwaters of Birch Creek; Birch Creek was heavily mined along the uppermost tributary (Eagle Creek) and along its mainstream above the confluence with the unmined control, Twelvemile Creek. Below this confluence, Birch Creek is designated a U. S. Wild River downstream to the point where it again passes under the Steese Highway about 202 km downriver, north of the town of Central.

Ptarmigan Creek (a tributary of upper Birch Creek), was unmined until late August 1983. Samples were taken there before and after the mining activity began. More detailed descriptions of these sites including additional flow characteristics and water quality data are presented in papers by Bjerklie & LaPerriere (1985), LaPerriere *et al.* (1985), Wagener & LaPerriere (1985) and Van Nieuwenhuyse & LaPerriere (1986).

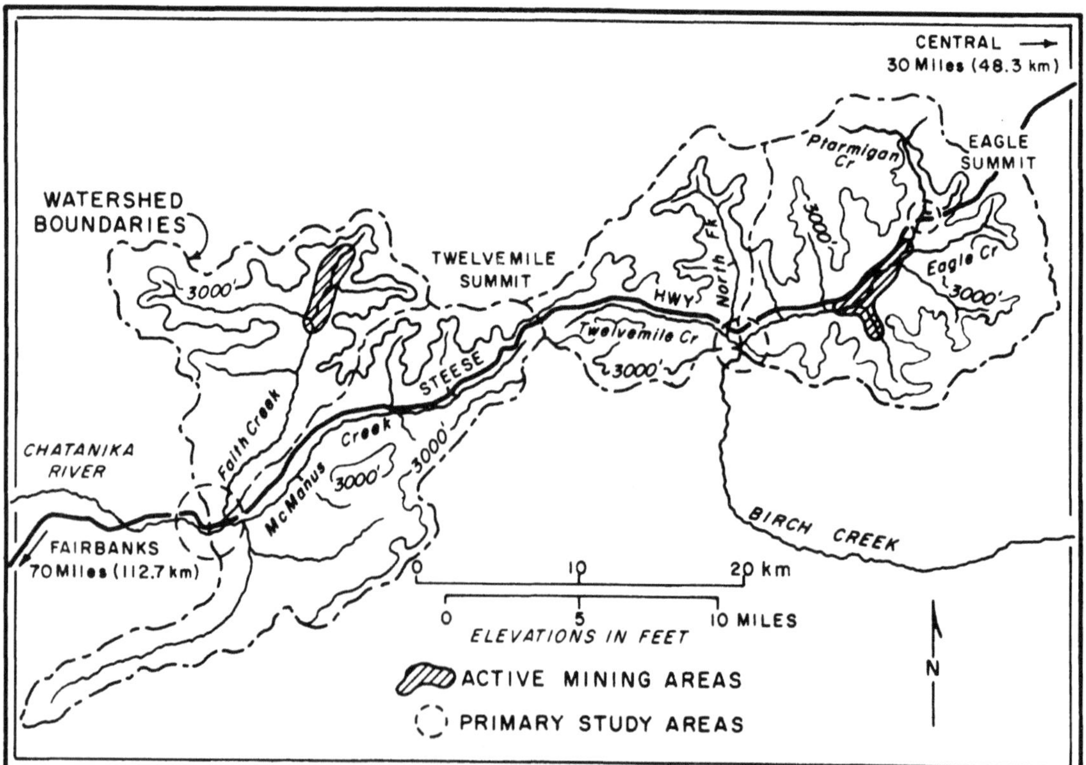

Fig. 2. Study sites on the headwaters of the Chatanika River and Birch Creek.

Methods

Water quantity and quality

Streamflow (discharge) was estimated by measuring water depth and current velocities at 0.6 of water depth at 10 to 20 even-distance intervals across the stream perpendicular to the flow, at sites with steady flow. Staff-gauges were placed in streams which were studied continuously in 1982 and 1983, and a stage-discharge relation was calculated for each stream. Flows not measured directly were estimated from these rating curves.

Field test kits or calibrated meters (HACH Chemical Company) were used for measurement of alkalinity and pH. Grab samples were taken for turbidity and read on a HACH 'Portalab' nephelometer. Replicate 50 ml water samples were frozen and stored (6 months or less) until total phosphorus was determined using the method of Gales *et al.* (1966) or as modified by Eisenreich *et al.* (1975).

Benthic algal cell densities

The stream substrate was divided into three size strata (Cummins, 1962): the largest, small boulders, ranged from Phi −7 to −8 (128–256 mm). The next, cobbles, ranged from Phi −5 to −6 (32–64 mm). The smallest substrate – fine silt to gravel (Phi 9 to −4; 0.0039 to 16 mm) – was not sampled because it represented only a small fraction of the total area and was highly unstable. A grid was laid out at both sites sampled (on the upper Chena River at Monument Creek and at West Fork), and three substrate samples were taken from each of the two size strata on each sampling date. A table of random numbers was used to select two numbers that defined an intersection on the grid for the location of each sample.

A large plastic bucket from which the bottom had been removed was placed around the sample before it was removed from the stream, to keep the periphyton intact. Each rock was placed into a double plastic bag and then into a dark insulated chest with ice for transport to the laboratory.

In the laboratory, the entire surface of each rock was scrubbed with a brass brush and rinsed with distilled water three times to remove the periphyton. The algal samples (20 mL) were preserved in a 3% formalin solution and allowed to settle for 1 week before the formalin was replaced with Caro-Safe, a nontoxic preservative (Carolina Biological Supply).

Surface area of each rock was determined by using heavy-gauge plastic food wrap. We wrapped the plastic snuggly around the rock and clipped off all the excess. The plastic was then weighed on a balance, and the rock surface area was determined by applying a linear regression equation derived from known areas and weights of the plastic.

A 0.1 ml subsample was used to determine cell densities. The microscope (×480) was fitted with a calibrated Whipple Disk and counts were made of fields in a Palmer Cell. Dilutions (made with a Pasteur pipette, by counting drops) were occassionally necessary when periphyton densities were high or foreign matter (e.g. sand) obscured the view. The counting procedure involved two separate observations because the periphyton samples included both small and large algal cells. During the first observation we counted every large algal cell in the Palmer Cell: *Didymosphenia geminata, Cladophora glomerata, Closterium* sp., *Mougeotia* sp., *Stigeoclonium tenue, Ulothrix zonata, Draparnaldia* sp., *Chaetophora elegans*, chains of *Fragilaria* sp., and filaments of unidentified large algae. In the second observation, we counted and identified the small algal cells in random microscopic fields until 100 cells from each of the six samples were counted. The algal unit approach (APHA *et al.*, 1975) was used for counting the tiny blue-green filamentous algae. These algae were rare in our samples, and were represented by only two types – *Lyngbya* sp. (average length 25 μm; about 14 cells) and *Oscillatoria* sp. (average length 25 μm; about 6 cells). Densities were calculated separately for the two observations and then summed. We rejected extreme values (those indicating a 95% or greater chance of being outliers) before calculating mean densities.

Chlorophyll a

For the upper Chena River (Monument Creek and the West Fork) the three substrate samples taken for periphytic chlorophyll *a* from each of the two substrate sizes were collected, transported, and scrubbed as described for periphyton identification samples. We added 3 ml of a saturated solution of magnesium carbonate ($MgCO_3$) to the sample in the bag before transport, to prevent acidification of chlorophyll *a* to phaeophytin *a*. For Delta Clearwater Creek, three to five cobbles (by far the dominant substrate) were taken randomly at each of three riffle stations, and a cylindrical sampler was attached to the upper surface with a bar-clamp. The area inside the circle (0.002 m^2) was scrubbed with a modified toothbrush to remove the attached algae. This was usually done in the field, but during inclement weather the samples were transported to the laboratory for processing. Samples from both rivers (upper Chena and Delta Clearwater Creek) were filtered through glass-fiber filters (1.6 μm effective pore size) and frozen for later analysis.

For the streams of the placer mining study (McManus, Faith, Twelvemile and Birch creeks) we used 1/8″ (3.18 mm) thick square plexiglass plates (0.01 m^2) as artificial substrates to characterize algal standing crop. A minimum of three plates were sampled periodically at random without replacement. A cylindrical plexiglass template was used to circumscribe an area of 0.003 m^2 in the center of each plate, which was then scraped bare with a modified toothbrush and filtered under slight pressure onto a glass-fiber filter described above. Several drops of saturated magnesium carbonate solution were added to prevent acidification. Filtered samples were stored dry and kept frozen for subsequent laboratory analysis.

Sestonic chlorophyll *a* samples (1–3 L) were taken in duplicate and immediately filtered through glass-fiber filters and frozen as described above. All filters were ground with 90% (v/v) acetone and allowed to extract for 24 h in a refrigerator. Analysis of chlorophyll *a* concentration was conducted on a narrow-beam spectrophotometer or on a fluorometer calibrated against a spectrophotometer using the methods and formulas of Golterman (1969). Outlier values were rejected (as described above) before mean values were calculated.

Primary productivity

To estimate algal productivity we used the standard (American Public Health Association *et al.*, 1980) freewater single-station diel oxygen curve of Odum (1956). A YSI Model 56 dissolved oxygen monitor, calibrated against five Winkler titrations before each measurement period, continuously recorded dissolved oxygen concentration (mg L^{-1}) and water temperature (°C) in the stream. We transferred these date by hand from the strip charts to data files, using an interval of 1 h (1982 and 1984) or 0.5 h (1983). We then calculated values for change in dissolved oxygen concentration, corrected them for diffusion, plotted them against time, and then solved for gross primary productivity, using the trapezoid rule. All calculations were based on mean temperature and dissolved oxygen values for the appropriate time interval. Accrual of oxygen by means other than from production and diffusion was considered negligible. Daytime respiration was quantified by connecting the pre-dawn and post-sunset minimums as suggested by Hall & Moll (1975). We used the tables of Mortimer (1981) to calculate oxygen saturation and the formula of O'Connor and Dobbins (1956) to derive the reaeration coefficient; Wilcock (1982) showed this formula to be best suited for application to relatively small headwater streams. The reaeration coefficient was corrected for temperature according to Elmore & West (1961). Atmospheric pressure was calculated from altitude according to Golterman (1969).

Results

Benthic algal standing crop

Cell densities of benthic algae were measured in Monument Creek and the West Fork of the upper Chena River over an entire year (Fig. 3). Late summer to early autumn was the period of maximum standing crop measured as number of cells.

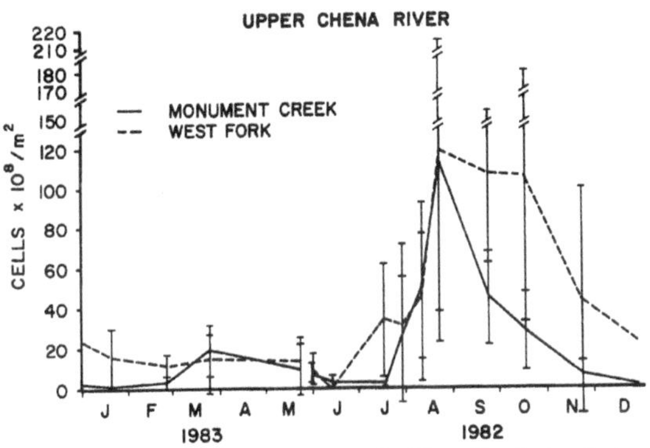

Fig. 3. Mean density (± SD; n = 6) of benthic algal cells in Monument Creek and the West Fork of the Chena River, May 1982 – May 1983.

Standing crop as chlorophyll *a* density was measured during the same year in the same streams (Fig. 4), and for nearly a full year in 1978 in Delta Clearwater Creek (Fig. 5). Again, late summer to early fall was the period of maximum standing crop, but another peak in spring was apparent in Delta Clearwater Creek. Note that the densities are about 10 times as high in Delta Clearwater Creek as in the upper Chena River.

Chlorophyll *a* densities from artificial substrates allowed to become colonized by benthic algae in the streams of the placer mining study during summer 1983 were maximum when the last one were sampled in September. The density in the unmined streams reached 2.87 ± 1.40 mg m^{-2} in Twelvemile Creek, and 9.49 ± 1.38 in McManus Creek. Moderately mined Faith Creek

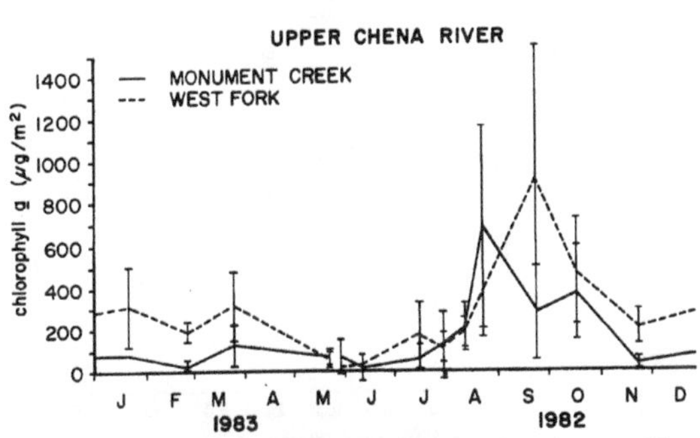

Fig. 4. Mean density (± SD; n = 6) of benthic algae as chlorophyll *a* in Monument Creek and the West Fork of the Chena River, May 1982 – May 1983.

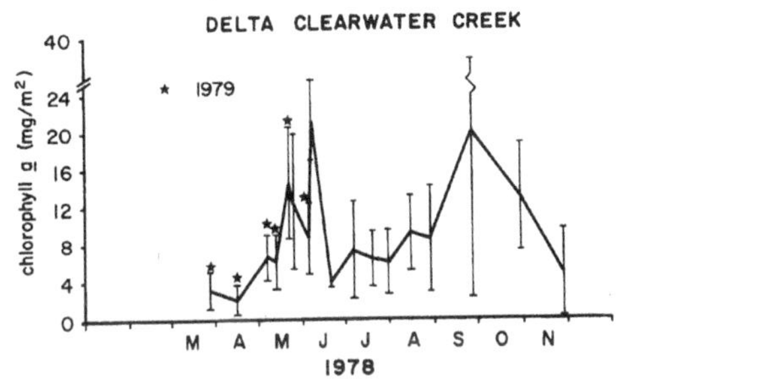

Fig. 5. Mean density (± SD; n = 3) of benthic algae as chlorophyll *a* in Delta Clearwater Creek, March 1978 – May 1979.

peaked at 3.83 ± 0.88 mg m^{-2}, but heavily mined Birch Creek was completely devoid of colonizing benthic algae.

Maximum standing crop as chlorophyll *a* varied positively and significantly with mean summer alkalinity ($r = 0.83$; $P \leq 0.05$) for the unmined, clear-water streams, but no relation was found between total phosphorus and standing crop.

Sestonic chlorophyll a

When not measured directly, the summer benthic algal standing crops were assessed by measuring the amounts of chlorophyll *a* suspended in the water column; a direct relation was assumed between the amount of suspended algae in the stream and the size of the standing crop of benthic algae (*sensu* Swanson & Bachmann, 1976). This assumption was tested for sestonic and benthic chlorophyll *a* data, paired for each sampling date of the Chena River study, by Spearman rank correlation and found acceptable ($P \leq 0.05$).

The relation between alkalinity and the chlorophyll *a* content of the suspended algae (Table 2) in the study streams was not significant. These study streams can be divided into two types, muskeg (or brown water) and boreal (or clear water) on the basis of watershed vegetation type and the associated water color. When separate analyses were conducted for streams of the two types, the relation between alkalinity and sestonic chlorophyll *a* was positive and significant ($r = 0.72$; $P \leq 0.05$) for clear-water streams

Table 2. Means and standard errors (n = 5) of selected physical, chemical and biological characteristics of the study streams.

Station	Sestonic chlorophyll *a* (mg m^{-3})	Av. velocity (m s^{-1})	Discharge (m^3 s^{-1})	Alkalinity (mg l^{-1}) (as $CaCO_3$)	Total phosphorus (mg m^{-3})
Summer 1979					
Monument Creek	0.31 ± 0.18	0.86 ± 0.17	1.48 ± 0.39	36 ± 5	3.6 ± 1.1
North Fork	0.33 ± 0.22	0.83 ± 0.14	2.71 ± 0.70	59 ± 7	4.0 ± 2.7
West Fork	0.46 ± 0.30	0.81 ± 0.13	4.18 ± 0.76	40 ± 5	5.0 ± 3.9
Big Bend	0.19 ± 0.13	1.05 ± 0.09	9.98 ± 1.77	47 ± 9	4.8 ± 1.6
Angel Creek	0.40 ± 0.19	0.94 ± 0.08	1.70 ± 0.44	29 ± 4	3.2 ± 1.9
Gauging Station	0.72 ± 0.29	– – –[a]	30.1 ± 4.3	66 ± 9	6.6 ± 4.4
Delta Clearwater	0.74 ± 0.47	0.90 ± 0.03	21.0 ± 2.1	142 ± 7	7.2 ± 3.1
Banner Creek	0.63 ± 0.63	0.49 ± 0.09	0.14 ± 0.06	82 ± 15	10 ± 4
Little Salcha	1.10 ± 0.50	0.75 ± 0.16	1.07 ± 0.74	50 ± 8	26 ± 8
O'Connor Creek	0.43 ± 0.59	0.60 ± 0.15	0.07 ± 0.04	113 ± 19	23 ± 3
Spinach Creek	0.34 ± 0.16	0.14 ± 0.01	0.04 ± 0.01	123 ± 16	9.0 ± 1.7
Goldstream Creek	1.12 ± 0.84	0.72 ± 0.12	0.75 ± 0.19	100 ± 4	32 ± 16
Kokomo Creek	0.93 ± 0.47	0.48 ± 0.12	0.56 ± 0.39	49 ± 14	16 ± 5
Summer 1983[b]					
Twelvemile Creek	0.18 ± 0.09	– – –	1.89 ± 2.90	40 ± 15	34 ± 14
McManus Creek	0.36 ± 0.31	– – –	1.39 ± 0.14	22 ± 6	23 ± 4

[a] – – – = Not measured.
[b] n = 9 for chlorophyll *a* and discharge, n = 14 for nutrients.

(Fig. 6); it was non-significant for brown-water streams however, and the negative trend of this relation was probably responsible for the lack of significance of the positive relation when data for all of the streams were combined.

The relation between sestonic chlorophyll *a* and total phosphorus (summer means) was positive and highly significant ($r = 0.74$; $P \leq 0.005$) for all study streams measured in 1979. However, the relation became non-significant when data for the two streams sampled in 1983 were included (Fig. 7). When data for brown-water streams were analyzed separately, a significant ($r = 0.68$; $P \leq 0.001$) positive relation was shown between sestonic chlorophyll *a* and total phosphorus.

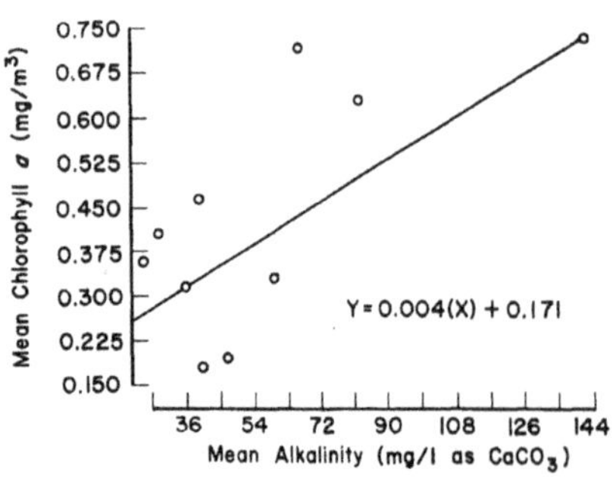

Fig. 6. Relation between alkalinity and chlorophyll *a* content of sestonic algae of ten high-subarctic, clear-water streams in Alaska.

Gross primary productivity

Primary productivity was low but measurable in streams in the high latitudes of the subarctic

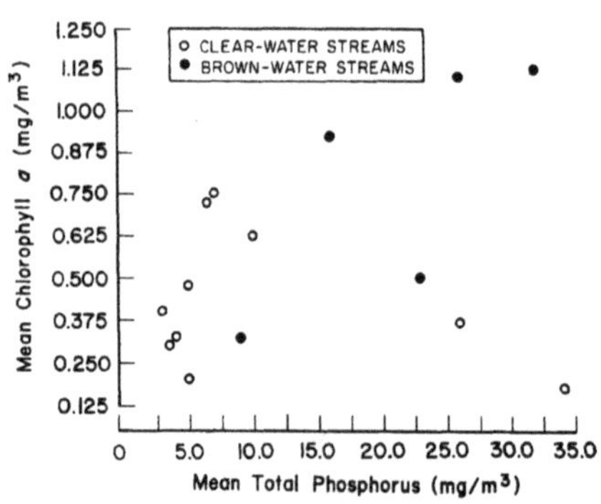

Fig. 7. Relation between total phosphorus and chlorophyll *a* content of sestonic algae of fifteen high-subarctic streams in Alaska.

Table 3. Selected primary productivity measurements and related physical data[a].

Stream	Gross P ($g\ O_2\ m^{-2}\ d^{-1}$)	Mean Temp. (°C)	Turbidity (NTU)	Discharge ($m^3\ s^{-1}$)	Av. depth (m)	Av. vel. ($m\ s^{-1}$)	Date
Monument Creek	0.2	4.3	---	0.87	0.22	0.55	27 May 1981
	0.5	0.1	---	0.39	0.19	0.31	8 Oct 1981
Delta Clearwater Creek	2.8	4.6	---	---	---	---	10 May 1980
	2.6	4.2	---	22.7	0.46	0.90	30 Jul 1983
	0.8	2.4	---	---	---	---	25 Sept 1979
	0.5	1.6	---	---	---	---	15 Oct 1979
Chatanika River Elliot H. Bridge	2.9	4.0	28	13.2	0.46	0.73	30 Aug 1984
Chatanika River Steese H. Bridge	1.0	5.6	6.8	6.09	0.27	0.84	18 Sept 1984
McManus Creek	0.2	6.0	0.31	1.38	0.32	---	14 June 1983
	1.1	9.0	0.92	1.17	0.31	---	10 Aug 1983
Faith Creek	0.2	9.6	136	1.49	0.22	---	5 Aug 1983
	0.6	9.0	130	1.33	0.22	---	10 Aug 1983
Twelvemile Creek	1.1	8.2	1.1	1.15	0.30	---	19 July 1982
	0.4	8.7	0.49	0.31	0.28	---	9 Aug 1982
Birch Creek	0.0[b]	8.1	3400	1.64	0.42	---	17 Aug 1982
	0.1[b]	9.7	1700	1.55	0.27	---	27 July 1983
Ptarmigan Creek							
above mine	0.9	5.1	0.41	0.22	0.20	---	11 Aug 1982
	0.5	4.1	0.52	0.56	0.26	---	19 Aug 1983
below mine	0.0[b]	4.2	680	0.67	0.27	---	19 Aug 1982

([a] Abbreviations: --- = not measured; [b] = below method detection limit)

(Table 3). Additional measurements for Monument Creek on 25 November and 21 December 1981, were both below the detection limit we calculated for the method ($0.2\ g\ O_2\ m^{-2}\ d^{-1}$). Turbidities were not measured at Monument Creek or at the Delta Clearwater Creek, but were most likely less than 1 NTU. Additional measurements were made for McManus, Faith, Twelvemile, Birch and Ptarmigan creeks (Van Nieuwenhuyse, 1983; Van Nieuwenhuyse & LaPerriere, 1986), but here we present mainly the high and low values measured for these streams.

All the streams in which we have measured primary productivity were clear-water streams. Although higher alkalinity was associated with higher measured primary productivity for unmined streams (i.e. excluding Faith, Birch and Ptarmigan creeks) ($r = 0.68$, $P \leq 0.01$) using the mean alkalinity and maximum productivity data for each stream, we lacked sufficient phosphorus data to examine the primary productivity to phosphorus relation.

Discussion

Annual mean numbers of benthic algal cells in the upper Chena River are comparable to those reported in other subarctic streams (McCart *et al.*, 1977; Ennis & Albright, 1982), and to similarly sized temperate streams (Moore, 1977; Wilhm *et al.*, 1978). We found no annual mean chlorophyll *a* values of benthic algae for other subarctic streams in the literature; however, our values were only about a tenth of those for temperate streams (Stockner & Shortreed, 1976; Wilhm *et al.*, 1978; Bott *et al.*, 1985). In fact, benthic chlorophyll *a* ranges for only a few streams in Quebec (Naiman, 1983) overlapped those of Monument Creek and the West Fork. This small amount of chlorophyll *a* despite the presence of moderate densities of algae, is probably a result of numerical dominance of small-celled diatoms (*Achnanthes* spp.) throughout the year in both Monument Creek and the West Fork of the Chena River (Anderson, 1984). This pattern of high density but low biomass in algae is similar to density-biomass relationships found in invertebrates in Monument Creek by Cowan and Oswood (1984), and the relation may be common in benthic communities of subarctic streams.

The higher standing crop of benthic algae in Delta Clearwater Creek was probably due to stable physical and chemical characteristics of this stream. Delta Clearwater Creek varies little in flow and, therefore, experiences little movement of the bottom substrate or scouring of epilithon. Water temperature also has a relatively small annual variation, from 0 °C to 7.5 °C (J. D. LaPerriere, 1978 & 1979, unpublished reports to Alaska Department of Natural Resources). Our measurements of standing crop were probably underestimates. The dominant alga in spring was a thallic chrysophyte, *Hydrurus foetidus*, that was large relative to our sampler, and, therefore, was probably not accurately sampled. This stream also had the highest alkalinity of all our streams, and relatively high total phosphorus concentrations.

Chlorophyll *a* densities in the upper Chena River stayed relatively high during winter. While these values are for total chlorophyll *a*, uncorrected for phaeophytin *a*, live-to-dead diatom ratios remained steady throughout the year (Anderson, 1984) indicating viable periphytic algae year-round. The winter community of benthic algae may be adapted to low-light level – we saw evidence of this adaptation in an increase in stalked diatoms during winter (Anderson, *op. cit.*), which forms probably have an advantage in receiving light by being raised above the prostrate forms.

The rationale that algal productivity is related to alkalinity is based on the fact that carbon, derived from carbonate alkalinity, is an algal macronutrient. Muskeg vegetation is underlain by permafrost and decomposes slowly, releasing tannins and other organic acids into the water flowing over and through the organic material. These organic acids give characteristic brown color to standing and flowing waters in muskeg-filled valleys. Alkalinity, being the measurement of the negative ions of the salts of weak acids, obviously includes those from tannins and lignins. These chemical species can be called 'false alkalinity' because they are non-carbonate alkalinity and the carbon they contain is not readily available to plants. This may explain why mean summer alkalinity in brown-water streams was not related to the concentration of sestonic chlorophyll *a* averaged over the summer.

Algae in lakes are expected to vary with phosphorus, in fact phosphorus is the major nutrient that is likely to be found in low concentration in lakes (Schindler, 1977). Kilkus *et al.* (1975) were not able to demonstrate a relation between phosphorus and chlorophyll *a* in Iowa streams where the soils in the watersheds were rich and contained relatively high phosphorus, but a direct relation has been reported for Missouri streams with low phosphorus concentrations (Jones *et al.*, 1984). Krewer & Harvey (1982) have demonstrated a positive linear or curvilinear relation for periphyton colonization (as chlorophyll *a*) with total dissolved phosphorus in artificial streams, and note that it would be expected to be a stronger relation for total phosphorus. We were able to show a relation between total phosphorus and sestonic chlorophyll *a* (as an indicator of standing crop of benthic algae) for 5 brown-water streams (and for 8 of 10 clear-water streams). Unfortunately, we did not have total phosphorus data for some of the streams where we measured primary productivity, and could not test the relation between total phosphorus and primary productivity.

Because our relation between alkalinity and primary productivity was based on as little as one measurement for a stream we must be cautious in our interpretation. Inorganic carbon stimulation of periphyton growth has been reported for a stream in Quebec (Dickman, 1973). More data will be needed to establish the relations between nutrients and primary productivity in these high-subarctic streams, and would probably be best gathered from experimental additions of nutrients (*sensu* Peterson *et al.*, 1983).

The gross primary productivity estimates we calculated from our measurements were in the same range as those reported for streams of similar size (as discharge) in the U. S. Pacific Northwest, Rocky Mountains and the Northeast (Bott *et al.*, 1985) – despite the occurrence of benthic chlorophyll *a* levels at least an order of magnitude lower. This relatively high productivity for low standing crop may be due to the relatively long daylight period during summer in the high subarctic, where no true darkness occurs for about 10 weeks.

The gross primary production of the Kuparuk River (Peterson *et al.*, 1986: 1264, Table 4) is considerably less than that of Delta Clearwater Creek where we measured chlorophyll *a* concentrations in the epilithon equivalent to those of the Kuparuk River, and also less than McManus Creek where we measured equivalent sestonic chlorophyll *a* concentrations. Delta Clearwater Creek as a spring-fed stream might be expected to have a higher productivity (Craig & McCart, 1975), but McManus Creek, is a surface-water fed stream which might be expected to have similar productivity to the Kuparuk. The Kuparuk also had similar alkalinity to McManus Creek, about one fifth or less that of Delta Clearwater Creek.

The Kuparuk River lies considerably north of our study area in the Arctic (~69° N). Though the Arctic has even longer daylight periods and for a longer duration in the summer than the high subarctic, it receives less integral radiation flux per day than our study area because of the lower sun angle (Hartman & Johnson, 1978; 48). The lower amount of solar radiation probably explains the lower primary production, and also is associated with a shorter growing season. Primary production was measured lower in the Kuparuk River in August than in July, but not in our streams where Van Nieuwenhuyse & LaPerriere (1986) noted higher efficiency of the benthic algae later in the season, as well as equivalent levels of primary production in August as compared to July, with highest productivity measured in late September.

Conclusion

Our research showed that two of the streams studied in the high subarctic of Alaska had moderate numbers of attached algae, but low biomass, as expressed by density of chlorophyll *a*, probably due to year-round dominance of the community by small-celled diatoms, *Achnanthes* spp. A significant positive relation was found between alkalinity and standing crop of benthic algal chlorophyll *a* in 5 subarctic clear-water streams, and between alkalinity and sestonic chlorophyll *a* in 10 subarctic clear-water streams.

The relation between sestonic chlorophyll *a* and total phosphorus was strongly positive for 13 subarctic streams studied in 1979, but additional data for two clear-water streams studied in 1983 did not fit this pattern. We hypothesize that this relation may be true only for brown-water streams in the high subarctic. Gross primary productivity measurements in the streams studied varied positively with alkalinity, and were comparable to those of northern temperate streams of similar size, despite lower biomass of benthic algae – perhaps due to the long day length during the subarctic summer.

Acknowledgements

We thank T. Winston Hobgood, Rob Holmgren, Arthur J. L. LaPerriere, IV, Huan Vinh Luong, and S. Mitchell Wagener for assistance in the field and laboratory. This work was supported by funds from the State of Alaska through the Delta Agricultural Project, and, the Alaska Department of Fish and Game. Other funds were provided by the Community Employment and Training Assistance Division of the State of Alaska, Department of Community and Regional Affairs; by the United States Department of the Interior as authorized under the Water Research and Development Act of 1978; and by the United States Environmental Protection Agency through Project No. 808340. This work was conducted through the Water Resources Center of the Institute of Northern Engineering, and the Alaska

Cooperative Fishery Research Unit, University of Alaska, Fairbanks. The Unit is a joint program of the US Fish and Wildlife Service, the Alaska Department of Fish and Game, and the University of Alaska, Fairbanks.

References

American Public Health Association, American Water Works Association & Water Pollution Control Federation, 1980. Standard methods for the examination of water and wastewater, 15th Ed. Am. Publ. Hlth Ass., Washington, D.C. 1072 pp.

Anderson, P. R., 1984. Seasonal changes of attached algae in two Alaskan subarctic streams. Unpublished master's thesis. University of Alaska, Fairbanks, Alaska. 112 pp.

Bjerklie, D. M. & J. D. LaPerriere, 1985. Gold-mining effects on stream hydrology and water quality, Circle Quadrangle, Alaska. Wat. Res. Bull. 21: 235–243.

Bott, T. L., J. T. Brock, C. S. Dunn, R. J. Naiman, R. W. Ovink & R. C. Peterson, 1985. Benthic community metabolism in four temperate stream systems: an inter-biome comparison and evaluation of the river continuum concept. Hydrobiologia 123: 3–45.

Cowan, C. A. & M. W. Oswood, 1984. Spatial and seasonal associations of benthic macroinvertebrates and detritus in an Alaskan subarctic stream. Pol. Biol. 3: 211–215.

Craig, P. C. & P. J. McCart, 1975. Classification of stream types in Beaufort Sea drainages between Prudhoe Bay, Alaska, and the Mackenzie Delta, N.W.T., Canada. Arct. Alp. Res. 7: 183–198.

Cummins, K. W., 1962. An evaluation of some techniques for the collection and analysis of benthic samples with special emphasis on lotic waters. Am. Midl. Nat. 67: 477–504.

Dickman, M., 1973. Changes in periphytic algae following bicarbonate additions to a small stream. J. Fish Res. Bd Can. 30: 1882–1884.

Eisenreich, S. J., R. T. Bannerman & D. E. Armstrong, 1975. A simplified phosphorus analysis technique. Envir. Lett. 9: 43–53.

Elmore, H. L. & W. F. West, 1961. Effect of water temperature on stream reaeration. J. San. Eng. Div. ASCE 87: 59–71.

Ennis, G. L. & L. J. Albright, 1982. Distribution and abundance of periphyton and phytoplankton species in two subarctic Canadian rivers. Can. J. Bot. 60: 224–236.

Gales, M. C., A. C. Julian & E. C. Kroner, 1966. A method for the quantitative determination of total phosphorus in filtered and unfiltered water. J. am. Wat. Wrks Ass. 58: 1363–1368.

Golterman, H. L., 1969. Methods for chemical analysis of fresh waters. International Biological Programme Handbook No. 8. Blackwell Scientific Publications, Oxford and Edinburgh. 166 pp.

Hall, C. A. & R. Moll, 1975. Methods of assessing aquatic primary productivity. In: H. Lieth & R. H. Whittaker (eds.) Primary productivity of the biosphere. Springer-Verlag, NY: 19–53.

Hartman, C. W. & P. R. Johnson, 1978. Environmental atlas for Alaska. Institute of Wat. Res., University of Alaska, Fairbanks. 95 pp.

Jones, J. R., M. Smart & J. N. Burroughs, 1984. Factors related to algal biomass in Missouri Ozark streams. Verh. int. Ver. Limnol. 22: 1867–1875.

Kilkus, S. P., J. D. LaPerriere & R. W. Bachmann, 1975. Nutrients and algae in some central Iowa streams. J. Wat. Pollut. Cont. Fed. 47: 1870–1879.

Krewer, J. A. & H. W. Holm. 1982. The phosphorus-chlorophyll *a* relationship in periphytic communities in a controlled ecosystem. Hydrobiologia 94: 173–176.

LaPerriere, J. D., 1983. Alkalinity, discharge, average velocity, and invertebrate drift concentration in subarctic Alaskan streams. J. Freshwat. Ecol. 2: 141–151.

LaPerriere, J. D., S. M. Wagener & D. M. Bjerklie, 1985. Gold-mining effects on heavy metals in streams, Circle Quadrangle, Alaska. Wat. Res. Bull. 21: 245–252.

McCart, P., P. Tsue, W. Grant & R. Green, 1977. Baseline studies of aquatic environments in the Athabasca River near lease 17. Environmental Research Monograph 1977–2. Syncrude Canada Limited, Vancouver, B. C., Canada.

Moore, J. W., 1977. Seasonal succession of algae in a eutrophic stream in southern England. Hydrobiologia 53: 181–192.

Mortimer, C. H., 1981. The oxygen content of air-saturated fresh waters over ranges of temperature and atmosphere pressure of limnological interest. Mitt. int. Ver. Limnol. 22. 23 pp.

Naiman, R. J., 1983. The annual pattern and spatial distribution of aquatic oxygen metabolism in boreal forest watersheds. Ecol. Monogr. 53: 73–94.

O'Connor, D. J. & W. Dobbins, 1956. The mechanism of reaeration in natural streams. J. San. Eng. Div. ASCE 82: 1115(1)–1115(30).

Odum, H. T., 1956. Primary production of flowing waters. Limnol. Oceanogr. 2: 85–97.

Peterson, B. J., J. E. Hobbie, T. L. Corliss & K. Kriet, 1983. A continuous-flow periphyton bioassay: tests of nutrient limitation in a tundra stream. Limnol. Oceanogr. 28: 583–591.

Peterson, B. J., J. E. Hobbie & T. L. Corliss, 1986. Carbon flow in a tundra stream ecosystem. Can. J. Fish. aquat. Sci. 43: 1259–1270.

Péwé, T. L., 1955. Origin of the upland silt near Fairbanks, Alaska. Geol. Soc. Am. Bull. 66: 699–724.

Schindler, D. W., 1977. Evolution of phosphorus limitation in lakes. Science 195: 260–262.

Stockner, J. G. & K. R. S. Shortreed, 1976. Autotrophic production on Carnation Creek, a coastal rainforest stream on Vancouver Island, British Columbia. J. Fish. Res. Bd Can. 33: 1553–1563.

Swanson, C. D. & R. W. Bachmann, 1976. A model of algal exports in some Iowa streams. Ecology 57: 1076–1080.
Van Nieuwenhuyse, E. E., 1983. Effects of placer mining on the primary productivity of interior Alaska streams. Unpublished master's thesis. University of Alaska, Fairbanks, Alaska. 120 pp.
Van Nieuwenhuyse, E. E. & J. D. LaPerriere, 1986. Effects of placer gold mining on primary production in subarctic streams of Alaska. Wat. Res. Bull. 22: 91–99.
Wagener, S. M. & J. D. LaPerriere, 1985. Effects of placer mining on the invertebrate communities of interior Alaska streams. Freshwat. Invertebr. Biol. 4: 208–214.
Wilcock, R. J., 1982. Simple predictive equations for calculating stream reaeration rate coefficients. N. Z. J. Sci. 25: 53–56.
Wilhm, J., J. Cooper & H. Naminga, 1978. Species composition, diversity, biomass, and chlorophyll of periphyton in Greasy Creek, Red Rock Creek, and the Arkansas River, Oklahoma. Hydrobiologia 57: 17–23.

Hydrobiologia 172: 77–95, 1989
W. F. Vincent and J. C. Ellis-Evans (eds) High Latitude Limnology

Broadscale patterns in the distribution of aquatic and terrestrial vegetation at three ice-free regions on Ross Island, Antarctica

Paul A. Broady
Department of Plant and Microbial Sciences, University of Canterbury, Christchurch, New Zealand

Key words: Antarctica, vegetation, ecology, aquatic, terrestrial

Abstract

Distribution patterns are presented for selected aquatic algae, and terrestrial algae, mosses and lichens, at three large, coastal ice-free regions on Ross Island. Each region is unique in certain aspects of its vegetation. The variation in areal and quantitative occurrence of different components of the vegetation in diverse ponds and streams, and over exposed ground surfaces, is related to the levels of marine salts in the environment, fertilisation by birds, water availability, substratum type and degree of exposure. A comparison is made with other antarctic, coastal, ice-free regions where similar broadscale patterns have been recognised.

Introduction

The purpose of this investigation was to compare broadscale patterns in the distribution of macroscopic and selected microscopic vegetation at three large, coastal, ice-free regions in the vicinity of Cape Bird, C. Royds and C. Crozier on Ross Island. The survey included terrestrial algae, lichens and mosses, and aquatic algae in ponds and streams. Much of the terrestrial vegetation could be termed 'semi-aquatic' as it was usually flushed with percolations of meltwater for periods during summer. No attempt was made to compile complete species lists.

Similar data has been collected at other ice-free regions in the Continental Antarctic Zone (sensu Pickard & Seppelt, 1984), for instance the Vestfold Hills (68° 35' S, 77° 58' E; Broady, 1986; Pickard, 1986; Seppelt, 1986a, b), Mawson Rock (67° 26′ S, 62° 53′ E; Seppelt & Ashton, 1978; Broady, 1982a) and East Ongul Island (69° S, 39° 30′ E; Matsuda, 1968; Karasawa & Fukushima, 1977; Nakanishi, 1977). As in the present study, the influence of salt on the distribution of vegetation was noted to be marked, together with effects due to water and nutrient availability. On Ross Island there are distinct contrasts in the environment and vegetation at each ice-free region which allows further insight into the factors controlling distribution. The broadscale, descriptive data also provides a foundation, and direction, for the development of more detailed ecological and physiological investigations.

Botanical investigations have been made previously at the three study areas. Algae at Cape Royds were first described by West & West (1911), and Fukushima (1964) reported further on the diatoms. Goldman *et al.* (1972) made brief observations on pond and lake algae and Holm-Hansen (1964) cultured algae from ponds and soils. The nutrient physiology of algae in ponds at

C. Bird and C. Royds was studied by Vincent & Vincent (1982), and Howard-Williams *et al.* (1986) examined the characteristics of a stream at C. Bird. Observations on algae in four ponds at C. Bird were provided by Zaneweld (1969) and in a single pond by Spurr (1975). Mosses have been described from Ross Island (Greene, 1967) and Dodge (1973) included lichen records from C. Crozier and C. Royds. Longton (1973) described terrestrial plant communities at C. Royds and C. Bird but did not investigate broad-scale distribution patterns.

Study sites

C. Bird and C. Royds are on the west coast and C. Crozier is at the eastern extremity of Ross Island (Fig. 1). The extent of ice-free land at each locality approximates 15, 13 and 18 km^2 respectively. They have received little human disturbance, in contrast to the next largest ice-free area at C. Armitage which has been considerably perturbed by station activities. There are a small number of other coastal rock outcrops, all smaller in extent than C. Evans which covers about 1 km^2.

The general topography of each area is shown in Fig. 2. In the absence of published contour maps of C. Bird and C. Crozier, the landform features shown are from sketch maps made in the field with the aid of an aneroid barometer. C. Bird is bounded to the east by the edge of the Mt. Bird ice-cap and, except for five or six prominent hills, slopes down to the coast. Similarly, C. Royds slopes in a generally westerly direction and the eastern boundary comprises the snow and ice-fields below Mt. Erebus. In contrast, C. Crozier comprises the lower, generally easterly-facing slopes of Mt. Terror. The approximate boundaries of ice-fields and major snowfields and snowdrifts which persist throughout summer are displayed in Fig. 3.

Adélie penguin rookeries are located at all three areas. At C. Bird, the three rookeries, Northern, Middle and Southern Rookery, have approximately 25 000, 1700 and 9600 breeding pairs respectively; at C. Royds there are 2 000 and at C. Crozier 168 000 pairs (Harper *et al.*, 1984; R. Taylor & P. Wilson, DSIR pers. comm.). There

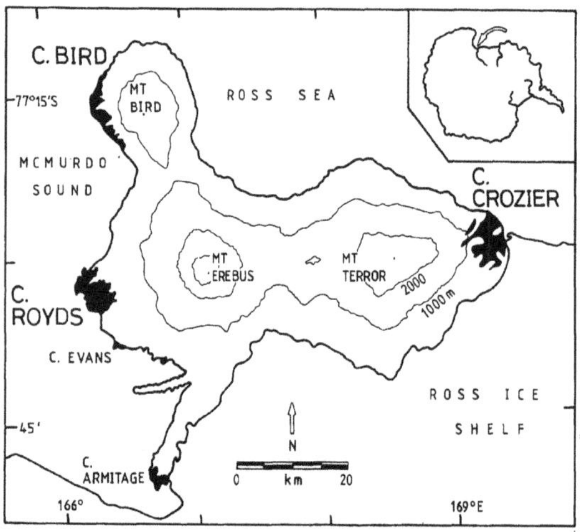

Fig. 1. Ross Island showing the location and approximate extent of the three major ice-free regions at Cape Bird, Cape Royds and Cape Crozier, together with other important features and contour lines at 1 000 m intervals. Inset, location of Ross Island.

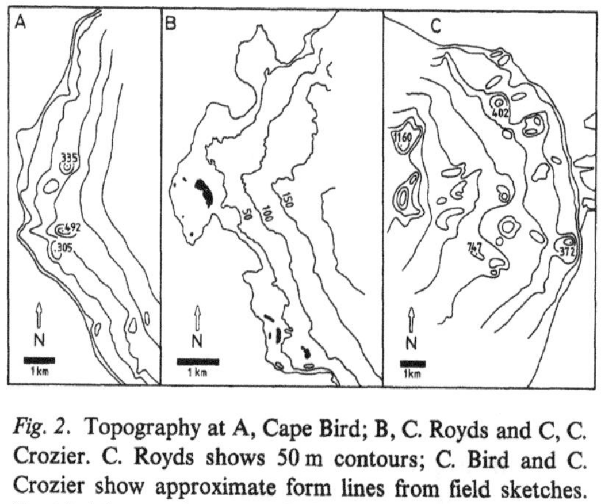

Fig. 2. Topography at A, Cape Bird; B, C. Royds and C, C. Crozier. C. Royds shows 50 m contours; C. Bird and C. Crozier show approximate form lines from field sketches. Spot heights are in metres. Lakes at C. Royds are shaded. Bases for maps are taken from: C. Bird, N.Z.M.S. 175/4, 1 : 25 000, 1961 and Ross Island, Antarctica, U.S. Geological Survey, 1 : 250000 Reconnaissance Series, 1970; C. Royds and C. Barne, N.Z. Lands and Survey, 37/108, 1 : 10 000, 1982; C. Crozier, N.Z.M.S. 175/16, 1:50 000, 1962.

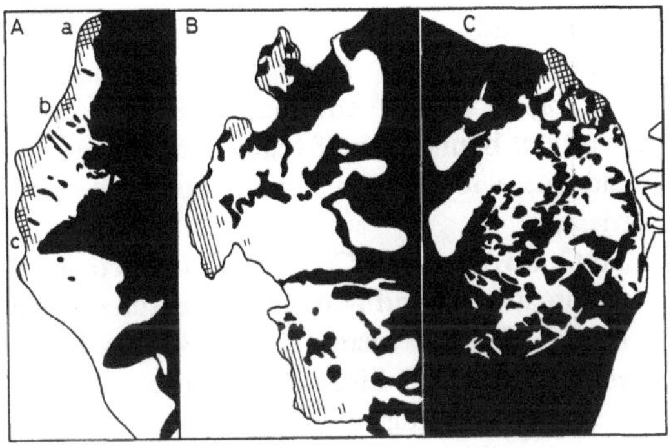

Fig. 3. Extent of permanent snow and ice-fields and locations of bird colonies at: A, Cape Bird; B, C. Royds; C, C. Crozier. Key: block-shaded areas are major snow and ice-fields; single hatching, skua nesting areas; cross-hatching, penguin rookeries; a, b, and c, locate Northern, Middle and Southern Rookery respectively at C. Bird.

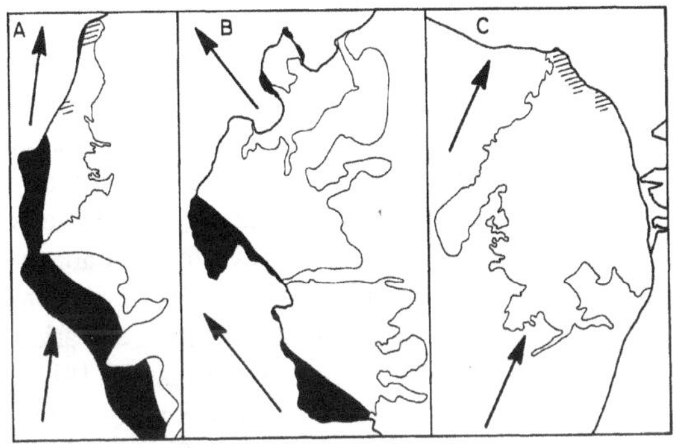

Fig. 6. Approximate directions of prevailing strong winds (arrows) and approximate areas of ground over which sea-spray is deposited (block-shaded) due to wind action on areas of open water at A, Cape Bird; B, C. Royds and C, C. Crozier. Line shading indicates areas of high salt levels due to bird concentrations. Fine lines enclose the major extent of ice-free ground (simplified from Fig. 3 and shown also in Figs 5, 6, 13–23).

are also substantial populations of skuas, breeding pairs being estimated as 399, 183 and 1000 at C. Bird, Royds and Crozier respectively (Ainley *et al.*, 1986).

There are high salt concentrations in a proportion of the soils and aquatic habitats. A comparison of heavily salt-affected, and relatively salt-free areas at C. Royds are shown in Figs. 4 and 5. Close to the coast the salts are largely derived from sea-spray blown during high winds across areas of land immediately downwind of open water (Fig. 6). Wind direction was inferred from snowdrift orientation in the lee of hills and from meteorological observations during the

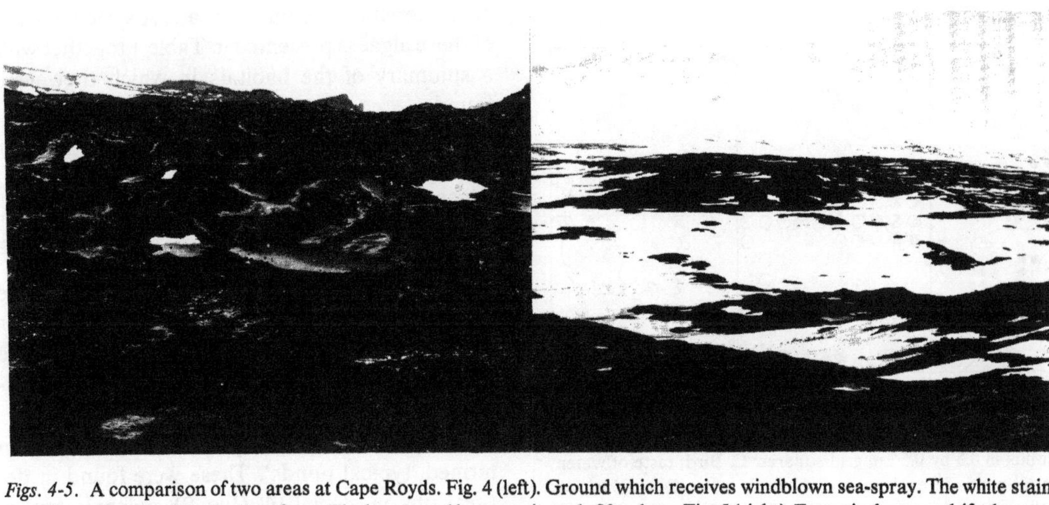

Figs. 4-5. A comparison of two areas at Cape Royds. Fig. 4 (left). Ground which receives windblown sea-spray. The white stains are salt crusts on rock and soil surfaces. The large pond is approximately 20 m long. Fig. 5 (right). Extensively snow-drifted ground on the lower slopes of Mt. Erebus (3270 m). The area beyond the frozen lake (Blue Lake) does not receive windblown sea-spray.

survey. These are in close agreement with earlier wind observations made by Simpson (1919). Evidence of sea-spray deposition was provided by a greatly increased incidence of white salt crusts on the surfaces of rocks and soils and in the generally increased salinities of pondwaters (Fig. 7). Salts of marine origin have been shown to be quantitatively most important in the McMurdo region, with lessening influences with increasing elevation and distance inland (Keys & Williams, 1981). Marine salts are also transferred to land by birds, in their faeces and excretions. This effect would be most marked in penguin rookeries.

Methods

Observations were made over one period of three to four weeks at each region during the summers of 1982-83 and 1984-85. Extensive notes were made describing the location and abundance of vegetation during traverses on foot over the large majority of each region. Traverse routes were varied in order to cover as wide an area as possible. Locations were plotted on large scale maps. These records were transformed into presence and absence date within 0.5 by 0.5 km squares of a grid covering each region. This has inevitably resulted in some loss of resolution, particularly at boundaries of vegetation change, but is adequate for demonstrating broadscale patterns.

At C. Royds and C. Crozier the conductivity of water in streams and ponds was measured using a Model 33 S-C-T meter (Yellow Springs Instument Co.) and at C. Bird salinity was assessed by taste. Samples of algae were removed for microscopic examination using light microscopy.

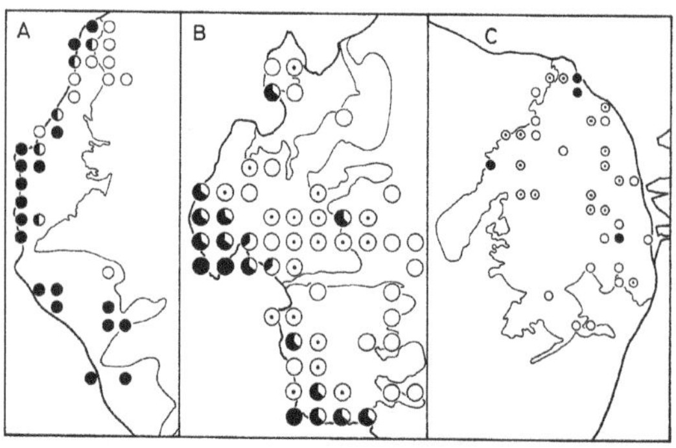

Fig. 7. Pond water taste and conductivity at A, Cape Bird; B, C. Royds and C, C. Crozier. Symbols indicate presence of ponds in 0.5 by 0.5 km grid squares: C. Bird, taste of water in all ponds in each square, ○ fresh taste; ● brackish to saline; ◐ both categories; C. Royds and C. Crozier, conductivity (mSm^{-1}) of all ponds in each square; ○ < 50; ⊙ < 300; ◐ O to > 300; ◐ 50 to > 300; ● > 300.

Results and discussion

Algae, lichens and mosses were found at all three regions in a wide range of aquatic and terrestrial habitats. However, the frequency of occurrence and abundance of different communities contrasted markedly from region to region. The broad distribution patterns in ponds and streams and over exposed ground are described below, together with discussion of controlling factors and results of studies elsewhere. Algae included in this account are those forming macroscopic growths visible to the unaided eye and species which, although not growing in such abundance, are characteristic of certain habitats. A systematic list of these algae is presented in Table 1 together with a summary of the habitats in which they were found.

Algae in ponds

Pond types

Four distinctly different categories of ponds were recognised on the basis of the nature of their catchments. Their distribution within the ice-free regions is shown in Fig. 8. and examples illustrated in Figs. 9–14.

Most numerous were ponds which have been termed 'typical ponds'. These were found in depressions in glacial till (Fig. 9), volcanic tephra and lava, and beach sands. They numbered 45, 145 and 45 at C. Bird, Royds and Crozier respec-

Table 1. Dominant and characteristic algae of terrestrial and aquatic habitats of ice-free regions on Ross Island.[1]

Systematic list of algae	Habitats[2]										
	1	2	3	4	5	6	7	8	9	10	11
CYANOPHYTA											
Chroococcales											
Gloeocapsa kuetzingiana Naeg.	+						+				
Chamaesiphonales											
Chroococcidiopsis sp.								+	+		
Oscillatoriales											
Oscillatoriaceae (including *Lyngbya, Oscillatoria* and *Phormidium* spp.)	+			+		+	+		+	+	+
Nostoc sp.	+						+				
Nostoc commune										+	
Scytonema sp.							+				
CHLOROPHYTA											
Volvocales											
Brachiomonas cf. *submarina* Boh.	+										
Chlamydomonas cf. *snowiae* Printz.			+								
C. cf. *subcaudata* Wille	+	+									
C. sp.					+						
Ulotrichales											
Klebsormidium sp.	+										
Ulothrix sp.	+	+				+	+				
Chaetophorales											
cf. *Coccobotrys* sp.								+	+		
cf. *Desmococcus* sp.						+					
Ulvales											
Prasiola calophylla (Carmich.) Menegh.						+				+	
P. crispa (Lightf.) Menegh.											+
Prasiococcus calcarius (Boye Pet.) Vischer								+		+?	+?
CHRYSOPHYTA											
Bacillariophyceae											
Centrales											
Melosira setosa Grev.	+										
Pennales											
Navicula cryptocephala Kuetz.	+	+									
N. muticopsis v. Heurck	+			+							
N. shackletoni West and West	+										
Pinnularia cymatopleura West and West				+							
Tropidoneis laevissima West and West	+	+									
CRYPTOPHYTA											
cf. *Chroomonas lacustris* Pa. and Rutt.	+										

[1] See text for a comparison of frequency of occurrence and abundance of growths at Cape Bird, C. Royds and C. Crozier.

[2] Habitats are as follows: 1–4, ponds; 1, typical; 2, shoreline; 3, rookery; 4, cryoconite; 5, snowfields and snowdrifts; 6, streams; 7–9, rock surfaces; 7, epilithic; 8, chasmoendolithic; 9, sublithic; 10–11, meltwater-flushed ground surfaces; 10, not significantly nutrient-enriched by birds; 11, in nutrient-enriched areas.

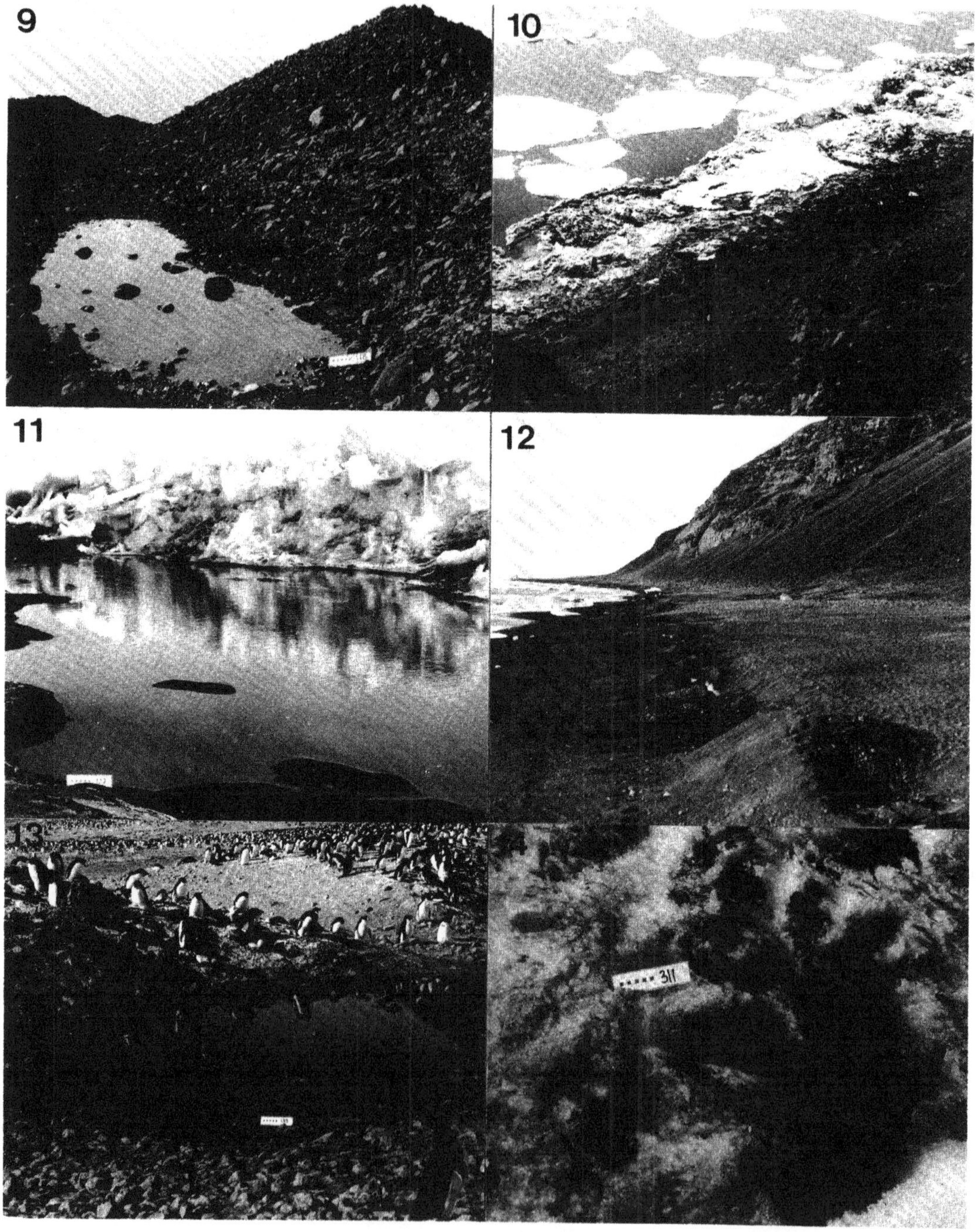

Figs. 9–14. Ponds at C. Bird. Fig. 9. A 'typical pond' in a depression in moraines in the north of the region. Fig. 10. Looking down onto a beach with sea-ice still attached along the shoreline; 'shoreline ponds' have formed along the landward edge of the ice. Fig. 11. A 'shoreline pond' fed by melt from coastal ice accumulations. Fig. 12. The same length of beach as shown in Fig. 11. after ice has broken away and melted completely. The 'shoreline ponds' have seeped away into the sand. Fig. 13. A 'rookery pond' in Northern Rookery. Fig. 14. Small 'cryoconite ponds' in the Mt. Bird ice-cap. The scale is in centimetres.

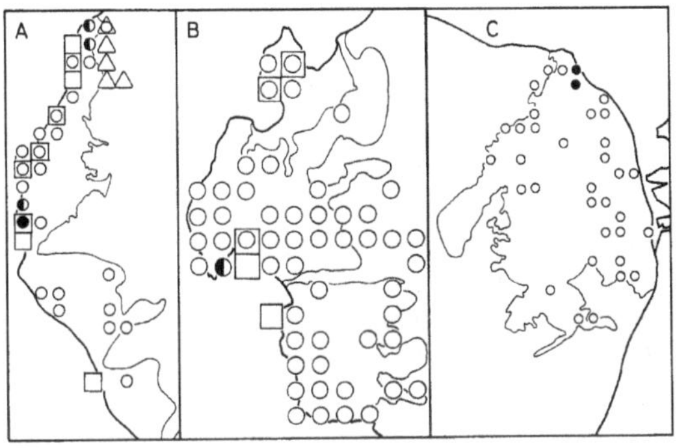

Fig. 8. Pond types at A, Cape Bird; B, C. Royds and C, C. Crozier. Symbols indicate presence of ponds in 0.5 by 0.5 km grid squares: △ 'cryoconite ponds'; □ 'shoreline ponds'; ● 'rookery ponds'; ○ 'typical ponds'; ◐ both 'rookery' and 'typical' ponds.

tively. They include all ponds other than those in the three remaining categories described below. Several were visited by penguins and in particular by skuas and received some nutrient enrichment from these sources. Their areas varied widely from 1 to 4 000 m^2. All were shallow with a maximum depth of <2 m. Some dried out during the study period. Ponds in areas affected by sea-spray were predominantly of high conductivity (300 to $>2\,000$ m Sm^{-1}) and of saline taste, whereas outside these areas the majority were of lower conductivity (50–300 m Sm^{-1}) and of fresh taste (Fig. 7).

Small ponds formed along the shoreline due to the melting of sea-ice which had become grounded on sandy beaches (Figs. 10–12). These 'shoreline ponds' were frequent in the northern half of C. Bird, with about 25 in total, infrequent at C. Royds, 5 in total, and none were observed at C. Crozier. Strong winds blow a layer of dark sand over the grounded ice. Solar heating of this causes the underlying ice to melt, forming ponds of 0.2–200 m^2 and up to 30 cm deep (see Fig. 9 in Vincent, 1987). On complete melting of the ice the water drains away and only a small depression remains as an indication of the previous presence of a pond. These ponds were generally brackish to saline.

Within and directly adjacent to the nesting areas of penguins (Fig. 3) there were turbid, highly nutrient and organically enriched ponds (Fig. 13). The catchments of these 'rookery ponds', and the ponds themselves, contained substantial guano deposits and scattered bird carcasses. They numbered 9, 1 and 4 at C. Bird, Royds and Crozier respectively. All were saline with conductivities >300 m Sm^{-1}.

Mineral sediments ('cryoconite') melting into glacier ice form 'cryoconite ponds' (Wharton *et al.*, 1985). These were found only in the ice bordering the north-eastern edge of C. Bird. Here the hard ice surface was darkened by substantial mineral deposits which had melted into the ice on flat areas (Fig. 14). The ponds were generally cylindrical, 20–40 cm diameter and 20–30 cm deep, although occasionally they attained 5 m^2 in surface area. The waters were of low conductivity, less than 50 m Sm^{-1}. Their absence at C. Royds and C. Crozier was probably due to the sloping topography of most ice-fields and the relatively clean ice surfaces.

Algal flora

The majority of 'typical ponds' contained benthic felts dominated by *Phormidium*, *Lyngbya* and *Oscillatoria* irrespective of conductivity and salinity (Fig. 15). In contrast, macroscopic growths of

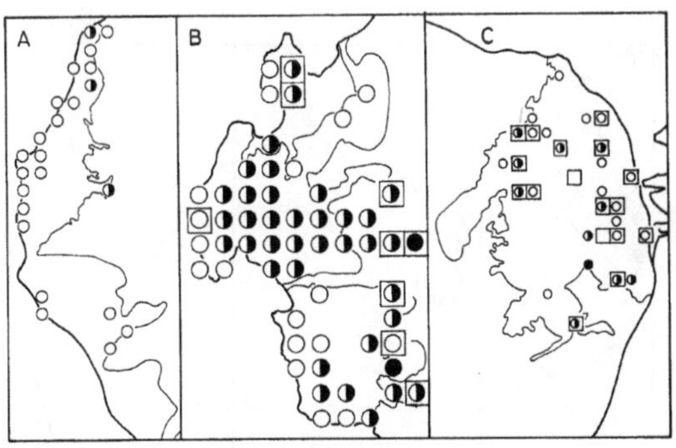

Fig. 15. Distribution of Cyanophyta in ponds at A, Cape Bird; B, C. Royds and C, C. Crozier. Symbols indicate presence in 0.5 by 0.5 km grid squares: ○ benthic felts dominated by Oscillatoriaceae; ● macroscopic colonies of *Nostoc* sp.; ◑ both Oscillatoriaceae and *Nostoc*; □ thin crusts and microscopic growths of *Gloeocapsa kuetzingiana*.

Nostoc sp. were encountered almost solely outside saline areas and were notably infrequent at C. Bird (Fig. 15). *Gloeocapsa kuetzingiana* was a frequent microscopic component within 'typical ponds' of low conductivity at C. Royds and C. Crozier alone (Fig. 15). However, it had a more restricted distribution at C. Royds where, except for a western coastal record, it was mostly confined to an eastern zone at the highest altitudes surveyed. In these ponds it dominated the relatively thin crusts which covered stones and gravel.

Other algae possessed distinct preferences for waters of high conductivity and were either absent or rarely found at C. Crozier. For instance in saline 'typical ponds', visibly yellow-green phytoplankton populations of cf. *Chroomonas lacustris* often developed (Fig. 16). These also frequently contained rich green growths of *Ulothrix* sp. (Fig. 16). This alga was frequent in 'shoreline ponds' associated with green benthic films of *Chlamydomonas* cf. *subcaudata* (Fig. 17) but the latter was infrequent in saline 'typical ponds'. A second species of *Chlamydomonas, C.* cf. *snowiae*, characterised the saline, nutrient-enriched 'rookery ponds' where dense unialgal populations developed (Fig. 17). It was not encountered in other habitats. Two saline 'typical ponds' at C. Royds contained small populations of the halophile *Brachiomonas* cf. *submarina* (Fig. 16).

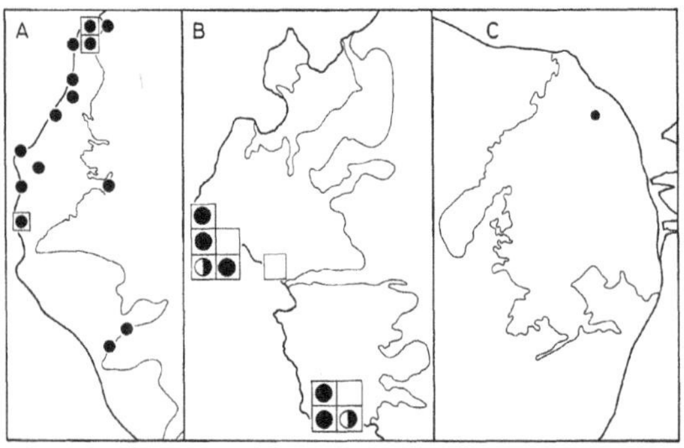

Fig. 16. Distribution of selected algae in ponds at A, Cape Bird; B, C. Royds and C, C. Crozier. Symbols indicate presence in 0.5 by 0.5 km grid squares; □ cf. *Chroomonas lacustris*; ● green filamentous growths of *Ulothrix* sp.; ◑ both *Ulothrix* sp. and *Brachiomonas* cf. *submarina*.

Diatoms were often abundant in saline ponds but infrequent elsewhere (Fig. 18). This is apparent in their widespread distribution at C. Bird, their almost total restriction to sea-spray affected areas at C. Royds and their rarity at C. Crozier. Three species were characteristic of saline 'typical ponds'; *Navicula shackletoni, Tropidoneis laevissima* and *Melosira setosa. T. laevissima* and *N. cryptocephala* dominated brown benthic films of diatoms in several 'shoreline ponds'. *N. muticopsis* displayed a wider range

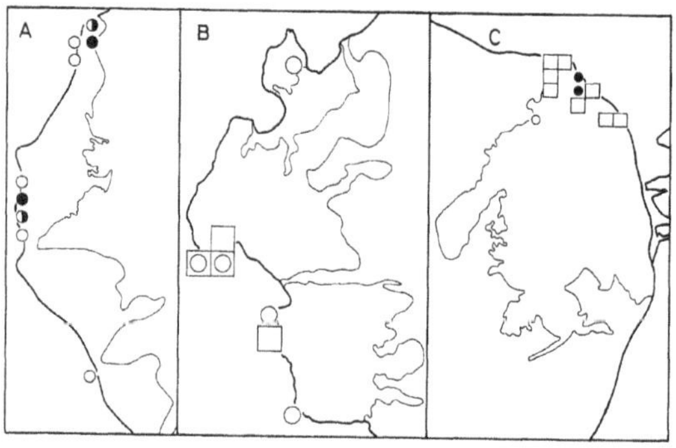

Fig. 17. Distribution of *Chlamydomonas* spp. in ponds and green snow at A, Cape Bird; B, C. Royds and C, C. Crozier. Symbols indicate presence in 0.5 by 0.5 km grid squares; ○ *C.* cf. *subcaudata*; ● *C.* cf. *snowiae*; ◑ both *Chlamydomonas* spp.; □ green snow dominated by an unidentified species of *Chlamydomonas.*

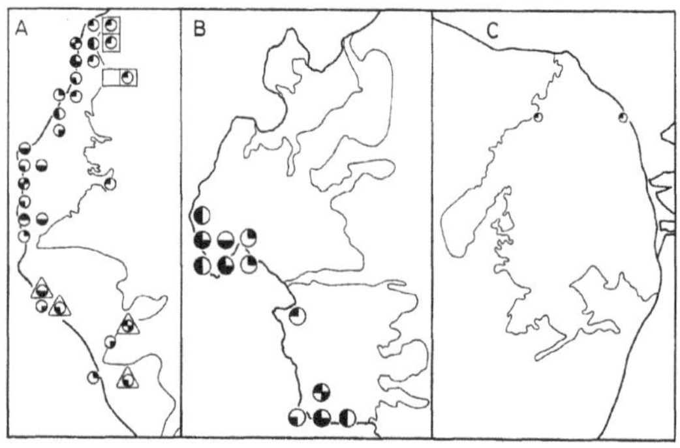

Fig. 18. Distribution of diatoms in ponds at A, Cape Bird; B, C. Royds and C, C. Crozier. Symbols indicate presence in 0.5 by 0.5 km grid squares; ◔ *Navicula cryptocephala*; ◒ *N. shackletoni*; ◓ *N. muticopsis*; ◐ *Tropidoneis laevissima*; □ *Pinnularia cymatopleura*; △ *Melosira setosa.*

of salinity tolerance being found at low abundance in 'cryoconite ponds' as well as in 'typical ponds'.

Pinnularia cymatopleura occurred solely in 'cryoconite ponds' at C. Bird (Fig. 18) and not in waters of similarly low conductivity in 'typical ponds'. 'Cryoconite ponds' also contained Oscillatoriaceae and unidentified chlorococcalean and chaetophoralean algae (Chlorophyta) at low abundance.

The abundance of Oscillatoriaceae and diatoms in lakes and ponds on Ross Island has been well known since reports by West & West (1911). Similar communities to those described above have been found elsewhere. In northern Victoria Land there were 'shoreline ponds' containing *Chlamydomonas* cf. *subcaudata* and *Ulothrix* sp., a 'rookery pond' with *C.* cf. *snowiae*, and *Nostoc* sp. as a frequent associate with Oscillatoriaceae in low conductivity 'typical ponds' (Broady, 1987). At Mawson Rock, ponds in saline areas were often dominated by diatoms, but also contained rich growths of green filaments resembling the *Ulothrix* sp. of this study (= *Urospora* sp., Broady, 1982).

The diatom flora of antarctic ponds has been shown to vary markedly with differences in water chemistry. Field studies by Fukushima (1961, 1964, 1967) and co-workers (Karasawa & Fukushima, 1977) showed *Tropidoneis laevissima, Navicula cryptocephala* and *N. muticopsis* to occur in saline ponds but that the latter prefered milder salinities. These observations were confirmed in culture for *T. laevissima* and *N. muticopsis* by Watanuki and Ohno (1976) and Watanuki (1979).

It is interesting to note that the first antarctic record of *Melosira setosa* was of marine specimens from the vicinity of the South Orkney Islands (Frenguelli, 1943). In northern Victoria Land it has been found in saline coastal pools (Broady, 1987).

Although regarding *Pinnularia cymatopleura* as a 'freshwater' alga, Fukushima (1964, 1967) found specimens in 'typical ponds' of high chlorinity at C. Royds and C. Evans. In northern Victoria Land (Broady 1987) records are from freshwater 'typical ponds' as well as 'cryoconite ponds'. It is one of the most abundant non-marine diatoms in the multitude of ponds on the McMurdo Ice Shelf (Kellogg & Kellogg, 1987).

Saline ponds and lakes in the vicinity of Lützow-Holm Bay (Watanuki & Fukushima, 1978)) and at the Vestfold Hills (Wright & Burton, 1981) contain substantial populations of *Dunaliella* sp. but this was not observed in similar habitats on Ross Island.

Snow algae

Green snow occurred at C. Royds and C. Crozier (Fig. 17). All locations were close to the coast where there was substantial snow melt in progress and fertilization by birds. Meltwater from the drifts was always of low conductivity (3–46 m Sm^{-1}). The absence of snow algae at C. Bird is surprising but may be related to the general lack of large snowdrifts, persisting throughout summer, in areas traversed by penguins.

At C. Royds the algae formed small patches in snow on the coastal ice-foot, adjacent to the penguin rookery. Also, an isolated patch occurred further south, downslope from an accumulation of feathers and guano deposited by moulting penguins. C. Crozier had much more extensive areas covering more than 4 ha adjacent to the penguin rookeries and skua colonies (Figs. 19–20). All samples were dominated by an unidentified species of *Chlamydomonas*, associated with occasional *Ulothrix* – like filaments and diatoms.

There is only a single previous note of snow algae in the Ross Sea regions, on Possession Is. in northern Victoria Land (Koob, 1966). The occurrences on Ross Is. are the furthest south on record. The growth of snow algae in the proximity of nesting birds has been noted by Akiyama (1979) and the requirement for melting snow has been recognised, e.g. Llano (1962). Extensive areas similar to those at C. Crozier have been remarked on only once before in the Continental Antarctic Zone, by Llano (1962) who saw 'many acres' of red snow on the Wilkes Land coast.

Figs.19-20. Snow algae at Cape Crozier. Fig. 19 (left). The location of the most extensive growths of snow algae at C. Crozier (indicated by the arrow) on a snowfield traversed by penguins and adjacent to skua nest sites. The field hut (3 m high) provides a scale. Fig. 20 (right). Rich green growths of snow algae (dark areas in foreground) revealed by breaking away a surface crust of icey snow. The scale is 10 cm long.

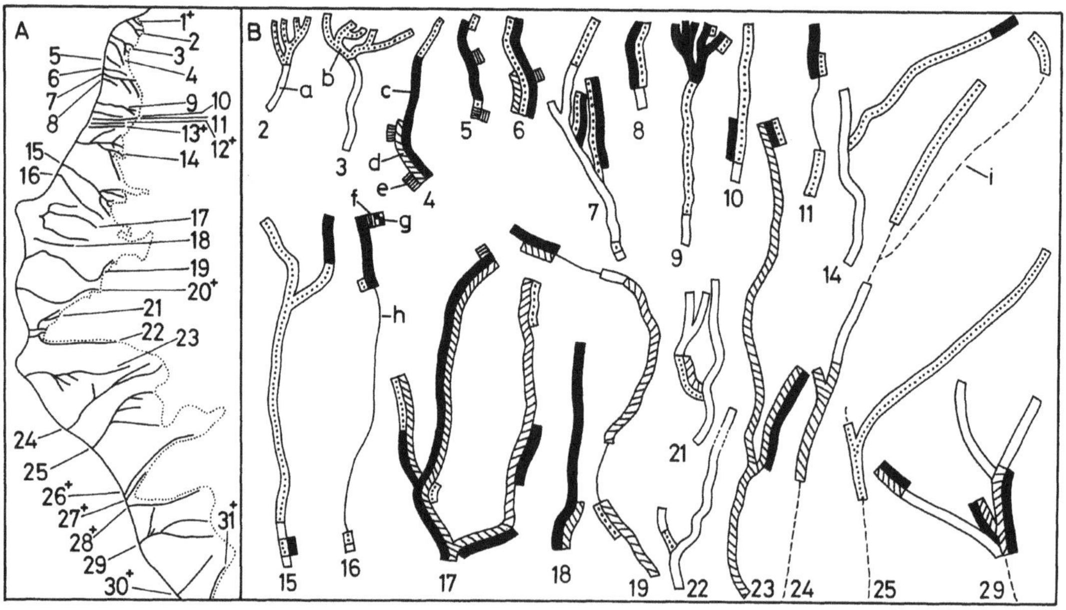

Fig. 21. Stream locations and their algal vegetation at Cape Bird. A, location of streams, + streams in which no vegetation was observed. B, distribution of macroscopic vegetation in streams: a, no vegetation observed; b, green epilithic crusts; c, oscillatoriacean felts; d, green filaments of *Ulothrix* sp.; e, foliose *Prasiola crispa*; f, colonies of *Nostoc commune*; g, ribbons of *Prasiola calophylla*; h, stream running below snowdrift; i, stretch of stream not examined.

Algae in streams

During summer, meltwater either percolates over the surface of wide areas of ground or flows within well-defined channels. Here a stream is regarded as the latter flow pattern. Streams are a major habitat for algae at C. Bird where there is a total length of approximately 48 km (Fig. 21A), but are of relatively minor importance at C. Royds and C. Crozier where they flow over only 8 and 4 km of ice-free ground respectively.

Streams vary greatly in physico-chemical and biological features. Notable differences occur in duration, volume and speed of flow, the loading of suspensoids and in water salinity. The abundance and type of vegetation varies from stream to stream and along the lengths of individual streams. The patterns found in streams at C. Bird were mapped in some detail (Fig. 21B).

Macroscopically visible vegetation was absent from 9 of the total 31 streams (Fig. 21A). These usually had rapid, relatively high volume peak flows, mostly derived from glacier ice rather than snowdrifts. They flowed over steep ground and contained an abundance of suspensoids which made the water turbid.

Thin green epilithic crusts were found on downstream faces of rocks and boulders and on the lower surfaces of large stones. Streams containing this type of growth alone, e.g. 2, 3 (Fig. 22) 22 and 25, were similar to those in which no vegetation was observed. Where crusts were found in streams which also contained other algae, they were dominant where flow rates were too great for establishment of the other algae. The crusts were dominated by cf. *Desmococcus* and *Phormidium fragile*.

Streams with a relatively slow, clear, peak water flow usually contained extensive reddish-brown felts, dominated by Oscillatoriaceae which covered the fines, gravel and small stones of the streambed. The most abundant felts were in streams 5, 17 and 18 and in the snowdrift-fed tributaries of stream 7. In other streams the felts were often sparse, thin and an indistinct grey.

Rich green epilithic filaments of *Ulothrix* sp. were dominant in some fast and gently flowing streams. They were encountered in most abundance in brackish to saline waters, e.g. streams 23 and 24 at C. Bird and some small flows in the sea-spray-swept areas at C. Royds (Fig. 23). However, at C. Bird the filaments were also abun-

Fig. 22. Stream 3 (see Fig. 21A) at Cape Bird. The vigorous, turbulent flow is derived from melt off the Mt. Bird ice-cap and contains a high loading of suspensoids. Sparse, epilithic green crusts occurred on downstream faces of rocks and boulders.

Fig. 23. Stream vegetation at C. Royds, indicating oscillatoriacean felts, green filaments of *Ulothrix* and stretches in which no algae observed; symbols as in Fig. 21.

dant in streams containing fresh water at the time of sampling, e.g. streams 17 and 19. They were not observed at C. Crozier.

Green ribbon-like plants of *Prasiola calophylla* were occasionally found attached to rocks and stones in fresh-water streams. However, this alga was most frequent in unchanneled melt flushes on saturated ground, as were *P. crispa* and *Nostoc commune* (see below).

There are distinct contrasts between the dominant algae in C. Bird streams and those described elsewhere. Although *Ulothrix* spp. were recorded by Hirano (1979) from a stream running into Lützow-Holm Bay, the species differ from the present specimens and were not noted to form luxuriant growths. Also, no specimens of the Ross Is. *Ulothrix* sp. have either been found in the ice-free valleys of southern Victoria Land (Broady, 1982b; Howard-Williams & Vincent, 1986; Howard-Williams *et al.*, 1986) or at Edmondson Point, northern Victoria Land (Broady, 1987). In those regions, rich growths of green filaments and ribbons comprise *Binuclearia tectorum*, *Prasiola calophylla* and *Tribonema elegans*. The former two species are rare in Ross Island streams and the latter has not been recorded. Additionally, species of *Gloeocapsa* and *Nostoc* have been commonly found in other areas but are rare in streams at C. Bird. Oscillatoriacean felts are ubiquitous. The major factor influencing these contrasts is considered to be the generally more saline C. Bird environment. Culture studies of the dominant taxa would help test this hypothesis.

At Signy Island, in maritime Antarctica, the stream flora (Hawes, 1988) contrasts markedly with that known from the Continental Zone, although oscillatoriacean felts are common in both. the most torrential streams are similar in containing sparse epilithic algae, however, these comprise solely diatoms. Additionally, the dominant filamentous chlorophytes are species of *Mougeotia, Zygnema* and a *Klebsormidium*-like alga. Despite these obvious contrasts, more revealing comparisons of stream floras will have to await the improvement of our taxonomic knowledge and the investigation of flowing waters in other regions.

Algae on exposed ground

Rock substrata

Epilithic algae were encountered where thin films of meltwater irrigated rock surfaces. Small growths numbered 3, 1 and 5 at Cape Bird, Royds and Crozier respectively. At the first of these, Oscillatoriaceae and *Ulothrix* sp. dominated the community, in contrast to *Gloeocapsa* spp., *Nostoc* sp. and *Scytonema* sp. in salt-free areas in the latter two regions.

Chasmoendolithic algae were present only in granitic, glacial erratic boulders. The predominant dark-coloured lavas were unsuitable for their development. Sixteen examples were found at C. Bird and just 2 at C. Crozier. No suitable boulders were present at C. Royds. Both green and blue-green growths occurred below thin flakes of rock, the former dominated by cf. *Coccobotrys* sp. and *Prasiococcus calcarius* and the latter by *Chroococcidiopsis* sp.

Sublithic (hypolithic) algae forming crusts on the undersurfaces of translucent stones lying on the soil were infrequent at C. Bird and absent from C. Royds and Crozier. Six quartz stones supported green and blue-green crusts. *Chroococcidiopsis* sp., *Phormidium* sp. and *Coccobotrys* sp. were dominant.

Compared with other regions, lithic algae were rare due to lack of suitable substrata. Epilithic communities were limited by the general absence of irrigated rock surfaces. At Mawson Rock (Broady, 1982a) and the Vestfold Hills (Broady, 1981a), 32 and 27 crusts of cyanophytes, dominated by *Gloeocapsa* spp. were restricted to salt-free areas. At the former region, meltwater- irrigated coastal rocks in salt-affected zones supported rich growths of *Ulothrix* sp. (= *Urospora* sp., Broady, 1982a) similar to the species found in saline ponds on Ross Island.

Endoliths are widespread where the geology of the country rocks is suitable, for instance in the southern Victoria Land dry valleys (Friedmann & Ocampo-Friedmann, 1984), the Vestfold Hills and Mawson Rock (Broady, 1981b). Similarly, it is only the paucity of translucent stones which restricts sublithic growths. If large numbers of stones had occurred, sublithic algae would have been frequent, as at the Vestfold Hills (Broady, 1986).

Particulate substrata

Soils are raw, mineral fines (lithosols) and ornithogenic soils with high guano content (Ugolini, 1970). Large areas of ground, particularly at C. Crozier, are covered by gravel and stones. Water availability at the surface varies greatly. At the maximum, melt water percolates over wide areas of ground, especially where covered by relatively poorly drained lithosols, and at the minimum moisture comes from the melt of occasional snowfall and snowdrift. Surfaces covered by gravel and stones are usually well-drained and relatively dry as water rapidly percolates down through the profile.

Distribution of Cyanophyta

Oscillatoriacean felts and macroscopic colonies of *Nostoc commune* (Fig. 24) were most abundant on poorly drained lithosols downslope from melting snowdrifts. Such ground was most extensive at C. Bird, where felt cover was almost continuous over areas up to 500 m^2, and in the central region of C. Royds coinciding with the terraced moraines recognised by Debenham (1923). The paucity of these algae at C. Crozier is related to the extensive occurrence of well-drained surfaces despite there being numerous large melting snowfields.

Similar to the ponds (Fig. 15), *Nostoc* was almost absent from saline areas whereas oscillatoriacean felts were present. This was most apparent at C. Bird where all but one record of *N. commune* was in the northern half.

Oscillatoriaceae and *N. commune* are well-known throughout the Antarctic Region on irrigated soils. Effects on *N. commune* similar to those recorded on Ross Island have been noted at Mawson Rock (Broady, 1982a) and the Vestfold Hills (Broady, 1986). At the latter region it occurred within the salt-affected zone only downslope from large melting snowdrifts where leaching by snowmelt probably removed restrictively high salt concentrations. This could also explain the southern record at C. Bird.

Distribution of Prasiola species

Two species of *Prasiola* exhibited distinctly different distribution patterns (Fig. 25). Green foliose thalli of the nitrophilous *P. crispa* were common on water-saturated ground in the vicinity of bird colonies. In contrast, *P. calophylla* was intolerant of such high nutrient and salt levels and was also excluded from areas receiving windblown seaspray. The relatively low frequency of occurrence

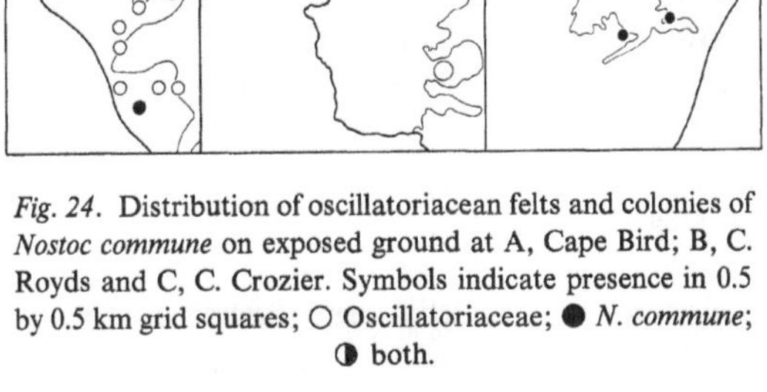

Fig. 24. Distribution of oscillatoriacean felts and colonies of *Nostoc commune* on exposed ground at A, Cape Bird; B, C. Royds and C, C. Crozier. Symbols indicate presence in 0.5 by 0.5 km grid squares; ○ Oscillatoriaceae; ● *N. commune*; ◑ both.

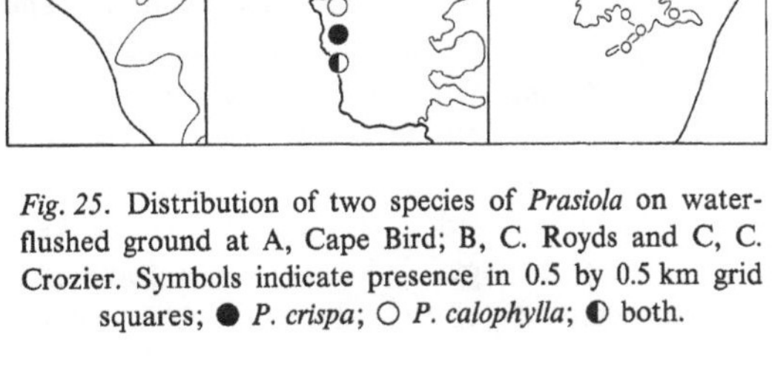

Fig. 25. Distribution of two species of *Prasiola* on water-flushed ground at A, Cape Bird; B, C. Royds and C, C. Crozier. Symbols indicate presence in 0.5 by 0.5 km grid squares; ● *P. crispa*; ○ *P. calophylla*; ◐ both.

at C. Bird is probably due to the generally higher salt levels over the region.

The ribbon-like thalli of *P. calophylla* were generally 1–4 cm long and attached by a holdfast to small stones and rocks in water seepage. The plants were often found immediately downslope from the downhill edges of snowfields where melt-water percolated between stones. Frequently they were not visible on upper stone surfaces but were revealed when the top layer of stones was overturned exposing the algae attached to their lower sides and undersurfaces.

P. crispa has been recognised as an important antarctic alga in nutrient-enriched, moist habitats since the earliest days of collection (West & West, 1911) and it is no surprise to find it localised around bird colonies on Ross Is. However, *P. calophylla* is much more poorly known although recent data show it to be widespread in southern Victoria Land (Broady, 1982b, 1987 and unpublished records). There it often grows luxuriantly in channelled streams but is also found amongst boulders, stones and gravel where lesser percolations flow over and through the substratum. The latter growth pattern was predominant on Ross Is, due to the few channelled water flows at C. Royds and Crozier.

Crusts of green algae

Green algal crusts coated the surfaces of soils, gravel and stones at all three areas but were particularly abundant at C. Crozier (Fig. 26). There they were mostly confined to the northern half of the region outside the perimeter of the penguin rookery. They were often found encrusting stones below the ground surface, in a pattern similar to *P. calophylla*. However, directly adjacent to the melting edges of snowfields, where a water supply was readily available, there were abundant powdery growths over exposed upper surfaces. The identity of the dominant alga in these crusts cannot be confidently determined. Although closely resembling the sarcinoid *Prasiococcus calcarius*, it is possible that growth forms and life-stages of *Prasiola crispa* and *P. calophylla* have a very similar morphology. The crusts were dominated less frequently by cf. *Coccobotrys* sp. At C. Royds,

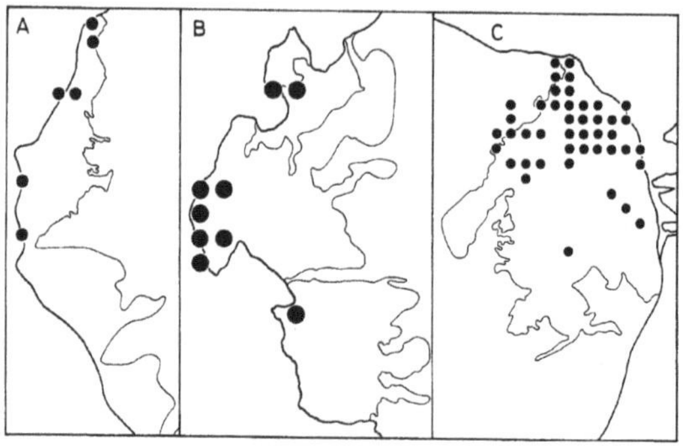

Fig. 26. Distribution of green algal crusts on exposed ground surfaces at A, Cape Bird; B, C. Royds and C, C. Crozier. Symbols indicate presence in 0.5 by 0.5 km grid squares.

similar crusts comprising cf. *P. calcarius* were predominantly encountered in areas receiving wind-blown sea-spray. At C. Bird green crusts were infrequent, and only on small areas of moist, mineral fines close to penguin colonies, but not where large quantities of guano were deposited. The flora of these was more diverse and although dominated by cf. *P. calcarius* and *Klebsormidium* sp. also contained over 20 species of other microalgae.

The ecology of *P. calcarius* in Antarctica has been discussed previously in some detail (Broady, 1983). In particular, the preference for salty environments was emphasized. It is possible that the specimens observed in salt-affected areas at C. Bird and C. Royds, and around the periphery of the penguin colony at C. Crozier, are this alga. However, in such habitats, they would also be exposed to nutrient-enrichment from birds and could be a growth form of the nitrophilous *P. crispa*. Similarly, the more southerly records in the relatively salt-free areas of C. Crozier overlap the distribution of *P. calophylla* (Fig. 16) and they could be a growth form of that alga. This confusion will be resolved only by critical studies on unialgal cultures of isolates of these algae.

Lichens

Distribution patterns are shown in Fig. 27. Crustose lichens included *Biatorella cerebriformis*,

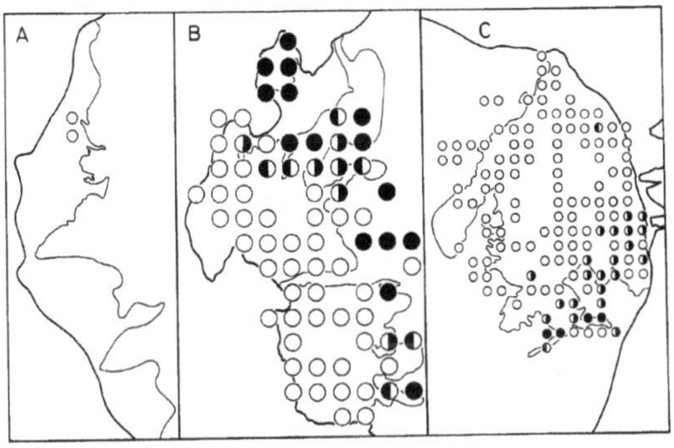

Fig. 27. Distribution of lichens at A, Cape Bird; B, C. Royds and C, C. Crozier. Symbols indicate presence in 0.5 by 0.5 km grid squares: ○ crustose lichens; ◑ *Usnea antarctica* and crustose lichens; ◐ *Umbilicaria aprina* and crustose lichens; ● *U. antarctica, U. aprina* and crustose lichens.

Candelariella antarctica and species of *Buellia, Caloplaca, Candelaria, Lecanora* and *Xanthoria*. Foliose lichens were represented by *Umbilicaria aprina* and fruticose by *Usnea antarctica*.

At C. Bird, lichens were extremely rare, being restricted to sparse growths of crustose *Caloplaca darbishirei*. This is possibly a response to the generally high levels of salt thought to occur over the region as a whole, even in the relatively protected northern half. In contrast crustose species were widespread and abundant at C. Royds and C. Crozier. At C. Royds they were absent from areas receiving maximum quantities of sea-spray, but outside these regions greatest abundance was at low altitudes close to the coast where nutrient input from skuas occurred. Similarly at C. Crozier maximum abundance was around the periphery of the penguin rookery, especially where nesting skaus were common. They were absent within the rookery where salt levels and mechanical disturbance were high. At higher altitudes, in both regions, most occurrences were of small, sparsely scattered *Buellia* and *Caloplaca*. A contrasting distribution was clearly seen with the foliose and fruticose species.

Usnea antarctica and *Umbilicaria aprina* were absent from C. Bird and at C. Royds and Crozier were present in a much more restricted area than crustose species. At C. Royds both occurred in a zone stretching from the northernmost coastal regions back in a southeasterly direction up onto the high slopes above the 150 m contour. At C. Crozier *U. antarctica* was found in a southeastern zone stretching from the coastal cliffs to a maximum altitude of about 300 m. *U. aprina* was more restricted and occurred in the southernmost part of the *Usnea* range except for one isolated record to the north.

These two species appeared to be restricted to areas of minimum salt levels and nutrient input, although northern occurrences at C. Royds occurred in the vicinity of nesting skuas. Observations made following periods of strong winds with snowdrifting indicated that it was in these areas that maximum accumulation of drift occurred, mostly on the downwind sides of boulders and knolls. It was in such microsites that the lichens were usually encountered. Subsequent melt would enhance moisture availability and could leach salts and nutrients from the substrata.

The inability of most lichens to establish in salt-laden environments has been described at Mawson Rock (Seppelt & Ashton, 1978) and Vestfold Hills (Seppelt, 1986a; Pickard, 1986). At both regions, strong winds blow salt spray over land areas downwind from open sea and saline lakes in a pattern similar to that at C. Royds. The same factor could be responsible for the impoverishment of lichen communities on East and West Ongul Is. compared with richer communities on rock areas along the neighbouring coast (Nakanishi, 1977). The latter receive salt-free katabatic winds straight off the continent. Similar winds also probably occur at Birthday Ridge, northern Victoria Land, where Kappen (1985a) encountered lichens both inland and at the coast.

At the Vestfold Hills, the distribution of different lichen species varies markedly (Seppelt, 1986a). Of particular note in the present context is the greatly restricted occurrence of *Usnea antarctica* compared with most crustose species, and especially *Buellia frigida*. In more detailed analyses (Pickard & Seppelt, 1984; Pickard, 1986), a set of lichens including *U. antarctica* has been recognised as occurring near the ice-sheet and on inland margins of small coastal rock outcrops.

This pattern is considered to be a response to higher moisture availability and considerably less wind-blow salt, as has been suggested above for the relatively narrow distribution of *U. antarctica* at Ross Is. This relationship might also be true for *U. sulphurea* which is absent from the Ongul Islands but present on nearby coastal ice-free areas (Nakanishi, 1977). Additionally, Kappen (1985) described *U. sulphurea* as more prominent in the vegetation further inland at Birthday Ridge, northern Victoria Land. He also suggested nutrient enrichment from birds to explain richer coastal communities. Nakanishi (1977) noted a similar response at coastal Skarvsnes in Lützow-Holm Bay.

Mosses

Bryum argenteum, B. antarcticum and *Sarconeurum glaciale* have been recorded from Ross Is. by Longton (1973). No attempt was made to distinguish the separate species during this study. Similar to crustose lichens, mosses were intolerant of high salt levels in sea-spray swept areas and penguin rookeries (Fig. 28). However, at C. Bird they were clearly more widespread than the lichens. Here the greatest abundance, and 83% of all stands, was in the area stretching 3 km south from Northern Rookery. Stands occurred as isolated cushions to confluent carpets up to 10–20 m^2. Generally they were on water-flushed substrata, particularly the relatively poorly drained mineral fines (Fig. 29). Sites were protected from strong winds and were in positions more likely to be supplied with melt percolating from snowdrifts as they occurred on more or less north-facing slopes, in depressions and gullies, and on ground downwind of hummocks and knolls.

At C. Royds no stands exceded a few square meters in extent and most occurrences were of one or a few, small scattered cushions. 56% of records, and maximum abundance, occurred below 50 m altitude. Above this height moss was much sparser, especially in the southern half of the region which is more exposed to the strong south-easterly winds than are the west to northwest facing slopes in the north. 69% of stands were in this northern area.

At C. Crozier, mosses were sparse and widely scattered. Most records were of single or a small number of adjacent cushions each less than 10 cm^2 in diameter. More abundant growths were found only on north to northeast-facing slopes in the northern part of the region where 71% of all stands were found. They were not seen over large

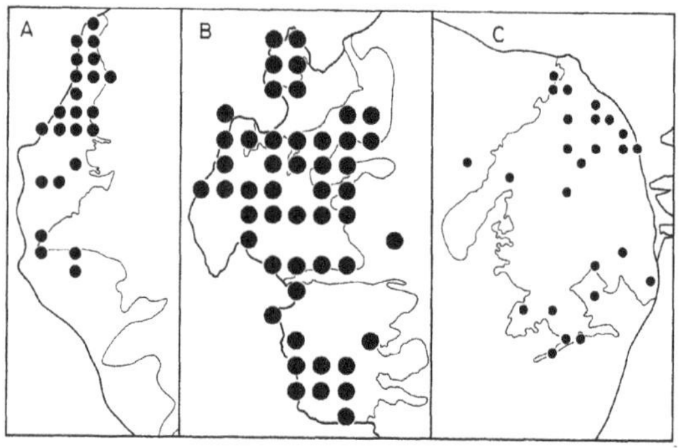

Fig. 28. Distribution of mosses at A, Cape Bird; B, C. Royds and C, C. Crozier. Symbols indicate presence in 0.5 by 0.5 km grid squares.

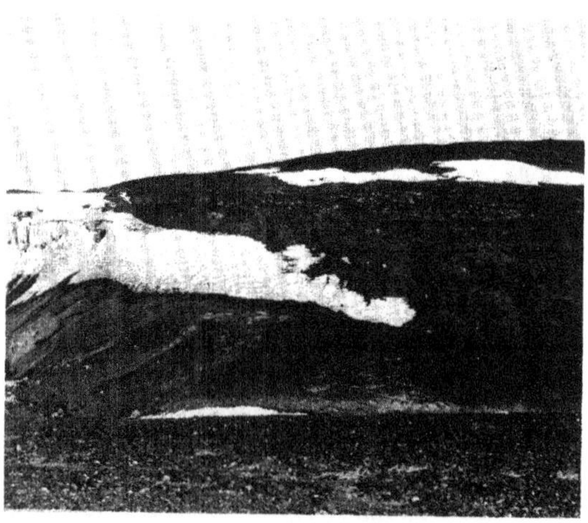

Fig. 29. Typical meltwater-flushed slopes, at Cape Bird, which supported moss growths. Cushions were found downslope from both snowdrifts, particularly in the shallow gullies. The foot of the hill is approximately 200 m distant.

areas inland, possibly due to the generally well-drained, stoney substratum not retaining sufficient moisture for their development. Isolated cushions were present up to a maximum elevation of 915 m but most were found below 300 m.

Mosses are absent from the most heavily salt-affected areas at the Vestfold Hills (Seppelt, 1986b; Pickard, 1986) and Mawson Rock (Seppelt & Ashton, 1978). This also appears to be the case at East Ongul Island although this is not stated by Matsuda (1968). The occurrence of mosses in the west and north-west of the island correlates with the presence of ponds of lowest chlorinity. This compares with an absence of mosses and presence of ponds of generally higher chlorinity to the east and south (Karasawa & Fukushima, 1977).

There are indications that mosses are more tolerant of salt than lichens. At the Vestfold Hills, Pickard (1986) distinguished certain areas as being basically occupied by only mosses, but considered that 'this is probably an artefact of collecting as a well-collected grid rarely contains only mosses'. Also, he noted with 'surprise' that grids with 'mosses only' co-occurred significantly with those containing saline lakes. At C. Bird mosses occurred widely outside the very confined distribution of lichens possibly due to the greater tolerance of mosses to the relatively high salt levels which occur throughout this region. Additionally, at East Ongul Island, mosses are frequent but the lichen flora is extremely poor (Nakanishi, 1977), possibly for the same reason.

Conclusions

It is apparent that major differences between the floras of different areas within individual ice-free regions on Ross Island, and between the regions themselves, are strongly influenced by the degree to which they are subjected to saline aerosols. Similar situations have been recognised in more northern, coastal continental, ice-free regions of East Antarctica, particularly at the Vestfold Hills (e.g. Pickard, 1986) and Mawson Rock (e.g. Seppelt & Ashton, 1978). However, such broadscale patterns have not been described in maritime Antarctica. There, the greater precipitation and subsequently greater volumes of fresh meltwater probably prevent the accumulation of such high levels of salt in soils and ponds.

This study has described distribution patterns at low resolution and those mostly of the dominant algae and broad groupings of lichens and mosses. Also, the environmental information is mostly subjective. The results serve as a base-line for more detailed studies with a greater quantitative component utilising statistical approaches to community analysis. In particular, the structure of algal communities in ponds and streams needs to be related to detailed data on water chemistry and temporal changes should be taken into account. It is probable that salinities of individual ponds change over the growing season as they melt out, evaporate and then re-freeze. The occurrence of species succession in response to such changes needs investigation. Also, there were indications that different crustose lichen species, and possibly moss species, exhibit different distribution patterns. No attempt has been made to distinguish them in the present report. These, and the distribution patterns of foliose and frutcose lichens, will be understood only when we can relate them to micrometeorological data and chemical analyses of their substrata and water supplies.

The data can also serve as a base-line for monitoring changes in vegetation over long periods. For instance, distribution patterns could alter in response to an amelioration of climate or to a change in the extent or persistence of sea-ice which would result in changes to the quantity of salt transferred to land. Additionally, an increase in human activity at these ice-free areas could result in damage to the plant communities. Knowledge of distribution patterns can help in the assessment of such effects and in the choice of areas requiring special protection.

Acknowledgements

It is a pleasure to thank Dr. L. G. Greenfield for inviting me to join the University of Canterbury expeditions of 1982-83 and 1984-85, and Antarctic Division, DSIR, Christchurch for logistic support. My companions in the field provided assistance and stimulating discussion and Dr. R. M. Crawford, University of Bristol identified *Melosira setosa*. The investigations were made whilst I was part of the Melbourne University Programme in Antarctic Studies.

References

Ainley, D. G., Morrell, S. H. & Wood, R. C., 1986. South Polar skua breeding colonies in the Ross Sea Region, Antarctica. Notornis 33: 155–163.

Akiyama, M., 1979. Some ecological and taxonomic observations on the coloured snow algae found in Rumpa and Skarvsnes, Antarctica. Mem. Nat. Inst. Polar Res., Special Issue No. 11: 27–34.

Bold, H. C. & Wynne, M. J., 1985. Introduction to the algae. Prentice-Hall, Englewood Cliffs, N.J.

Broady, P. A., 1981a. Ecological and taxonomic observations on subaerial epilithic algae from Princess Elizabeth Land and Mac. Robertson Land, Antarctica. Br. phycol. J. 16: 257–266.

Broady, P. A., 1981b. The ecology of chasmolithic algae at coastal locations of Antarctica. Phycologia, 20: 259–272.

Broady, P. A., 1982a. Ecology of non-marine algae at Mawson Rock. Antarctica. Nova Hedw. 36: 209–229.

Broady, P. A., 1982b. Taxonomy and ecology of algae in a freshwater stream in Taylor Valley, Victoria Land, Antarctica. Arch. Hydrobiol. Suppl. 63.3, Algological Stud., 32: 331–349.

Broady, P. A., 1983. The antarctic distribution and ecology of the terrestrial, chlorophytan alga *Prasiococcus calcarius* (Boye Petersen) Vischer. Polar Biol. 1: 211–216.

Broady, P. A., 1986. Ecology and taxonomy of the terrestrial algae. In: J. Pickard (ed.), Antarctic Oasis. Academic Press, Sydney: 165–202.

Broady, P. A., 1987. A floristic survey of algae at four locations in northern Victoria Land. N.Z. Antarctic Record, 7: 8–19.

Debenham, F., 1923. Report on the maps and surveys. British (Terra Nova) Antarctic Expedition 1910–1913. Harrison and Sons Ltd., London: 1–94, ± 14 maps.

Dodge, C. W., 1973. Lichen flora of the Antarctic Continent and adjacent islands. Phoenix Publications, Canaan.

Frenguelli, J., 1943. Diatomeas de las Orcadas del Sur. Revta. Mus. La Plata, 5: 221–265.

Friedmann, E. I. & Ocampo-Friedmann, R., 1984. Endolithic microorganisms in extreme dry environments – analysis of a lithobionitic microbial habitat. In: M. J. Klug & C. A. Reddy (eds) Current Perspectives in Microbial Ecology. American Society for Microbiology, Washington, D.C.: 177–185.

Fukushima, H., 1961. The brief notes on the diatoms vegetation at the Prince Olav Coast, Antarctica. Bull. Mar. Biol. Sta. Asamushi, 10: 237–240.

Fukushima, H., 1964. Diatoms vegetation on ice-free areas of Cape Royds, Antarctica. Antarctic Record, 22: 1–13.

Fukushima, H., 1967. A brief note on diatom flora of Antarctic inland waters. In: T. Nagata (ed.) Proceeding of the Symposium on Pacific-Antarctic Sciences. Eleventh Pacific Science Congress, Tokyo: 253–264.

Goldman, C. R., Mason, D. T. & Wood, B. J. B., 1972. Comparative study of the limnology of two small lakes on Ross Island, Antarctica. In: G. A. Llano (ed.) Antarctic Terrestrial Biology. American Geophysical Union, Washington, D.C.: 1–50.

Greene, S. W., 1967. Bryophyte distribution. In: B. C. Bushnell (ed.) Antarctic Map Folio Series, Folio 5, Terrestrial Life of Antarctica. American Geographical Society, New York: 11–13, plate 5.

Harper, P. C., Knox, G. A., Spurr, E. B., Taylor, R. H., Wilson, G. J. & Young, E. C., 1984. The status and conservation of birds in the Ross Sea sector of Antarctica. ICBP Tech. Publ. No. 2: 593–608.

Hawes, I., 1988. Filamentous green algae in freshwater streams on Signy Island, Antarctica. (This volume).

Hirano, M., 1979. Freshwater algae from Yukidori Zawa, near Syowa Station, Antarctica. Mem. Nat. Inst. Polar Res., Special Issue No. 11: 1–25.

Holm-Hansen, O., 1964. Isolation and culture of terrestrial and fresh-water algae of Antarctica. Phycologia, 4: 43–51.

Howard-Williams, C., Vincent, W. F. & Wratt, G. S., 1986. The Alph River Ecosystem: a major freshwater environment in Southern Victorialand. N.Z. Antarctic Record, 7: 21–33.

Howard-Williams, C., Vincent, C. L., Broady, P. A. & Vincent, W. F., 1986. Antarctic stream ecosystems: variability in environmental properties and algal community structure. Int. Revue ges. Hydrobiol. 71: 511–544.

Kappen, L., 1985. Vegetation and ecology of ice-free areas of northern Victoria Land, Antarctica 1. The lichen vegetation of Birthday Ridge and an inland mountain. Polar Biol. 4: 213–225.

Karasawa, S. & Fukushima, H., 1977. Diatom flora and environmental factors in some fresh water ponds of East Ougul Island. Antarctic Record, No 59: 46–53.

Kellogg, D. E. & Kellogg, T. B., 1987. Diatoms of the McMurdo Ice Shelf, Antarctica: implications for sediment and biotic reworking. Palaeogeog. Paleaeoclimat. Palaeoecol. 60: 77–96.

Keys, J. R. & Williams, K., 1981. Origin of crystalline, cold desert salts in the McMurdo region, Antarctica. Geochim. Cosmochim. Acta, 45: 2299–2309.

Koob, D. D., 1966. Ecology of some antarctic algae. Antarct. J. U.S. 1: 144–145.

Llano, G. A., 1962. The terrestrial life of the Antarctic. Sci. Am. 207: 213–230.

Longton, R. E., 1973. A classification of terrestrial vegetation near McMurdo Sound, continental Antarctica. Can. J. Bot. 51: 2339–2346.

Matsuda, T., 1968. Ecological study of the moss community and micro-organisms in the vicinity of Syowa Station, Antarctica. JARE Sci. Rep. Ser. E. 29: 1–58.

Nakanishi, S., 1977. Ecological study of the moss and lichen communities in the ice-free areas near Syowa Station, Antarctica. Antarctic Record, 59: 68–96.

Pickard, J., 1986. Spatial relationships of the vegetation of the Vestfold Hills, In: J. Pickard (ed.) Antarctic Oasis. Academic Press, Sydney: 275–308.

Pickard, J. & Seppelt, R. D., 1984. Phytogeography of Antarctica. J. Biogeog. 11: 83–102.

Seppelt, R. D., 1986a. Lichens of the Vestfold Hills. In: J. Pickard (ed.) Antarctic Oasis. Academic Press, Sydney: 247–274.

Seppelt, R. D., 1986b. Bryophytes of the Vestfold Hills. In: J. Pickard (ed.) Antarctic Oasis. Academic Press, Sydney: 221–245.

Seppelt, R. D. & Ashton, D. H., 1978. Studies on the ecology of the vegetation at Mawson Station, Antarctica. Aus. J. Ecol. 3: 373–388.

Simpson, G. C., 1919. Meteorology. Vol. 1 Discussion. British (Terra Nova) Antarctic Expedition 1910-1913. Committee for the Publication of the Scientific Results: 1–326.

Spurr, B., 1975. Limnology of Bird Pond, Ross Island, Antarctica. N.Z.J. Mar. Freshw. Res. 9: 547–562.

Ugolini, F. C., 1970. Antarctic soils and their ecology. In: M. W. Holdgate (ed.) Antarctic Ecology. Academic Press, London: vol. 2: 673–692.

Vincent, W. F. & Vincent, C. L., 1982. Response to nutrient enrichment by the plankton of antarctic coastal lakes and the inshore Ross Sea. Polar Biol. 1: 159–165.

Vincent, W. F., 1987. Antarctic limnology. In: A. B. Viner (ed.) Inland Waters of New Zealand. DSIR, Wellington: 379–412.

Watanuki, T., 1979. Isolation and culture of antarctic diatoms from the saline lakes in the Sôya Coast, East Antarctica. Mem. Nat. Inst. Polar Res., Special Issue No. 11: 35–41.

Watanuki, T. & Fukushima, H., 1978. Saline lake algae in Antarctica. Polar News (Kyokuchi), No. 27: 30–33.

Watanuki, T. & Ohno, M., 1973. Cultivation of antarctic microalgae (3). Isolation and culture of Antarctic endemic diatom *Tropidoneis laevissima* W. & G. S. West from the bottom sand of the salt lake Suribati at Skarvsnes in Lützow-Holm Bay, Antarctica. Antarctic Record, No. 56: 33–36.

West, W. & West, G. S., 1911. Freshwater algae. In: J. Murray (ed.) British Antarctic Expedition, 1907–1909: Reports on the Scientific Investigations; Biology, Vol. 1 (Pt. VIII): 263–298.

Wharton, R. A. Jr., Vinyard, W. C., Parker, B. C., Simmons, G. M. Jr. & Seaburg, K. G., 1981. Algae in cryoconite holes on Canada Glacier in Southern Victorialand, Antarctica. Phycologia, 20: 208–211.

Wright, S. W. & Burton, H. R., 1981. The biology of Antarctic saline lakes. Hydrobiologia, 82: 319–338.

Zaneveld, J. S., 1969. Cyanophytan mat communities in some meltwater lakes at Ross Island, Antarctica. Proc. Ak. Wetenschappen, Amsterdam, Series C, 72: 299–305.

Hydrobiologia 172: 97–110, 1989
W. F. Vincent and J. C. Ellis-Evans (eds) High Latitude Limnology

Community structure of benthic invertebrates in interior Alaskan (USA) streams and rivers

Mark W. Oswood
Institute of Arctic Biology, University of Alaska, Fairbanks, Alaska 99775, USA

Key words: Alaska, arctic, biogeography, high latitude, insects, macroinvertebrates, rivers, streams

Abstract

Taxonomic composition of benthic invertebrates in interior Alaskan streams and rivers is summarized from published and unpublished data. Diptera dominate the Alaskan stream fauna and constitute a larger proportion of the benthos in Alaskan streams than in streams of temperate North America. Plecoptera and Ephemeroptera are the next most abundant in Alaskan streams with Trichoptera generally very scarce. Several orders that occur regularly in streams of temperate North America are absent (or in very low abundance) in interior Alaskan streams: Hemiptera, Odonata, Megaloptera, Coleoptera. Net-spinning caddisflies, burrowing mayflies, and several families of stoneflies (Pteronarcyidae, Peltoperlidae and Perlidae) are conspicuous by their absence or extreme scarcity. Taxonomic composition varies significantly among hydrologic regions (major watersheds) and among stream types (springs, headwater streams, small rivers, and large rivers). Only two taxa (Chironomidae and Nemouridae) significantly increase in proportional contribution from south to north while many taxa decrease.

Introduction

Much of our ecological knowledge is derived from studies in regions with moderate climates. Studies from the polar extremes and tropical regions (not to mention the deeps of the oceans) have been hindered by problems of accessibility, logistical costs and the difficulties of conducting research under extreme climatic conditions. Studies of lotic systems are no exception, with the consequence that we have only scant information on the geographical limits of generalities derived from studies of streams and rivers of temperate regions (Harper, 1981).

Studies over large spatial or temporal scales (especially in remote regions) can seldom be carried out by one researcher (or team of researchers) so that the use of 'historic data sets' is necessary (Bowser, 1986). Most often, these data sets will be the result of short-term research projects with goals and procedures differing from project to project. Many valuable data sets reside in the 'gray literature' (e.g. theses, reports of governmental agencies, and environmental impact assessments). The critical problem with compilation and use of this collective 'research memory' (Risser & Treworgy, 1986) is reliability of the data and conclusions. Nearly every study of stream macroinvertebrates involves a unique sampling protocol (devices and sampling schedule) and identifications of macroinvertebrates are carried out to different taxonomic levels in different studies. Identifications undertaken prior to the availability of modern taxonomic works (e.g. Merritt & Cummins, 1978; Pennak, 1978) are often uncertain. Use of data sets from the literature requires that one find a robust 'lowest common denominator' among the data sets and use caution in analyses and interpretations.

In spite of these difficulties, there are a number of examples of successful 'literature-based' studies in freshwater ecology. Wiggins & Mackay (1978) compared trophic and habitat ecology for the genera of major orders (especially Trichoptera) of Nearctic aquatic insects. Logan & Brooker (1983) used data from stream studies in North America and the United Kingdom to examine differences in the faunas of riffles and pools. Clifford (1980) used published data on mayfly (Ephemeroptera) abundance to characterize patterns in taxonomic structure and mayfly abundance in the Holarctic region. Several works have tabulated and summarized literature values for various aspects of lotic ecology. Examples include seston concentrations (Webster *et al.*, 1979), primary production (Bott *et al.*, 1985; Minshall, 1978), allochthonous inputs (Anderson & Sedell, 1979), and organic matter budgets (Cummins *et al.*, 1983). The growing literature in stream ecology represents a valuable database for the examination of spatial and temporal patterns beyond the scope of individual investigators.

Alaska is isolated from the contiguous United States (Fig. 1) with a size and geographic diversity that exceeds that of many countries. Climates range from maritime in the southeastern panhandle to arctic in the north with vegetation likewise ranging from temperate rain forest through boreal forest to alpine and arctic tundra. Much of the research on rivers and streams has been associated with recent resource development, such as the trans-Alaska Pipeline (Alexander & Van Cleve, 1983). Although there is a relatively large literature on the lotic fishes of Alaska, the literature on lower trophic levels (primary producers and invertebrates) and basic processes (primary production and detrital ecology) is meagre. Further, most of the literature on stream macroinvertebrates is in the 'gray literature' and not readily accessible.

In this study, I have summarized the data on lotic macroinvertebrates of Alaska based upon an extensive literature search. This study includes data from mainland ('interior') Alaska but the panhandle (Southeast Alaska) is not included (Fig. 1). Southeast Alaska is geographically distinct from interior Alaska and best considered as part of a latitudinal transect including the continental United States. Data on numerical abundance of benthic macroinvertebrates from 114 data sets were analyzed in terms of the proportional (percent) contribution of various taxa. These analyses are used to provide an overall summary of the macroinvertebrate composition of Alaska's rivers and streams. The taxonomic composition of lotic invertebrates is also analyzed in terms of Alaska's hydrologic regions (major watersheds) as shown in Fig. 1, in terms of stream type (spring, headwater stream, small river, or large river), and as a function of the latitude of the study site. Finally, I compare the average taxonomic composition of small (headwater) streams of Alaska with several sites in temperate North America.

Methods

Appendix 1 provides a summary of the 114 data sets used in this study. Characteristics of the data sets summarized in the appendix include the following: data source, latitude (to the nearest degree), hydrologic region, stream type, sampling method(s), and seasonal extent of sampling (one-time vs. repeated sampling). Sampling methods, sampling schedules, and the taxonomic level of identifications varied widely between studies. The minimum data required for inclusion in this study were numerical abundance values with at least ordinal level identifications of insects. From these abundance values, I calculated the proportional contributions of the insect orders and non-insects to the total number of organisms, and if available, the proportions of insect families within each insect order and the proportions of major taxa of non-insects within the total non-insects. Analyses of generic composition were not done because few studies provided generic identifications and

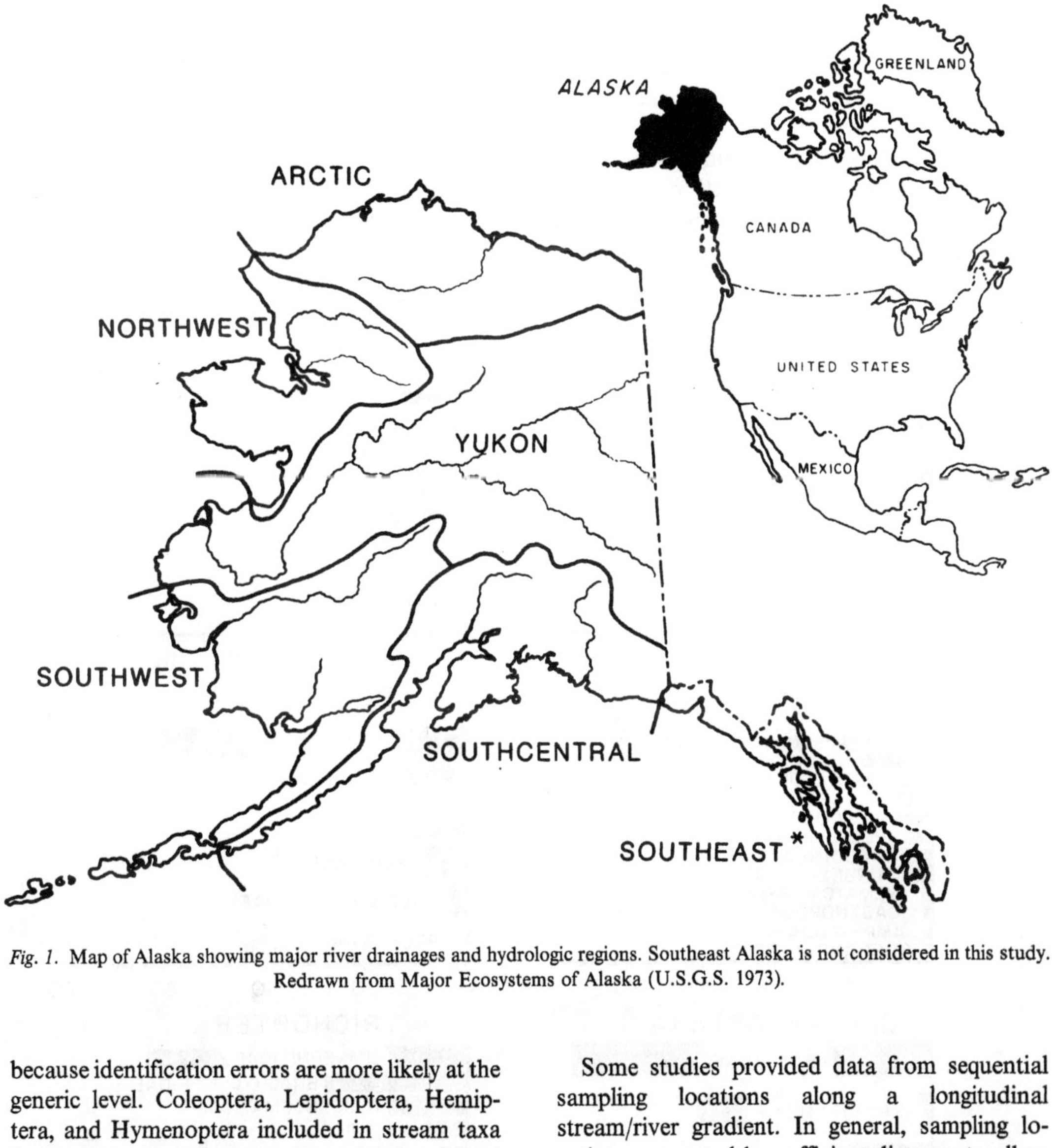

Fig. 1. Map of Alaska showing major river drainages and hydrologic regions. Southeast Alaska is not considered in this study. Redrawn from Major Ecosystems of Alaska (U.S.G.S. 1973).

because identification errors are more likely at the generic level. Coleoptera, Lepidoptera, Hemiptera, and Hymenoptera included in stream taxa lists were assumed to be terrestrial forms and were not included in analyses unless such representatives were explicitly indicated by authors to be aquatic forms or unless unambiguously aquatic taxa (e.g. Gerridae) were indicated. The majority of data sets provided identifications of families within Diptera (and identifications of major taxa of non-insect invertebrates) but did not provide family identifications for other insect orders.

Some studies provided data from sequential sampling locations along a longitudinal stream/river gradient. In general, sampling locations separated by sufficient distance to allow a change in stream order or marked change in stream size were treated as separate (i.e. independent) data sets while data from contiguous sample sites were combined. Three biogeographic variables were associated with each data set: latitude (to the nearest degree), hydrologic region, and stream type. Hydrologic regions (Fig. 1) are natural divisions representing major watersheds of

Alaska. Hydrologic regions roughly correspond to regions of major climatic differences with cold, dry conditions in the Arctic region, increasing average temperatures and precipitation in the Northwest, Yukon, and Southwest regions and maritime climate (warm, moist) in the Southwest region. Four categories of stream types were used: springs, headwater streams, small rivers, and large rivers. Approximate boundaries for stream types were derived from Cummins (1975, Figure 2): headwaters (orders 1–3, < 10 m width), small rivers (orders 4–6, 10–75 m width), large rivers (orders ≥ 7, width > 75 m). Some studies did not provide information on either stream order or stream width; in these cases stream type was estimated from photos, discharge data, or other site descriptors.

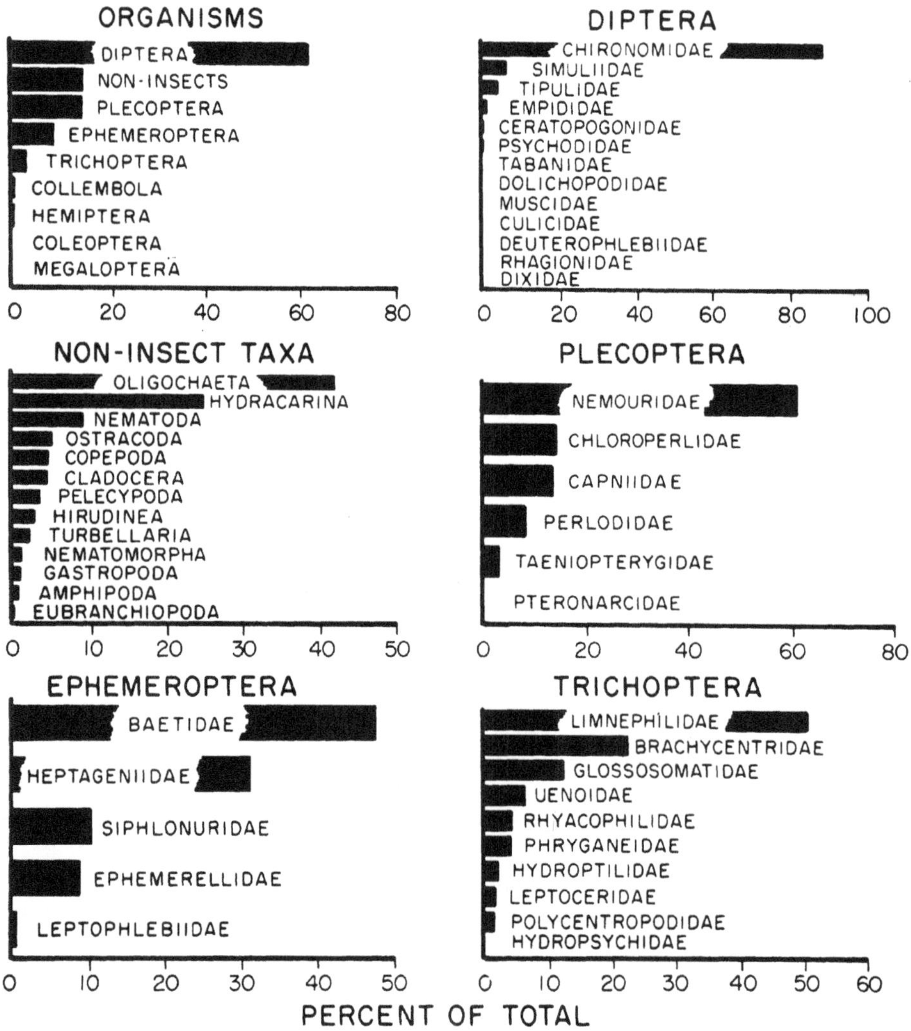

Fig. 2. Composition of benthic macroinvertebrates of Alaskan rivers and streams. Each graph lists taxa from greatest to least percent contribution by numerical abundance.

All data are expressed as a proportion or percentage contribution of a particular taxon to a higher taxonomic category. Such percentage data (especially for data values outside the range of 30–70%) generally do not meet the assumption of normality required for use of parametric statistics (Sokal & Rohlf, 1981). Therefore, all analyses were carried out using non-parametric tests. Relationships between percent composition of taxa and latitude were examined via Spearman rank correlation and differences in taxonomic composition among stream types and among hydrologic regions were examined using Kruskal-Wallis non-parametric analyses of variance.

Results

In interior Alaskan lotic communities, Diptera dominated the stream and river faunas with Plecoptera, Trichoptera, and Ephemeroptera collectively accounting for less than 25% of the fauna (Fig. 2). Collembola, Hemiptera, Coleoptera, and Megaloptera each constituted less than 1% of the fauna. The Diptera were represented by 13 families but midge larvae (Chironomidae) accounted for nearly 90% of the lotic Diptera. The Plecoptera were dominated by Nemouridae, and the Ephemeroptera by Baetidae and Heptageniidae. Although the Limnephilidae constituted over half of the Trichoptera, many other families were common. Net-spinning caddisflies (notably Hydropsychidae) were in very low abundance. Non-insect invertebrates were dominated by oligochaete worms and water mites (Hydracarina) but a long list of other taxa contributed to the non-insects.

Study sites spanned approximately 10° latitude and were located from near Valdez on the Gulf of Alaska to the Arctic Coastal Plain bordering the Arctic Ocean. Taxonomic composition changed markedly over this latitudinal range (Table 1). Nemouridae and Chironomidae were the only taxa whose relative proportions increased with increasing latitude, while the general trend for most invertebrate taxa (except Diptera) was a decrease in relative proportion. This indicates a simplification of the lotic fauna along a south to north gradient with increasing dominance by nemourids and chironomids. The Diptera (as an order) showed no significant change in proportion with latitude presumably because of contradictory trends of the dipteran families i.e. a proportional increase in Chironomidae and decreases in Ceratopogonidae, Dixidae, Empididae, Psychodidae, Simuliidae, and Tabanidae.

Alaska has six major hydrological regions (Fig. 1). The Southeast region (i.e. the panhandle) is not considered in this study and no data sets were located from the Southwest region so that comparisons of the lotic fauna are limited to the Arctic, Yukon, Northwest, and Southcentral regions. The proportional contributions of major taxa for each hydrologic region are shown in

Table 1. Taxa showing a significant relationship between percent composition and latitude (percent of total organisms for insect orders, percent of order for insect families, percent of total non-insects for non-insect taxa). Significance of relationships based upon Spearman rank correlation analyses. * = $P < p.05$, ** = $P < 0.01$.

Increasing representation with increasing latitude	Decreasing representation with increasing latitude
Nemouridae**	Collembola*
Chironomidae**	Ephemeroptera*
	Plecoptera*
	Trichoptera**
	Ephemerellidae*
	Leptophlebiidae*
	Siphlonuridae**
	Chloroperlidae**
	Perlodidae**
	Taeniopterygidae**
	Glossosomatidae*
	Ceratopogonidae**
	Dixidae*
	Empididae**
	Psychodidae**
	Simuliidae**
	Tabanidae*
	Turbellaria**
	Ostracoda*
	Pelecypoda**

Table 2. Only the Plecoptera and Trichoptera showed significant differences in proportional contributions between hydrologic regions. The Trichoptera showed their highest contribution to the lotic fauna in the Southwest region and lowest contribution in the Arctic region. Too few studies provided family level identifications within the Ephemeroptera, Plecoptera, and Trichoptera to allow analyses of differences in contributions of families (to orders) among hydrologic regions. However, these analyses are shown in Table 3 for Diptera and non-insect invertebrates. As in the previous analyses of latitudinal distributions, the families within the Diptera exhibited different biogeographic patterns. Three families (Psychodidae, Simuliidae, Tabanidae) reached their highest proportional contribution and one family (Chironomidae) its lowest in the Southcentral region.

Proportional contributions of all major insect orders showed significant differences between stream types (Table 4). The Ephemeroptera and Trichoptera reached their maximum contribution to the lotic fauna in small rivers, the Plecoptera in

Table 2. Percentage contribution of major taxa to the stream fauna of each hydrologic region. Taxa constituting a small percentage of the macroinvertebrate fauna (see Figure 2) not included. See Figure 1 for boundaries of hydrologic regions. Probabilities of no difference between hydrologic regions based upon Kruskal-Wallis non-parametric analyses of variance. NS = not significant (P > 0.05), * = P < 0.05, ** = P < 0.01.

	Arctic	Northwest	Yukon	Southcentral	Differences between regions
n =	53	10	39	12	
Ephemeroptera	8.4	6.4	7.6	9.4	NS
Plecoptera	13.2	10.1	17.2	7.8	*
Trichoptera	0.7	0.4	2.9	8.4	**
Diptera	60.0	65.0	62.0	65.1	NS
Non-Insects	17.4	18.1	9.4	8.7	NS

Table 3. Taxa showing significant differences in proportional representation among hydrologic regions. Boundaries of hydrologic regions given in Figure 1. Probabilities of no difference between hydrologic regions based upon Kruskal-Wallis non-parametric analyses of variance. * = P < 0.05, ** = P < 0.01. AR = Arctic, NW = Northwest, YK = Yukon, SC = Southcentral. Sample sizes for comparisons among dipteran families and non-insect taxa (in parentheses) as follows: AR = 53 (47), NW = 9 (7), YK = 35 (37), SC = 12 (12).

Taxon	Significance level	Hydrologic region highest proportional representation	Hydrologic region of lowest proportional representation
DIPTERA			
Ceratopogonidae	*	NW	YK
Chironomidae	*	YK	SC
Psychodidae	**	SC	NW
Simuliidae	**	SC	NW
Tabanidae	**	SC	NW
Tipulidae	*	NW	YK
NON-INSECTS			
Turbellaria	*	NW	YK
Ostracoda	*	SC	NW
Hydracarina	*	YK	NW
Gastropoda	*	AR	NW
Pelecypoda	**	SC	NW

headwater streams, and the Diptera in springs. Once again the dipteran families showed quite different trends (Table 5). The Chironomidae reach their maximum contribution in the spring fauna, the Psychodidae in headwater streams, Simuliidae and Tipulidae in small rivers, and the Ceratopogonidae and Empididae in the large river fauna. The amphipods (like the Chironomidae) reached their maximum in the spring fauna (Table 5).

Discussion

Hynes (1970) points out the remarkable similarity in the macroinvertebrate fauna of stony streams throughout the world. Hynes (1970; pp. 115–116) and Cummins (1972; pp. 48–50) list typical major macroinvertebrate taxa of running waters. A comparison of these lists with Fig. 2 reveals the relatively depauperate nature of the Alaskan lotic fauna. Several orders, which occur regularly in temperate regions of North America, are absent or at very low abundance in interior Alaskan rivers and streams: Hemiptera, Odonata, Mega-

Table 4. Percentage contribution of major taxa to each stream type. Taxa constituting a small percentage of the macroinvertebrate fauna (see Figure 1) not included. Definitions of stream types in text. Probabilities of no difference between stream types based upon Kruskal-Wallis non-parametric analyses of variance. NS = not significant (P > 0.05), * = P < 0.05, ** = P < 0.01.

	Springs	Headwaters	Small rivers	Large rivers	Differences between stream types
n =	12	44	44	14	
Ephemeroptera	3.7	7.4	10.7	5.3	**
Plecoptera	4.3	20.1	14.1	0.5	**
Trichoptera	1.0	2.0	3.4	0.4	*
Diptera	75.0	59.0	58.2	69.2	*
Non-Insects	15.8	11.1	12.6	24.4	NS

Table 5. Taxa showing significant differences in proportional representation among stream types. Stream types discussed in text. Probabilities of no difference between stream types based upon Kruskal-Wallis non-parametric analyses of variance. * = P < 0.05, ** = P < 0.01. SP = spring, HW = headwater streams. SR = small rivers, LR = large rivers. Sample sizes for comparisons among dipteran families and non-insect taxa (in parentheses) as follows: SP = 12 (9), HW = 40 (44, 42 for Oligochaeta), SR = 40 (41), LR = 14 (9).

Taxon	Significance level	Stream type of highest proportional representation	Stream type of lowest proportional representation
DIPTERA			
Ceratopogonidae	*	LR	SP
Chironomidae	**	SP	SR
Empididae	*	LR	SP
Psychodidae	*	HW	LR
Simuliidae	**	SR	LR
Tipulidae	**	SR	SP
NON-INSECTS			
Oligochaeta	*	HW	SR
Amphipoda	*	SP	SR/LR
Hydracarina	*	SR	SP

loptera, Neuroptera and Coleoptera. Net-spinning caddisflies, burrowing mayflies, several families of stoneflies (Pteronarcyidae, Peltoperlidae and Perlidae), and truly lotic Coleoptera (e.g. Psephenidae and Elmidae) are conspicuous by their absence or extreme scarcity. The average taxonomic composition of Alaskan headwater streams (n = 44) is compared with published data from three streams in temperate North America in Fig. 3. Although these stream sites represent a range of stream types (relatively low- and high-elevation Rocky Mountain streams and a northern prairie stream) they obviously do not represent the diversity of stream types of North America and firm conclusions concerning large scale latitudinal trends in North American streams must await a much more extensive comparison. The most striking feature of the Alaskan fauna is the dominance by Diptera compared to temperate streams (Fig. 3). Diptera constitute an average of 59% of the invertebrate fauna of Alaskan headwater streams while Andrews & Minshall (1979) report a range of 5.3–23.8% Diptera for 11 unpolluted Rocky Mountain streams. Plecoptera also appear to generally constitute a larger proportion of the Alaskan stream fauna than of temperate streams while Ephemeroptera and Trichoptera are proportionally less abundant in Alaskan streams (Fig. 3). Coleoptera are significant elements of the fauna of temperate streams (Fig. 3) but only sporadically present at very low densities in Alaskan streams. MacLean (1975) and Danks (1981), in comparing the taxonomic composition of high latitude insect faunas with the North American or World insect faunas, likewise show the increased dominance by Diptera and decreased contribution of Coleoptera at high latitudes. However, these comparisons are based upon percentage contributions of species within each order to the total number of species of a region rather than percent contributions to total numerical abundance.

In contrast to this apparent faunal impoverishment of Alaskan streams, densities of lotic invertebrates do not appear to be markedly lower in Alaskan streams compared to streams of temperate regions (Cowan & Oswood, 1984). Spring-fed streams in the Alaskan arctic are characterized by very high densities of Chironomidae (Craig & McCart, 1974; Glesne & Deschermeier,

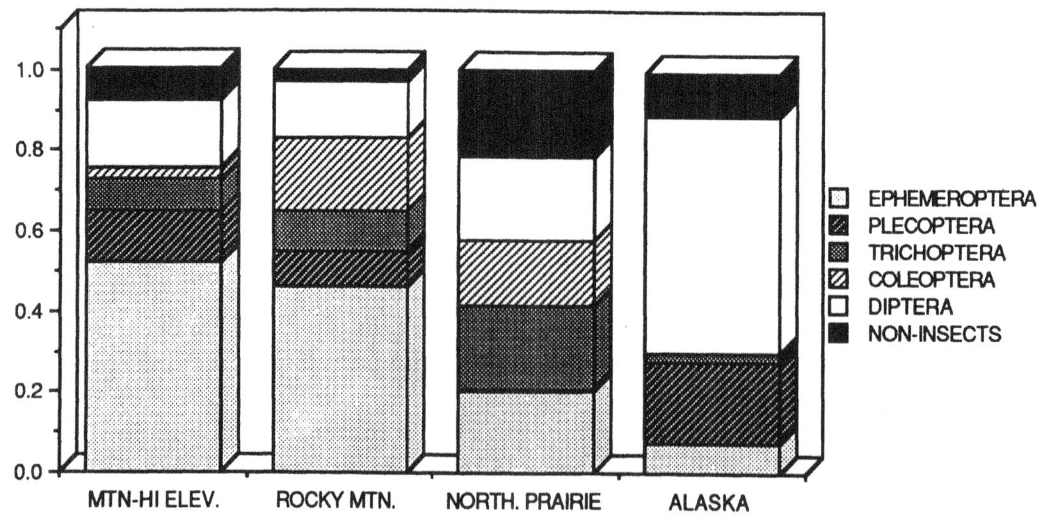

Fig. 3. Comparison of the taxonomic composition of interior Alaskan headwater streams (n = 44) with three streams in temperate North America. MTN-HI ELEV. = a high altitude Rocky Mountain stream in Colorado (Cline *et al.*, 1982), ROCKY MTN. = a relatively low elevation Rocky Mountain stream in Idaho (Minshall, 1981), and NORTH. PRAIRIE = a prairie stream in North Dakota (Neel, 1985).

1984) and macroinvertebrate densities of non-spring streams may be high as well (Cowan & Oswood, 1984; Wedemeyer, unpublished).

Latitudinal changes in taxonomic composition were evident within Alaska (Table 1) as well as in comparisons with the faunas of temperate regions. Only Nemouridae and Chironomidae showed significant increases in percent contribution with increasing latitude while Collembola, Ephemeroptera, Plecoptera, and Trichoptera (as well as many lower taxa) decreased with increasing latitude. Barton (1986) found similar latitudinal (53° N–67° N) trends in stream riffle faunas of the Mackenzie River basin in Canada.

Stream type (Tables 4 and 5) and hydrologic regions (Tables 2 and 3) appear to be useful landscape features for understanding patterns in the relative abundance of Alaskan lotic invertebrates. However, other features (e.g. physiographic regions, geology, soils, climate, or vegetation may also prove useful in understanding and predicting distributional patterns of Alaskan invertebrates. Scudder (1979) suggests that drainage patterns are the most important feature in understanding the zoogeography of Canadian lotic insects. Corkum (in ms.) examined patterns in the relative abundance of lotic invertebrates over a large region of northwestern North America. She found that biological patterns were associated with major landscape features (major drainage basins, physiographic regions, bedrock geology). Larsen *et al.* (1986) defined aquatic ecoregions for Ohio, based upon land-surface form, land use, potential natural vegetation, and soil parent material and found that fish assemblages corresponded to the defined ecoregions. Using the converse approach, Hawkes *et al.* (1986) used multivariate techniques and presence/absence data for common stream fishes to define fish ecoregions for the State of Kansas. These ecoregions (based upon fish assemblages) corresponded to major landscape features (physiographic regions, river basins, geology, soil, and potential natural vegetation).

Biogeographic patterns immediately raise questions of causality. Present distributions of Alaskan lotic insects are undoubtedly the result of both historical factors (e.g. Pleistocene glaciation and invasions of river systems) and ecological factors. Alaska's thermal environment varies from relatively warm southern coastal regions with maritime influences to the extreme rigor of the arctic. Physiological abilities (e.g. tolerance of freezing) of stream invertebrates are little known. Trophic resources (organic detritus and periphyton) must certainly differ in abundance, quality, and seasonality among stream types and geographical regions.

One of the benefits of literature-based syntheses is the identification of data gaps and research needs. No data were located from the Southwest region (a large region which includes the Kuskokwim River drainage and the Aleutians – see Fig. 1) and data from springs were available only from the Arctic region. I would appreciate learning of additional sources of data. Relatively few studies identified members of the Ephemeroptera, Plecoptera, or Trichoptera to family. A systematic survey of Alaska's rivers and streams involving a uniform sampling protocol, identification of all insects to family, and a sampling scheme based upon major landscape features (hydrologic regions, stream types, physiographic regions etc.) would provide a firmer basis for both descriptive biogeography and for formulating questions of causality. However, this study has shown significant patterns in the faunal composition of Alaskan rivers and streams in spite of the great diversity of sampling procedures used by researchers.

Acknowledgements

I thank Drs. Steve MacLean, David Murray, James Reynolds, Phyllis Weber, and Ms. Kate Wedemeyer for comments on the manuscript. Ms. Tina Picolo assisted with data entry and Judy Oswood with proof-reading. Mr. Don Borchert (Institute of Arctic Biology, Univ. Alaska) assisted greatly in graphics preparation. Research support for collection of unpublished data provided by The U.S. National Park Service and The U.S. Department of Energy.

Appendix 1. Characteristics of data sets used in analyses. Hydrologic regions shown in Figure 1. Abbreviations for stream types as follows: SP = springs, HW = headwater streams, SR = small river, LR = large river (definitions in text). Abbreviations for sampling devices as follows: BX = box sampler, HS = Hess-type sampler, SR = Surber-type sampler, TR = 'ten-rock', DR = drift net, RB = rock-filled baskets, MP = multi-plate artificial substrate, AN = aquatic net, GS = grab samples, EG = Ekman Grab, PS = pump sampler (details of sampling methods in original references). Abbreviations for extent of sampling as follows: R = reconnaisance (one sample time), E = extended sampling (repeated sampling).

	Reference	Site	Latitude	Stream Type	Hydrologic Region	Sampling Devices	Extent of Sampling
1	Slack *et al.*, 1979	Kuyuktuvuk Cr.	68	HW	YK	DR, TR, AN	R
2	Slack *et al.*, 1979	Kuyuktuvuk Cr.	68	HW	YK	DR, TR, AN	R
3	Slack *et al.*, 1979	Kuyuktuvuk Cr.	68	SR	YK	DR, TR, AN	R
4	Slack *et al.*, 1979	Kuyuktuvuk Cr.	68	SR	YK	DR, TR, AN	R
5	Slack *et al.*, 1979	Dietrich R.	68	SR	YK	DR, TR, AN	R
6	Slack *et al.*, 1979	Atigun R.	68	HW	AR	DR, TR, AN	R
7	Slack *et al.*, 1979	Atigun R.	68	HW	AR	DR, TR, AN	R
8	Slack *et al.*, 1979	Atigun R.	68	HW	AR	DR, TR, AN	R
9	Slack *et al.*, 1979	Atigun R.	68	SR	AR	DR, TR, AN	R
10	Slack *et al.*, 1979	Atigun R.	68	SR	AR	DR, TR, AN	R
11	Slack *et al.*, 1979	Unnamed trib. to Atigun R.	68	HW	AR	DR, TR, AN	R
12	Maurer & Toland, 1984 & Maurer 1986	Middle Cr.	61	HW	SC	HS	E
13	Maurer & Toland, 1984 & Maurer 1986	Lone Cr.	61	HW	SC	HS	E
14	Craig & McCart, 1974	Echooka Sp.	69	SP	AR	SR	R
15	Craig & McCart, 1974	Ribdon Sp.	69	SP	AR	SR	R
16	Craig & McCart, 1974	Lupine Sp.	69	SP	AR	SR	R
17	Craig & McCart, 1974	Kuparuk R.	69	SR	AR	SR	R
18	Craig & McCart, 1974	Toolik R.	69	SR	AR	SR	R
19	Craig & McCart, 1974	Happy Valley Cr.	69	SR	AR	SR	R
20	Craig & McCart, 1974	Accomplishment Cr.	69	SR	AR	SR	R
21	Craig & McCart, 1974	Ribdon R.	69	SR	AR	SR	R
22	Craig & McCart, 1974	Ribdon trib.	69	HW	AR	SR	R
23	Craig & McCart, 1974	Echooka R.	69	HW	AR	SR	R
24	Craig & McCart, 1974	Lupine R.	69	SR	AR	SR	R
25	Cowan, 1983	Monument Cr.	65	HW	YK	BX	E
26	Frey *et al.*, 1970	Chena R.	65	SR	YK	SR	E
27	McCoy, 1974	Little Chena R.	65	SR	YK	RB	E
28	McCoy, 1974	Chena R.	65	SR	YK	RB	E
29	Nauman & Kernodle, 1975	Happy Valley Cr.	69	HW	AR	SR, TR	R
30	Brown, 1987	Stampede Cr.	64	HW	YK	BX	E
31	Brown, 1987	Clearwater Fk. Toklat R.	64	SR	YK	BX	E
32	Viavant & Oswood unpublished	Imnavait Cr.	69	HW	AR	PS	E

33	Watson *et al.*, 1966	Ogotoruk Cr.	68	SR	NW	?	E
34	Oswood *et al.*, 1984	Little Poker Cr.	65	HW	YK	SR	E
35	Wagener, 1984	Twelve Mile Cr.	65	HW	YK	BX	E
36	Wagener, 1984	McManus Cr.	65	HW	YK	BX	E
37	Miller *et al.*, 1986	Imnavait Cr.	69	HW	AR	HS	R
38	Miller *et al.*, 1986	Poker and Caribou Cr.	65	HW	YK	HS	R
39	Nauman & Kernodle, 1974	Allison Cr.	61	HW	SC	AN, TR, MP, RB	E
40	Wedemeyer, unpublished	Eldorado Cr.	64	HW	YK	BX	E
41	Wedemeyer, unpublished	Jumbo Cr.	64	HW	YK	BX	E
42	Lotspeich *et al.*, 1970	Dennison Fk.	64	SR	YK	SR	E
43	Lotspeich *et al.*, 1970	West Fk.	64	SR	YK	SR	E
44	Lotspeich *et al.*, 1970	Big Timber	64	HW	YK	SR	R
45	Elliott G.V., 1982	Mainline Sp.	68	HW	AR	SR	E
46	Elliott G.V., 1982	Holden Cr.	68	HW	AR	SR	E
47	Elliott G.V., 1982	Spike Camp Cr.	68	HW	AR	SR	E
48	Childers *et al.*, 1979	Etivluk R.	69	SR	AR	AN	R
49	Childers *et al.*, 1979	Colville R.	69	LR	AR	AN	R
50	Childers *et al.*, 1979	Killik R.	69	LR	AR	AN	R
51	Childers *et al.*, 1979	Kovrorak Sp.	68	SP	NW	AN	R
52	Childers *et al.*, 1979	Ikpikpuk R.	70	LR	AR	AN	R
53	Childers *et al.*, 1979	Meade R.	70	LR	AR	AN	R
54	Childers *et al.*, 1979	Avalik R.	70	LR	AR	AN	R
55	Childers *et al.*, 1979	Utukok R.	70	LR	AR	AN	R
56	Childers *et al.*, 1979	Kokolik R.	70	SR	AR	AN	R
57	Childers *et al.*, 1979	Kukpowruk R.	69	SR	AR	AN	R
58	Childers *et al.*, 1979	Pitmega R.	69	SR	AR	AN	R
59	Childers *et al.*, 1979	Ipewik R.	68	LR	NW	AN	R
60	Childers *et al.*, 1979	Kukpuk R.	68	LR	NW	AN	R
61	Childers *et al.*, 1979	Ogoturuk Cr.	68	SR	NW	AN	R
62	Childers *et al.*, 1979	Kivalina R.	68	LR	NW	AN	R
63	Childers *et al.*, 1979	Wulik R.	68	LR	NW	AN	R
64	Childers *et al.*, 1979	Kugururok R.	68	SR	NW	AN	R
65	Childers *et al.*, 1979	Squirrel R.	67	LR	NW	AN	R
66	Childers *et al.*, 1979	Kobuk R.	67	LR	NW	AN	R
67	Childers *et al.*, 1977	Flood Creek Sp.	69	SP	AR	TR	R
68	Childers *et al.*, 1977	Red Hill Sp.	70	SP	AR	TR	R
69	Childers *et al.*, 1977	Katakturuk R trib. spring	70	SP	AR	TR	R
70	Childers *et al.*, 1977	Sadlerochit Sp.	70	SP	AR	TR, SR	E
71	Childers *et al.*, 1977	Hulahula R. Sp.	70	SP	AR	TR	R
72	Childers *et al.*, 1977	Aichilik R. Sp.	70	SP	AR	TR	R
73	Childers *et al.*, 1977	Ekaluakat R. Sp.	70	SP	AR	TR	R
74	Childers *et al.*, 1977	Kongakut R. Sp.	70	SP	AR	TR	R
75	Childers *et al.*, 1977	Ekaluakat R.	70	LR	AR	SR	R
76	Elliott, 1980	Gulkana R.	63	HW	SC	SR	E
77	Elliott, 1980	Dann Cr.	63	HW	SC	SR	E
78	Elliott, 1980	Million Dollar Cr.	65	HW	YK	SR	E

Appendix 1. (Continued)

	Reference	Site	Latitude	Stream Type	Hydrologic Region	Sampling Devices	Extent of Sampling
79	Nauman & Kernodle, 1974	Kuparuk R.	70	SR	AR	SR, MP, RB	E
80	Nauman & Kernodle, 1974	Putuligayuk R.	70	SR	AR	SR, AN, GS, MP, RB	E
81	Nauman & Kernodle, 1974	W. Channel Sagavanirktok R.	70	LR	AR	SR	R
82	Nauman & Kernodle, 1974	Sagavanirktok R.	69	SR	AR	GS, MP, RB	E
83	Nauman & Kernodle, 1974	Happey Valley Cr.	69	HW	AR	AN, TR, RB, SR, MP	E
84	Nauman & Kernodle, 1974	Atigun R. trib.	69	HW	AR	AN, TR	R
85	Nauman & Kernodle, 1974	Atigun R.	68	SR	AR	AN, TR	E
86	Nauman & Kernodle, 1974	Atigun R. trib.	68	HW	AR	AN, TR	R
87	Nauman & Kernodle, 1974	Atigun R. trib.	68	HW	AR	AN, TR	R
88	Nauman & Kernodle, 1974	Atigun R. trib.	68	HW	AR	AN, TR	R
89	Nauman & Kernodle, 1974	Atigun R. trib.	68	HW	AR	AN	R
90	Nauman & Kernodle, 1974	Kuyuktuvuk Cr. trib.	68	HW	YK	AN, TR	R
91	Nauman & Kernodle, 1974	Kuyuktuvuk Cr. trib.	68	HW	YK	AN, TR	R
92	Nauman & Kernodle, 1974	Kuyuktuvuk Cr. trib.	68	SR	YK	AN, TR, GS	E
93	Nauman & Kernodle, 1974	Kuyuktuvuk Cr. trib.	68	SR	YK	AN, TR	R
94	Nauman & Kernodle, 1974	Dietrich R.	68	SR	YK	AN	R
95	Nauman & Kernodle, 1974	Middle Fk. Koyukuk R.	67	SR	YK	GS, AN, SR	E
96	Nauman & Kernodle, 1974	Minnie Cr.	67	HW	YK	MP, RB, AN, TR	E
97	Nauman & Kernodle, 1974	Wiseman Cr.	67	HW	YK	SR, MP, RB	E
98	Nauman & Kernodle, 1974	Slate Cr.	67	SR	YK	GS, AN, RB	E
99	Nauman & Kernodle, 1974	Prospect Cr.	67	SR	YK	AN, TR, SR	E
100	Nauman & Kernodle, 1974	Hess Cr.	66	SR	YK	MP, RB, TR, SR, AN	E
101	Nauman & Kernodle, 1974	Tolovana R.	65	SR	YK	SR, RB	E
102	Nauman & Kernodle, 1974	Chatanika R.	65	SR	YK	AN, TR, SR	E
103	Nauman & Kernodle, 1974	Caribou Cr.	65	HW	YK	SR, TR, MP, RB	E
104	Nauman & Kernodle, 1974	Chena R.	65	SR	YK	AN, MP, SR, RB, TR	E
105	Nauman & Kernodle, 1974	Salcha R.	64	SR	YK	TR, MP, RB, GS, AN	E
106	Nauman & Kernodle, 1974	S. Fk. Minton Cr.	64	HW	YK	GS	R
107	Nauman & Kernodle, 1974	Shaw Cr.	64	SR	YK	GS, AN, TR, RB, EG	E
108	Nauman & Kernodle, 1974	Fish Cr.	63	HW	SC	SR, MP, RB, AN, TR	E
109	Nauman & Kernodle, 1974	Gulkana R.	63	SR	SC	SR, AN, MP, RB, GS, TR	E
110	Nauman & Kernodle, 1974	Tazlina R.	62	SR	SC	SR	R
111	Nauman & Kernodle, 1974	Squirrel Cr.	62	SR	SC	SR, MP, RB, AN, TR	E
112	Nauman & Kernodle, 1974	Little Tonsina R.	62	SR	SC	SR, RB, GS, AN, TR, EG, MP	E
113	Nauman & Kernodle, 1974	Stuart Cr.	61	HW	SC	SR, RB	E
114	Nauman & Kernodle, 1974	Solomon Gulch	61	HW	SC	GS, AN, TR, MP, RB	E

References

Alexander, V. & K. Van Cleve, 1983. The Alaska pipeline: a success story. Ann. Rev. Ecol. Syst. 14: 443–463.

Anderson, N. H. & J. R. Sedell, 1979. Detritus processing by macroinvertebrates in stream ecosystems. Ann. Rev. Entomol. 24: 351–377.

Andrews, D. A. & G. W. Minshall, 1979. Distribution of benthic invertebrates in the Lost Streams of Idaho. Amer. Midl. Nat. 102: 140–148.

Barton, D. R., 1986. Invertebrates of the Mackenzie system. In B. R. Davies & K. F. Walker (eds), The ecology of river systems. Junk Publishers, Dordrecht, The Netherlands.: 473–492.

Bott, T. L., J. T. Brock, C. S. Dunn, R. J. Naiman, R. W. Ovink & R. C. Petersen, 1985. Benthic community metabolism in four temperate stream systems: an inter-biome comparison and evaluation of the river continuum concept. Hydrobiologia 123: 3–45.

Bowser, C. J., 1986. Historic data sets: lessons from the past, lessons for the future. In W. K. Michener (ed.), Research data management in the ecological sciences. University of South Carolina Press, Columbia, SC.: 155–179.

Brown, B. N., 1987. Community structure and longitudinal patterns of benthic invertebrates in a heavy metal contaminated Alaskan river system. M. S. Thesis. University of Alaska, Fairbanks. USA.

Childers, J. M., D. R. Kernodle & R. M. Loeffler, 1979. Hydrological reconnaissance of Western Arctic Alaska, 1976 and 1977. Open-File Report 79–699. U.S. Geol. Survey. Anchorage, Alaska.

Childers, J. M., C. E. Sloan, J. P. Meckel & J. W. Nauman, 1977. Hydrological reconnaissance of the Eastern North Slope, Alaska, 1975. Open-File Report 77–492. U.S. Geol. Survey. Anchorage, Alaska.

Clifford, H. F., 1980. Numerical abundance values of mayfly nymphs from the holarctic region. In J. F. Flannagan & K. E. Marshall (eds), Advances in Ephemeroptera biology. Plenum Press, NY: 503–509.

Cline, L. D., R. A. Short & J. V. Ward, 1982. The influence of highway construction on the macroinvertebrates and epilithic algae of a high mountain stream. Hydrobiologia 96: 149–159.

Cowan, C. A., 1983. Phenology of benthic detritus input, storage and processing in an Alaskan subarctic stream. M. S. Thesis. University of Alaska, Fairbanks. USA.

Cowan, C. A. & M. W. Oswood, 1984. Spatial and seasonal associations of benthic macroinvertebrates and detritus in an Alaskan subarctic stream. Polar Biol. 3: 211–215.

Craig, P. C. & P. J. McCart, 1974. Classification of stream types in Beaufort Sea drainages between Prudhoe Bay, Alaska and the Mackenzie Delta. In P. J. McCart (ed.), Classification of streams in Beaufort Sea drainages and distribution of fish in arctic and sub-arctic drainages. Canadian Arctic Gas Study Limited, Biological Report Series. Vol. 17, Chap. 1. 47 pp.

Cummins, K. W., 1972. What is a river? – a zoological description. In R. T. Oglesby, C. A. Carlson & J. A. McCann (eds), River ecology and man. Academic Press, NY: 33–52.

Cummins, K. W., 1975. The ecology of running waters; theory and practice. In D. B. Baker, W. B. Jackson & B. L. Prater (eds), Proceedings Sandusky River Basin Symposium. Tiffin, Ohio.: 278–293.

Cummins, K. W., J. R. Sedell, F. J. Swanson, G. W. Minshall, S. G. Fisher, C. E. Cushing, R. C. Petersen & R. L. Vannote, 1983. Organic matter budgets for stream ecosystems: problems in their evaluation. In J. R. Barnes & G. W. Minshall (eds), Stream ecology: application and testing of general ecological theory. Plenum Press, NY: 299–353.

Danks, H. V., 1981. Arctic arthropods: a review of systematics and ecology with particular reference to the North American fauna. Entomological Society of Canada, Ottawa. 608 pp.

Elliott, G. V., 1980. First interim report on the evaluation of stream crossings and effects of channel modifications on fishery resources along the route of the trans-Alaska pipeline. U.S. Fish and Wildlife Service, Special Studies, Anchorage, Alaska, 77 pp.

Elliott, G. V., 1982. Final report on the evaluation of stream crossings and effects of channel modifications on fishery resources along the route of the trans-Alaska pipeline. U.S. Fish and Wildlife Service, Special Studies, Anchorage, Alaska, 110 pp.

Frey, P. J., E. W. Mueller & E. C. Berry, 1970. The Chena River: the study of a subarctic stream. Federal Water Quality Administration, Northwest Region, U.S. Dept. of the Interior, Alaska Water Laboratory, College, Alaska. Project No. 1610 – 10/70. 96 pp.

Glesne, R. S. & S. J. Deschermeier, 1984. Abundance, distribution and diversity of aquatic macroinvertebrates of the North Slope of the Arctic National Wildlife Refuge, 1982 and 1983. U.S. Fish and Wildlife Service. Fairbanks Fishery Resources Progress Report Number FY84-2. Fairbanks, Alaska.

Harper, P. P., 1981. Ecology of streams at high latitudes. In M. A. Lock & D. D. Williams (eds), Perspectives in running water ecology. Plenum Press, NY: 313–337.

Hawkes, C. L., D. L. Miller & W. G. Layher, 1986. Fish ecoregions of Kansas: stream fish assemblage patterns and associated environmental correlates. Envir. Biol. Fishes 17: 267–279.

Hynes, H. B. N., 1970. The ecology of running waters. University of Toronto Press. 555 pp.

Larsen, D. P., J. M. Omernik, R. M. Hughes, C. M. Rohm, T. R. Whittier, A. J. Kinney, A. L. Gallant & D. R. Dudley, 1986. Correspondence between spatial patterns in fish assemblages in Ohio streams and aquatic ecoregions. Envir. Manag. 10: 815–828.

Logan, P. & M. P. Brooker, 1983. The macroinvertebrate faunas of riffles and pools. Wat. Res. 17: 263–270.

Lotspeich, F. B., E. W. Mueller & P. J. Frey, 1970. Effects of large scale forest fires on water quality in interior Alaska. U.S. Dept. Interior, Federal Water Pollution Control Administration, Alaska Water Laboratory, College, Alaska. 115 pp.

McCoy, G. A., 1974. Preconstruction assessment of biological quality of the Chena and Little Chena rivers in the vicinity of the Chena Lakes Flood Control Project near Fairbanks. U.S. Geological Survey. Water – Resources Investigations 29–74. 84 pp.

MacLean, S. F. Jr., 1975. Ecological adaptations of tundra invertebrates. In F. J. Verburg (ed.), Physiological adaptation to the environment. Intext Educational Publishers, NY: 269–299.

Maurer, M. A., 1986. Chemical and biological water quality of selected streams in the Beluga Coal Area, Alaska. Alaska Division of Geological and Geophysical Surveys. Public-data File 86–51.

Maurer, M. A. & D. C. Toland, 1984. Water-quality data from the Beluga coal-field area, Alaska, June 1982 through March 1983. Report of Investigation 84–27. Alaska Division of Geological and Geophysical Surveys.

Merritt, R. W. & K. W. Cummins, 1978. An introduction to the aquatic insects of North America. Kendall/Hunt Publishing Co. Dubuque, Iowa. 441 pp.

Miller, M. C., J. R. Stout & V. Alexander, 1986. Effects of a controlled under-ice oil spill on invertebrates of an arctic and a subarctic stream. Environ. Poll. (Series A): 42: 99–132.

Minshall, G. W., 1978. Autotrophy in stream ecosystems. BioScience 28: 767–771.

Minshall, G. W., 1981. Structure and temporal variations of the benthic macroinvertebrate community inhabiting Mink Creek, Idaho, USA, a 3rd order Rocky Mountain stream. J. Freshwat. Ecol. 1: 13–26.

Nauman, J. W. & D. R. Kernodle, 1974. Aquatic organisms from selected sites along the proposed trans-Alaska pipeline corridor September 1970 to September 1972. Basic-Data Report. U.S. Geol. Survey. Anchorage, Alaska. 23 pp.

Nauman, J. W. & D. R. Kernodle, 1975. The effect of a fuel oil spill on benthic invertebrates and water quality on the Alaskan arctic slope, Happy Valley Creek near Sagwan, Alaska. J. Res. U.S. Geol. Survey 3: 495–500.

Neel, J. K., Sr., 1985. A northern prairie stream. University of North Dakota Press, Grand Forks, N.D. 274 pp.

Oswood, M. W., C. W. Slaughter & J. W. Hilgert, 1984. Effects of vegetation removal in an Alaskan subarctic watershed: preimpact benthic invertebrate studies. In S. R. Bredthauer (Chairman), Alaska's water: a critical resource. Proc. Alaska Section, American Water Resources Association. Inst. Water Resources, Univ. Alaska, Fairbanks. IWR Report 106: 183–193.

Pennak, R. W., 1978. Fresh-water invertebrates of the United States. 2nd Edn. John Wiley and Sons., NY: 803 pp.

Risser, P. G. & C. G. Treworgy, 1986. Overview of ecological research data management. In W. K. Michener (ed.), Research data management in the ecological sciences. University of South Carolina Press, Columbia, S.C.: 9–22.

Scudder, G. G. E., 1979. Present patterns in the fauna and flora of Canada. In H. V. Danks (ed.), Canada and its insect fauna. Memoirs Entomol. Soc. Canada No. 108. Ottawa: 87–179.

Slack, K. V., J. W. Nauman & L. J. Tilley, 1979. Benthic invertebrates in a north-flowing stream and a south-flowing stream, Brooks Range, Alaska. Water Resources Bull. 15: 108–135.

Sokal, R. R. & F. J. Rohlf, 1981. Biometry. 2nd Edn. W. H. Freeman and Co. San Francisco. 859 pp.

Wagener, S. M., 1984. Effects of gold placer mining on stream macroinvertebrates of interior Alaska. M. S. Thesis. University of Alaska, Fairbanks. USA.

Watson, D. G., W. C. Hanson, J. J., Davis & C. E. Cushing, 1966. Limnology of tundra ponds and Ogoturuk Creek. In N. J. Wilimovsky & J. N. Wolfe (eds), Environment of the Cape Thompson Region, Alaska. U.S. Atomic Energy Commision.: 415–435.

Webster, J. R., E. F. Benfield & J. Cairns, Jr., 1979. Model predictions of effects of impoundment on particulate organic matter transport in a river system. In J. V. Ward & J. A. Stanford (eds), The ecology of regulated streams. Plenum Press, NY: 339–364.

Wiggins, G. B. & R. J. Mackay, 1978. Some relationships between systematics and trophic ecology in nearctic aquatic insects, with special reference to Trichoptera. Ecology 59: 1211–1220.

Hydrobiologia 172: 111–127, 1989
W. F. Vincent and J. C. Ellis-Evans (eds) High Latitude Limnology

Variability of macroinvertebrate community composition in an arctic and subarctic stream

Michael C. Miller & James R. Stout[1]
Department of Biological Sciences, University of Cincinnati, ML 06, Cincinnati, Ohio 45221, USA;
[1] *Present address: CECOS International Inc., 4879 Spring Grove Ave, Cincinnati, Ohio 45323, USA*

Key words: Alaska, arctic, benthos, drift, ordination, species richness, species turnover

Abstract

The macroinvertebrate community composition was compared in two Alaskan streams (USA) for numeric and species constancy during the ice-free period from 1981 to 1983. Imnavait Creek is a first order arctic stream (60° 39′ N, 149° 21′ W) draining upland tundra in the foothills of the Brooks Range. Caribou-Poker Creek is a 4th order subarctic stream (65° 08′ N, 147° 28′ W) draining the taiga forest north of Fairbanks, Alaska. The aquatic insect larvae and other macroinvertebrates were sampled with drift nets and Hess bottom samplers for four periods, each 1 week long in the ice free season of three years. We found 112 species in the arctic stream and 138 species in the subarctic stream in a chironomid-dominated community. In any sample period the communities contained 51–60 species in the arctic and 49–92 species in the subarctic. Between the four sample periods on average 39% and 50% of the species were present in two sequential samples in the arctic and subarctic stream, respectively. New immigrants, never before found in the system, averaged 37% and 31% of the community, respectively. These systems are exposed to several intermediate disturbances: prolonged and variable freeze-up, extreme variation in discharge, wide diel and seasonal changes in temperature, and erosion by frazil and anchor ice. The dipterans that compose the most numerous and variable taxa must have variable diapause, ability to grow in cold waters, and good dispersal powers, even migrating across drainages in the arctic. Much of the seasonal dominance pattern appears therefore to be stochastic.

Introduction

In arctic streams compared to temperate streams, we anticipate finding many fewer species of benthic invertebrates (aquatic insects, mollusks, acarina, oligochaetes, etc.) as the climate becomes more severe. Presumably invertebrate species might be reduced by intolerance to prolonged freezing, by their inability to diapause at any stage of the lifecycle and by lack of dispersal mechanisms to colonize depauperate or highly seasonal habitats. In Alaskan streams compared to temperate streams, there is some reduction in insect diversity at the ordinal and familial level and changes in the relative contribution of the fauna by order and family; however, not necessarily any reduction in richness at the species level (Oswood, 1988).

In streams macroinvertebrate diversity may be more directly affected by stream order than latitude (Coffman, unpublished MS; Naiman *et al.*, 1987; Slack *et al.*, 1977, 1979). Intermediate-sized streams have more diverse insect communities than do smaller streams and larger rivers (Brooker & Morris, 1980; Bruns *et al.*, 1982; Culp & Davies, 1982; Minshall *et al.*, 1985;

Naiman *et al.*, 1987; Schaeffer & Perry, 1986; Vannote *et al.*, 1980). Presumably as river size increases so does the number of habitats, the diversity of food sources and the number of colonization sources, all tending to enhance the taxonomic composition of the stream/river. In addition, intermediate sized-streams have a more variable seasonal range of temperature and discharge as compared to smaller streams (Sanford & Ward, 1981). Discharge and temperature variation can act as intermediate disturbances, enhancing species richness by maintaining species density below saturation (Connell, 1978; Pickett & White, 1985). Variation in species composition should be less in small streams with fewer species and with rapid colonization by drift (Fox, 1977).

Increasing trophic richness and severe disturbance can reduce species richness by selecting for the fastest growing, best colonizing species (Fisher *et al.*, 1982). Trophic richness leads to reduced species diversity and increased dominance in many types of systems including grasslands (Tilman, 1982, 1987), in streams (Reger and Kevern, 1981), in plankton of lakes (Tilman, 1982) and in mathematical models (Houston, 1979; Tilman, 1982) However, most rivers studied in the arctic are unproductive (Peterson *et al.*, 1985) and only a few cases have invertebrate densities comparable to those in temperate streams (Craig & McCart, 1974; Cowan & Oswood, 1984).

The role of marked seasonal disturbance and patch dynamics have received much attention in stream insect communities in maintaining species richness (McAuliffe, 1983, 1984; Sousa, 1984; Stout & Vandermeer, 1975). Natural disturbances in arctic streams which may open up local patches or generate system wide disruption include summer flooding, winter freezing and spring thawing, moving ice and rapidly fluctuating temperatures. Summer flooding loosens attached algae and invertebrates by sheer stress, erodes attached forms with suspended detritus or sand grains blasting the upstream surface of rocks, rolls over rocks dislodging insect larvae, and erodes large blocks of terrestrial peat from the side and bottom of the stream. Early or late winter freeze-up may catch physiologically unhardened species in a vulnerable stage. Ice blocks released by the spring thaw, especially in the subarctic may dislodge rocks. Frazil ice and anchor ice formed nightly in the early winter just before freeze-up effectively devastate any attached forms on the surface of the rocks. As the ice crystals collect on the rocks at night or as the anchor ice is released in large pieces with some diurnal warming, the surface-dwelling organisms are vulnerable to physical damage.

The frequency of intermediate disturbances interacts with stream order. Statzner & Higler (1985, 1986) contend that the compositional diversity is highest in streams at the first and second transition zone of hydraulic stress or where there is inconsistancy in this stess because of variance in flow or channel characteristics. The joining of larger order tributary streams in mid-reaches that make these systems more diverse and perhaps in a state of flux (Minshall *et al.*, 1985). High and low discharge events change the area of the wetted perimeter and available habitat of the river many times seasonally, and wide variation in temperature both daily and seasonally have been interpreted as being disturbances to many species with aquatic life cycle stages (Stanford & Ward, 1983). Many of these disturbances are predictable in a stochastic sense over years of observation. Thus globally, the life cycles, diapause and dispersal abilities of the macroinvertebrates have been selected for their ability to endure these disturbances. However, at the habitat level, changes in abundance and composition might be extreme and resultant community composition appear stochastic (Riece, 1981, 1984). It is variation in small stretches (25–100 m) of stream over parts of several years that we will address here.

Climate interacts with life history selecting for stream insect species that can be frozen and survive, that can remain active at cold temperatures, and that retain high dispersal characters or mechanisms to ensure local recolonization. In contrast to these expectations in stream communities, Butler *et al.* (1980) found that in permanent tundra pond systems on the coastal plain of

Alaska some sibling species of chironomids (Diptera) have lost their ability to fly to disperse or even to recognize their own mates except by the synchrony of their emergence. The emergence after pupation is apparently set by the days since ice thaw in the summer usually, June 7–19 at 71° N (Miller *et al.*, 1980). Life spans ranged from 1–7 years for chironomid species in fishless, coastal tundra ponds, where risk of loss between years was low. In Toolik Lake which lies in the same watershed as our arctic stream, the abundance of chironomids in the benthos varied dramatically between winter and spring (S. Mozley, N.C.S.U. personal communication) with life spans of 1 to 4 years, depending upon species (Hershey, 1985). In the Kuparuk River (3 km to the West), life histories of macroinvertebrates are almost all univoltine with only a large *Brachycentrus* (Trichoptera) having a two year life cycle (Hershey, personal communication; Peterson *et al.*, 1985). Hence selection pressures on life cycle length are different in lakes than in streams based on predictability and freedom from severe disturbances. In smaller, high-gradient streams, especially those in the mountains of the Brooks Range, freeze up causes flow to stop and streams remain dry all winter. Spring snow melt remakes stream channels moving boulders in the process. The cold and severity of climate should select for species that colonize and develop annually.

The present study is one of few with extensive sampling in an arctic stream and a subarctic stream over parts of several years (Frey, Muller & Berry, 1970). We examined the seasonal variability in macroinvertebrates in both an arctic and subarctic stream. Large samples were used to be sure of collecting moderately rare species. The ice free season in the arctic and subarctic is short, so that almost any species that will exist in the community will have to be present and growing during most of the flowing water portion of the year. Following pupation and before young of the year reach sufficient size to sample with our methods, they may appear to become quite rare. Drift nets, bottom grabs, and rock basket samples were used to sample several microhabitats.

Methods

Macroinvertebrates were collected both early and late in each thaw season over a week of sampling by taking drift samples, bottom samples (Hess samples) and colonized rocks in baskets. Both a first order arctic and fourth order subarctic stream were sampled over parts of three years. Drift nets were either collected once after an hour and then after 24 hours over several days, or were collected twice daily after 12 hours of day or night exposure. Discharge through each net was measured at the beginning and at the end of each collecting period with a current meter (General Oceanics Model 2035-MK111). Discharge for the streams was determined at the same time using cross-sectional area and discharge at three points across channel (Wetzel and Likens, 1979). Drift nets were either 10 × 30 cm opening with 1 meter-long net of 363 um mesh Nitex nylon or 30 × 45 cm Wildco drift nets (Saginaw, MI, model 15-C20). Eight bottom samples were taken at each time using the Hess sampler containing 0.086 m^2 (Wildco, Sagninaw Michigan). In most cases samples, spread in white enamel pans, were sorted for live invertebrates and preserved in 70% ethanol. In a few cases they were floated off with sucrose (30% by weight), cleaned of detritus and preserved. In the latter case we picked the residue for 5 minutes to check for missed individuals.

Rock baskets (35 × 35 × 5 cm) constructed of hardware cloth were filled with 18 scrubbed, dried, flat rocks ranging in size from 66–960 g and having 62–171 cm^2 surface area on their largest flat side. Twelve baskets were set in the river in riffles along the sample reaches. The rocks were harvested in the next season by setting the recovered basket in a plastic pan underwater, scrubbing the rocks and collecting the insects with a 100 μm sieve.

All larger macroinvertebrates were enumerated and identified under dissecting microscope (Wild M5 or M8). In larger samples, the chironomids were split using a Folsom Plankton Splitter (WILDCO, Saginaw, MI) to reduce the number counted to between 75 and 200. The chironomids were mounted in CMCP mounting medium

(Masters Chem. Co., Arlington Hts. Illinois) according to the method of Beckett & Lewis (1982). All identifications used published works taking them to the lowest taxon using an Olympus BH-2 compound microscope (see Miller *et al.*, 1986). Unidentifiable morphospecies were drawn in detail, given species designations (eg. 1, 2, 3 etc.) and cataloged for reference. However, each morphospecies identification was used consistently between samples at other times and in the other stream. Data were entered in LOTUS 123, edited, and sent to AMDAHL 370 mainframe for multivariate analysis by DECORANA (*De*trended *Cor*respondence *Ana*lysis) (Gaugh, 1982) in order to examine the differences between sampled communities among seasons. DECORANA ordinates seasonal samples in species space with minimal overlap in common causality between the first and second ordinate. For DECORANA, we used the numeric abundance of species in drift and Hess samples, deleting species with <2 occurrences in order to minimize the impact of very rare species. The coefficient of community, Sorenson's Index (CC), was used to compare the similarity in species composition based on presence and absence in paired sample communities, where $CC = 2c/(a + b)$ (where c is the number of species in common between two samples, and a and b are the number of species in each of the two communities) (Whittacker, 1973).

The suspended chlorophyll was quantified by filtering water samples on glass fiber filters, extracting it in 90% acetone, and reading absorbance or fluorescence on a Perkin Elmer ratio recording spectrophotometer or Turner 111 fluorometer, respectively (Wetzel and Likens, 1979). Epilithic chlorophyll was collected with a 5.2 cm diameter rubber-ended core tube through which the attached material was dislodged with a rotary brush and collected with a large suction pipette several times. Coarse particulate matter was measured by the dry weight of organic debris caught in the drift nets (>363 μm) divided by the volume of water (m^3) that passed through the net in the collection interval.

The arctic stream, we call Imnavait Creek, is an eastern fork of the Kuparuk River in the northern foothills of the Brooks Range, Alaska (68° 39′ N, 149° 21′ W) at 823 m elevation. The subarctic stream, called Caribou-Poker Creek is a tributary to the Chantanika River formed by the confluence of Poker Creek and Caribou Creek, North of Fairbanks, Alaska (65° 08′ N, 147° 28′ W) in the Caribou-Poker Creek Experimental Ecological Preserve operated by University of Alaska, Institute of Northern Forests and US Dept. of Forestry. The former drains upland tundra vegetation underlain by continuous permafrost; the latter drains taiga forest of black spruce and aspen/birch/alder stands underlain by discontinuous permafrost. Hence the arctic stream is open to direct and indirect radiation 24 hours per day during the arctic summer. Thus despite being a primary stream, it is not shaded as in the classic river continuum hypothesis (Vannote *et al.*, 1980). The fourth order taiga stream is shaded much of the day over most of its length by the spruce, alder, aspen, and birch trees that line its banks. These trees contribute to the high concentration of coarse particulate organic matter collected here (Table 1). The arctic stream drained 4.3 km^2 above our sampling reach, collecting its water from parallel ridges (960 m elevation) on each side, falling 110 m in 4.2 km to our stations. It joined the Kuparuk River 7 km down stream, never receiving drainage from another permanent stream. The subarctic stream drained 124 km^2 from hills reaching 800 m, falling 580 m to the Chantanika River (Slaughter, 1971) (Fig. 1).

The two streams were each sampled extensively a total of four times, over the ice-free parts of three years. Imnavait Creek was sampled in the autumn 1981 (17–24 Sept.) as it froze, in the spring 1982 (3–9 June) just as it melted, in autumn 1982 (24–28 August), and in early summer 1983 (13–18 June). Caribou-Poker Creek was sampled in autumn 1981 (7 Oct–20 Nov.) as it froze, in spring 1982 (17–31 May & 12 June) during the spring melt, in autumn 1982 (30 Aug.–3 Sept.) and in midsummer 1983 (23 June–20 July).

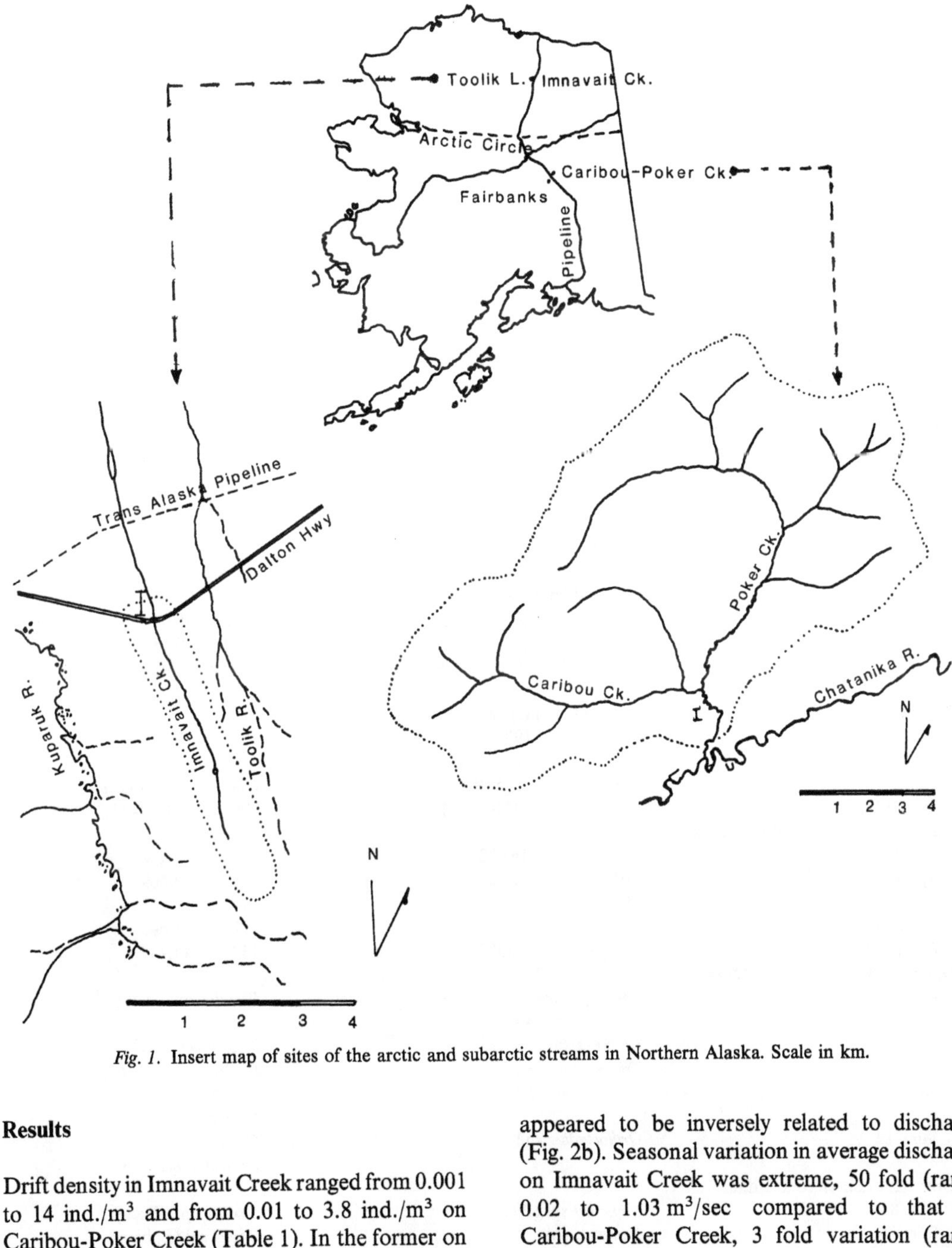

Fig. 1. Insert map of sites of the arctic and subarctic streams in Northern Alaska. Scale in km.

Results

Drift density in Imnavait Creek ranged from 0.001 to 14 ind./m^3 and from 0.01 to 3.8 ind./m^3 on Caribou-Poker Creek (Table 1). In the former on three of four sample dates the drift density appeared to decline with increased discharge (Fig. 2a). In the latter on all dates, drift density appeared to be inversely related to discharge (Fig. 2b). Seasonal variation in average discharge on Imnavait Creek was extreme, 50 fold (range 0.02 to 1.03 m^3/sec compared to that in Caribou-Poker Creek, 3 fold variation (range 0.66–1.82 m^3/sec) (Table 1). In the arctic stream, summer discharges comparable to the spring snow melt have been noted in several years (1983,

Table 1. Physical, chemical, and biological differences between Imnavait Creek and Poker-Caribou Creek, Alaska.

Parameter	Imnavait Ck.		Caribou-Caribou Ck	
Average Discharge (m^3/sec)				
Aut. 81	0.02		0.86	
Spr. 82	1.03		1.82	
Aut. 82	0.04		0.66	
Sum. 83	0.04		1.11	
Average Velocity (m/sec)				
Aut. 81	0.34		0.17	
Spr. 82	0.49		0.40	
Aut. 82	0.42		0.28	
Sum. 83	0.37		0.23	
Suspended Chlorophyll (mg/m^3)	2.69		9.44	
Coarse Particulate Organic Matter				
CPOM (mg/m^3)	9.4		393.8	
CPOM/Chl. a Ratio				
(mg CPOM/mg chl)	21.2		48.4	
Epilithic Chlorophyll (μg/cm^2)				
Aut. 81	0.34		1.46	
Spr. 82	3.48		0.25	
Aut. 82	1.04		N.A.	
Sum. 83	0.45		0.81	
Drift Density-	avg. (ind./m^3)	(range)		
Aut. 81	3.5	(0.21–13.1)	0.77	(0.03–1.9)
Spr. 82	0.06	(0.004–0.27)	0.94	(0.16–3.8)
Aut. 82	0.02	(0.001–0.034)	0.26	(0.07–0.7)
Sum. 83	0.79	(0.011–3.7)	0.76	(0.01–2.6)
Benthic Density	avg. (ind./m^3)	(range)		
Aut. 81	917	(93–2174)	394	(70–593)
Spr. 82	334	(23–1401)	626	(397–1238)
Aut. 82	489	(245–946)	654	(315–1495)
Sum. 83	791	(432–1378)	799	(304–1413)
Number of Species and individuals counted				
Aut. 81	60	16,932	62	5,166
Spr. 82	50	2,032	76	3,569
Aut. 82	51	1,038	49	767
Sum. 83	53	2,335	92	5,643
Composite Total	112	22,337	138	15,235
Dominant Chironomidae (Diptera)				
Aut. 81	*Phaenopsectra* spl *Trichotanypus* sp.		*Pseudodiamesa* spl	
Spr. 82	*Diamesa* sp. *Trichotanypus* sp.		*Micropsectra* sp3 *Heterotrissocladius* sp.	
Aut. 82	*Orthocladius* sp2		*Cricotopus* spl *Micropsectra* spl	
Sum. 83	*Orthocladius* sp2		*Cricotopus* spl *Heterotrissocladius*	
Dominant non-chironomidae				
Aut. 81	*Nemoura* spl (Plecop.)		*Skwala* sp. (Plecop.)	
Spr. 82	*Nemoura* spl		*Baetis* spl (Ephem.)	
Aut. 82	*Nemoura* spl		Lumbriculidae (Oligoch.)	
Sum. 83	*Nemoura* spl *Prosimulium* (Diptera)		*Cynigmula* spl (Ephem.)	

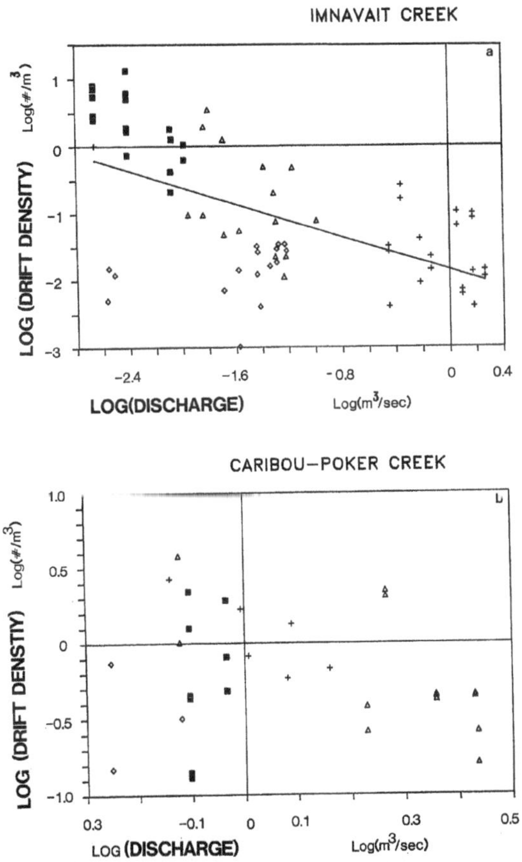

Fig. 2. a) Drift density as a function of discharge of four seasonal samples in Imnavait Creek as a double log plot. Squares = aut. 81; pluses = spr. 82; diamonds = aut. 82; and triangles = sum. 83. Line is fit by least squares, $r^2 = 31.8\%$ of variation, df = 69, $p < 0.05$. b) Drift density as a function of discharge of four seasonal samples in Caribou-Poker Creek as a double log plot. Symbols as above.

1984, 1986 unpublished data). Hence the scour and potential damage done by spring melt and summer precipitation runoff might be considerably different in the two systems. Much of the pulsed snow melt in the subarctic stream flowed over the ice long before the river ice thawed, thereby protecting the benthos in the main channel from the full force of the spring melt water. Similarly, in the arctic stream the snow melt water flowed over the surface of the ground-fast ice eroding rapidly through the ice but minimizing damage to the benthos over those first few critical days of high discharge (Kane & Carlson, 1973).

Density in Hess samples averaged from 334 to 917 ind./m^2 in Imnavait Creek and from 349 to 799 ind./m^2 in Caribou-Poker Creek. The substrate in Imnavait Creek was cobble to gravel, often mixed with large amounts of organic matter from decayed terrestrial peat. The substrate in Caribou-Poker creek was cobble and larger, with sand-sized particles at depth under the rocks in our reach. Coarse organic matter, which was recent detritus and litter fallen from the riparian zone, was 40 X as great as in the arctic stream (Table 1). Epilithic algae (as measured as chlorophyll) as potential food for grazers was higher in the more open arctic stream on most dates (Table 1).

The macroinvertebrate species collected in each of the seasonal benthic samples in Imnavait Creek were 37, 22, 20 and 23 in aut. '81, spr. '82, aut. '82, and sum. '83, respectively (Table 2). In Caribou-Poker Creek there were 36, 39, 28 and 49 species collected in those seasons, respectively. In both systems the Hess samples contained fewer species than found in the drift (avg. = 65.9% and 65.7% in Imnavait and C-P Creeks, respectively). The average coefficient of community between arrays of species in the benthic and drift samples was 45% in Imnavait Creek and 64% in C-P Creek over the four seasons (Table 2). Apparently the drift in the northern stream was derived less from the benthos than in the southern stream.

Rock baskets incubated over a season contained many fewer species than other samples. The scrubbed rocks obviously developed little periphyton during over-winter exposures in darkness. In the arctic stream the baskets were frozen into solid ice 8 months from when they were set in place til spring thaw. In Imnavait Creek the baskets were colonized by 6, 13, and 18 species over the periods from late aut.-spr., spr.-sum. and early aut.-early sum. incubations, respectively. The coefficient of community of rock basket fauna compared to their source community in Hess samples was 21.4%, 42.4%, and 48.7% over the same incubation periods. In the subarctic stream the rocks baskets were subject to several

Table 2. Sorenson's coefficient of community between benthic and drift invertebrate communities in Imnavait Creek and Caribou-Poker Creek over four seasons.

Creek	Aut. 81 benthos	Aut. 81 drift	Spr. 82 benthos	Spr. 82 drift	Aut. 82 benthos	Aut. 82 drift	Sum. 83 benthos	Sum. 83 drift
Imnavait Ck.								
no. spp.	37	40	22	39	20	32	23	44
no. in common	19		16		8		16	
Coeff. Community	0.49		0.52		0.31		0.48	
					AVERAGE =		0.45	
Caribou-Poker Ck.								
no. spp.	36	52	39	64	28	38	49	83
no. in common	26		31		23		42	
Coeff. Community	0.59		0.60		0.70		0.64	
					AVERAGE =		0.63	

episodes of being covered with anchor ice and bombarded with crystals of frazil ice at freeze-up over the two winters. Anchor ice forms under a clear cold winter sky when longwave radiation cools the rocks under water sufficiently that ice crytals can form attached to the rocks. During the day and under a permanent ice cover, the anchor ice thaws, often breaking loose in large clumps moving down river. Hence the colonization of rock baskets set in the winter before permanent ice cover seems unlikely. In C-P Creek, the baskets were colonized by 26, 31, and 35 species over the aut.-spr., spr.-aut, and aut.-sum. intervals, respectively. The coefficient of community between species in rock baskets and in benthos was 55.4%, 69.6%, and 59.5% over those intervals, respectively. In both streams colonization was highest over the summer period.

Summing all sample types over the four seasonal samples the most numerous orders in Imnavait Creek were Diptera (Chironomidae 30–70%), Plecoptera (Nemouridae 5–47%) and Coleoperta (Helphoridae 0.5–2%), with only 3 other insect orders being present. In Caribou-Poker Creek the most common were the Diptera (Simulidae and Chironomidae 37–68%), Ephemeroptera (0.6–57%) or Plecoptera (Perlodidae 0.6–24%). Minor contributions of three other insect orders (Trichoptera (0.5–6%), Coleoptera (0.05–10%) and Hemiptera (0–.8%)) and four other non-insect orders (Stout, 1986).

Community composition, particularly dominance and density, might reveal the potential for a biological competition for space (Bruns and Minshall, 1983). Using the collected set of species from each seasonal sampling, dominance diversity curves for each stream all appeared to be log normal as expected (log(# ind./species) vs ranked species). The low density and the lack of dominance may indicate the inability of any single species to exclude others by competition. But the lack of dominance does not eliminate the possibility that pairwise species interactions do occur that affect species abundances and distribution (Bruns & Minshall, 1983).

Changes in community composition by season were visualized by using detrended correspondence analysis (DECORANA, Gaugh, 1982)

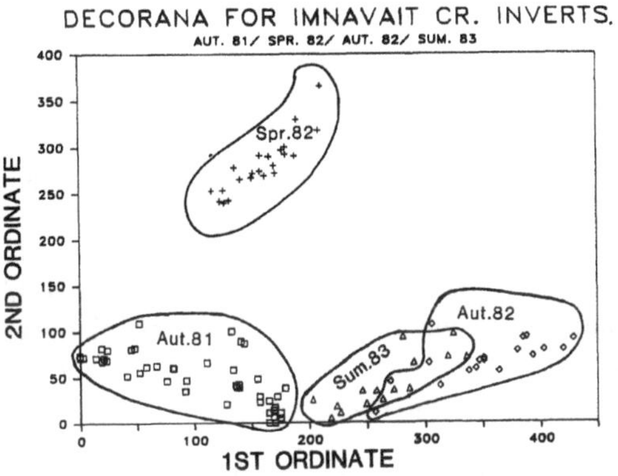

Fig. 3. DECORANA separation of seasonal samples using species composition data from Imnavait Creek. Species with <2 occurences were deleted. The line enclosing each set of seasonal samples is hand drawn to include all samples from that season while minimizing overlap with other seasons.

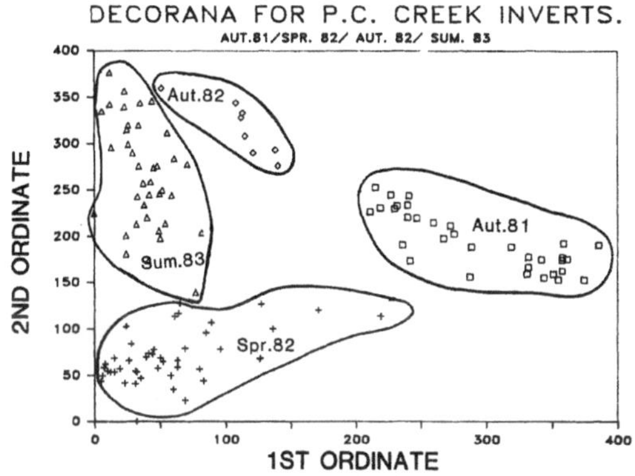

Fig. 4. DECORANA separation of seasonal samples using species composition data from Caribou-Poker Creek. See explanation from Fig. 3.

using individual drift and bottom samples from each site for each season. Using this ordination technique we can place the communities on two primary ordinate axes in a location determined by species composition, where their distance from the coordinate center are standard deviation units. We might have expected the distribution of samples to overlie one another, their being from the same stream. However in Imnavait Creek three non overlapping groups were detected (Fig. 3): Autumn 1981, Spring 1982, Autumn 1982-Summer 1983. In Caribou-Poker Creek none of the four seasonal samples overlapped at all (Fig. 4). Similarly, complete separation was found in the benthic community (Hess samples) of C-P Creek over the four seasonal samples (not shown). In Imnavait Creek benthic samples, three groupings separated with overlap between Autumn '82 and Spring '83 benthic samples as found in drift samples.

We examined the numbers of species that were lost or not sampled between successive seasonal samples in each stream using a composite of drift, benthic and rock basket collections. In Imnavait Creek the number of morphospecies enumerated for the seasons of autumn 1981, spring 1982, autumn 1982 and summer 1983 were 60, 50, 51, and 53, respectively, from a cumulative pool of 112 species. Each seasonal sample amounted to 54%, 45%, 45% and 47% of the species pool, respectively. In Caribou-Poker Creek 62, 76, 49 and 92 species were found over the seasons, respectively, representing 45%, 55%, 36% and 67% of the cumulative species pool of 132 species (Table 1).

For our analysis, the terms local extinctions or emigrations are used to designate morphospecies which were not found in successive seasonal samples. Similarly, immigrants are those species that appear or reappear in a seasonal sample, not in the previous sample. For Imnavait Creek, of the 60 initial species in autumn 1981, 23 species (38%) persisted overwinter to the spring of 1982. Therefore, 37 species (62%) became locally extinct (Fig. 5). In the spring of 1982, 27 novel species immigrated (54% of the community). By autumn 1982 only 16 (27%) of the initial pool persisted, but 24 (48%) of the spring species pool persisted til that autumn. In the autumn of '82 the community was colonized by 15 novel species (29% of season total) and recolonized by 12 species (23%) that were last seen in Autumn 1981, a year earlier. These 12 recolonizing species might be the near the limit of those species with annual life cycles recolonizing the stream and growing into a collectable size classes only in the autumn of the year. In the summer of 1983, of the 53 species present only 9 (15%) were present

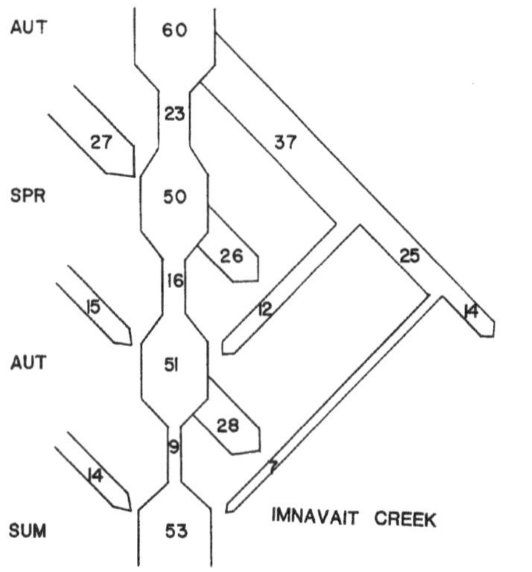

Fig. 5. Emigration, immigration and recolonization of invertebrate species in Imnavait Creek. The area of each block and the width of each arrow are proportional to the number of species sampled by all methods in each season. The vertical connecting tube is the number of species from the autumn 1981 community that were carried over from season to season. The arrows away from the blocks are those species that were not found in the subsequent season. The arrows entering blocks from the left are new immigrants found for the first time in the study. The arrows entering the blocks from the right are species which went locally extinct after autumn 1981, but recolonized the community in a third or fourth season.

continuously from autumn 1981. The community in summer '83 was colonized by 14 species never found in any sample and 7 species that disappeared in 1981. By summer '83, 4 of the 12 species that recolonized in autumn 1982 persisted, 7 recolonized from the original community and 9 persisted throughout. Thus, 20 species (33%) of the original 60 species were present two years later. In the three seasonal transfers 37/60 (62%), 26/50 (52%), and 28/51 (55%) went locally extinct or became so rare that they were not sampled. Of the original 60 species in autumn 1981, 38% were present the following season, 47% one year later, and 33% two years later, an average carryover of 39% from season to season. Immigration of novel species, not collected in any previous sample, composed an average of 37% of the season's species pool.

For Caribou-Poker Creek, of the 62 initial species in autumn '81, 36 species (58%) persisted to the spring of 1982 (Fig. 6). Therefore, 26 species (42%) became locally extinct by spring. In spring '82, 40 novel species colonized the stream (53% of the spring community). By autumn '82 only 23 (37%) of the initial pool persisted, but 41 (54%) of the spring community persisted til autumn. In autumn 1982 the community was colonized by 9 new species (18% of autumn community) and recolonized by 5 (10%) that were last collected in autumn '81. In summer '83, of the 92 species present only 23 (25%) were present continuously from autumn '81. The community in summer '83 was colonized by 21 novel species and 4 species that disappeared in autumn '81. Only 2 of the 5 species recolonizing in autumn '82 persisted til summer '83. Thus 29 species (47%) of the original 62 species were present in the sampled community two years later. In the three successive sample periods 26/42 (42%), 41/76 (54%), 9/49 (18%) of the community became locally extinct or were not sampled (avg. 38%). Of the original 62 species present in autumn '81, 58% were present in the following season, 45% were present one year later, and 47% were present two years later, an average

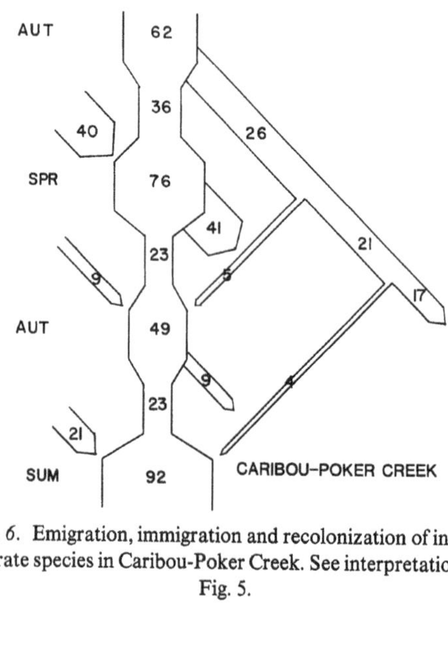

Fig. 6. Emigration, immigration and recolonization of invertebrate species in Caribou-Poker Creek. See interpretation in Fig. 5.

carryover of 50%. Immigration of novel species not collected in any previous sample season averaged 31%.

The dominant species changed in 3 of 4 seasonal samples in both streams (Table 1). The dominant chironomidae in Imnavait Creek changed from the red *Phaenopsectra* sp. 1 in autumn '81 to *Trichotanypus posticalis* and *Zalutschia pusa* in spring '82. In autumn '82 and summer '83, *Orthocladius* sp2 and *Micropsectra* sp. 2, respectively, were most common. The dominant chironomidae in Caribou-Poker Creek changed from *Pseudodiamesa* sp. 1 to *Microspectra* sp. 3 & *Hetertrissocladius* sp. 2 & 3, to *Cricotopus* sp. 1 and to *Cricotopus* sp. 1 & *Heterotrissocladius* sp. 3 over the same 4 periods, respectively. In Imnavait Creek the non-chironomid fauna was dominated numerically by *Neumora arcticus* (Plecoptera) in all four seasonal samples, although their numbers dropped almost 1.5 orders of magnitude over the winter of 81–82 (Stout 86). In Caribou-Poker Creek, *Skwala* sp. 1 (Plecoptera) and *Prosimulium* sp. H (Diptera); *Baetis* sp. 1 (Ephemeroptera), Lumbriculidae (Annelida) and *Cynigmula* sp. 1 (Ephemeroptera) were the non-chironomid dominants over the four seasons, respectively. (Table 1).

On all dates the chironomids were the most diverse family, often with small numbers of sampled individuals. Our expectation was that rare species would be those that were lost and replaced most frequently as predicted by the equilibrium theory of island biogeography (MacArthur & Wilson, 1967). In theory, small populations localized in specialized habitats are at risk of stochastic catastrophes causing extinction. This was not the case, because the distribution of extant species ranked by their abundance was not significantly different than the same distribution of species that disappeared between two sequential seasonal samples. A Kolmogorov-Smirnov statistic of maximum deviation for testing goodness of fit of distributions was applied to compare the ranked numerical abundance of species in each seasonal community and the ranked numerical abundance of those going extinct before the next sampling period (Bradley 1968, Sokol & Rohlf 1969). For Imnavait Creek the K statistic was not significant for any sampling interval (Aut. '81 × extinctions aut. '81-spr. '82 $K_{df\ 9} = 0.09$, NS; Spr.'82 × extinctions spr. '82-aut. '82 $K_{df\ 9} = 0.06$, NS; Aut. '82 × extinctions aut. '82-sum. '83 $K_{df\ 9} = 0.23$, NS). In Caribou-Poker Creek the Kolmogorov-Smirnov test statistic 'K' similarly showed that there was no difference between abundance frequency distribution of the species pool and those that were to go locally extinct over three seasons (aut. '81 × extinctions aut. '81-spr. '82 $K_{df\ 9} = .11$, NS; spr. '82 × extinctions spr. '82-aut. '82 $K_{df\ 9} = .13$, NS; aut. '82 × spr. '83 $K_{df\ 9} = .23$, NS). Thus based upon local abundance in samples, the probability of going extinct or being unsampled in the next season was nearly constant regardless of species abundance.

For example, in autumn 1981 *Neumora* sp. 1 (Plecoptera), probably *arcticus*, was the numeric dominant represented by late instars that drifted in large numbers as the stream froze (Miller *et al.*, 1986). By spring '82 that species had all but disappeared. It reappeared as a common species on all other dates. These larvae were frozen under the ice before we left in winter, 1981 and could not have emerged before the stream melted to the bottom in early June, while we were sampling. When they emerged, if they did, remains a mystery. Of the 34 local extinctions over the winter of '81–'82, 27 were dipterans, only 10 of which were represented by 1 or 2 individuals in our composite samples. Over the summer of '82 13 of 27 species which disappeared were dipterans. Over the winter of '82–'83, 15 of the 22 species which disappeared were dipterans. In Caribou-Poker Creek the most common taxa going extinct were dipterans as well (20 of 29, 40 of 48 and 7 of 10 species over winter '81–'82, summer of '82, and winter of '82–'83, respectively).

Discussion

Four hypotheses are considered viable for explaining the compositional diversity of fauna in arctic streams; 1) duration and predictability of ice-free season; 2) habitat heterogeneity of stream course; 3) stream order; 4) disturbance-frequency. Flow stability was first proposed as an explanation for the variation in invertebrate composition and biomass by McCart and Bain (1972) in surveying streams/river across the north slope of the Brooks Range. Among streams sampled by Surber samplers, three which were spring-fed had a mean of 30 800 invertebrates/m^2 as compared to 1 950 ind./m^2 in four foothills streams and 491 ind./m^2 in five montane rivers. A spring fed stream, called Echooka Spring, had more than five times the invertebrate density (avg. = 84 000 ind./m^2) and biomass (avg. = 43 g/m^2). Over the ice-free summer, the foothills streams were the warmer than montane streams and both were warmer than spring-fed streams. Moreover, the variation in temperature between dates decreased in the same order. The discharge of Echooka Spring was constant throughout the winter (1.13–1.54 m^3/sec) while the others froze solid, first the montane and then the foothills streams. The number of taxa identified by order and family only were similar in spring and foothills streams, both being more diverse than montane streams. The most numerous family in their streams were the midges (Chironomidae, Diptera), which were also the most diverse taxonomically in our surveys and in the review by Oswood (1988). Constancy of flow and water availability into the winter are important criteria for enhancing density and species richness in arctic streams. In our study, the arctic stream froze solid early (Sept–Oct.) and melted late (mid May-early June); while the subarctic stream froze late (Nov.) and melted earlier (mid-April), maintaining flow all winter from groundwater seepage in an area of discontinuous permafrost. In fact, winter stoneflies emerged under the ice in the south while climatic conditions were still rigorous.

Slack *et al.* (1977, 1979) found that species richness increased in third to fourth order, mid-elevation streams in surveys of invertebrates on the North Slope. Analysis of their data show that the number of species appeared to be inversely related to the log of the slope of the river bed, but the slope of the river bed was strongly correlated with stream order. Hence middle elevation rivers are mid-order rivers and high elevation streams are low order streams in the Brooks Range. In temperate systems, species richness is often highest in mid-order rivers (order 5–8) where maximal range in discharge and temperature is also found (Stanford & Ward, 1983; Vannote *et al.*, 1980). In arctic rivers, McCart *et al.* (1972) found the highest variance in temperature occured in mid-order rivers. Lower order streams maintained a more stable temperature regime because their source water pecolated over the permafrost. Of the two streams in the present study, variation in temperature was greater in our fourth order subarctic river (Table 1) and the variation in flow was greater in our first order stream.

Drift of invertebrates in our arctic stream ranged 200 fold (0.017–3.51 ind./m^3) between seasons, but only 3.5 fold (0.265–0.94 ind./m^3) in the subarctic stream (Table 1). The drift density was inversely related to discharge on three of our four seasonal samples (Figs. 2a & 2b). Drift density may have been directly related to benthic density in Imnavait Creek ($r = 0.83$, $df = 2$, NS), but almost constant, unrelated to benthic density in Caribou-Poker Creek with only four dates. In a series of streams in the Tanana River drainage (near our subarctic stream), LaPerriere (1982) reported an inverse relationship between stream size and drift in summer samples. Stream gradients in her sample ranged from 1–14 m/km, within the optimal range for diversity in McCart *et al.* (1973), 7–23 m/km. Drift densities might be related to the benthic densities, which tend to be greater in lower order streams. Craig and McCart (1975) noticed an apparent increase of benthic density in small rivers in their survey of rivers reaching the Beaufort Sea, between Sagavonirktok River and the Mackenzie River to the east. Such a trend was found on the Dietrich River by stream order, but not in the Atigun River series in their 1971 study (McCart *et al.*, 1973). Assuming that mid to low flow invertebrate drift is behavioral and not catastrophic, then invertebrate drift is probably related to density of the benthos.

LaPerriere (1983) believed that invertebrates were released or dislodged per unit area from the bottom at a constant rate, so that drift density was an inverse function of discharge in subarctic streams near Caribou-Poker Creek. We feel that the apparent increase in density in invertebrates in low order rivers in the arctic and the close compositional similarity between drift and benthos (Table 2) means that drift density will be greater in her smaller streams under near steady state conditions in mid summer.

Species richness has long been associated with the number and variety of microhabitats. The mid-order streams potentially have more diversity of habitat if one does not consider the cumulative habitat of inflowing streams (Minshall *et al.*, 1985). However, our arctic tundra stream was a beaded stream composed of melted ice-wedge ponds connected by narrow rocky channels in our reach and by peat-covered channels above our study section. Thus, there were pond specialists and stream specialists supported in the same primary stream. The dominant zooplankter in the drift (1984–1985) was *Eurycercus lamellatus* (Cladocera), not normally a stream invertebrate at all. This large chydorid probably lives feeding on epiphytes attached to the grass *Arctophila fulva* and the sedges growing in the margins of the beads. Depth of beaded pools could be 2 meters at low flow and diameter could be up to 20 meters across. Thus habitats in beaded ponds with soft organic bottoms and on rooted aquatic plants and in the rocky and peat-lined riffles represented high habitat heterogenity in our arctic stream. At high flow the water might be 1 meter over the tops of the beaded ponds and the channel 10–15 meters wide, flooding over the riparian vegetation. Substrates in the stream included terrestrial eroding peat, organic debris, sand, gravel, cobble, and huge glacial boulders exposed in the channel. Our subarctic taiga-forest stream was more conventional in our reach, composed of riffles and pools below the convergence of Caribou and Poker Creeks. Some pools included debris dams and quiescent sections that allowed organic debris to build up. Below our study sections a new beaver dams changed the morphology of the lower stream dramatically during the study period (Miller *et al.*, 1986). All substrata were cobble dominated material which the stream had deposited at some time. The apparently, large species pool in the arctic stream may be partially explained by unusual habitat diversity of its beaded morphology.

Since species-area curves predict the inclusion of more and more species with sampling more habitats and more guilds, we could question what is the appropriate sample size for comparing species richness. Within Alaska on the Atigun and Dietrich Rivers, 20 to 100 km south of our arctic stream, the species richness varied from 14 to 26 taxa at a given station. Collectively that was 49 species in each river and 68 taxa from both rivers over a two day collecting period using 10 rock samples, drift nets, and dip net (Slack *et al.*, 1979). In all cases more species were collected with dip net than with drift net or 10 rock samples (Slack *et al.*, 1976) with 4648 and 3813 individuals enumerated in each system, respectively. In any given seasonal sample at four adjacent stations we found between 51–60 and 49–92 morphospecies with enumeration of 1000–16900 and 760–5640 individuals in the arctic and subarctic stream, respectively (Table 1). Our finding of 112 and 138 taxa in the arctic and subarctic streams, respectively, was the cumulative enumeration of 22000 and 15000 individuals over two years. By inference only, the surveys of Slack *et al.* (1976, 1979) and Nauman & Kernodle (1975) would appear to fit the same relationship to new species added per 1000 individuals counted, although one survey was taken over seasons and the other over several stations up a river on a single date ($\log(\text{spp.}) = .44 * \log(\text{ind.}) + 0.07$, $r^2 = 0.79$, $df = 47$). Thus the species richness/individual were comparable in all surveys, at least on a double log plot, such that the apparent differences in species richness in these surveys in the same general areas of Alaska are different because of the increased number of specimens enumerated.

There was a high turnover of species in our samples in both streams. If this were not a real phenomenon, the obvious reasons for this apparent species replacement in samples might include inconsistent taxonomy, unequal sample sizes between dates and an inability to capture all life history stages with standard survey techniques. We attempted to minimize the mis-identifications by cataloging. Sample sizes did vary by season as one might have expected from 1000 to 16000 in the arctic and 760 to 5600 in the subarctic (Table 1). Therefore, number of species identified could be, in part, a function of sample size. Our samples were larger than most others from North Slope collections. However, the source of variation in the species pool is the most puzzling finding of this study to us. Turnover in rare species might be an artifact expected in variable sample sizes; however, the replacement was found both in common and rare species in proportion to their abundance in the sample (see above). Lenat (1983) observed a 50% seasonal variation in chironomid species in an Oregon river. In a stream habitat stressed by discharge variation and sediment load, Winget (1984) found that the seasonality and abundance of *Brachycentrus* sp. (Trichoptera) varied widely, disappearing for an entire year or for multiple years in samples from any given month.

Could local populations be effectively unsampled because much of the population in a patch is destroyed? Freezing for long periods or variance in the timing and permanence of the freezing period was the biggest stress inferred from the regional data of McCart & Bain (1972). Olsson (1981) reported large mortalities of most macroinvertebrates (except chironomids) from experimental freezing at 4 °C for five months. However, when buried and frozen naturally in the river bed, most frozen macroinvertebrates revived, with chironomids surviving in greatest proportion (86–97%). Therefore, it is plausible that the large changes in numerical abundance of macroinvertebrate species were caused by the more severe and longer periods of freezing in Imnavait Creek as compared to Caribou-Poker Creek. The variation in time of freeze-up varied by 1.5 months between 1981 (second week of Sept.) and 1982 (last week of October). The variance in the spring melt is also variable from mid-May in most years to as late as June 1 in 1982. Instances of autumn freeze beginning and then thawing have been observed in the arctic. The larger size of the subarctic stream and the seepage water enhanced the constancy and timing between years of freezing and thawing. This difference supports our finding that the average percentage replacement by novel species was 37% vs 31%, and the survival of original species pool was 39% vs 50% in the arctic and subarctic stream respectively.

Disturbance of intermediate severity and frequency is a mechanism, popularized by Connell (1978) which has been substantiated in many systems as a probable mechanism for maintaining local diversity, by eliminating the best competitors on a random basis (Sanford & Ward, 1983). If prolonged freezing, anchor and frazil ice formation, and summer floods and temperature variation are the major disturbances causing local extinctions, then the arctic species array must be disturbed often and selected for high dispersal capability. Temperature variation and discharge variation were greatest in the arctic stream compared to the subarctic stream. This would have enhanced the variance and magnitude of the hydraulic stress on the organism living in the laminar sublayer on rocks (Statzner & Hilger, 1986).

Source areas for colonization by adult females may be more diverse in the arctic drainages that the subarctic. The rivers located near Imnavait Creek (Fig. 1), namely the Kuparuk River, the Toolik River and Oksrukuyik Creek, all drain parallel to one another to the Arctic Ocean about 200 km to the north. These drainages are separated by a long distance water route that must be navigated by their fishes but the drainages come within 10 km by air at our sample area. The differences in source water (foothills vs mountain glaciers), differences in their history during the Holocene and differences in selective regimes might allow marked divergence in the communities they support. The drainage of Caribou-Poker Creek is dendritic within each river system rather than linear. The watersheds of the equal sized drainages are separated by mountain passes, thereby limiting potential colonization between drainages. After one or more of the potential catastrophes of freeze-up or flood, affected streams may contain sections of lower density than they might support under constant conditions. Hence the streams may be invasible by novel, dispersing species. For this kind of dispersal to occur as our data imply, many species, particularly the chironomids, might be selected for facultative diapause at many stages in the life cycle for continuous growth even at low temperatures, and for high dispersal characteristics both as larvae (by drift) and adults. Thus the species that will proliferate each spring might be very stochastic, depending upon which species emerges and colonizes first in the spring or which colonizer in the autumn had instars that achieved sufficient size to survive the winter successfully in large numbers.

Conclusions

1. Species richness in a given season was higher in the a subarctic stream (138 spp.) compared to an arctic stream (112 sp) over four seasonal samples in parts of three years.
2. Drift density appeared from this and other studies to be related to benthic density in the arctic, but not the subarctic stream. The species composition of the drift was more closely related to the benthic composition in the subarctic stream. More kinds of habitats in the beaded (ponded) arctic stream probably contributed to the more diverse drift in the arctic stream.
3. The community composition of the drift and benthic invertebrates varied so much between seasons, that there was very little overlap between in an ordination of sampled communities using species (Decorana).
4. The local colonization of stream fauna between seasons was high and noncyclic over parts of three years. In the arctic stream new immigrants averaged 37% of the species present in any seasonal set of samples and only 39% of the species were present from the previous seasonal samples. In the subarctic stream, new immigrants averaged 31% of the species present in any season and 50% of the species were present from the previous seasonal sample.
5. Independent drainages were linear and more densely packed at the headwaters of the arctic stream across the North Slope of the Brooks Range than in the subarctic stream, where drainage basins are dendritic. Dispersal between different drainages is more likely for the foothills stream.
6. Shorter ice free seasons, more unpredictability about the timing of freeze-up and spring melt, sequential icing and thawing, nature of snow melt flood, timing and severity of summer runoff events, and greater temperature variation contributed to the intermediate disturbances that kept habitats open or undersaturated in the arctic compared to the subarctic drainage.

Acknowledgements

This work was sponsored by a grant from the US Environmental Protection Agency, Environmental Biology Program, Office of Exploratory Research (R808779-01) to Vera Alexander, Institute of Marine Science, University of Alaska. Much of the laboratory support, equipment and field support was made available by her. The authors would like to thank Frank Muller Karger, Curt Meininger and Bernard Moller for help in collecting samples under rigorous conditions. Curt Meininger did the multivariate statistical analysis. Scott Hartsock helped with laboratory preparation of invertebrates, mounting and illustration.

References

Beckett, D. C. & P. A. Lewis, 1982. An efficient procedure of slide mounting of larval chironomids. Trans. Am. Microsc. Soc., 101: 96–99.

Bradley, J. V., 1968. Distribution-free statistical tests. Prentice-Hall, Inc. Englewood Cliffs, N.J. 388 p.

Bright, G. R., 1982. Secondary benthic production in a tropical island stream. Limnol. Oceanogr. 27: 472–480.

Bruns, D. A. & G. W. Minshall, 1983. Macroscopic models of community organization: analyses of diversity, dominance, and stability in guilds of predaceous stream insects. pp 231–264. In Barnes, J. R. & G. W. Minshall (eds.). Stream Ecology. Plenum Press, N.Y.

—, —, J. T. Brock, C. E. Cushing, K. W. Cummins & R. L. Vannote, Ordinations of functional groups and organic matter parameters from the middle fork of the Salmon River, Idaho. Freshwat. Biol. 1: 2–12.

Butler, M., M. C. Miller and S. Mozley, 1980. Macrobenthos. pp 297–339. Hobbie, J. E. (ed.) Limnology of Tundra Ponds. U.S. IBP Synthesis Series 13, Dowden, Hutchinson and Ross, Inc. Stoudsburg, Pa.

Connell, J. H., 1978. Diversity in tropical rainforests and coral reefs. Science 199: 1302–1320.

Cowan, C. A. & M. W. Oswood, 1984. Spatial and seasonal associations of benthic macroinvertebrates and detritus in an Alaskan subarctic stream. Polar Biol. 3: 211–215.

Craig, P. C. & P. McCart, 1974. Classification of stream types in Beaufort Sea drainages between Prudhoe Bay, Alaska and the Mackenzie Delta. *In* P. J. McCart (ed). Classification of streams in Beaufort Sea Drainages and distribution of fish in arctic and subarctic drainages. Canadian Arctic Gas Study Limited, Biological Report Series 17(1): 1–47.

Culp, J. M. & R. W. Davies, 1982. Analysis of longitudinal zonation and the river continuum concept in the Oldman-South Saskatchewan River system. Can. J. Fish. aquat. Sci. 39: 1258–1266.

Fisher, S. G., L. J. Gray, N. B. Grimm & D. E. Busch, 1982. Temporal succession in a desert stream ecosystem following flash flooding. Ecol. Monogr. 52: 93–110.

Fox, L. R., 1977. Species richness in streams – An alternative mechanism. Am. Nat. 111: 1017–1021.

Frey, P. J., E. W. Mueller & E. C. Berry, 1970. The Chena River, the study of a subarctic stream. Federal Water Quality Administration Project Rept. 1610. 96 p.

Gauch, H. G. Jr., 1982. Multivariate analysis in community ecology. Cambridge U. Press. Cambridge, UK.

Hershey, A. E., 1985. Littoral chironomid communities in an arctic Alaskan lake. Holarct. Ecol. 8: 39–48.

Houston, M., 1979. A general hypothesis of species diversity. Am. Nat. 113: 81–101.

Kane, D. L. & R. F. Carlson, 1973. Hydrology of the central arctic river basins of Alaska. Institute of Water Resources, U. Alaska. Rept. No. IWR-41. 50 p.

LaPerriere, J. D., 1983. Alkalinity, discharge, average velocity and invertebrate drift concentrations in subarctic Alaskan streams. J. Freshwat. Ecol. 2: 141–151.

Lenat, D. R., 1983. Chironomid taxa richness: Natural variation and use in pollution assessment. Freshwat. Invert. Biol. 2: 192–194.

McArthur, R. H. & E. O. Wilson, 1967. The theory of island biogeography. Monographs in Population Biology 1. Princeton U. Press, Princeton, NJ.

McAuliffe, J. R., 1983. Competition, colonization patterns, and disturbance in stream benthic communities. In Barnes, J. R. & G. W. Minshall. Stream Ecology. Plenum Press, N.Y. pp 137–156.

—, 1984. Competition for space, disturbance, and the structure of a benthic stream community. Ecology 65: 894–908.

McCart, P., P. Craig & H. Bain, 1972. Report on fisheries investigations in the Sagavanirktok River and neighboring drainages. Alyeska Pipeline Service Co., Bellevue Washington. 186 p.

Miller, M. C., G. R. Hater, P. Spatt, P. Westlake and D. Yeakel, 1986. Primary production and its control in Toolik Lake, Alaska. Arch. Hydrobiol./Suppl. 74(1): 97–131.

—, R. T. Prentki & R. J. Barsdate, 1980. Physics. In Hobbie, J. E. (ed.) Limnology of Tundra Ponds. U.S. IBP Synthesis Series 13, Dowden, Hutchinson and Ross, Inc. Stoudsburg, Pa. pp 51–75.

—, J. R. Stout & V. A. Alexander, 1986. Effects of a controlled under-ice oil spill on invertebrates of an arctic and a subarctic stream. Environ. Poll. (Series A) 42: 99–132.

Minshall, G. W., K. W. Cummins, R. C. Petersen, C. E. Cushing, D. A. Bruns, J. R. Sedell & R. L. Vannote, 1985. Developments in stream ecosystem theory. Can. J. Fish. aquat. Sci. 42: 1045–1055.

—, R. C. Petersen, Jr. & C. F. Nimz, 1985. Species richness in streams of different size from the same drainage basin. Am. Nat. 125: 16–38.

Naiman, R. J., J. M. Melillo, M. A. Lock, T. E. Ford & S. Reice, 1987. Longitudinal patterns of ecosystem processes and community structure in a subarctic river. Ecology 68: 1139–1156.

Nauman, J. W. & D. R. Kernoodle, 1975. The effect of a fuel oil spill on benthic invertebrates and water quality on the Alaska Arctic Slope, Happy Valley Creek near Sagwon, Alaska. J. Res. U.S. Geol. Survey 3: 495–500.

Olsson, T. I., 1981. Overwintering of benthic macroinvertebrates in ice and frozen sediment in a North Swedish river. Holarctic Ecology 4: 161–166.

Oswood, M. W., 1988. Community structure of benthic invertebrates in interior Alaskan (USA) streams and rivers. Hydrobiologia (this volume).

Peterson, B. J., J. E. Hobbie, A. Hershey, M. Lock, T. Ford, J. R. Vestal, M. Hullar, M. C. Miller, R. Ventullo & G. Volk, 1985. Transformation of a tundra stream from heterotrophy to autotrophy by addition of phosphorus. Science 229: 1383–1386.

—, —, & T. L. Corliss, 1986. Carbon flow in a tundra stream ecosystem. Can. J. Fish. aquat. Sci. 43: 1259–1270.

Pickett, S. T. A. & P. S. White (eds.), 1985. The ecology of natural disturbance and pathc dynamics. Academic Press, Inc. N.Y.

Reger, S. J. & N. R. Kevern, 1981. Benthic macroinvertebrates of three Michigan streams. J. Freshwat. Ecol. 1: 179–188.

Reice, S. R., 1981. Interspecific associations in a woodland stream. Can. J. Fish. aquat. Sci. 38: 1271–1280.

—, 1984. The impact of disturbance frequency on the structure of a lotic riffle community. Ver. int. Ver. Limnol. 22: 1906–1910.

Sanford, J. A. & J. V. Ward, 1981. Insect species diversity as a function of environmental variability and disturbance in stream systems. In Barnes, J. R. & G. W. Minshall (eds.). Stream Ecology. Plenum Press, N.Y. p 265–278.

Schaeffer, D. J., J. A. Perry, 1986. Gradients in the distribution of riverine benthos. Freshwat. Biol. 16: 745–757.

Slack, K. V., J. W. Nauman & L. J. Tilley, 1976. Evaluation of three collecting methods for a reconnaissance of stream benthic invertebrates. J. Res. U.S. Geol. Survey 4: 491–495.

—, — & —, 1977. Benthic invertebrates in an arctic mountain stream, Brooks Range, Alaska. J. Res. U.S. Geol. Survey 5: 519–527.

—, — & —, 1979. Benthic invertebrates in a north-flowing stream and a south-flowing stream, Brooks Range, Alaska. Water Resources Bull. 15: 108–135.

Slaughter, C. W., 1971. Caribou-Poker Creeks Research Watershed, interior Alaska, Background and current status. Cold Regions Research and Engineering Laboratory, Hanover, N.H. Special Report 157. 10 p.

Sokal, R. R. & F. J. Rohlf, 1969. Biometry. W. H. Freeman & Comp., San Francisco.

Sousa, W. P., 1984. The role of disturbance in natural communities. Ann. Rev. Ecol. Syst. 15: 353–391.

Stanford, J. A. & J. V. Ward, 1983. Insect species diversity as a function of environmental variability and disturbance in stream systems. In Barnes, J. R. & G. W. Minshall (eds.) Stream Ecology. Plenum Press, N.Y. p 265–278.

Statzner, B. & B. Hilger, 1985. Questions and comments on the river continuum concept. Can. J. Fish. aquat. Sci. 42: 1038–1044.

—, 1986. Stream hydraulics as a major determinant of benthic invertebrate zonation patterns. Freshwat. Biol. 16: 127–139.

Stout, J. R., 1986. Macroinvertebrate drift and community composition in an arctic and subarctic stream in Alaska. Thesis. University of Cincinnati, OH. 72 p.

Tilman, D., 1982. Resource competition and community structure. Princeton University Press, Princeton, New Jersey.

—, 1984. Plant dominance along an experimental nutrient gradient. Ecology 65: 1445–1453.

—, 1987. Secondary succession and the pattern of plant dominance along experimental nitrogen gradients. Ecol. Monogr. 57: 189–214.

Vannote, R. L., G. W. Minshall, K. W. Cummins, J. R. Sedell & C. E. Cushing, 1980. The river continuum concept. Can. J. Fish. aquat. Sci. 37: 130–137.

Ward, A. F. & D. D. Williams, 1986. Longitudinal zonation and food of larval chironomids (Insecta: Diptera) along the course of a river in temperate Canada. Holarct. Ecol. 9: 48–57. stream.

Wetzel, R. G. & G. Likens, 1979. Limnological Methods. Saunders, N.Y.

Whittacker, R. H., 1973. Handbook of vegetation science, Part V. Ordination and classification of communities. W. Junk Publishers, The Hague.

Winget, R. N., 1984. *Brachycentrus americanus* and *B. occidentalis* (Trichoptera) in a regulated stream. J. Freshwat. Ecology 2: 373–381.

Hydrobiologia 172: 129–148, 1989
W. F. Vincent and J. C. Ellis-Evans (eds) High Latitude Limnology

Geochemical processes in the Lake Fryxell Basin (Victoria Land, Antarctica)

William J. Green, Thomas J. Gardner, Timothy G. Ferdelman, Michael P. Angle, Lawrence C. Varner & Philip Nixon
School of Interdisciplinary Studies, Miami University, Oxford, Ohio 45056

Key words: anoxic basin, Antarctic streams, geochemical processes, Lake Fryxell, metal cycling, nitrogen deficiency, reverse weathering

Abstract

Major ion, nutrient, transition metal, and cadmium concentrations are presented for nine meltwater streams flowing into Lake Fryxell, a permanently stratified lake with an anoxic hypolimnion in Taylor Valley, Antarctica. For the major ions, stream compositions are considered in terms of dissolution of marine-derived salts and chemical weathering of local rocks. Although Lake Fryxell has undergone significant evaporative concentration, only calcite, of the simple salts, is predicted to precipitate. Geochemical budgets indicate, however, that large quantities of K, Mg, and SO_4 have also been removed from the lake. Reverse weathering may be an important sink for K and Mg, although magnesium removal with calcium carbonate phases is also likely. Assuming constancy of composition over recent geologic time, all of the salts in the Fryxell water column could have been delivered under present flows in about three thousand years (chloride age).

Comparison of nutrient concentrations in these meltwater streams with other flowing waters in the world reveals that the Fryxell streams are strikingly deficient in NO_3-N but not PO_4-P. The apparent nitrogen deficiency in Lake Fryxell itself can be attributed to the low annual stream loadings of this nutrient.

Stream concentrations and loadings are also presented for Mn, Fe, Co, Ni, Cu, and Cd. 'Dissolved' metal concentrations correlate roughly with average crustal abundances, suggesting that chemical weathering is the major source for these elements. Vertical metal profiles within Lake Fryxell itself appear to be governed by the formation of insoluble sulfide phases, or, in the case of Mn, by $MnHPO_4$. However, dissolved nickel levels in sulfide-bearing waters are much higher than can be explained in terms of metal-sulfide equilibria, and we suspect that significant organic complexing of this metal is occurring in the deeper waters.

Introduction

The ice-free valleys of Antarctica's coastal areas constitute only a small (< 5%) and highly atypical fraction of the continent's surface. Extreme desert environments, these valleys are unique because they are both polar and completely rainless. All precipitation, falling as snow, is quickly lost by sublimation. The katabatic winds, which sweep down from the Polar Plateau, maintain the relative humidity at less than 2%.

Taylor Valley, the site of Lake Fryxell, is situated in southern Victoria Land, on the western coast of the Ross Sea in the McMurdo Sound region of the Transantarctic Mountains. This valley, like the other major valleys of the McMurdo Oasis, cuts in an east-west direction from McMurdo Sound to the Polar Plateau. Claridge & Campbell (1977) have observed that the major rock types found in the valley include dolerites, franite, basalts, gneisses, schists, sandstones, and metasediments.

Lake Fryxell is situated toward the broad eastern end of the Taylor valley and is 5 km long, 1.6 km wide, and has a maximum depth of about 19 meters. At least nine streams feed the lake, and these are derived from the Canada, Commonwealth, and Howard Glaciers. The locations of the streams are shown in Fig. 1.

The chemical limnology of Lake Fryxell was first examined by Angino *et al.* (1962), who found the principal ions to be sodium, chloride, and bicarbonate. These workers noted that although there were abundant aragonite coatings on the rocks near Lake Fryxell, calcium concentrations in Fryxell waters were somewhat lower than expected. They suggested that $CaCO_3$ or $CaSO_4$ might be precipitating from the lake. Angino *et al.* (1962) also commented on the possible sources of the lake's salts, suggesting that ocean spray, patches of NaCl and Na_2SO_4 associated with

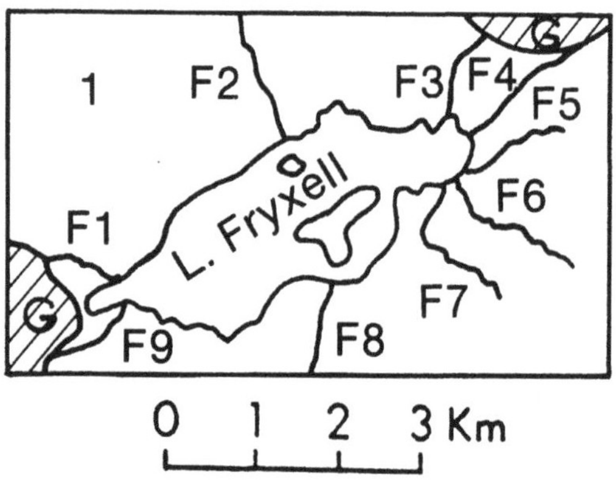

Fig. 1. Sketch of Lake Fryxell showing the locations of the nine meltwater streams discussed. G = glacier.

local moraines, relict seawater, and volcanic hotsprings might all be contributing ions to the lake. Because of the low magnesium content in Lake Fryxell, relict seawater was viewed as an 'improbable' source. The slight warming (2.5 °C) of deep waters was attributed to solar heating.

Torii *et al.* (1975) were the first to measure nutrient concentrations in the lake. Their study revealed rapidly increasing phosphorus concentrations with depth and a paucity of nitrate throughout the water column. In addition, they observed that H_2S appeared at about 8 m, making the anoxic zone a significant fraction of the lake's volume.

Vincent (1981) studied the phytoplankton ecology and physiology of the lake and considered the relative importance of light, temperature, and nutrient supply in the primary production strategies of Antarctic phytoplankton. He suggested that nutrient supply, rather than *in situ* light or temperature, was the primary factor controlling the variations in productivity among dry valley lakes.

Parker *et al.* (1982) observed that a benthic blue-green algal community covers the shallower areas of Lake Fryxell. Approximately 1.8×10^6 m^2 of benthic surface was considered amenable to habitation by these organisms. However, the role which these play in the overall ecology of the lake, specifically their ability to extract nutrients and other solutes directly from the water column, has not been determined.

The present study considers Lake Fryxell in the context of its nine meltwater streams. The objectives are to present a reasonably comprehensive geochemical description of the streams and of the lake itself and to identify the major processes involved in lake evolution and in element cycling and fate. Where appropriate, comparisons are made between this system and other flowing and standing waters of interest.

Methods

Most stream and lake samples were collected during the 1982–83 Antarctic field season. However, a separate, more detailed lake sampling was conducted for Mn, Co, Ni, Cu, Cd, and sulfides during the 1983–84 field season. Major ions were analyzed according to the methods presented in Green & Canfield (1984). Nutrients were determined using procedures discussed by Canfield & Green (1985). Hydrogen sulfide was measured in the field using the methylene blue technique (Cline, 1969).

Mn, Fe, Co, Ni, Cu, and Cd concentrations were determined in the nine meltwater streams and in Lake Fryxell. Metal analysis involved the use of clean procedures during all stages of the field and laboratory work, as discussed by Green *et al.* (1986b). Both 'total' and 'dissolved' concentrations are reported. 'Total' refers to analyses made on samples collected directly in acid-soaked (20% nitric for one week; 1% Ultrex for one day; DIW rinsed) linear polyethylene bottles, acidified with ultrapure nitric acid to $pH < 2$, stored in the dark, and analyzed several months later using graphite furnace atomic absorption spectrophotometry (Perkin-Elmer 3030), preceded in most cases by an extraction procedure (Danielson *et al.*, 1978). 'Dissolved' (values appear in parentheses in the table) refers to samples processed by pressure filtering through 0.45 micron acid-washed filters. Evidence of contamination in some samples is attributed to handling of filters in the field under conditions of high wind. Detection limits were (in nM): Mn = 0.9; Fe = 1.8; Co = 0.3; Ni = 0.3; Cu = 0.8; and Cd = 0.009.

Nine discharges to Lake Fryxell were monitored several times during the season using a pygmy-type flow meter. Three streams (Fig. 1) clearly dominate the hydrologic budget of the lake. F_1 originates at the Canada Glacier, flows southeast down a steep rocky slope, and then issues onto a delta before entering the lake. This stream contributes an estimated 25% of the annual inflow. F_3 derives from the Commonwealth Glacier and empties into the northeastern basin. It is responsible for 26% of the annual water input and is the major source of inorganic solutes. F_6 cuts a narrow channel and enters from the southeastern shore, bringing roughly 16% of the annual water supply. Since Antarctic streams have high daily and annual variability (Chinn, 1981), loadings based on a single season must be considered as rough estimates. We are encouraged, however, by the fact that the annual discharge measured for the 1982–83 season was within a few percent of the water loss expected from a surface ablation (Chinn, 1981) rate of $30\ cm\ a^{-1}$.

The annual load of a given ion, nutrient species, or metal was obtained as the sum of the loads from each stream. The load from a given stream was computed as the product of the average concentration and the annual stream flux. The total quantity of a particular species in the lake was determined from the lake's bathymetry and from vertical concentration profiles (see Green *et al.*, 1986a).

Major ion, nutrient, silica, pH and temperature, and selected metal data were analyzed using the equilibrium computer program WATEQ2 (Ball *et al.*, 1979). The calcite saturation indices (SI) reported later were determined in this way, as were the equilibrium states of metal sulfides.

Results and discussion

Major ions: stream and lake concentrations and budgets

The streams flowing to Lake Fryxell are small by temperate zone standards. They have channels which range from one to ten meters in width, and water depths generally not in excess of 60 cm. The streams discussed in this paper have lengths between roughly 0.5 and 10 km, and discharge 10^7–10^8 liters of water annually to Lake Fryxell. All of the streams were considerably smaller, both in terms of length and discharge, than the Onyx River which feeds Lake Vanda. The chemistry of this river has been discussed recently by Green & Canfield (1984) and by Downes *et al.* (1986), and its hydrology has been detailed by Chinn (1981).

Table 1 presents the major ion compositions and the 1982–83 annual discharges of streams associated with Lake Fryxell. Cation mole fractions in each stream are presented in the triangular diagrams of Figure 2. The composition

Table 1. Major element concentration (μM) in meltwater streams and annual stream discharges ($l\ a^{-1}$)

Stream	Ca	Mg	Na	K	HCO_3	SO_4	Cl	Discharge
F_1 (3)*	76	15	66	18	115	24	100	4.0×10^8
F_2 (3)	400	68	282	79	804	95	268	1.4×10^8
F_3 (4)	183	62	308	55	441	28	320	4.2×10^8
F_4 (1)	142	41	372	51	513	11	217	0.6×10^8
F_5 (1)	88	64	403	67	559	3	212	1.4×10^8
F_6 (1)	293	42	199	62	679	5	147	2.5×10^8
F_7 (1)	568	136	521	82	1410	69	508	0.3×10^8
F_8 (2)	546	123	453	80	1140	96	501	0.8×10^8
F_9 (1)	94	94	147	26	223	6	169	0.6×10^8

* Number in parentheses indicates the number of samples averaged to get reported values.

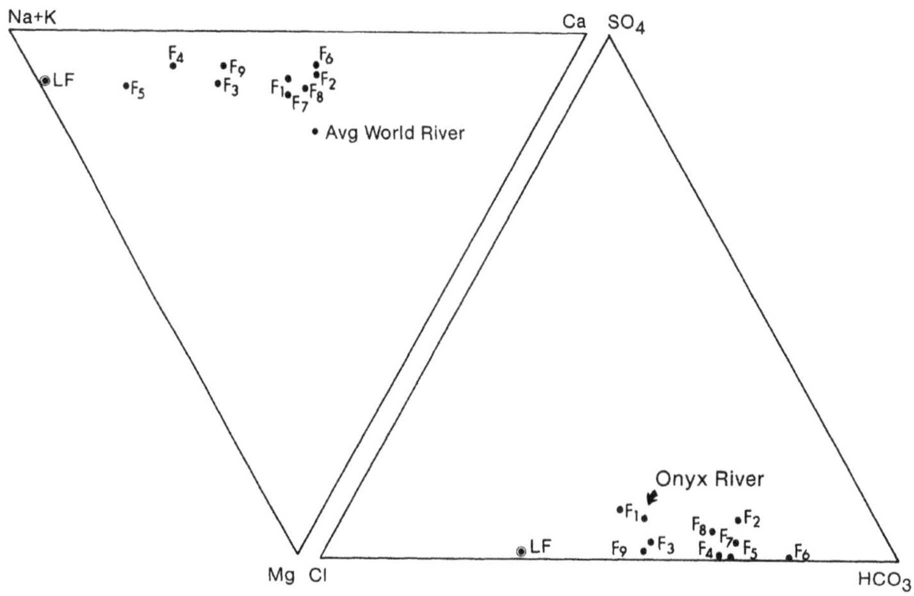

Fig. 2. Cation and anion trilateral diagrams giving locations of Fryxell streams. Axes are mole %.

of average world river water, taken from Livingstone (1963), is shown for comparison. Cations fall exclusively along the Na + K-Ca leg of the figure, with Mg and K appearing as minor constituents of all waters. In all streams, Na or Ca is the dominant cation.

Following the work of Keys & Williams (1981) and Claridge & Campbell (1977), it is possible to view the sodium concentrations of these streams as an index of the degree to which salts of predominantly marine origin are being leached from soils and morainal materials. Some fraction of the sodium may also come from the weathering of feldspars. Calcium, by contrast, can be taken as an index of chemical weathering of an assortment of materials, including ferromagnesian minerals in the Ferrar Dolerites, and calcite coatings derived from marble. However, since Ca/Mg molar ratios in these streams are much larger than would be predicted by dolerite weathering, we suspect that much of the calcium derives ultimately from the weathering of marble (Claridge & Campbell, 1977).

Magnesium is probably derived both from the direct weathering of dolerites and from sea salt aerosols. Using the Mg/Cl molar ratio for average seawater and the measured chloride concentrations in the streams, we estimate that 36–89% of the Mg in the Fryxell streams could have come from marine aerosols. The relatively low magnesium content of these waters coupled with the apparent importance of the marine source term for this element suggest that dolerite weathering is not a major contributor to the dissolved (major ion) load of Antarctic meltwaters. We note, in support of this claim, that where we have silica data for Fryxell streams, this constituent is low (roughly 3% of the dissolved load), indicating that the weathering of silicate rocks is quantitatively less important than (1) the dissolution of marine salts and (2) the dissolution of calcite crusts.

SO_4 in Fryxell streams is probably acquired by dissolution of Na_2SO_4 salts in the region which were first reported by Angino *et al.* (1962). Using the argument given above for magnesium, we calculate that a significant fraction of this stream sulfate derives from marine aerosols. K can be attributed to dissolution of muscovite micas, as suggested by Claridge & Campbell (1977).

Streams of the Fryxell System which have higher calcium mole fractions tend to be those which flow relatively long distances from glaciers in the Kukri Hills and Asgard Mountains. These include F_2, F_6, F_7, and F_8. This observation is consistent with the finding of Green & Canfield (1984) that the Ca/Na ratio for the Onyx River gradually increases with distance from the source (Wright Lower Glacier).

Another perspective can be gained by placing the Fryxell streams within the envelope of world waters developed by Gibbs. Gibbs (1970) classified surface waters according to whether their chemical composition was regulated principally by atmospheric inputs, by interaction with rocks in the drainage basin, or by evaporation and resulting deposition of insoluble minerals. Figure 3 shows total dissolved solids (TDS) vs. Na/(Na + Ca). Here the Taylor Valley streams exhibit a rather broad spectrum of compositions. F_2, F_6, F_7, and F_8, all of which flow from glaciers in the Kukri Hills and Asgard Mountains, plot toward the region of 'rock dominance.' Streams flowing a short distance from either the Canada or Commonwealth Glacier (F_1, F_9, F_3, F_4, F_5) tend to have lower dissolved solids and higher sodium to calcium and chloride to bicarbonate ratios.

When compared with other world surface waters, these streams are generally dilute waters whose compositions, in Gibb's (1970) terms, are influenced by atmospheric precipitation (marine aerosols) and by rock (or rock-derived salt) weathering, though not by evaporation and crystallization. Length of flow appears to be an important variable, with longer streams having higher ionic strengths and greater calcium to sodium and bicarbonate to chloride ratios. The positions of these streams in Fig. 3 are consistent with a 'multiple source' hypothesis for dissolved solutes.

In turning our attention to Lake Fryxell itself, it is worth remarking on several general features of the lakes of the McMurdo Oasis. First, solar heating is an important physical phenomenon in the region. Highly transparent ice covers transmit radiation to the underlying waters and insure that, even during periods of cold evaporation, water temperatures do not fall much below 0 °C. This is important because it renders unlikely any temperature-related reversals in the normal sequence of mineral precipitation. Second, most lakes of the region exhibit only intermediate degrees of salt concentration relative to inputs. Lake Fryxell is more similar to the lakes of the Ethiopian Rift than to the highly saline Great Salt Lake (Spencer *et al.*, 1985), or to many of the closed-basin lakes discussed by Hardie & Eugster (1970), Eugster & Hardie (1978), and Eugster & Jones (1979). Third, evaporation in these lakes, including Lake Fryxell, occurs by sublimation of surface ice. Under steady state conditions, equilibrium is maintained by addition of ice to the bottom of the

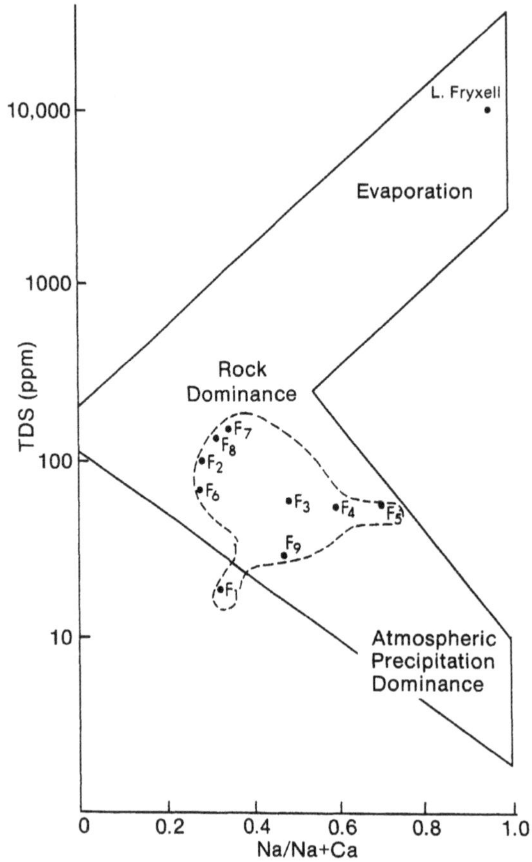

Fig. 3. Gibbs diagram showing location of Fryxell streams in relation to regions of 'rock dominance' and 'atmospheric precipitation dominance.'

ice cover and by the release of latent heat (McKay *et al.*, 1985). During periods of unusual cold, when sublimation exceeds water input, the lakes shrink in volume and the salinity of their waters increases. It should be noted that this concentration mechanism is thermodynamically equivalent to simple evaporation.

Chemically, Lake Fryxell is a highly stratified water body, having a Na-Cl-HCO_3 brine. The water column concentrations of major ions and silica are given in Table 2 and show that the brine is of the Na-Cl-HCO_3 type, with an ionic strength of 0.175 at 17 meters. Some of the major ions follow a diffusion-driven distribution, increasing monotonically with depth (Fig. 4). Sulfate values, however, peak at twelve meters and decrease in deeper waters, as a result of sulfate reduction. Anoxia occurs at 9 meters, and sulfide values finally exceed 1 mM at 17 meters.

The evolution of a Na-Cl-HCO_3 composition is readily understood in terms of the dilute inflows feeding the lake. Hardie & Eugster (1970) have emphasized the importance of early calcite deposition in the evolution of closed-basin lakes from dilute inflows. Lakes forming from inflows having bicarbonate to calcium molar ratios in excess of two tend to evolve toward calcium-deficient, bicarbonate-rich waters. Examination of stream concentrations for Lake Fryxell in Table 1 shows that $M_{HCO3}/M_{Ca} > 2$ for all inflows except F_1. In one of the major inflows, F_5, this ratio is 6.3. The relatively low calcium concentrations in Lake Fryxell, first discussed by Angino *et al.* (1962), are seen on this analysis to be a natural consequence of the evaporation sequence followed by inflows with high bicarbonate to calcium molar ratios.

Data in Table 2 were analyzed using the equilibrium computer program WATEQ (Ball *et al.*, 1979). All waters were supersaturated with respect to calcite, but undersaturated with respect to gypsum. Calcite (and possibly magnesian calcite) deposition would appear to be a major geochemical process in the contemporary water column.

Rough geochemical budgets for the seven major ions were estimated by computing the annual stream inputs, multiplying these by the chloride age of the lake to get total inputs over time, and then comparing these with the amounts in the lake. Data are presented in Table 3.

Budget calculations suggest large Ca and HCO_3 losses; SO_4 losses are nearly three times greater than the quantity of sulfate presently in the lake. Also, large K and Mg losses are indicated. Na is conservative.

The three processes which reasonably account for the observed deficits are calcite (and magnesian calcite) deposition, sulfate reduction, and,

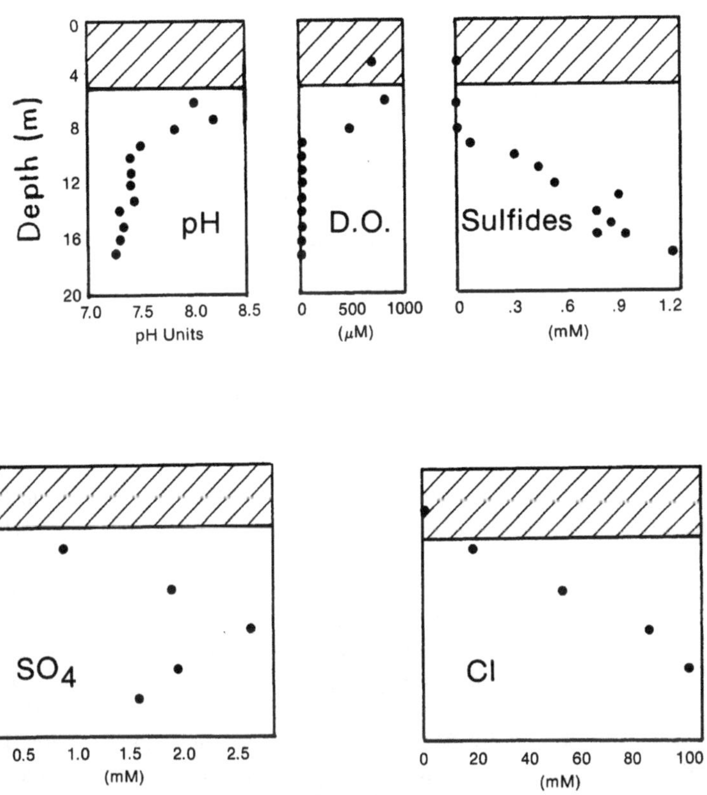

Fig. 4. Water column profiles of pH, D.O., sulfides, sulfate, and chloride.

Table 2. Data for Fryxell water column. (All concentrations in mM).

Depth (m)	T °C	Ca	Mg	Na	K	HCO_3	SO_4	Cl	SiO_2	Log (SI) calcite	pH
3	0.50	0.028	0.011	0.083	0.028	0.05	0.01	0.144	-	−2.92	8.2
6	1.0	1.40	2.30	22.2	2.26	10.4	0.90	19.0	0.22	0.56	8.0
9	2.0	2.22	6.48	57.7	4.62	26.3	1.89	52.8	0.45	0.53	7.5
12	2.0	3.18	10.1	104	6.15	41.0	2.65	84.7	0.48	0.74	7.4
15	2.0	3.52	12.1	120	6.15	53.9	1.95	99.3	0.50	0.81	7.4
17	2.0	2.75	12.9	129	8.00	55.0	1.58	102	0.54	0.61	7.3

more tentatively, reverse weathering, as shown by the equations below:

1. Calcite deposition:
   $Ca^{2+} + HCO_3^- = CaCO_3 + H^+$
2. Sulfate reduction:
   $8SO_4^{2-} + 2Fe_2O_3 + 15C + 7H_2O =$
   $4FeS_2 + 14HCO_3^- + CO_3^{2-}$
3. Reverse weathering:
   cation-poor Al silicates
   $+ HCO_3^- + SiO_2 +$ cations $=$
   cation Al silicates $+ CO_2 + H_2O$

Table 3. Flux estimates and morphometric parameters for the Lake Fryxell system.

	Ca	Mg	Na	K	HCO_3	SO_4	Cl
Flux (10^3 moles a^{-1})	324	78.3	384	70.6	774	46.7	357
In lake (10^5 moles)	456	1300	12570	878	5466	352	10700
2997* × flux (10^5 moles)	9710	2347	11508	2115	23197	1399	10700
Residuals (10^5 moles)	−9254	−1047	+1062	−1237	−17731	−1047	0

Lake volume	4.32×10^{10} l
Surface area	7.4×10^6 m²
Total inflow	1.58×10^9 l yr^{-1}

* Cl age in years (in lake: flux)

The first two processes are common and well documented in the evolution of closed-basin lakes. Calcite removal in this system is supported both by budget calculations and by thermodynamic evidence (see Table 2 for calcite SI as a function of depth), and sulfate reduction is indicated by the sulfate profile and by the high concentration of sulfides. The third, reverse weathering, was originally proposed by Mackenzie & Garrels (1966) to explain marine K and Mg balances. More recently, Von Damm & Edmond (1984) have used the concept to account for K and Mg losses in lakes of the African Rift Valley. The plausibility of mineral reconstitution reactions in Lake Fryxell is suggested both by the geochemical budgets and (more tentatively, given uncertainties in the free energy data) by thermodynamic calculations, which indicate that K-feldspars, Mg-montmorillonite, and Mg-beidellites may be forming as authigenic phases. Mineralogical data on the sediments are clearly needed to test this claim.

Nutrients: stream and lake concentrations and budgets

We will present a more detailed discussion of the behavior of nutrients in Lake Fryxell at a later date, but it is worthwhile in this context to mention some of the important features of nutrient chemistry in the system.

Figure 5 gives average concentrations of NO_3-N in the nine streams feeding the lake. Values were highest in F_1, F_2, and F_3, but in no case did they exceed 3 μM (42 μg l^{-1}). Compared with other flowing waters in the world, these nitrate values are extremely low, and are less than half of Maybeck's (1982) average world value of 7.1 μM (100 μg l^{-1}). Dissolved phosphorus levels, by contrast, are generally within the range (0.064 μM to 0.77 μM) given by Maybeck (1982) for unpolluted rivers. Fig. 6 shows that orthophosphate phosphorus concentrations are between 0.84 μM for F_3 and 0.21 μM for F_8.

Based on our earlier study of the Onyx River – the major input to Lake Vanda (Canfield & Green, 1985) – nutrient acquisition in Antarctic meltwater streams appears to involve two sources; atmospherically derived nitrate incorporated into glacial ice and soil-derived phosphate.

Nutrient profiles along the Onyx River (Canfield & Green, 1985) in Wright Valley reveal large initial nitrate values associated with the Wright Lower Glacier. Farther downstream, particularly beyond Bull Lake, nitrate decreases rapidly until near Lake Vanda it is only about 0.2 μM. Canfield & Green (1985) attributed this profile to nitrate uptake by stream organisms. Howard-Williams *et al.* (1986) have recorded biological removal of inorganic nutrients in the Onyx River and have noted a drop of 86% in NO_3-N levels during passage of water through one 1200-meter stretch of river bed. Similar decreases in NO_3-N downstream from a glacier source have been observed for one of the inflows

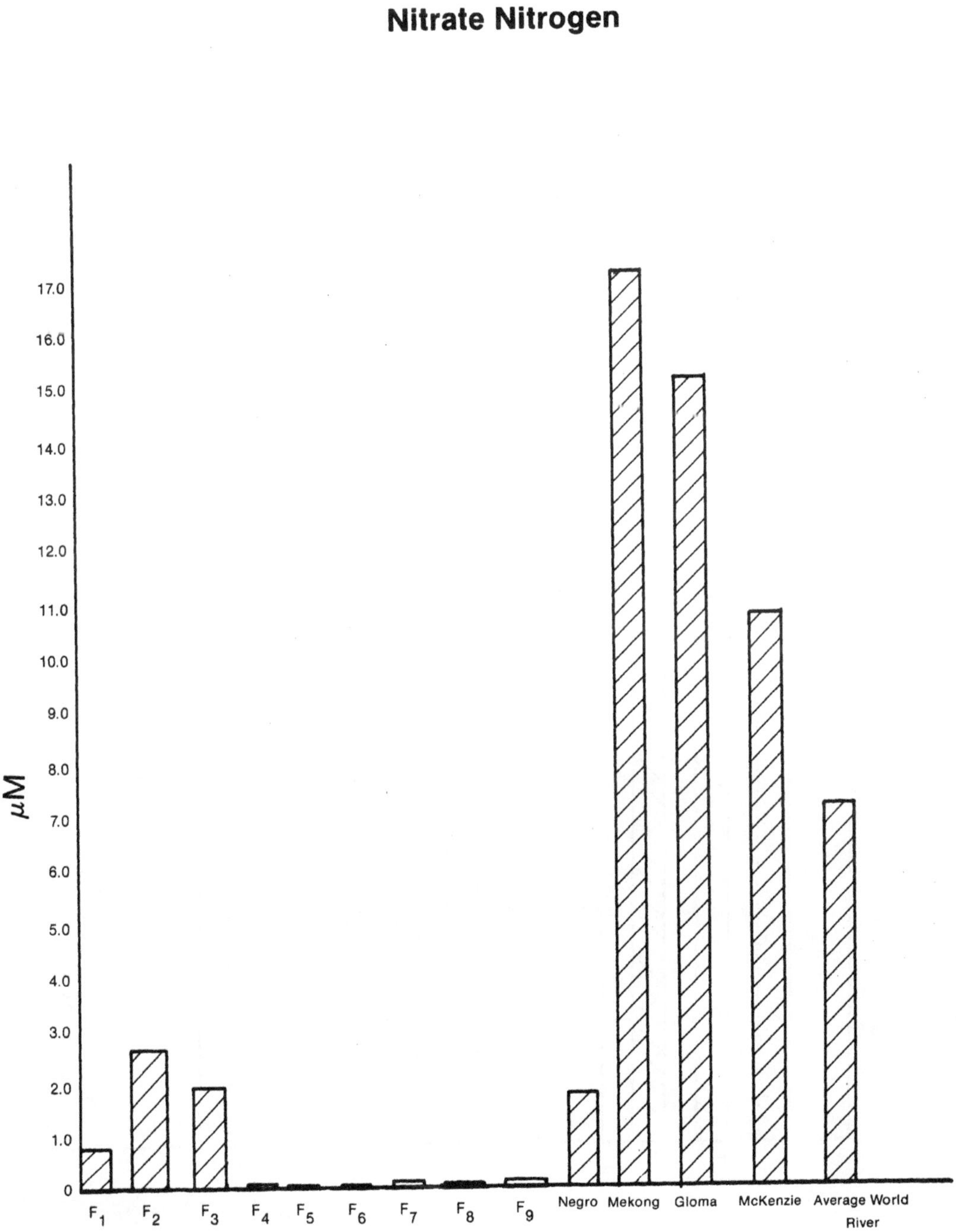

Fig. 5. NO_3-N levels in Fryxell streams as compared with rivers worldwide. World river data are from Maybeck (1982).

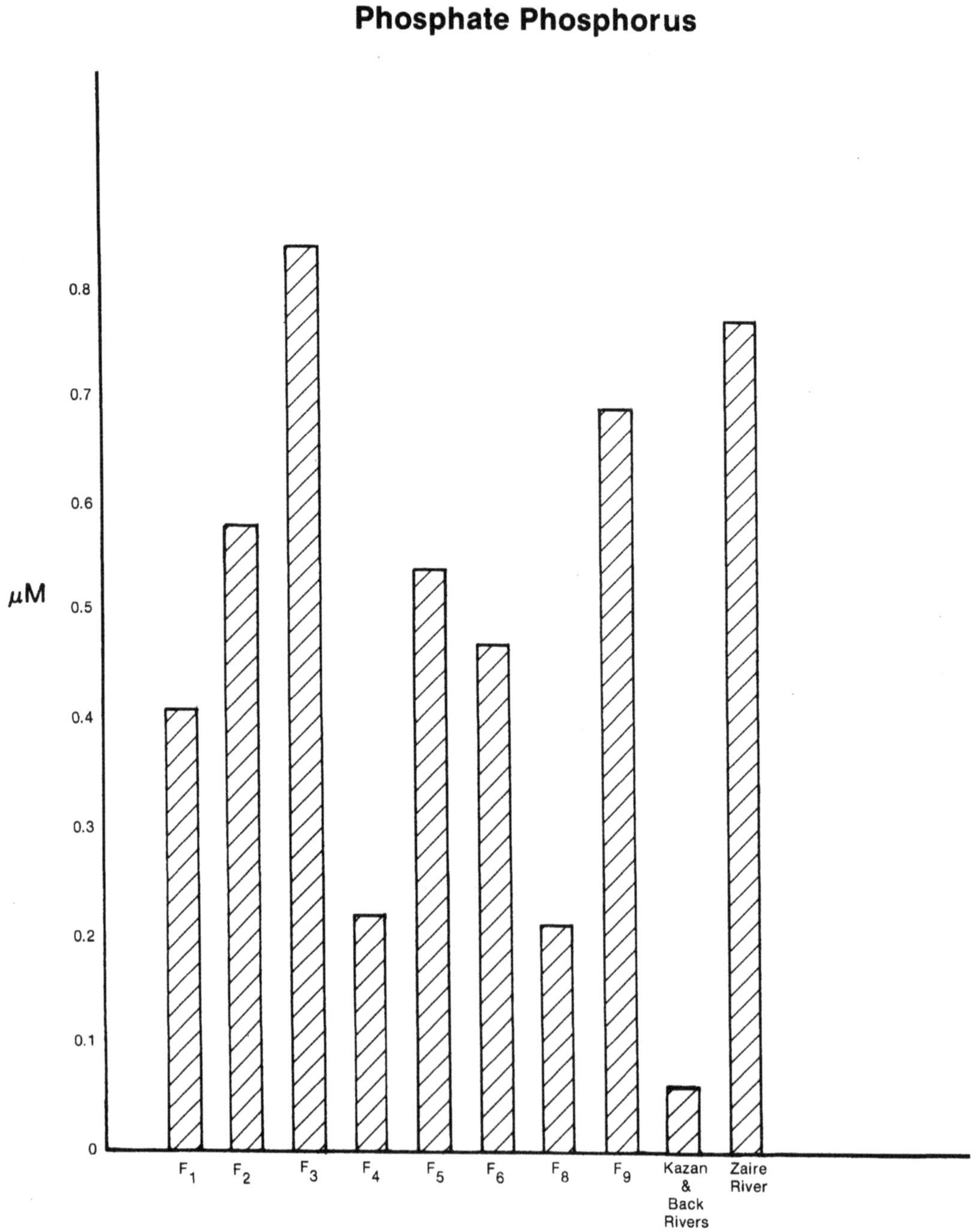

Fig. 6. Dissolved reactive phosphorus in Fryxell streams compared with Kazan, Back, and Zaire rivers (Maybeck, 1982).

to Lake Fryxell (Howard-William *et al.*, 1986), again suggesting the importance, in some cases, of glacier-derived nitrogen in these systems.

Nutrient budgets can be summarized for the 1982–83 season as follows. We estimate that 88 kg of total P, 22 kg of phosphate phosphorus, and 22 kg of NO_3-N reached the lake during the flow period, and that the molar ratio of incoming inorganic nitrogen to phosphorus was only 2.2, suggesting nitrogen limitation. However, this claim must be qualified in view of the recent observation of Downes *et al.* (1986) that Antarctic streams contain significant quantities of dissolved organic nitrogen and phosphorus, much of which may be in biologically available form.

Nutrient profiles and concentrations within the Fryxell water column itself reflect both strong lake stratification and extremely low nitrogen inputs. Fig. 7 shows that the major nutrient species all increase rapidly with depth. Inorganic N/P ratios increase with depth and range over values 4.7 to 9.8. N/P ratios everywhere in the Fryxell water column are higher than N/P ratios for inflows, indicating more efficient cycling of nitrogen in this closed system. In this connection, we compute that approximately 11 kg a^{-1} of NH_4-N are being diffusionally transported across the lake's 9-meter contour. Nitrogen regenerated along the NH_4-cline is, as first suggested by Vincent (1981), an important source of this nutrient to the lower euphotic zone.

We note, finally, that the areal loading ($g\,N\,m^{-2}\,a^{-1}$) of inorganic nitrogen to Lake Fryxell is some two to three orders of magnitude lower than the loading to the temperate N-limited lakes discussed by Rast & Lee (1978). For comparative purposes we show Lake Fryxell on the Rast-Lee nitrogen loading diagram in Fig. 8.

Metals: concentrations, fluxes, lake distributions

Stream concentrations for Mn, Fe, Co, Ni, Cu, and Cd are presented in Table 4. Total metal concentrations were highest, often by an order of magnitude or more, in the stream F_3. We estimate that this stream alone contributes 71% of the incoming Mn, 72% of the Fe, 81% of the Co, 56% of the Ni, and 45% of the influent Cu and Cd. Clearly, future budget studies will wish to emphasize the role played by this single meltwater stream of the Commonwealth Glacier in the chemistry of the lake.

Since total metals vary dramatically with discharge rate (Green *et al.*, 1986b), comparison of metals in these streams with world river water should be based on filtered values. Table 4 presents data for the Onyx River and for world river dissolved metals (Martin & Whitfield, 1983). Iron concentrations in the Taylor Valley streams and in the Onyx River are significantly higher than the world average (see Fig. 9), suggesting that ferromagnesian minerals associated with the Ferrar Dolorites are being chemically weathered.

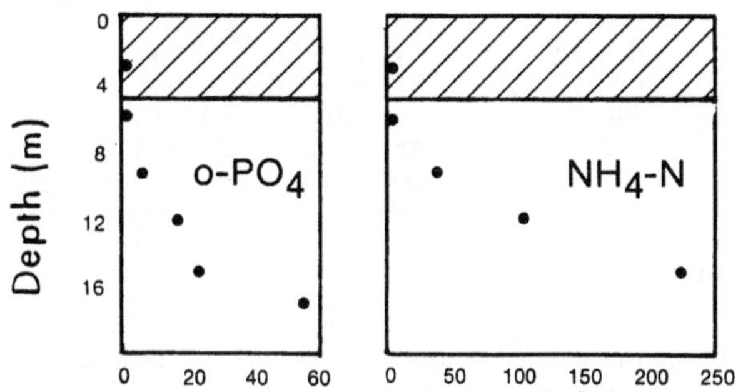

Fig. 7. Dissolved reactive phosphorus and ammonium nitrogen in Lake Fryxell.

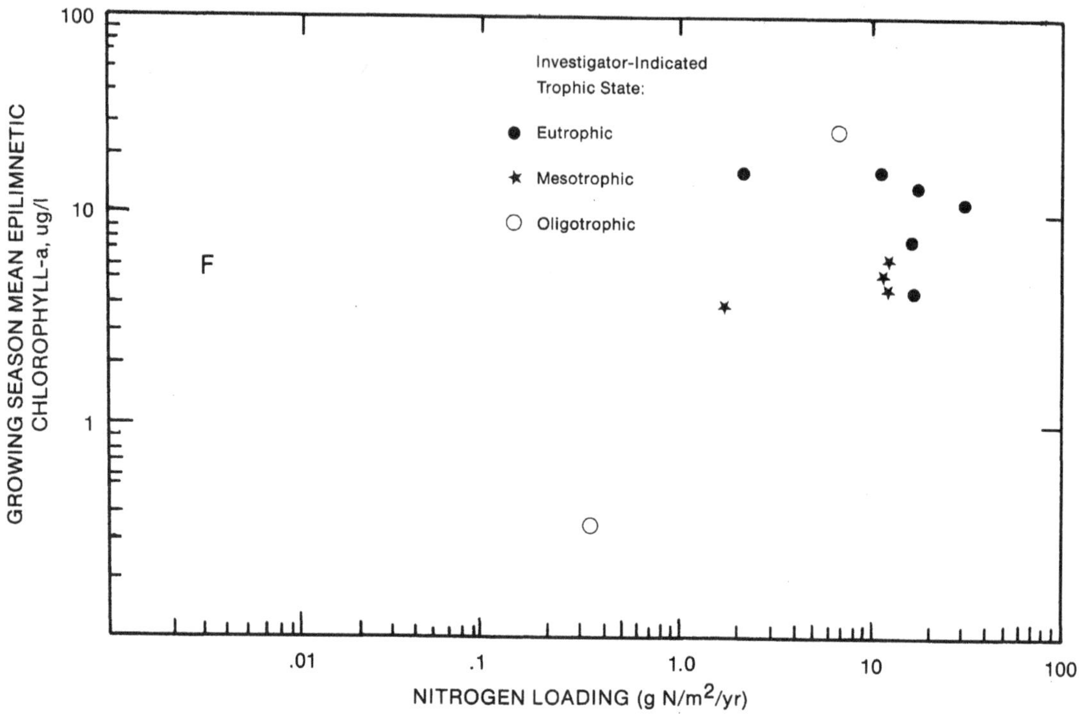

Fig. 8. Location of Lake Fryxell on a nitrogen loading diagram. (After Rast & Lee, 1978).

Table 4. Average metal concentrations in Fryxell streams (nM). Values in parentheses are 'dissolved' concentrations.

Stream	Mn	Fe	Co	Ni	Cu	Cd
F_1	90 (30)	6200 (1500)	3.0 (0.8)	5.6 (4.2)	31 (–)	0.69 (0.62)
F_2	590 (330)	38100 (20400)	6.1 (6.1)	11.6 (–)	9 (–)	0.71 (–)
F_3	2190 (50)	154100 (1500)	6.3 (0.3)	98.5 (3.4)	81 (–)	2.61 (0.36)
F_4	430 (20)	22400 (n.d.)	3.4 (n.d.)	6.6 (1.4)	19 (–)	1.16 (0.54)
F_5	990 (30)	42300 (500)	1.9 (0.5)	9.7 (0.5)	111 (–)	0.71 (0.27)
F_6	350 (30)	39300 (770)	13.8 (0.5)	30 (0.17)	14 (–)	0.58 (0.54)
F_7	90 (5)	5180 (520)	1.4 (n.d.)	2.8 (–)	19 (–)	2.0 (–)
F_8	90 (70)	6530 (4400)	2.5 (–)	3.9 (–)	8.5 (–)	0.62 (–)
F_9	40 (20)	940 (820)	0.34 (0.51)	n.d. (–)	7.9 (–)	0.36 (–)
Onyx R.*	451 (286)	16902 (1146)	10.6 (<0.3)	22.3 (6.6)	8.1 (1.7)	0.03 (0.03)
World**	(150)	(720)	(3.4)	(8.5)	(23.6)	(0.18)

* Green *et al.* (unpublished)
** Martin and Whitfield (1983)

Dissolved cadmium concentrations also appear to be higher in Fryxell streams than in average river water (Fig. 10). While in the case of cadmium there is the possibility of contamination arising from filtration we suspect that these values reflect genuinely high levels of this element in the region. Geochemical pathways are likely to involve aeolian deposition of Cd onto glacial ice and

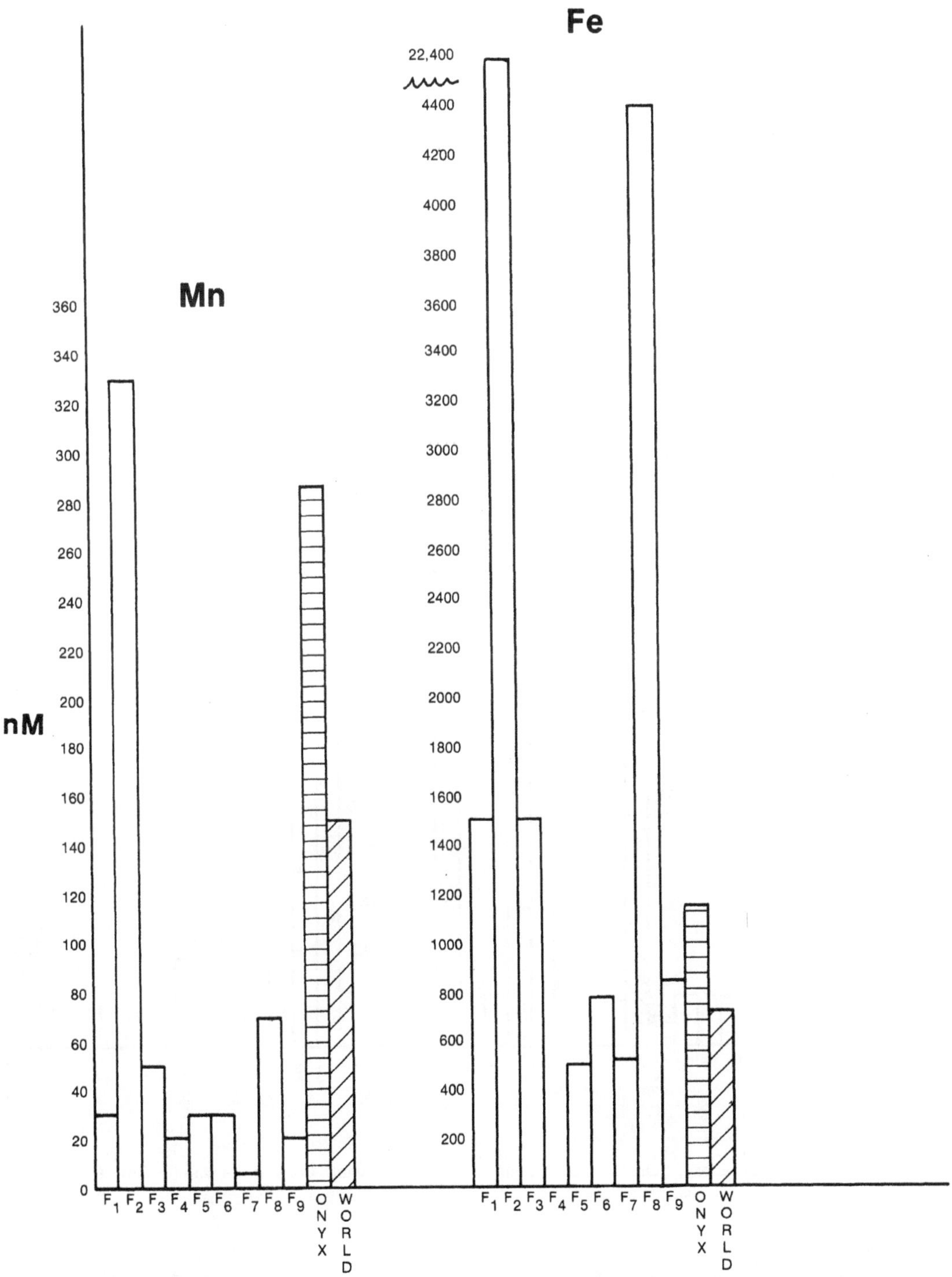

Fig. 9. Nanomolar concentrations of dissolved Mn and Fe in Fryxell streams. Comparison with the Onyx River (Wright Valley) and average river water (Martin & Whitfield, 1981).

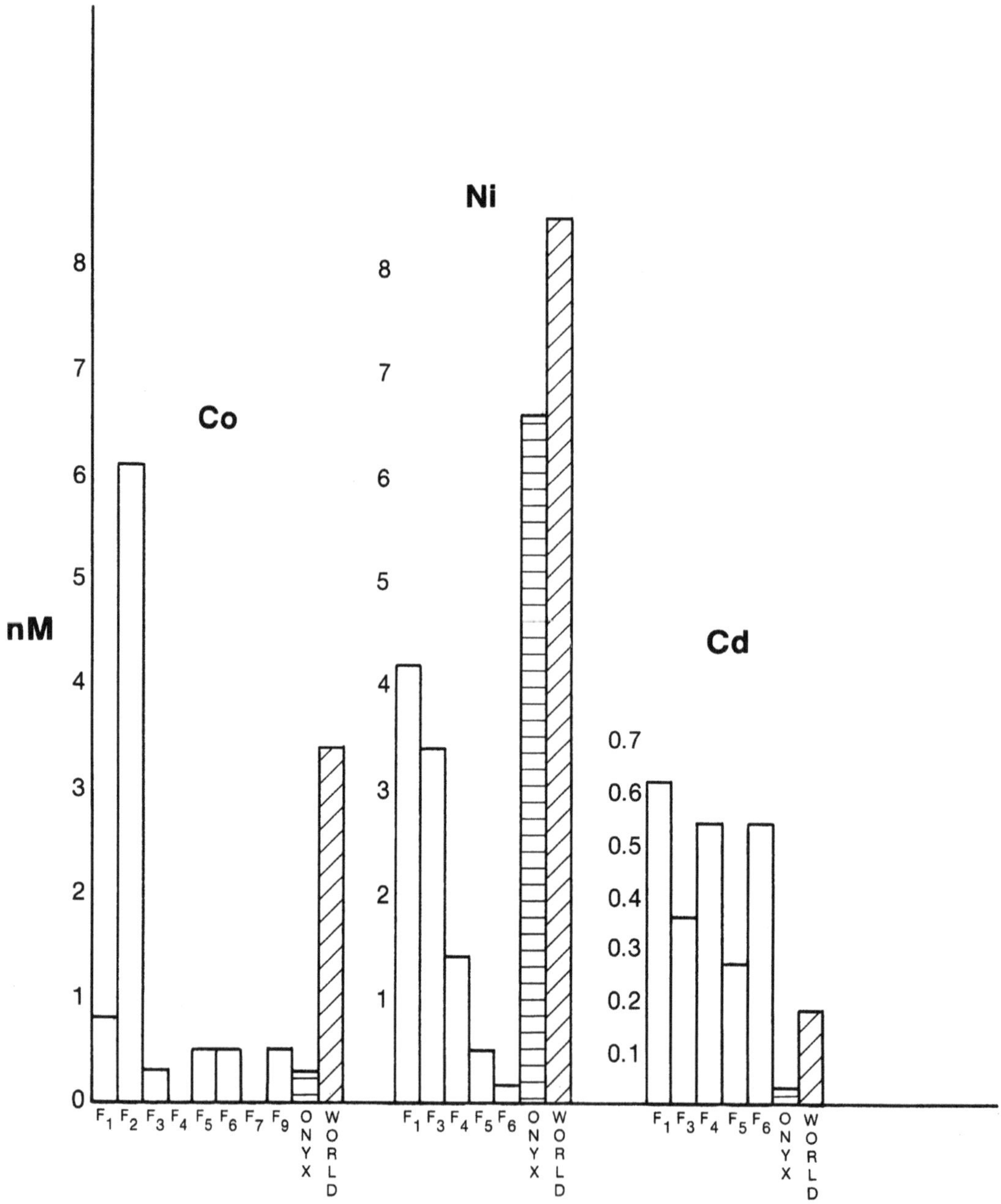

Fig. 10. Nanomolar concentrations of dissolved Co, Ni, and Cd in Fryxell streams. Comparison with the Onyx River (Wright Valley) and average river water (Martin & Whitfield, 1983).

subsequent transport by flowing waters. Weathering of local rock may also contribute to some degree to high Cd concentrations in the Fryxell streams.

Mn, Co, and Ni are generally present in lower concentrations in the Fryxell streams (Figs. 9 and 10) than in average world river water. Based on total copper concentrations, dissolved copper values for most streams are also probably lower than in average river water.

Chemical weathering of local rocks and soils appears to be the major source of most heavy metals in the Fryxell streams and in the Onyx River. This is indicated also by the correlation between dissolved metal concentrations and metal abundance in average crustal rock (Fig. 11), taken from Mason's compilation (1966). On this plot, it should be noted that Cd concentrations again appear elevated over those to be expected from dissolution of crustal rock alone. Atmospheric inputs of this element, possibly from volcanic sources, are likely.

Stream loads to Lake Fryxell are delivered to a basin, 35% of whose water volume is anoxic. This fact determines metal concentrations, profiles, and dynamics within the lake (Table 5 and Fig. 12).

Mn concentrations (Table 5) increase with depth between 6 m and 8 m. At the oxic/anoxic interface, between 8 m and 9 m, concentrations increase by a factor of 32, and then decrease rapidly with depth. The profile suggests that sinking oxide particles are being reduced in sub-oxic and anoxic waters, giving rise to the boundary maximums. Decreasing concentrations in deeper waters are a response to mineral formation reactions. In this connection, computations using WATEQ2 show that deep Fryxell waters below 12 meters are supersaturated with respect to $MnHPO_4$. Precipitation of this mineral from the lake would provide the sinks necessary to account both for the manganese profile and for the apparent removal of phosphorus relative to nitrogen. Alternatively, the inclusion of maganese as a minor component in precipitating calcite could also account for significant manganese deposition.

There is a maximum in the 'total' iron concen-

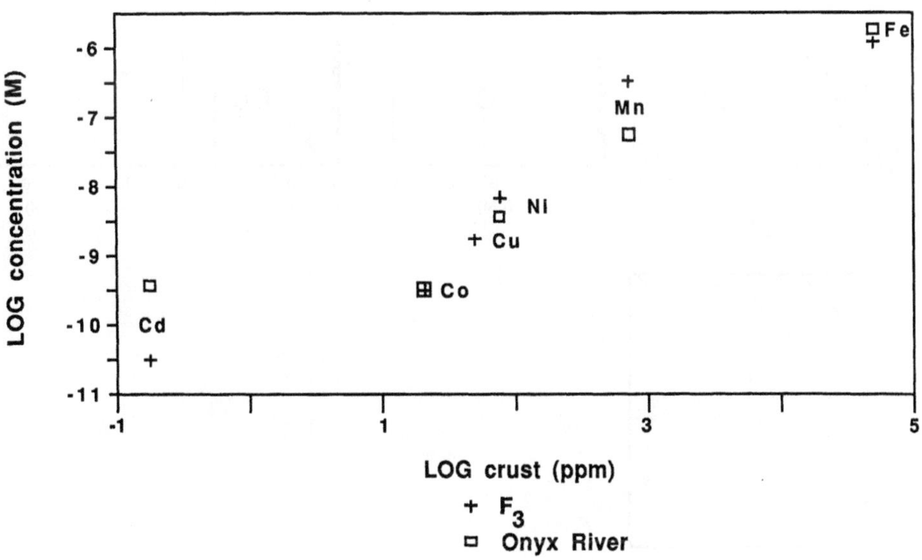

Fig. 11. Plot of dissolved metal concentrations in F_3 and the Onyx River against average crustal rock. Crustal rock values from Mason (1966).

Table 5. Metals and selected chemical parameters in Lake Fryxell. (Metals in nM; O_2 and sulfides in μM).)

Depth	pH	O_2	Mn	Fe	Co	Ni	Cu	Cd	Sulfides
6	8.05	808	(10.6)*	71.7 (122)	nd (nd)	91.7 (99.7)	71.0 (80.0)	0.53	nd
7	8.22	---	(23.3)	---		36.2 (52.0)	44.7 (42.4)	0.43	nd
8	7.81	470	(118)	---		33.4 (---)	46.3 (---)	0.33	nd
9	7.50	nd	(3754)	1398 (192)	nd (nd)	29.2 (---)	4.1 (3.0)	0.32	69.9
10	7.38	nd	(2852)	---		17.0 (33.6)	1.1 (1.6)	0.05	319
11	7.39	nd	(1822)	---		29.0 (36.3)	3.2 (nd)	0.02	457
12	7.40	nd	(1179)	753 (197)	nd (nd)	32.1 (35.3)	nd (nd)	nd	550
13	7.45	nd	(762)	---		30.3 (21.8)	nd (nd)	nd	889
14	7.32	nd	(565)	---		30.7 (34.4)	2.4 (nd)	0.14	798
15	7.35	nd	(552)	771 (1058)	nd (nd)	33.9 (39.0)	0.95 (nd)	0.09	847
16	7.30	nd	(510)	---		28.0 (26.9)	nd (nd)	0.10	937
17	7.26	nd	---	757 (549)	nd (nd)	---	---	---	1210
17.5	---	---	(479)	---		21.8 (22.3)	nd (nd)	0.04	---

* Number in parentheses indicates 'dissolved' values with filtration through acid-washed 0.45 micron filters.
--- indicates that no measurement was made.
nd means 'not detected', or below detection limits.

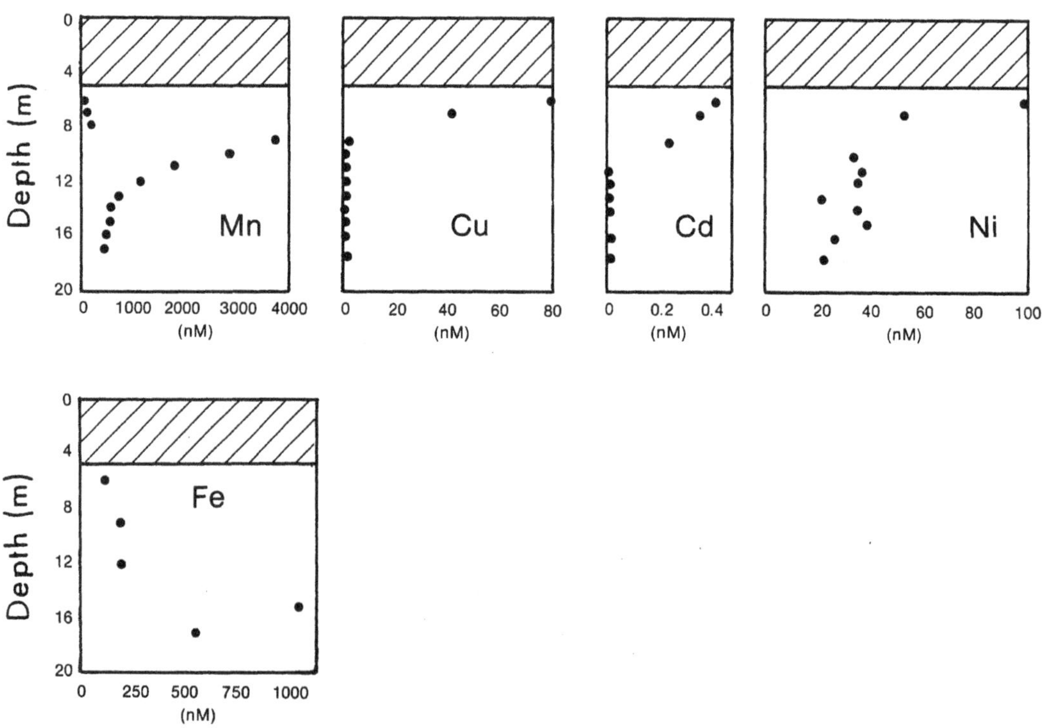

Fig. 12. Water column profiles for Mn, Cu, Cd, Ni, and Fe in Lake Fryxell.

tration at nine meters, with much of this being present in the particulate form. Calculations indicate supersaturation with respect to greigite, and this is consistent with geochemical observations in other anoxic basins (Emerson *et al.*, 1983; Jacobs *et al.*, 1985; Green *et al.*, 1986a). The high concentrations indicated for 'dissolved' iron at 15 and 17 m may be an artifact associated with using a 0.45 μm filter instead of 0.2 μm. Iron control by sulfide phases is likely below 9 meters.

Concentrations of copper and cadmium decrease from oxic to anoxic waters and we could detect no 'dissolved' copper below 11 meters. As in Lake Vanda (Green *et al.*, 1986a), chalcocite precipitation is indicated by thermodynamic calculations. Although Cd is clearly being removed from anoxic waters, insufficient data are available on the Cd-sulfur system to allow useful saturation estimates to be made.

Ni exhibits perhaps the most intriguing and, in a sense, most unexpected behavior of any metal in anoxic basins. Jacobs *et al.* (1985) called attention to the fact that in Framvaren Fjord nickel is supersaturated with respect to known sulfide phases. In Lake Fryxell, we calculate supersaturation with respect to millerite at 9 meters and at all depths below. Unlike copper and cadmium, nickel levels are not significantly lower in sulfide-bearing waters than they are under oxic conditions.

It is worthwhile to speculate what governs nickel concentrations in anoxic basins. One possibility follows from consideration of the Irving-Williams series. Equilibrium constants for organo-metallic complexes, involving divalent metal ions and ligands containing nitrogen donor atoms, are in the following order:

$$\mathrm{Mn} < \mathrm{Fe} < \mathrm{Co} < \mathrm{Ni} < \mathrm{Cu} > \mathrm{Zn}.$$

Since the oxidation state of copper in anoxic basins is Cu(I), of the elements studied here Ni(II) should form the most stable organo-metallic complexes. As Porterfield (1984) has emphasized, this order is a response to decreasing ionic radii (and hence increasing polarizing ability of the metal ion) across the transition metal series, and to increasing crystal field stabilization energy. We suggest, then, that much of the dissolved nickel in Lake Fryxell, and in other anoxic basins, may be organically complexed by N-containing ligands.

Cobalt was unfortunately below detection at all depths using our extraction procedure. We have found, however, that in Lake Vanda (Green *et al.*, unpublished) Co exhibits a maximum at the oxic/anoxic interface and that concentrations fall off rapidly with depth. Our cobalt profiles in Lake Vanda closely resemble those observed by Jacobs *et al.* (1985) in Framvaren Fjord, and cobalt in Lake Fryxell can be expected to behave in a similar manner.

Table 6 presents whole lake residence times for metals based on 'total' metal inputs and 'total' anoxic zone concentrations. The long residence times of Mn and Ni are yet another index of the tendency of these elements to be retained in anoxic basins.

Conclusions

Nine meltwater streams deliver dissolved and suspended materials to Lake Fryxell. Even though these streams have erratic flow patterns and are likely to have widely varying discharge rates from year to year (Chinn, 1981), this short-term study of their major ion, nutrient, and metal chemistry provides, in broad outline, a picture of solute acquisition processes in the Lower Taylor Valley.

The inflows to Lake Fryxell have disproportionately high concentrations of sodium and

Table 6. Metal loadings, lake content, and whole lake residence times.

Metal	In lake (moles)	Loading (moles a^{-1})	Residence time (years)
Mn	29842	1302	22.9
Fe	14550	90137	0.16
Co	< 8.2	32.7	<0.25
Ni	1379	55.6	24.8
Cu	794	74.6	10.6
Cd	7.92	2.46	3.2

chloride as compared with average world river water. Livingstone (1963) reports sodium values of 12.7 mole percent and chloride values of 10.2 mole percent in average river water. By contrast, stream F_3, the largest contributor of solutes to Lake Fryxell, has a sodium mole fraction of 22% and a chloride mole fraction of 23%. Other streams in the basin are also enriched in these ions. This points out the importance of the marine source term for these elements and for some fraction of the magnesium.

The acquisition of marine salts can only be part of the story, however, because these streams contain significant quantities of dissolved calcium and bicarbonate, well in excess of those to be expected from marine aerosols alone. Dissolution of calcite coatings derived from bedrock (e.g., marble) weathering is likely. Low magnesium and (where data are available) low silica values suggest that dissolution of ferromagnesian minerals may be of less importance in explaining the major ion chemistry of these streams.

The 'multiple source hypothesis' for soluble salts is supported further by the location of Fryxell streams on the Gibb's plot in Fig. 3. On that diagram, compositions tend to fall between 'atmospheric dominance' and 'rock dominance.' Those streams flowing the longest distances (F_6, F_2, F_8, F_7) plot in the vicinity of the great continental rivers, toward the region of 'rock dominance.'

Budgets for the 1982–83 field season can only be suggestive, in that they are snapshots in time. But on the *assumption* that compositions have remained relatively constant with respect to the major ions, we conclude that large quantities of calcium, bicarbonate, and sulfate have been lost in the evaporative evolution of Lake Fryxell, and that potassium and magnesium have also been incorporated into the sediments. Calcium carbonate (and Mg-calcite) precipitation, sulfate reduction, and, plausibly, reverse weathering are suggested both by budget calculations and by thermodynamic arguments – and these processes are sufficient to account for the evolution of the present lake waters from inflows having the compositions observed.

Perhaps surprisingly, the age of the contemporary lake, as indicated by the chloride budget, is relatively young – on the order of a few thousand years of present flows. We are convinced that Lake Fryxell has evolved in a manner similar to other closed-basin lakes worldwide (Eugster & Jones, 1979) and that its location in polar terrain is not sufficient reason to invoke exotic evolutionary pathways.

Although more detailed and longer term studies will probably alter the nutrient flux estimates given here – since these elements are notoriously coupled to stream flow – we believe that the major conclusions are likely to be confirmed: namely, that inorganic nitrogen concentrations are extremely low as compared with other flowing waters in the world; and that areal nitrogen loading to Lake Fryxell is well below that of temperate lakes. When coupled with our observations on the Onyx River, we conclude that inorganic nitrogen has its proximate origin in glacial ice and that phosphorus is derived from soils.

Within Lake Fryxell itself, nitrogen is more efficiently cycled than phosphorus, and diffusion of ammonia from deeper waters is an important source of this nutrient to the euphotic zone. Our calculations suggest that phosphorus may be lost to the sediments by precipitation of $MnHPO_4$.

With the possible exception of cadmium, metals are derived from stream weathering of local rocks and soils. The relative abundance of 'dissolved' iron is further evidence that ferromagnesian minerals are being attacked and solubilized. Elevated cadmium levels point toward an aeolian source for some fraction of this metal.

The distributions of iron and manganese in Lake Fryxell are both influenced by reductive dissolution of these metals, although the peak in 'dissolved' iron at 15 meters is not consistent with the high sulfide concentrations at that depth. Sinks for these metals, as indicated by thermodynamic calculations, include Fe_3S_4 (greigite) and $MnHPO_4$. Manganese may also be co-precipitating with calcite.

Profiles of copper and cadmium in Lake Fryxell are determined by the formation of insoluble sulfide phases. Nickel, by contrast, is highly soluble in anoxic, sulfide-bearing waters and we suspect that its geochemistry is influenced by the formation of soluble organo-metallic complexes. Its position in the Irving-Williams series is consistent with this claim.

Whole lake residence times underscore the mobility of nickel and manganese in anoxic basins and highlight the geochemical contrasts between the adjacent transition metals manganese and iron. Anoxic basins provide a setting conducive for the separation of these elements in nature.

Acknowledgments

We are grateful to the National Science Foundation's Office of Polar Programs for support of this research, and to Provost Bud Williamson of Miami University for his general encouragement and assistance over the years. We appreciate numerous conversations with Keith Chave and Don Canfield on the geochemistry of Antarctic lakes and are grateful to W. Vincent for many helpful comments. Finally, we are indebted to Betty Marak for typing the manuscript.

References

Angino, E. E., K. B. Armitage & J. C. Tash, 1962. Chemical stratification in Lake Fryxell, Victorialand, Antarctica. Science 138: 34–36.

Ball, J. W., E. A. Jenne & D. H. Nordstrom, 1979. WATEQ2 – a computerized chemical model for trace and major element speciation and mineral equilibria in natural waters. In E. A. Jenne (ed.), Chemical modeling in aqueous systems. Am. Chem. Soc., Washington, DC: 815–836.

Canfield, D. E. & W. J. Green, 1985. The cycling of nutrients in a closed-basin Antarctic lake: Lake Vanda. Biogeochemistry 1: 233–256.

Chinn, T. J. H., 1981. Hydrology and climate in the Ross Sea area. J. Roy. Soc. New Zealand 11: 373–386.

Claridge, G. G. C. & I. B. Campbell, 1977. The salts in Antarctic soils, their distribution and relationship to soil processes. Soil Science 123: 377–384.

Cline, J. D., 1969. Spectrophotometric determination of hydrogen sulfide in natural waters. Limnol. Ocean. 14: 454–458.

Danielson, L. G. & S. Westerlund, 1978. An improved metal extraction procedure for the determination of trace metals in seawater by atomic absorption spectrometry with electrothermal atomization. Anal. Chim. Acta 98: 47–57.

Downes, M. T., Howard-Williams & W. F. Vincent, 1986. Sources of organic nitrogen, phosphorus, and carbon in Antarctic streams. Hydrobiologia 134: 215–225.

Emerson, S., L. Jacobs & B. Tebo, 1983. The behavior of trace metals in marine anoxic waters: Solubilities at the oxygen-hydrogen sulfide interface. In C. S. Wong, E. Boyle, K. W. Bruland, J. D. Burton & E. D. Goldberg (eds.), Trace metals in seawater. Plenum, NY: 579–608.

Eugster, H. P. & L. A. Hardie, 1978. Saline lakes. In A. Lerman (ed.), Lakes–chemistry, geology, physics. Springer-Verlag, NY: 237–293.

Eugster, H. P. & B. F. Jones, 1979. Behavior of major solutes during closed-basin brine evolution. Am. J. Sci. 279: 609–931.

Gibbs, R. J., 1970. Mechanisms controlling world water chemistry. Science 170: 1088–1090.

Green, W. J. & D. E. Canfield, 1984. Geochemistry of the Onyx River (Wright Valley, Antarctica) and its role in the chemical evolution of Lake Vanda. Geochim. Cosmochim. Acta 48: 2457–2467.

Green, W. J., D. E. Canfield, G. F. Lee & R. A. Jones, 1986a. Mn, Fe, Cu, and Cd distributions and residence times in closed basin Lake Vanda (Wright Valley, Antarctica). Hydrobiologia 134: 237–248.

Green, W. J., T. G. Ferdelman, T. J. Gardner, L. C. Varner & M. P. Angle, 1986b. The residence times of eight trace metals in a closed basin Antarctic lake. Hydrobiologia 134: 249–255.

Hardie, L. A. & H. P. Eugster, 1970. The evolution of closed-basin brines. Miner. Soc. Am. Spec. Publ. 3: 273–290.

Howard-Williams, C., C. L. Vincent, P. A. Broady & W. F. Vincent, 1986. Antarctic stream ecosystems: Variability in environmental properties and algal community structure. Int. Revue ges. Hydrobiol. 71: 511–544.

Jacobs, L., S. Emerson & J. Skei, 1985. Partitioning and transport of metals across the O_2/H_2S interface in a permanently anoxic basin: Framvaren Fjord, Norway. Geochim. Cosmochim. Acta 49: 1433–1444.

Keys, J. R. & K. Williams, 1981. Origin of crystalline, cold desert salts in the McMurdo region, Antarctica. Geochim. Cosmochim. Acta 45: 2299–2309.

Livingstone, D. H., 1963. Chemical composition of rivers and lakes. U.S. Geol. Surv. Prof. Paper 440-G.

Mackenzie, F. T. & R. M. Garrels, 1966. Chemical mass balance between rivers and oceans. Am. J. Sci. 264: 507–525.

Martin, J. M. & M. Whitfield, 1983. The significance of river input of chemical elements to the ocean. In C. S. Wong, E. Boyle, K. W. Bruland, J. D. Burton & E. D. Goldberg (eds.), Trace metals in seawater. Plenum Press, NY: 265–296.

Mason, B., 1966. Principles of geochemistry (3rd ed.). John Wiley & Sons, Inc., NY: 329 pp.

Maybeck, M., 1982. Carbon nitrogen and phosphorus transport by world rivers. Amer. J. Sci. 282: 401–450.

McKay, C. P., G. D. Clow, R. A. Wharton & S. W. Squyres, 1985. Thickness of ice on perennially frozen lakes. Nature 313: 561–562.

Parker, B. C., G. M. Simmons, K. G. Seaburg, D. Cathey & F. C. T. Allnut, 1982. Comparative ecology of plankton communities in seven Antarctic oasis lakes. J. Plank. Res. 4(2): 271–286.

Porterfield, W. W., 1984. Inorganic chemistry: A unified approach. Addison-Wesley Publishing Co., Reading, MA: 688 pp.

Rast, W. F. & G. F. Lee, 1978. Summary analysis of the U.S. portion of the North American OECD eutrophication study results emphasizing nutrient loading lake response relationships and trophic state indices. U.S. EPA. EPA 600/3-78-008, Corvallis Environmental Research Laboratory, Corvallis, OR: 454 pp.

Spencer, R. J., H. P. Eugster, B. F. Jones & S. L. Rettig, 1985. Geochemistry of Great Salt Lake, Utah. I: Hydrochemistry since 1850. Geochim. Cosmochim. Acta 49: 727–737.

Torii, T., N. Yamagata, S. Nakaya, S. Murata, T. Hashimoto, O. Matsubaya & H. Sakai, 1975. Geochemical aspects of the McMurdo saline lakes with special emphasis on the distribution of nutrient matters. In T. Torii (ed.), Geochemical and geophysical studies in Dry Valleys, Victoria Land, Antarctica. Memoirs Natn. Inst. Polar Res. (Japan) Spec. Issue No. 4: 5–29.

Vincent, W. F., 1981. Production strategies in Antarctic inland waters: Phytoplankton eco-physiology in a permanently ice-covered lake. Ecology 6: 1215–1224.

Von Damm, K. L. & J. M. Edmond, 1984. Reverse weathering in the closed-basin lakes of the Ethiopian Rift. Amer. J. Sci. 284: 835–862.

Hydrobiologia 172: 149–164, 1989
W. F. Vincent and J. C. Ellis-Evans (eds) High Latitude Limnology

Some aspects of iron cycling in maritime antarctic lakes

J. C. Ellis-Evans & E. C. G. Lemon
British Antarctic Survey, Natural Environment Research Council, High Cross, Madingley Road, Cambridge, England, CB3 OET

Key words: Antarctic lake, iron, iron oxidation, iron reduction, phosphorus, sulphide

Abstract

Iron occurs in extremely high concentrations in certain maritime Antarctic freshwater lakes which seasonally develop an anoxic zone. In oligotrophic Sombre Lake the data show that Fe(II) precipitates as Fe(III) oxyhydroxides which bind phosphorus and return it to the sediments. In nutrient-enriched Amos Lake, significant quantities of sulphide are also produced and this binds a proportion of the released Fe(II) so reducing the ratio of total iron to phosphorus at the redox boundary where the oxyhydroxides are formed. A proportion of the sediment-released phosphorus therefore reaches the upper waters of this lake (unlike in Sombre Lake) and provides the initial nutrient source for under-ice phytoplankton development in spring. Iron-reducing bacteria have been isolated, from Sombre Lake sediments, which apparently utilise the abundant Fe(III) oxyhydroxides. From thermodynamic considerations (assuming Fe(III) is not limiting) these should outcompete sulphate reducers and methanogens (both previously reported from Sombre and Amos Lakes) and could therefore constitute an important component of the anaerobic mineralisation of organic carbon in such lakes.

Introduction

It has long been recognised that iron is an important component of aquatic sediment geochemistry but only recently has iron reduction been considered a potentially significant process in the mineralisation of organic matter (Froelich *et al.*, 1979; Sorensen, 1982; Jones *et al.*, 1983; Brannon *et al.*, 1984; Lovley & Philips, 1986a). The freshwater lakes of Signy Island, South Orkney Islands, are typical maritime Antarctic systems which have been extensively studied by the British Antarctic Survey for over 15 years (Heywood, 1985). Work by Gallagher (1985) indicated that large concentrations of iron accumulated in the anoxic waters of these lakes each winter and that this appears to influence nutrient release to overlying oxic waters. Studies have shown that a significant anaerobic microflora, including methanogens, is active in the anoxic regions of these permanently cold water bodies during the same period (Ellis-Evans, 1985). Recent work (Froelich *et al.*, 1979; Sorensen, 1982; Jones *et al.*, 1984; Lovley & Philips, 1986a) suggests that iron reduction is microbially-mediated and that in some environments iron-reducing activity may be linked to organic matter mineralisation. The present study examines the interaction of iron, phosphorus and sulphide in two contrasting lake systems during the long period of winter ice cover and presents evidence for a population of iron-reducing bacteria potentially competing with other microbial groups for the available organic matter.

Study sites

Signy Island (60° 43′ S, 45° 38′ W) lies at the northern edge of the Weddell Sea (Fig. 1a) and falls within the maritime Antarctic climatic zone with mean monthly air temperatures between +2.0 and −18.0 °C. The seventeen lakes of Signy Island share a common geology but encompass a wide range of trophic status from ultra-oligotrophic to grossly eutrophic (Heywood *et al.*, 1980). The period of open water differs from lake to lake over the summer period of January – early April and during this time the lakes are subject to varying degrees of wind-induced mixing. For the

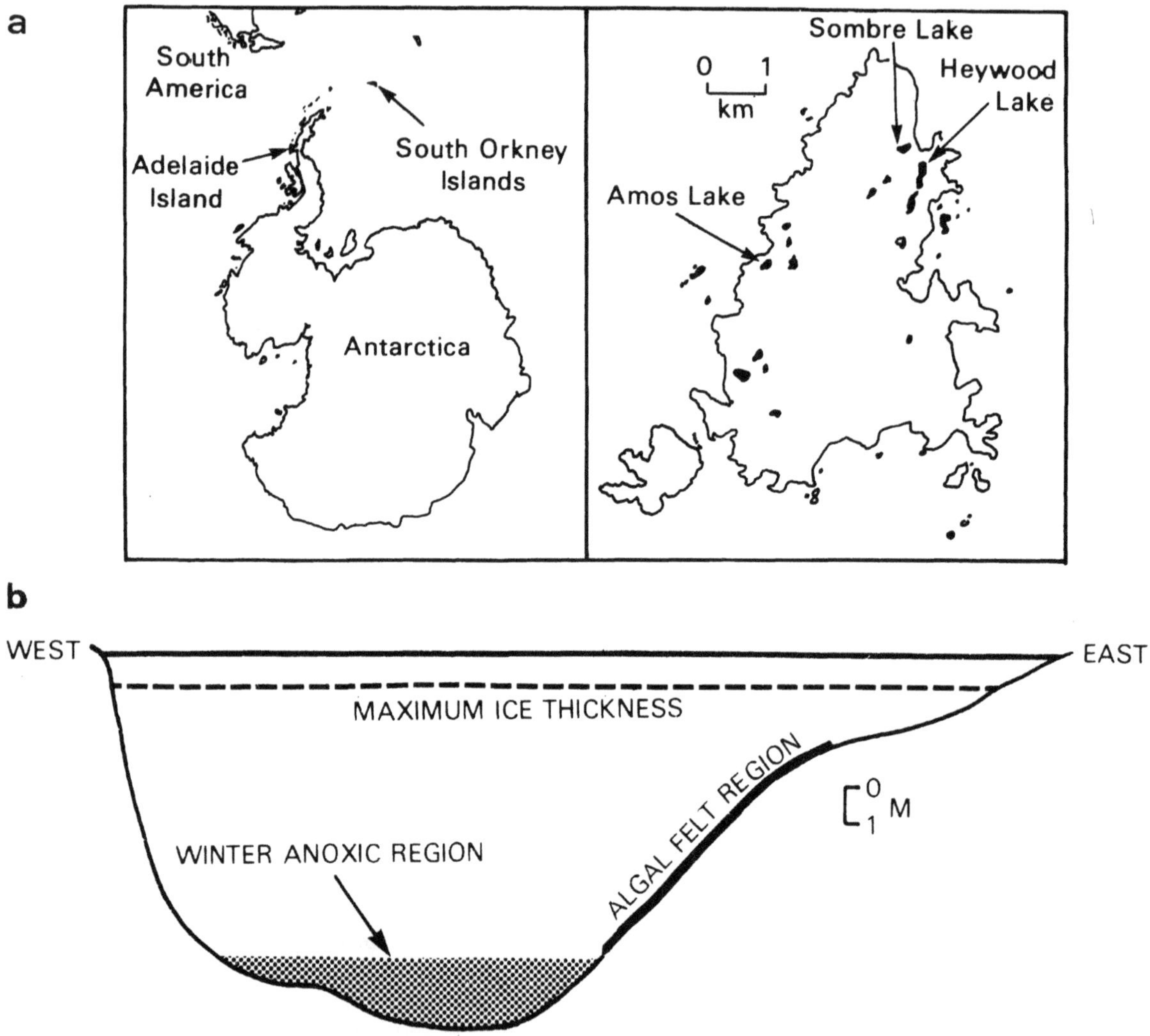

Fig. 1. (a) Maps to show the location of Signy Island (South Orkney Islands) in relation to the Antarctic continent and the position of the three main study lakes at Signy Island.
(b) Diagram to indicate the position of and extent of ice cover, benthic algal mats and the anoxic sump in Sombre Lake. Only the vertical dimensions are to scale.

remainder of the year all the lakes are ice-covered to a depth of 1 m and develop stable density gradients. The annual temperature range is 0–6 °C and is even more stable in deep water (1–3 °C).

The two lakes used in the present study (Fig. 1b) are markedly different in biological and chemical terms. Sombre Lake (max. depth 10.8 m) is a clear, oligotrophic system receiving no significant enrichment and with a biota dominated by benthic cyanobacterial mats. The inflows are essentially snow melt water and are frozen for 8–9 months each year. Amos Lake is a small, shallow (max. depth 4.1 m) eutrophic lake and receives considerable nutrient enrichment from the surrounding catchment which is frequented by seals and seabirds in summer (Ellis-Evans, 1982). A dense phytoplankton develops during spring and summer in response to the high nutrient levels but this sediments from the water column when the ice reforms in April/May.

Methods

In summer, samples were taken from a rubber dinghy moored to a line stretched across the lake. In winter, a petrol engine-driven 250 mm diameter ice auger was employed to cut a hole at the sampling site. Samples for the water chemistry were obtained using the siphon sampling device described in Gallagher (1985). This apparatus is designed to sample through ice cover and allow samples to be taken without significant exposure to air. The device was fitted with a sediment detector (Cunningham & Davison, 1980) to permit sampling within 5 cm of the sediment surface. Samples were taken at 0.1 m intervals up the water column, starting at the sediment surface, and water from each interval was stored in 150 ml polyethylene bottles or 250 ml glass bottles for chemical and polarographic analyses respectively. All bottles were cleaned in Decon 90, acid-soaked overnight (10% HCl) and rinsed several times with double distilled deionised water before use. Bottles for total iron were predosed with 3.2 ml of 5 M HCl whilst those for total sulphide were dosed with 1 ml of 0.3 M zinc acetate after taking the water sample. Predosing, in the latter case, would have resulted in the acetate solution freezing in the bottle at prevailing air temperatures. Samples for polarographic analysis were taken observing the recommendations of Davison (1977a) and using bottles with modified stoppers to minimize contact with oxygen during manipulations.

All water samples were returned to the laboratory within 2–3 h in an insulated box. Where filtered samples were required, an Antlia hand pump fitted with a three way valve was used to draw up samples through acid-cleaned GF/F filters into a collection bottle. Care was taken to avoid inclusion of air in this process. Measurements of oxygen and temperature were made in situ with a YSI Model 57 meter and stirrer-assisted polarographic electrode.

Total Fe was analyzed in UK by atomic absorption spectrophotometry using the standardized technique in Methods for the Examination of Water and Associated Materials (1983). Total P was analyzed following the methodology of Eisenreich *et al.* (1975). The molybdenum blue colour reaction was measured at 882 nm by uv/vis spectrophotometry using a 4 cm cell. Total sulphide was measured using the N, N-dimethyl-p-phenylenediamine method described in Golterman *et al.* (1978), modified to increase sensitivity. The colour reaction was measured at 745 nm, again using a 4 cm cell. Sulphate was analyzed turbidimetrically (Tabatabai, 1974) but subtracting a reagent blank to remove possible interference from humics and ferric iron. Recent work by Davison and Finlay (1986) suggest a maximum overestimate of 10 μM may be anticipated in the presence of high ferric iron levels. Polarographic analysis of Fe(II) and dissolved oxygen in lake bottom waters was determined, according to Davison (1977b), by means of an EG & G (Princeton Applied Research) Model 303 static mercury dropping electrode and Model 264 Polarographic Analyzer. Oxygen was analyzed by sampled DC polarography, as was Fe(II) at concentrations > 2 mg l^{-1}. Below this concentration differential pulse polarography was employed for

Fe(II). Alkalinity was calculated from a Gran titration (Talling 1973) and pH was measured using a combined glass electrode preconditioned with lake water.

Water samples for microbiological analysis were obtained using the siphon sampler described above. Sediment samples were obtained using the surface sediment sampler described in Ellis-Evans (1982). Both water and sediment samples were returned to the laboratory within 2 h and all manipulations were undertaken in a constant temperature room at +5 °C unless otherwise stated. Where necessary, cores were extruded under a stream of nitrogen.

Iron-reducing bacteria were initially obtained in semi-solid enrichments using the medium described by Jones (1983). Subsequently, the liquid medium described in Lovley & Philips (1986a) was employed, sometimes in semi-solid form, with considerable success.

The distribution of iron reduction and methanogenesis in sediments was undertaken with extruded sediment slices representing 0.0–0.5 cm, 0.5–1.0 cm, 1–2 cm, 2–3 cm and 3–4 cm depths. These were slurried in serum vials, carefully excluding oxygen by employing the methods of Latham & Wolin (1978) as far as possible. All vials were incubated in the dark for 15 days at 30 °C which earlier work had suggested would be appropriate for maximum activity of anaerobes (Ellis-Evans, 1981b). As in all experiments described here, samples were incubated under an atmosphere of 95% nitrogen/5% carbon dioxide. Fe(II) production was monitored by removing 0.2 ml samples and analyzing for Fe(II) with ferrozine (Sorensen, 1982). Methane production in the vials was monitored by headspace analysis using a Pye Unicam PU4500 gas chromatograph fitted with a gas sampling valve and FID. The details are given in Ellis-Evans (1984).

Surface sediments from Sombre Lake were incubated for 3 weeks at 25 °C to try to ensure that Fe(III) was depleted. Amendments with amorphous ferric oxyhydroxides, either saturated with phosphate or phosphate-free, were made to achieve concentrations of 200–300 μmol of added Fe(III) per gm dry sediment. The oxyhydroxide was formed by neutralizing a 0.4 M solution of ferric chloride with sodium hydroxide. This was then washed five times with deionised water to remove chlorides. Phosphate was adsorbed onto samples of oxyhydroxides by adding the slurry to 0.25 M potassium phosphate at neutral pH. The slurry was then centrifuged and washed as before. Acetate was used as a further amendment at a final concentration of 25 mM. Again, incubations were undertaken at 30 °C under an atmosphere of 95% nitrogen/5% carbon dioxide.

A third set of experiments with surface sediment slurries examined the effect of various electron donors on Fe(II) production in the presence or absence of Fe(III) oxyhydroxides. Amendments to serum vials were as follows: – 50 μmol of Fe(III), 15 μmol of acetate, 10 μmol of glucose and 50 μmol of hydrogen. An incubation period of 15 days at 30 °C was used. Slurries were also incubated in the presence or absence of nitrate, an alternative electron acceptor to Fe(III), and over a range of temperatures to test for temperature response.

Some experiments used intact cores in which the overlying water was replaced by a nutrient solution as described by Jones *et al.* (1983). Unstoppered cores were incubated for 12 h at 5 °C and then the solution replaced with fresh solution and the cores stoppered. This hopefully re-established a chemical gradient within the sediments but in the presence of overlying water of known composition. Small samples removed from the incubated cores at intervals for analysis were replaced from a reservoir of the mineral solution. The cores were incubated at 30 °C.

Results

Oxygen profiles. During summer open water (Fig. 2), both lake water bodies were fully oxygenated due to the influence of wind-induced turbulence. Towards the end of April both lakes began to develop an ice cover and the effects of wind action were thus removed. By June when day length was at its annual minimum, ice cover

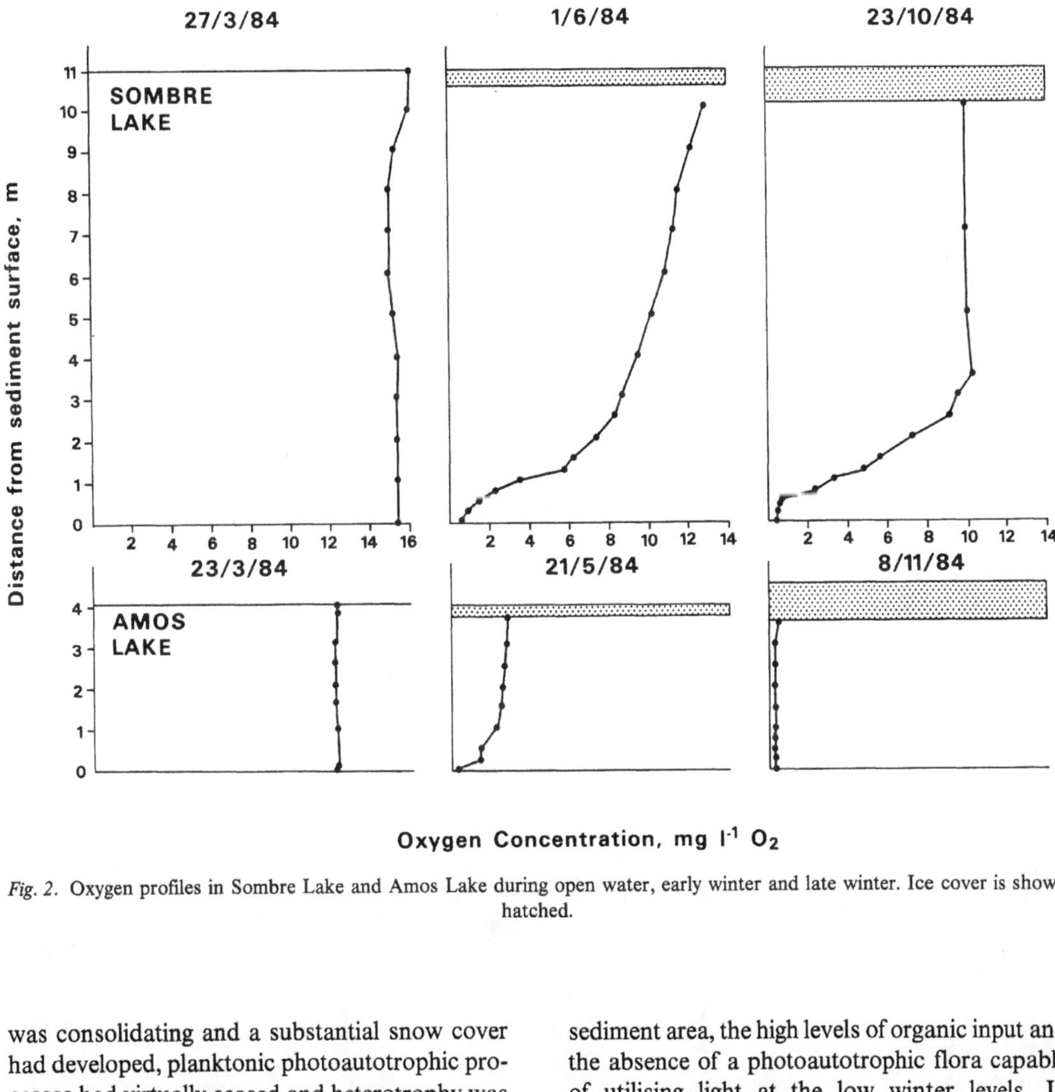

Fig. 2. Oxygen profiles in Sombre Lake and Amos Lake during open water, early winter and late winter. Ice cover is shown hatched.

was consolidating and a substantial snow cover had developed, planktonic photoautotrophic processes had virtually ceased and heterotrophy was pre-eminent. This is reflected in the Amos Lake profile (Fig. 2) where rapid deoxygenation (47g O_2 m^{-2} sediment d^{-1}) was measured on a whole lake basis. By July, the entire water column of Amos Lake had become fully anoxic (from field oxygen sensor measurements) though polarographic analyses did reveal trace amounts of oxygen in the upper regions of the water column throughout the winter. The high rate of oxygen consumption can be explained in terms of the small volume of Amos Lake in relation to its sediment area, the high levels of organic input and the absence of a photoautotrophic flora capable of utilising light at the low winter levels. In Sombre Lake, phytoplankton were scarce but an extensive cyanobacterial mat extended to 8 m depth. The only portion of the oxygen profile to show a marked decrease was that below 8 m where little or no algal mat occurred. In June, when incident light levels were lowest, a decline in oxygen content was evident with depth in the region 2–7 m where mat growth occurred. However, by October, light levels had risen and the trend towards lower oxygen concentrations had been countered, if not reversed, as indicated by the vertical oxygen profile shown in Fig. 2 (23/10/84). The slow decline observed in the top two to three metres can be attributed to the virtual absence of any substantial sediments in this region. Below 8 m oxygen concentrations declined sharply and in the last half metre, at the deep spot, no oxygen was detected throughout the period August – November.

Sombre Lake chemistry. In summer, concentrations of iron and various nutrients were low (Fig. 3) throughout the water column. However, as ice cover developed and wind influences were removed so aerobic mineralisation became increasingly significant. With the depletion of oxygen in the bottom waters the oxic/anoxic boundary moved upwards out of the sediments into the overlying water. Both iron and phosphorus mobilized within the anoxic sediments diffused upwards into the water column, the iron occurring at almost milli-molar concentrations. Sulphide was also detectable, just above the sediment surface but never attained high concentrations throughout the winter. Earlier work (J. B. Gallagher, pers. comm) suggested that conditions might be appropriate for $FeCO_3$ formation, but in

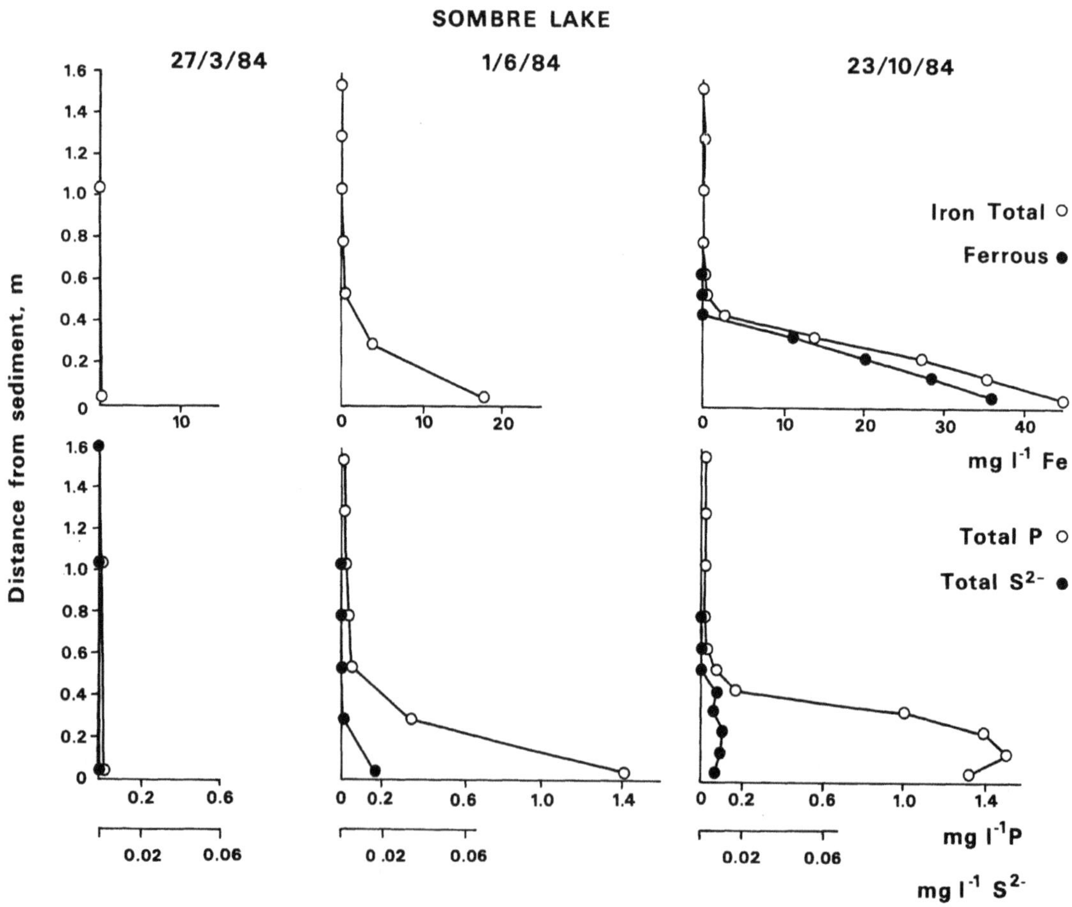

Fig. 3. Profiles of total and ferrous iron, total phosphorus and total sulphide in bottom waters of Sombre Lake during open water, early winter and late winter.

the present study pH values were consistently less than 7.0 in the bottom waters and carbonate formation was therefore unlikely (W. Davison, pers. comm.). Calculated solubility products for FeS indicated that sulphide concentrations were too low to exert much effect on the Fe profile in Sombre Lake.

Phosphorus (Fig. 3) never diffused more than 0.6 m above the sediment surface at the deep station and the profile broadly followed the profiles for total iron and ferrous iron. The profiles suggest that ferrous iron diffused up towards an oxic boundary where oxidation occurred. Direct observation (using SCUBA) revealed the presence of a brown turbid suspension extending, on occasion, up to 0.8 m above the deep spot sediments. The sedimenting material appeared to be a complex of amorphous ferric oxyhydroxides and phosphate, for which such oxyhydroxides have a high affinity. In the presence of such high concentrations of Fe(III), clearly little or no phosphorus transfer to the oxic waters occurred over winter. Calculations from data supplied by J. B. Gallagher (in prep.) suggest that the bulk of phosphorus supply to the oxic waters of this lake was instead derived from the limited nutrient content of spring snow-melt water. The reversal of the total phosphorus profile in the bottom sample, together with the relevant pIAP and solubility product calculations, indicate that a second process of phosphorus removal, as ferrous phosphate, was also occurring at the highest phosphorus concentrations.

Amos Lake chemistry. Examples of the seasonal changes in vertical profiles of iron, phosphorus and sulphide are given in Fig. 4. Phosphate concentrations in Amos Lake were always at least an order of magnitude higher than in Sombre Lake. In contrast, iron concentrations were consistently lower in the enriched lake, even with the development of winter anoxia. One of the most significant features of the data presented here was the high concentrations of sulphide present in Amos Lake over winter when compared to Sombre Lake. Peak dissolved sulphide concentrations were attained in October/November when two orders of magnitude separated the maximum concentrations recorded in Amos Lake and Sombre Lake (0.84 and 0.014 $mg\,l^{-1}$ respectively). The concave form of the sulphide profile indicates that sulphide production was confined to the sediments and that at the deep spot, saturation with respect to ferrous sulphide, occurred late in the winter resulting in precipitation of ferrous sulphide. pK_{sp} calculations for FeS indicated values of < 3.0 in the bottom 0.5 m of the vertical profile and that values then increased to 7.0 at 0.8 m. Analysis of sulphate indicated that the late summer mean concentrations of 4 $mg\,l^{-1}$ quickly declined in the bottom 0.5 m of the vertical profile as anoxia developed under ice cover. Less than 0.2 $mg\,l^{-1}$ sulphate was detectable in the bottom waters of Amos Lake from early September to late December.

The vertical profiles of iron and phosphorus which had developed by November (Fig. 4) were again essentially concave indicating a sediment-based control but between 0.2 and 0.6 m above the sediment the profile shape of both these measurands and Fe(II) was distorted by a lateral interaction which was attributed to the lake morphometry. The phosphorus and total iron profiles were very similar but departed from that of Fe(II) approximately 0.7 m above the sediment surface. This coincided with higher levels of particulate iron oxyhydroxides which peaked at about 0.8 m. Analysis of these oxyhydroxides demonstrated the presence of phosphorus, so clearly some degree of co-precipitation had occurred. However, the phosphorus profile and that of total iron showed no sharp decline at the oxic-anoxic boundary and the overall profile suggests that either a quantity of the phosphorus was not adsorbed and diffused into the upper waters or, less likely, that phosphorus release occurred in sediments located above the oxic-anoxic interface. Comparison with earlier sampling dates shows a slow accumulation of phosphorus in the upper waters of the lake in the second half of the winter, confirming that freeze-out during formation of the ice cover was not a significant source for this accumulation.

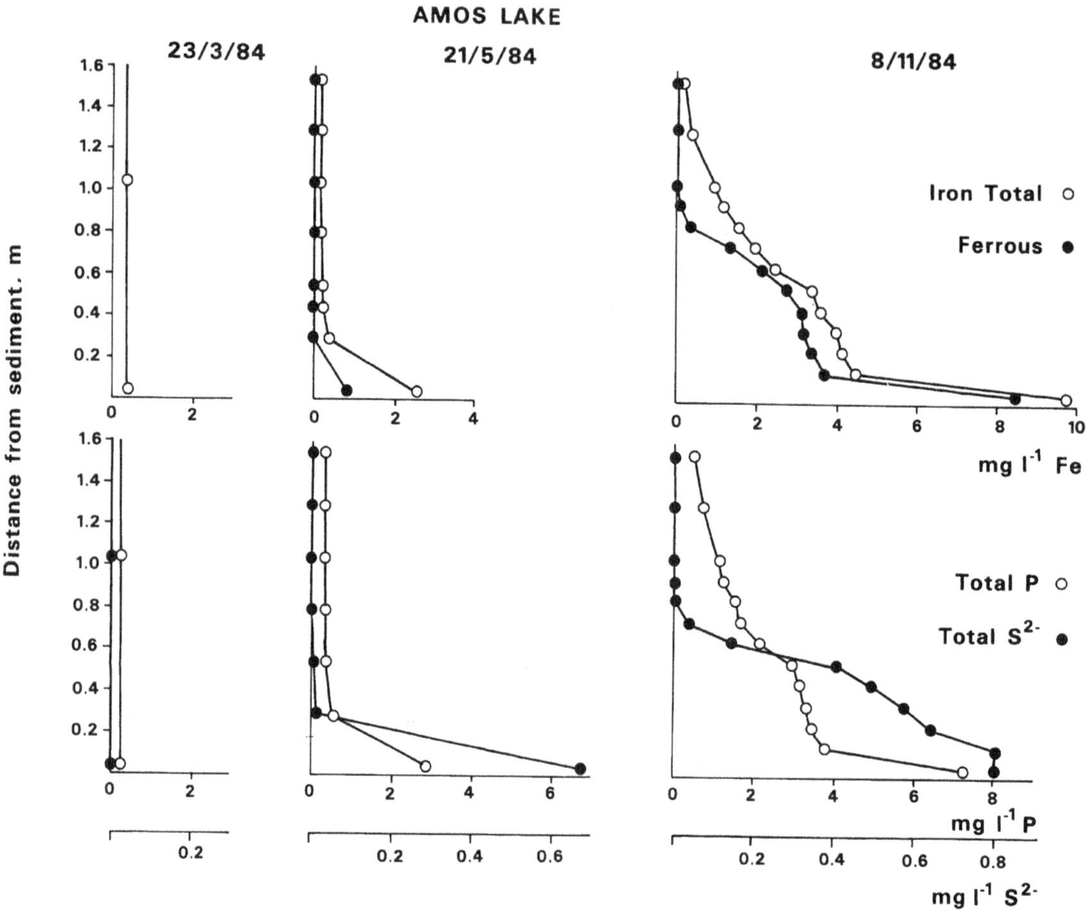

Fig. 4. Profiles of total and ferrous iron, total phosphorus and total sulphide in bottom waters of Amos Lake during open water, early winter and late winter. Compare axes with Fig. 3.

Examination of the filters from anoxic filtration experiments undertaken between Sept 27 and Nov 26 indicated that this was the period when anoxia peaked in Amos Lake although anoxic conditions were evident in bottom waters for 6 months in all. Three zones of particulate material were recognised (Fig. 5) based on colour and particle size, namely a brown layer comprising amorphous iron oxyhydroxides and two black zones comprising fine particulate or coarse particulate FeS. In both FeS zones, the concentration of the relevant particle decreased with height above the sediment and there was some suggestion of a steady transition to larger particles with depth. The brown zone tended to peak in concentration towards the centre and decline towards either perimeter of the zone. Towards the middle of November reoxygenation of the upper water column caused the oxic boundary to move down towards the sediment and all three zones retreated accordingly.

Iron reduction in Sombre Lake sediments. Profiles of ferrous iron in Sombre Lake indicated that

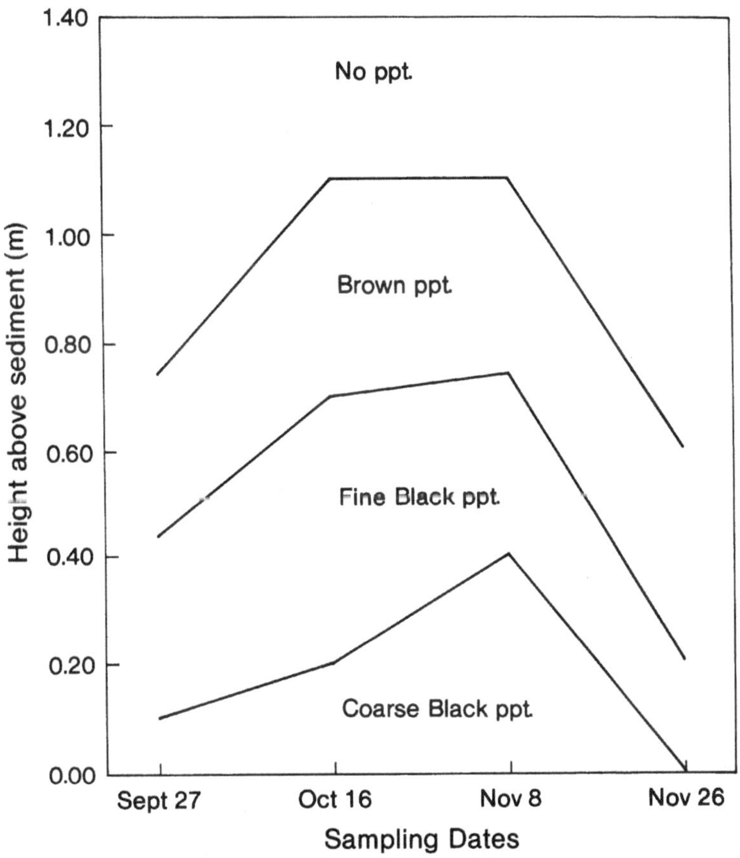

Fig. 5. Diagram to indicate the position of the three zones (defined in text) of particulate material in the bottom waters of Amos Lake in late winter. Only four sampling dates are shown here for clarity.

reduction was occurring in the sediments rather than the overlying water. Measurements of iron reduction using extruded core samples (Fig. 6) confirmed the absence of iron reduction in water immediately above the sediment surface and indicated that peak activity was associated with the 0.0–0.5 cm layer. Experiments were undertaken at 30 °C for up to 15 days and during this period it was noted that the rate of iron reduction in the surface sediment declined after five days. A similar trend was seen at the other depths but was not as pronounced as considerable variability was found in these latter samples.

Analysis of methane production in these same core samples (Fig. 7) revealed similar levels of production at all depths but that the 0.0–0.5 cm layer, in particular, significantly increased its methane production rate over the 15 day period suggesting a degree of interaction between methanogens and iron reducers. Examination of Amos Lake water and sediment samples indicated that, in early winter, iron reduction was located in the surface sediments but that the location steadily moved up the water column as anoxia developed. By November, peak activity was located at the lower edge of the brown preci-

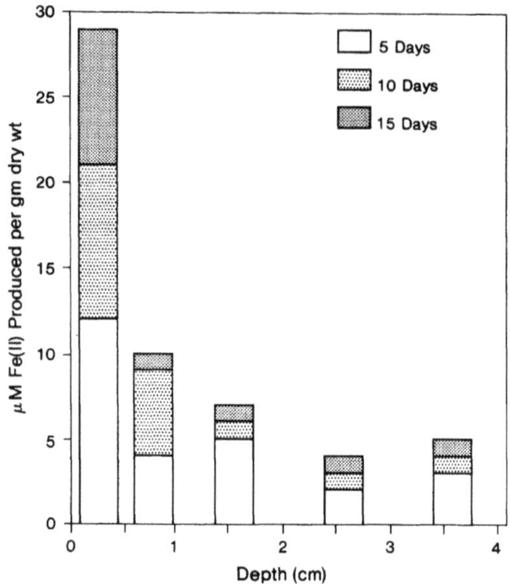

Fig. 6. The distribution of iron reduction at five depths within the sediment profile of Sombre Lake. Production values after 5, 10 and 15 days are shown.

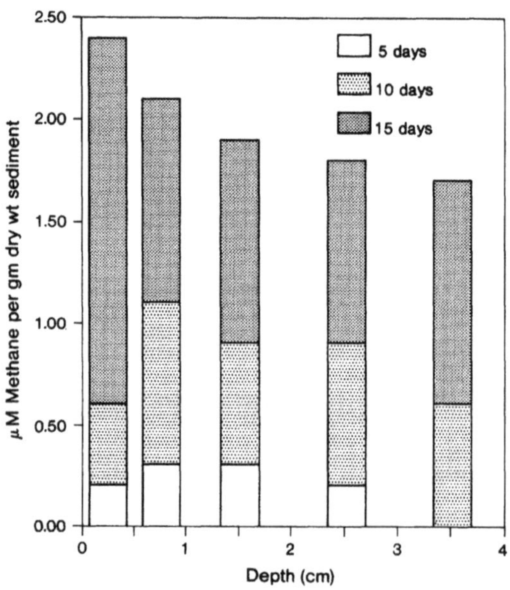

Fig. 7. The distribution of methane production at five depths in Sombre Lake sediments. Production values after 5, 10 and 15 days are shown.

pitate zone with little or no activity evident (from MPN estimates) in the coarse black precipitate zone.

Amendment experiments. The fact that iron reduction was largely located in the surface sediments of Sombre Lake suggests that iron reduction was favoured by the presence of iron oxyhydroxides since Fe(III) was detected up to 5 cm into the sediment profiles examined though as yet the form it takes is unknown. The effect of additions of amorphous ferric oxyhydroxides on the activity of Fe(III)-depleted surface sediment samples is shown in Fig. 8. Initially, iron reduction occurred in the control but with time the rate decreased. In contrast the rate increased with time when either amorphous iron oxyhydroxides amended or unamended with phosphate was added. There was no significant difference between either treatment so the presence of phos-

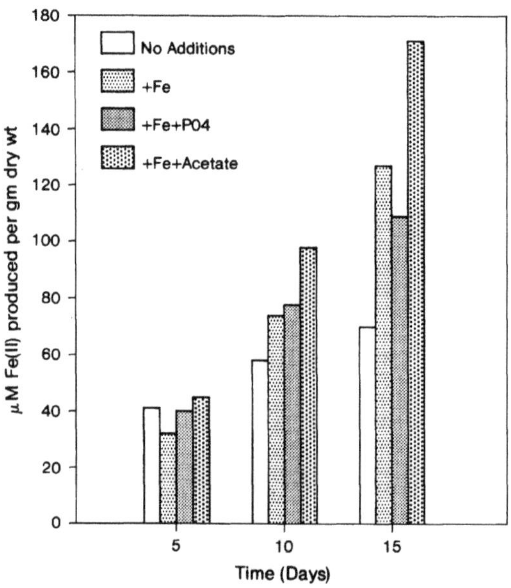

Fig. 8. Amendment experiments with Sombre Lake surface sediments. Samples were either unamended or amended with amorphous ferric oxyhydroxide (+Fe), amorphous ferric oxyhydroxide saturated with phosphate (+Fe+PO_4) or amorphous ferric oxyhydroxide and acetate (+Fe+Acetate).

phorus was clearly irrelevant to iron reduction. The third amendment of particulate iron plus acetate gave a significantly higher rate after 15 days than the other amendments indicating that these iron reducers could respond to fermentation end-products. Incubation with acetate alone gave rates only slightly higher than the control.

Methanogenesis was measured in a parallel set of amendment experiments (data not shown) which revealed that methanogenesis was inhibited by 31–65% of the control in the +Fe and +Fe+PO_4 amendments but stimulated 17% above the control in the presence of iron and acetate and 39% in the presence of acetate alone. This suggested that the iron reducers were controlled primarily by the supply of Fe(III) and secondarily by the availability of assimilable carbon. In a situation where acetate was in short supply but Fe(III) oxyhydroxides plentiful they could apparently out-compete methanogens for electron donors. A subsequent experiment compared Fe(II) production in sediment samples amended with acetate, glucose or hydrogen in the presence or absence of amorphous iron oxyhydroxides (Table 1). The results indicate that highest production rates in both control and amended samples occurred in the presence of Fe(III) but that the addition of electron donors also had a positive effect. There was no significant difference between acetate and glucose amendments but both resulted in somewhat higher production than the hydrogen amendment. Addition

Table 1. Production of Fe (II) in Sombre Lake surface sediments in response to amendment with various electron donors and amorphous Fe (III) oxyhydroxides

Fe (III) addition	Donor treatment			
	Control	+ Acetate	+ Glucose	+ Hydrogen
– Fe (III)	14 ± 3.8	25 ± 2.9	22 ± 1.7	25 ± 4.3
+ Fe (III)	27 ± 4.7	48 ± 5.1	43 ± 5.2	39 ± 5.0

Incubation temperature 30 °C for 15 days.
Amendments were as follows: – 50 μmol of Fe (III), 15 μmol of acetate, 50 μmol of hydrogen and 10 μmol of glucose per gm wet sediment.
Results expressed as μmoles of Fe (II) produced per gm dry sediment. Mean (n = 5) ± 1 standard deviation.

of various fermentable substrates, such as cellulose, also resulted in increased production over the control value, particularly in the presence of Fe(III), indicating that the iron-reducing bacteria population in this lake could potentially utilize a wide range of fermentation products.

Effect of temperature and nitrate. Sediment cores were incubated at a range of temperatures in the presence or absence of mercuric chloride. The response of iron reduction to temperature is shown in Fig. 9 and indicates that highest production occurred between 30–40 °C (no significant difference between the two values) and followed a classic biological response curve. The lowest incubation temperature presented in the figure is 10 °C and the calculated rate of production at this temperature was exceedingly slow. Production at +3 °C was approximately 60% less again but high variability in the results was obtained at these very low production rates. Addition of mercuric chloride prevented biological activity and revealed that 17–29% of measured activity could be attributed to chemical activity. Nitrate concentrations declined rapidly in the bottom waters of both lakes as anoxia developed (data not shown)

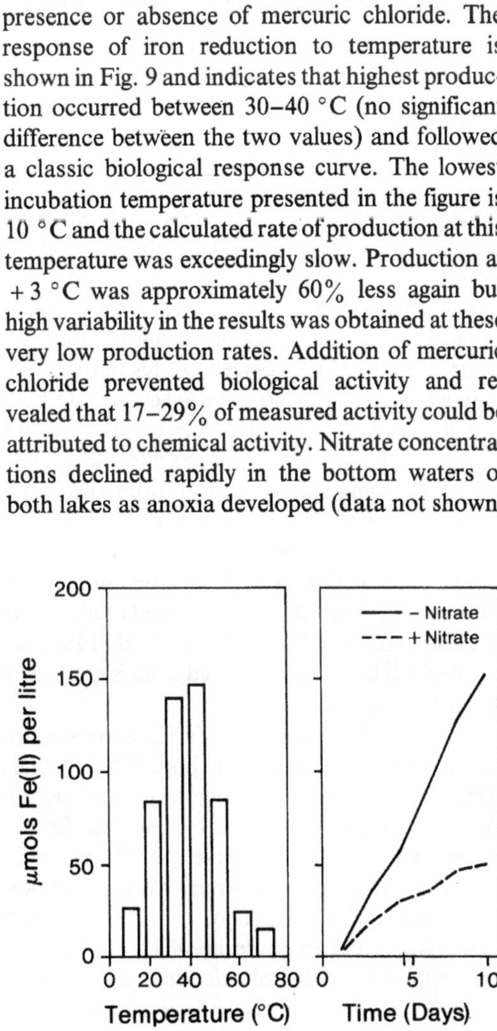

Fig. 9. (a) The effect of temperature on Fe(II) production in surface sediments from Sombre Lake.
(b) The effect of addition of nitrate on Fe(II) production in surface sediments from Sombre Lake.

but was nevertheless present albeit in low concentrations in the region of iron reduction in both lakes. In Sombre Lake sediment cores, the addition of nitrate (Fig. 9) resulted in a decrease of 42–66% in Fe(II) production over the control. Iron reduction in enrichments of Amos Lake water samples showed 80–90% inhibition in the presence of nitrate concentrations comparable to those measured in the upper waters of the lake in winter.

Discussion

Iron cycling – chemical aspects

Both lakes are seasonally anoxic for extended periods, at least in the bottom waters, due to the presence of a substantial ice cover which effectively seals the lake environment from external influences. Anoxic conditions result from the delivery of reduced substances to sites of oxidation activity. These reduced substances include both organic particulate material derived from the lake biota and allochthonous sources and various dissolved reducing species derived primarily, if not exclusively, from microbial metabolism (Jones, 1985). The species include methane, various sulphide forms, Fe(II), Mn(II) and ammonium, most of which are formed in the sediments rather than the water column of lakes (Mayer *et al.*, 1982) and all of which have been reported from Signy Island lakes (Gallagher, 1985).

Substantial quantities of Fe(II) were released under anoxic conditions in both lakes studied here and migrated up the water column by diffusion. This process, termed by some the 'ferrous wheel' (Campbell & Torgersen, 1980), has been well documented in temperate lakes and is now fairly well understood (Davison *et al.*, 1980). It comprises a closed cycle in which the reduced ions diffuse to a zone of oxidation, precipitate as oxyhydroxides and then settle gravitationally to become reduced once again deep within the anoxic zone. The oxidation process occurs spontaneously at neutral pH values and appears to be entirely abiotic in most freshwater lakes (Davison & Seed, 1983). The entire cycle may be confined within the sediments throughout the winter period at shallower sites, such as the algal mat zone of Sombre Lake, where an oxic surface layer exists year-round (Ellis-Evans, 1982, 1984). In the anoxic sump of Sombre Lake only iron oxidation was located outside the sediments, whilst in Amos Lake the entire cycle may have been water column-based by late winter.

The tight coupling of iron cycling with phosphorus, which is also released from sediments under anoxic conditions, was readily demonstrated in this study. Phosphate adsorbs strongly to the freshly precipitated Fe(III) oxyhydroxides (Stumm & Morgan, 1981) and so, like iron, is largely trapped within the anoxic zone. The coprecipitation prevents phosphorus from entering the oxic waters of Sombre Lake and thus helps maintain the lakes' oligotrophic character.

In Amos Lake, the profiles suggest that whilst a large proportion of the released phosphorus was trapped by oxyhydroxides at the oxic/anoxic interface, a small proportion was released into the overlying waters. Concentrations of sediment-derived phosphorus were high relative to Fe(II) in this lake, compared to Sombre Lake, reflecting its higher nutrient status. High concentrations of sulphide were also present in Amos Lake bottom waters, particularly later in the winter, and it is well documented that sulphide and Fe(II) readily interact around neutral pH to form insoluble ferrous sulphide (Davison & Heaney, 1978). Seasonally anoxic systems have been shown (Davison, 1980) to have solubility product constants (pK_{sp}) that are close to the laboratory determined value for freshly precipitated amorphous ferrous sulphide and that these lie in the range 2.59–3.25. In the present study, values of between 2.55 and 2.85 were calculated for all water samples up to 0.5 m above the sediment surface during the period September to November. This suggested that these waters were saturated with respect to ferrous sulphide and was confirmed in anoxic filtration experiments by the presence of a black precipitate which increased in particle size with depth. The 0.0–0.5 m interval above the sediment in Amos Lake was therefore a region where a substantial proportion of soluble iron was removed from solution. The shape of the iron profiles however suggest that soluble iron still maintained high concentrations beyond 0.5 m. Sulphide-mediated reduction of Fe(III) oxyhydroxides also probably occurred in Amos Lake since a distinct intermediate brown/black zone was distinguishable at times between the brown oxyhydroxide and black sulphide zones.

Sulphide was never present in high concentrations in Sombre Lake water column, though acidification of sediment samples did indicate the presence of substantial amounts of acid-volatile sulphides at depth in the sediment. Both the low concentrations of sulphide present and pK_{sp} calculations indicate that soluble iron was not influenced to any significant degree by sulphide precipitation.

Iron cycling – microbial aspects

Roughly $2\,g\,C\,m^{-2}$ of primary production-derived C reaches the sediment surface of Sombre Lake each year (Hawes, 1985) compared to almost $200\,g\,C\,m^{-2}$ in Amos Lake (Ellis-Evans & Sanders, 1988). This greater availability of organic material in the latter is attributable to the presence of high concentrations of nutrients, including phosphate, supplying primary production over the summer period. It is reflected in greater microbial activity within Amos Lake (Ellis-Evans, 1982) as the high sulphide concentrations in the present data set demonstrate. Both lake sediments contain sulphate-reducing and putrefactive bacteria but far higher numbers of both groups occur in Amos Lake (Herbert & Bell, 1974; Ellis-Evans, 1982). Most Probable Number estimates of sulphate reducers indicated a rapid increase in numbers in surface sediments during early winter but numbers soon plateaued and subsequently remained unchanged for the remainder of the anoxic period. This can be explained in terms of sulphate limitation ($<25\,\mu g\,l^{-1}$ in pore water) from the data presented here. Evidence from earlier work by Ellis-Evans (1984) suggest that sulphate limitation is of widespread occurrence in Signy Island lakes as in freshwater systems elsewhere (Jones, 1985). The fact that sulphide production was greatest in the second half of the winter would seem to suggest that putrefactive bacteria perhaps play a significant role in Amos Lake sulphur cycling as Molongoski & Klug (1980) have proposed for hyper-eutrophic Wintergreen Lake.

Until recently, interest in geomicrobiological cycling in lakes under anoxic conditions was restricted largely to studies of sulphate reduction and methanogenesis. Both processes can be terminal stages in the decomposition of complex organic matter to carbon dioxide or methane with associated accumulation of low stable concentrations of various organic intermediates (Lovley & Klug, 1986). However a similar processing of organic matter with Fe(III) as the electron acceptor is also feasible and this has been investigated by various workers including Jones *et al.* (1983, 1984) and, more recently, Lovley & Philips (1986a, b).

The present study has demonstrated that iron reduction in Signy Island lakes is largely a biotic process restricted by the permanently low lake temperatures and primarily located at sites where the supply of Fe(III) is greatest. Munch and Ottow (1982) have shown that physical contact between the bacterium and the iron particles is necessary for maximum reduction rates and this contact requirement in part explains the location of peak iron reduction/population numbers in the lakes studied here. However, in the case of Sombre Lake the fact that most organic breakdown occurred at the sediment surface (Ellis-Evans, 1985) must also have been an important influence.

Both chemolithotrophic and chemo-organotrophic bacteria have been implicated in iron reduction but given the extremely slow rate of the lithotrophic process (Balashova & Zavarzin, 1980) and the diversity of chemo-organotrophs, capable of more rapid Fe(III) reduction, present in natural systems (Jones, 1985) it seems likely that the latter group is predominant. Nitrate reductase is thought to be responsible for Fe(III) reduction observed in many soil bacteria (Ottow & Munch, 1978) and links with the finding here that the presence of nitrate lowered the rate of

Fe(III) reduction. Further work is necessary to confirm this possibility.

Both Jones *et al.* (1983) and Lovley & Philips (1986a) have examined the effect of various forms of iron on iron reduction and found that iron oxyhydroxides gave the highest reduction rates. The former group actually used $FeCl_3$ but under the pH conditions prevailing in their experiments it would have immediately hydrolysed and slowly flocculated as oxyhydroxides. Oxyhydroxides, saturated with respect to phosphorus, were no less suitable than phosphorus-free material in promoting iron reduction in Sombre Lake enrichment studies. Studies by Lovley & Philips (1986b) of oxyhydroxides saturated with humics or adsorbed onto clay similarly showed no significant effect on the reduction rate compared to unamended oxyhydroxides. In this most recent work however, Lovley & Philips concluded that, whilst Fe(III) occurred to considerable depth in their river sediments, below 3–4 cm depth Fe(III) occurred mainly as mixed Fe(III)-Fe(II) compounds resistant to reduction. This was not investigated in the present study but similar mixed iron compounds have been reported from Lake 227 sediments (Coey *et al.*, 1974). If generally applicable to seasonally anoxic lake sediments it could represent a significant spatial limitation for iron reduction but since assimilable organic matter is largely confined to the upper sediments in Signy Island lakes (Ellis-Evans, 1981a), this was probably not critical.

Of particular significance in this study was the observation that the addition of fermentation end-products such as acetate and hydrogen or fermentable substrates such as cellulose, in the presence of Fe(III), increased the reduction rate over that of Fe(III) addition alone. In contrast, Jones *et al.* (1983) considered that most chemoorganotrophic iron reducers gained no significant energetic advantage from the process, using it apparently as a hydrogen sink by which NAD could be regenerated. Indeed the addition of fermentation end products such as acetate and butyrate were found to be inhibitory to Fe(II) production and even though evidence of energy conservation was found, the scale of iron reduction was considered small overall in comparison to the reducing potential of the available substrates. The preliminary work presented here is in agreement with the findings of Lovley & Philips ((1986a) that, given the presence of amorphous Fe(III) oxyhydroxides, iron reducers are capable not only of significant transfer of electrons to Fe(III) during organic matter mineralisation but can also cause inhibition of methane production by utilization of acetate and hydrogen. These workers used a variety of Fe forms to demonstrate that amorphous oxyhydroxides caused the greatest inhibition of methane production and was also the most suitable electron acceptor for acetate metabolism. Acetate-utilizing methanogens appear to predominantly occur in the upper sediments of Sombre Lake (Ellis-Evans 1984) where iron reducers also predominate. Given that amorphous Fe(III) oxyhydroxides are apparently the main source of Fe(III) to these sediments during anoxia, it would appear that methanogenesis could be inhibited by iron reduction in the surface sediments. Methane profiles of these sediments (Ellis-Evans, 1984) have indicated that maximum methane production does indeed occur beneath this surface layer.

Acknowledgements

We are grateful to the British Antarctic Survey for facilities to undertake this work and to the personnel of Signy Station for their assistance and companionship. We thank J. B. Gallagher for constructive criticism of the draft manuscript.

References

Brannon, J. M., D. Gunnison, R. M. Smart & R. L. Chen, 1984. Effects of added organic matter on iron and manganese redox systems in sediment. Geomicrobiol. J. 3: 319–341.

Balashova, V. V. & G. A. Zavarzin, 1980. Anaerobic reduction of ferric iron by hydrogen bacteria. Microbiology 48: 635–639.

Campbell, P. & T. Torgersen, 1980. Maintenance of iron meromixis by iron redeposition in a rapidly flushed monimolimnion. Can. J. Fish. Aquat. Sci. 37: 1303–1313.

Coey, J. M. D., D. W. Schindler & F. Weber, 1974. Iron compounds in lake sediments. Can. J. Earth Sci. 11: 1489–1493.

Cunningham, C. C. & W. Davison, 1980. An opto-electronic sediment detector and its use in the chemical microprofiling of lakes. Freshwater Biol. 10: 413–418.

Davison, W., 1977a. Sampling and handling procedures for polarographic measurements in hypolimnetic water. Freshwater Biol. 7: 393–401.

Davison, W., 1977b. The polarographic measurements of O_2, Fe^{2+}, Mn^{2+} and S^{2-} in hypolimnetic water. Limnol and Oceanogr. 22(4): 743–753.

Davison, W., 1980. A critical comparison of the measured solubilities of ferrous sulphide in natural waters. Geochim. Cosmochim. Acta. 44: 803–808.

Davison, W. & B. J. Finlay, 1986. Ferrous iron and phototrophy as alternative sinks for sulphide in the anoxic hypolimnia of two adjacent lakes. J. Ecol. 74: 663–673.

Davison, W. & S. I. Heaney, 1978. Ferrous-iron-sulphide interactions in anoxic hypolimnetic waters. Limnol. Oceanogr. 23: 1194–1200.

Davison, W., S. I. Heaney, J. F. Talling & E. Rigg, 1980. Seasonal transformations and movements of iron in a productive English lake with deep water anoxia. Schweiz. Z. Hydrol. 42: 196–224.

Davison, W. & G. Seed, 1983. The kinetics of the oxidation of ferrous iron in synthetic and natural waters. Geochim. Cosmochim. Acta. 47: 67–79.

Eisenreich, S. J., R. T. Bannerman & D. E. Armstrong, 1975. A simplified phosphorus analysis technique. Environ. Letts. 9(1): 43–53.

Ellis-Evans, J. C., 1981a. Freshwater microbiology in the Antarctic I-Microbial numbers and activity in oligotrophic Moss Lake. Brit. Antarct. Surv. Bull. 54: 85–104.

Ellis-Evans, J. C., 1981b. Freshwater microbiology at Signy Island, South Orkney Islands, Antarctica. PhD thesis (CNAA) 283 pp.

Ellis-Evans, J. C., 1982. Seasonal microbial activity in Antarctic freshwater lake sediments. Polar Biol. 1: 129–140.

Ellis-Evans, J. C., 1984. Methane in maritime Antarctic freshwater lakes. Polar Biol. 3: 63–71.

Ellis-Evans, J. C., 1985. Decomposition processes in maritime Antarctic lakes. In Seigfreid, W. R., P. R. Condy and R. M. Laws (eds.) Antarctic Nutrient Cycles and Food Webs. Springer Verlag: 253–260.

Ellis-Evans, J. C. & M. W. Sanders, 1988. Observations on microbial activity in a seasonally anoxic, nutrient-enriched maritime Antarctic lake. Polar Biol. 8: 311–318.

Froelich, P. N., G. P. Klinkhammer, M. L. Bender, N. A. Luedtke, G. R. Heath, D. Cullen, P. Dauphin, D. Hammond, B. Hartman & V. Maynard, 1979. Early oxidation of organic matter in pelagic sediments of the eastern equatorial Atlantic: suboxic diagenesis. Geochim. Cosmochim. Acta. 43: 1075–1090.

Gallagher, J. B., 1985. The influence of iron and manganese on nutrient cycling in shallow freshwater Antarctic lakes. In Seigfried, W. R. & R. M. Laws (eds.) Antarctic Nutrient Cycles and Food Webs, Springer Verlag: 234–237.

Golterman, H. L., R. S. Clymo & M. A. M. Ohnstad, 1978. Methods for physical and chemical analysis of fresh waters. IBP Handbook No. 8 (2nd ed.). Blackwell: 89–91.

Hawes, I., 1985. Factors controlling phytoplankton populations in maritime Antarctic lakes. In Seigfreid, W. R., P. R. Condy & R. M. Laws (eds.) Antarctic Nutrient Cycles and Food Webs. Springer Verlag: 245–252.

Herbert, R. A. & C. R. Bell, 1973. Nutrient cycling in freshwater lakes on Signy Island, South Orkney Islands. Brit. Antarct. Surv. Bull. 37: 15–20.

Heywood, R. B., 1985. Antarctic inland waters. In Laws, R. M. (ed.) Antarctic Ecology Vol 1 Academic Press: 279–343.

Jones, J. G., 1983. A note on the isolation and enumeration of bacteria which deposit and reduce ferric iron. J. Appl. Bact. 54: 305–310.

Jones, J. G., 1985. Microbes and microbial processes in sediments. Phil. Trans. R. Soc. Lond. A 315: 3–17.

Jones, J. G., S. Gardner & B. M. Simon, 1983. Bacterial reduction of ferric iron in a stratified eutrophic lake. J. Gen. Microbiol. 129: 131–139.

Jones, J. G., S. Gardner & B. M. Simon, 1984. Reduction of ferric iron by heterotrophic bacteria in lake sediments. J. Gen. Microbiol. 130: 45–51.

Latham, M. J. & M. J. Wolin, 1978. Use of a serum bottle technique to study interactions between strict anaerobes in mixed culture. In Lovelock, D. W. & R. Davies (eds.) Techniques for the Study of Mixed Populations. SAB Technical Series No. 11 Academic Press: 113–124.

Lovley, D. R. & M. J. Klug, 1986. Model for the distribution of methane production and sulphate reduction in freshwater sediments. Geochim. Cosmochim. Acta. 50: 11–18.

Lovley, D. R. & E. J. P. Philips, 1986a. Organic matter mineralization with reduction of ferric iron in anaerobic sediments. Appl. Environ. Microbiol. 51(4): 683–689.

Lovley, D. R. & E. J. P. Philips, 1986b. Availability of ferric iron for microbial reduction in bottom sediments of the freshwater tidal Potomac River. Appl. Environ. Microbiol. 52(4): 751–757.

Mayer, L. M., F. P. Liotta & S. A. Norton, 1982. Hypolimnetic redox and phosphorus cycling in hypereutrophic Lake Sebasticook, Maine. Water Res. 16: 1189–1196.

Methods for the Examination of Water and Associated Materials, 1983. Iron in raw and potable waters by atomic absorption spectrophotometry. HMSO, London, 20 pp.

Molongoski, J. J. & M. J. Klug, 1980. Quantification and characterization of sedimenting particulate organic matter in a shallow hypereutrophic lake. Freshwat. Biol. 10: 497–506.

Mortimer, C. H., 1941. The exchange of dissolved substances between mud and water in lakes. I and II. J. Ecol. 29: 280–329.

Mortimer, C. H., 1942. The exchange of dissolved substances between mud and water in lakes. III and IV. J. Ecol. 30: 147–201.

Munch, J. C. & J. C. G. Ottow, 1982. Effect of cell contact and iron(III) oxide form on bacterial iron reduction. Zeit. Pflanz., Dung. Boden. 145: 66–77.
Ottow, J. C. G. & J. C. Munch, 1978. Mechanisms of reductive transformations in the anaerobic microenvironment of hydromorphic soils. In Krumbein, W. E. (ed.) Environmental Biogeochemistry and Geomicrobiology Vol 2. Ann Arbor: 483–491.
Sorensen, J., 1982. Reduction of ferric iron in anaerobic, marine sediment and interaction with nitrate and sulphate. Appl. Environ. Microbiol. 43: 319–324.
Stumm, W. & J. J. Morgan, 1981. Aquatic Chemistry, 2nd Edition. Wiley 780 pp.
Tabatabai, M. A., 1974. Determination of sulphate in water samples. Sulphur Inst. J. 10: 11–13.
Talling, J. F., 1973. The application of some electrochemical methods to the measurement of photosynthesis and respiration in freshwater. Freshwater Biol. 3: 335–362.

Hydrobiologia 172: 165–172, 1989
W. F. Vincent and J. C. Ellis-Evans (eds) High Latitude Limnology

Nitrogen cycling in Arctic lakes and ponds

V. Alexander, S. C. Whalen & K. M. Klingensmith[1]
Institute of Marine Science, University of Alaska-Fairbanks, Fairbanks, Alaska 99775, USA; [1] *Present address: Forest Soils Laboratory, Agricultural and Forestry Experiment Station, University of Alaska – Fairbanks, USA*

Abstract

Nitrogen cycle studies were carried out on two distinct systems in arctic Alaska. Research on the first, a series of small tundra thaw ponds near Barrow, was carried out as part of the International Biological Program Tundra Biome program. Research on the contrasting system, Toolik Lake, was done during the year following completion of the Barrow pond research. Toolik Lake is located on the north flank of the Brooks Range, approximately 130 km south of the Beaufort Sea coast.

The two systems had some similarities, but also some distinct differences which appeared to be associated with their different depths. Tundra ponds are very shallow, freeze to the bottom in winter and the waters are continuously in contact with the sediment. The sediment, therefore, serves as a major source of inorganic nitrogen for the phytoplankton. As a result, phosphorus rather than nitrogen deficiency is evident. In the deep lake, *in situ* nitrogen regeneration and allochthonous sources are important and nitrogen deficiency can occur.

An examination of the nitrogenous nutrient regime of these two systems and the major cycle processes is included.

Introduction

Until the last fifteen years, knowledge about Alaskan arctic lakes was based primarily on brief surveys. The available information has been summarized by Livingstone (1963). Subsequently, Kalff (1970) and Hobbie (1973) have also presented reviews of the earlier published literature. A major change took place with the advent of the International Biological Program (IBP) Tundra Biome program, which included a strong aquatic component. This focused on ecosystem function and driving forces for polygonal thaw ponds which are abundant on the coastal region in the Barrow area.

Following the completion of the IBP research and its synthesis and publication in the mid-1970s, emphasis shifted to a study of rates and processes in deep arctic lakes, which also occur throughout arctic Alaska. Our nitrogen cycling portion of this research was carried out on Toolik Lake and adjacent waters over a ten-year period from 1976 to 1986. Although the regional limnological interest was a primary motivation for studying Toolik Lake, there are some general advantages in conducting nitrogen cycle research in such a simple system and the rather pristine conditions present an opportunity to study a system with minimal human influence. There are few areas in the world which have been subjected to less human impact in the past, although current activity along the pipeline corridor, where Toolik and other major arctic lakes are located, is changing this. Here, we present an overview and comparison of our results on the nitrogen regime in both of these aquatic systems. The scope of the nitrogen cycle work included assessing the sources and fate of nitrogen, the phytoplankton nitrogen assimilation regime, the *in situ* nitrogen cycle processes and the overall nitrogen budget.

Study sites

The location of the two major study sites is shown in Fig. 1 and a brief description follows.

The thaw ponds at Barrow are small (30 × 40 m) and shallow, with a maximum depth of 0.5 m. They are surrounded by wet tundra dominated by grasses and sedges and are isolated from adjacent ponds by ridges formed by the uplifting of underlying ice wedges. Toolik Lake is typical of large lakes in the foothills of the Brooks Range. It has a surface area of 1.5 km^2, a mean depth of 7 m and a maximum depth of 25 m. Surrounded by tussock tundra, Toolik Lake has two inlets and a single major outlet.

Neither of these water bodies becomes anoxic at the sediment-water interface. Both Toolik Lake and the ponds are ice covered from late September until late June and in winter the ponds become solidly frozen to the sediment surface. The weather is similar in the two areas, often overcast and cool, with summer air temperatures around 10 °C. In both, the surface water temperature can reach 15 °C but is typically cooler, about 6 °C in the Barrow ponds and 8° to 10 °C in Toolik Lake. Toolik Lake stratifies thermally for five to six weeks in mid-summer with the maximum extent of the thermocline to about 9 m. The

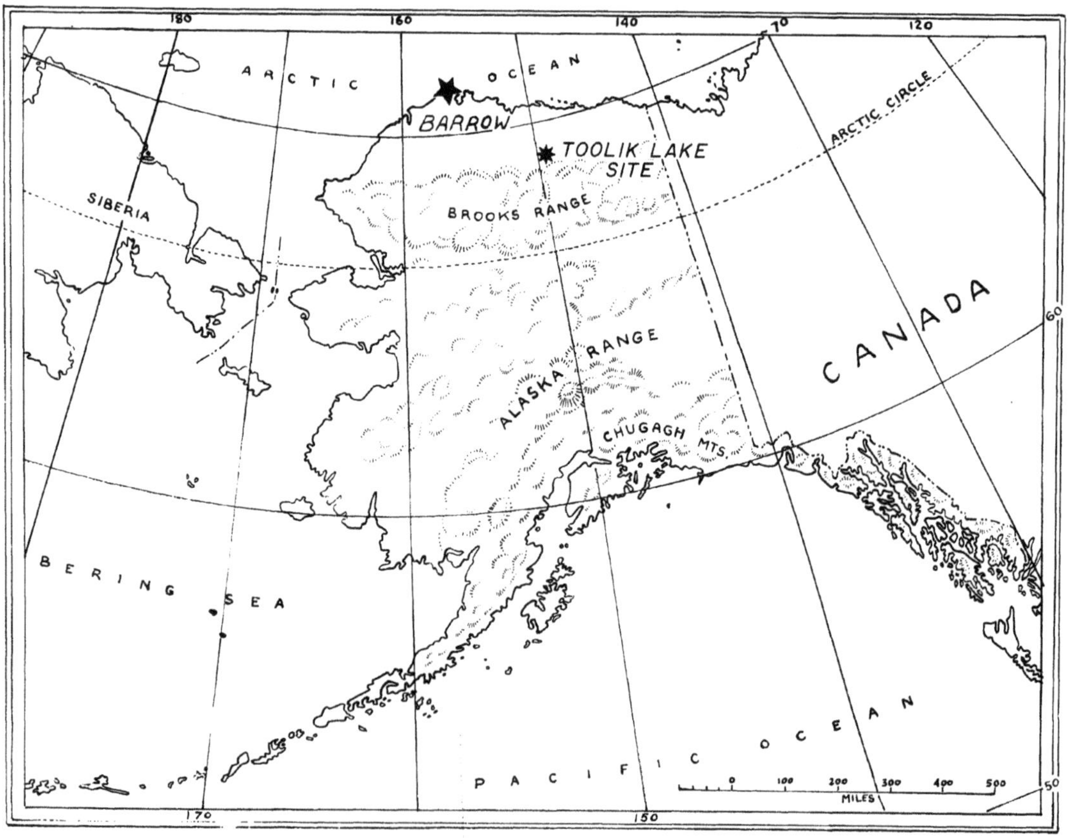

Fig. 1. Map showing the two major research sites – the Barrow thaw ponds and Toolik Lake.

Barrow ponds are subject to constant wind mixing and do not stratify. Both areas are underlain by permafrost with a summer thaw depth of 25 to 100 cm for the Barrow tundra (Walker, 1973) and 0.5 m for the Toolik Lake watershed.

Methods

All the methods used in these studies have been described fully in Prentki *et al.* (1980), Alexander *et al.* (1980), Klingensmith (1981) and Whalen (1986). They will be outlined only briefly here.

Analyses for ammonium, nitrate, dissolved organic nitrogen and particulate nitrogen followed standard chemical techniques. Phytoplankton primary production was measured using the ^{14}C method with liquid scintillation. Phytoplankton nitrate and ammonium uptake was measured using ^{15}N-labeled compounds with analysis using mass spectrometry. Chlorophyll *a* was measured on acetone extracts of material retained by a micropore glass fiber filter, either spectrophotometrically or fluorometrically. Nitrogen fixation was measured using the acetylene reduction technique. Denitrification was measured using the acetylene block method (Toolik Lake) or using $^{15}NO_3^-$ isotope tracers (Barrow) and nitrification using $^{15}NO_3^-$ tracer techniques.

Results and discussion

Nitrogenous chemical composition of waters

Both thaw pond and Toolik Lake water are dominated by the dissolved organic fraction. Inorganic nitrogen concentrations are low in both environments, although the levels were generally higher in the ponds (Table 1). This was probably due to agitation of the sediments by winds. In both systems, a runoff pulse dominates the annual allochthonous nutrient input.

The nitrogen chemistry of sediments

Ammonium dominates the inorganic fraction of the interstitial water in both environments, although the concentration in the ponds is one order of magnitude greater than that in Toolik Lake (Table 2). These Toolik Lake data were taken close to the average depth of the waters and vegetation was avoided in the sampling area. The ammonium concentration in the pond sediment decreased by two orders of magnitude at the margins and in areas with rooted vegetation. From this and based on uptake measurements we have concluded that for the Barrow pond sediments nitrogen cycling is controlled by macrophyte uptake, resulting in the low concentrations of interstitial nutrients at the pond margins (McRoy & Alexander, 1975). In Toolik Lake, 66–76% of the primary production is remineralized in the water column based on current ^{15}N uptake measurements and sediment trap studies. This pelagic mineralization decreases particulate flux to the sediment and bacteria have little organic matter to decompose.

Nitrogen fixation

Barrow pond sediments were tested for nitrogen fixation: Of four ponds tested in 1972, two showed no activity at all, one showed acetylene reduction on one of three occasions. The fourth pond exhibited activity on all dates. The average nitrogen fixation for the active pond was 0.02 mmol N m^{-2} d^{-1} which translates to about 2 mmol N m^{-2} yr^{-1} for a 100-day growing season. The average nitrogen fixation rate for terrestrial tundra surrounding the pond was about

Table 1. Comparative water chemistry for Toolik Lake (1980) and Barrow ponds (1970–1971; 1975). All data as μmol N l^{-1}, range and mean.

Nutrient	Toolik Lake	Barrow Ponds
NO_3^-	0–3.29[a]	0–6.4
	0.46	1.44
NH_4^+	0–0.28	0.68–6.07
	0.17	2.32
DON	11–30	57–100
	17	60
PN	0.8–6.0	2.2–11.4
	2.1	4.3

[a] Mean during open water period = 0.10 μmol l^{-1}. See text for details.

Table 2. Comparative sediment chemistry for Toolik Lake (July 1979) and Barrow ponds (July 1973). All data as μmol l^{-1}.

Sediment depth (cm)	Water depth (m)	NH_4^+-N	NO_3^--N
	Pond J		
0–8	0.18	190	0.4 (Between rooted
8–16	0.18	235	0.3 vegetation beds)
	Toolik Lake		
0–9	10	13	1.7
9–15	10	27	0

3.4 mmol N m^{-2} yr^{-1} (Barsdate & Alexander, 1975), although much higher activity was found in wet mossy areas (Schell & Alexander, 1973). Considering the average dissolved inorganic nitrogen content in the sediment, we conclude that *in situ* nitrogen fixation is a minor source of nitrogen to the pond nitrogen budget, unlike the terrestrial budget where Barsdate & Alexander (1975) and Alexander *et al.* (1978) have shown nitrogen fixation to be important.

The role of nitrogen fixation in Toolik Lake is not clear. Surveys for nitrogen fixation were conducted at the start of the season, June 27–29, and later in the summer, August 9–10, 1976. The watershed, inlets, sediment and water column were tested. The highest activity in the aquatic environment was found in the stream benthos, with a maximum rate of 152 μmol C_2H_4 m^{-2} d^{-1} or about 0.1 mmol N m^{-2} d^{-1} in a stream with mats of *Nostoc* sp. A sediment core also had measurable rates of acetylene reduction (Fig. 2). Detectable but generally low levels of acetylene reduction were found at most depths in the water column, with the total nitrogen fixed in the water column about 0.4 mmol m^{-2} d^{-1} for the early summer period and 0.07 mmol N m^{-2} d^{-1} in August. The interesting outcome of this survey was that, despite very low temperatures, almost every sample showed some nitrogenase activity. Heterocystous cyanophytes were never a significant component of the phytoplankton.

Subsequent nitrogen fixation surveys of the water column in 1979 and 1980 did not show any nitrogen fixation and, similarly, no sediment nitrogen fixation was found in 1979 and 1980, except in a bed of *Nitella* sp. Since there is only one *Nitella* patch in Toolik Lake, and much of the bottom is rocky, these later results suggest that *in situ* nitrogen fixation in the lake is not significant. This is probably due to the low phosphorus levels in both water and sediment. Cornwell (1985) reported that soluble reactive phosphorus is undetectable in Toolik Lake water and that vivianite formation and adsorption strongly regulate the inorganic phosphorus in the sediment pore-water.

Clearly, the latter results contradicted our earlier findings and, therefore, we are unable to assign a relative importance to nitrogen fixation in the Toolik Lake budget with any reasonable degree of confidence. However, *in situ* fixation is probably not very important.

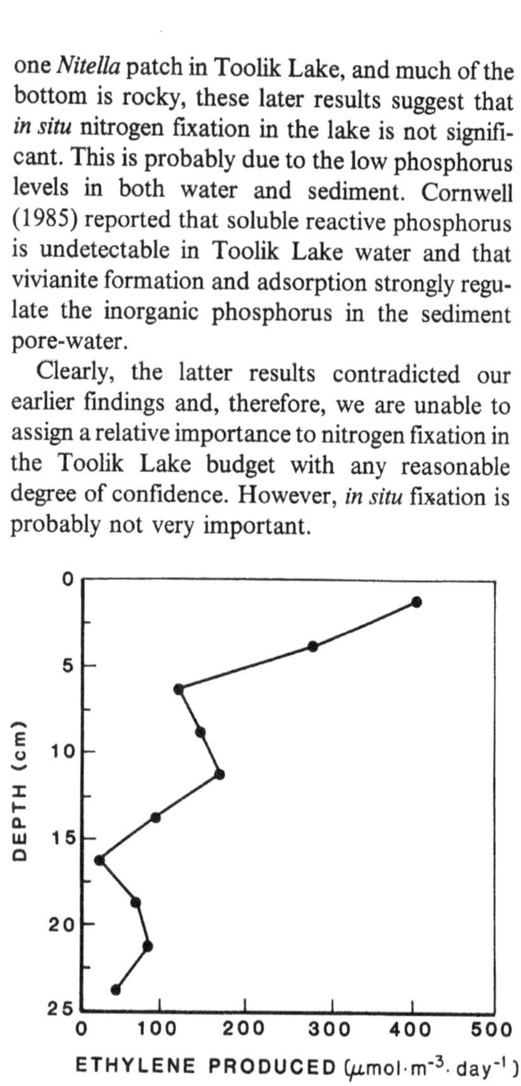

Fig. 2. Nitrogen fixation, as measured by acetylene reduction, in Toolik Lake sediment.

Denitrification and nitrification

Rate measurements of denitrification in the Barrow ponds were conducted on a single core to which $^{15}NO_3^-$ was added. Subsequent appearance of isotope in the N_2 after 16 d was measured. Data were considered to give the 'potential' rate of activity due to inoculation with excess isotope-labeled substrate. The measured rate was 0.01 nmol N cm^{-3} d^{-1}. No nitrification experiments were conducted.

Potential denitrification rates were considerably higher (by two orders of magnitude) in Toolik Lake than in the ponds. The average potential rate for three dates in 1980, from several locations in the lake, was 3.1 nmol cm^{-3} day^{-1}. It was interesting to note higher N_2O accumulation during the acetylene blockage experiments without C_2H_2 addition than with added C_2H_2. The usual assumption in these measurements is that the rate of N_2O accumulation is greater in the presence than in the absence of acetylene since acetylene blocks the further reduction of N_2O which has been produced from NO_3^- by denitrification. We have postulated that our unusual results may mean that a substantial proportion of the N_2O was due to nitrification, and, further, that the addition of acetylene may have reduced the amount of nitrate produced by nitrification, thus eliminating the supply of substrate for denitrification. Previous work (Walter *et al.*, 1973) had showed that C_2H_2 inhibits nitrification. Our hypothesis is also supported by our finding that when NO_3^- was added to cores there was a 9-fold increase in N_2O production in the presence C_2H_2 and only 1.5-fold increase without added C_2H_2.

Measurements of potential nitrification rates in Toolik Lake sediment produced an average rate of 3.5 nmol cm^{-3} d^{-1}. These are similar to the denitrification rates, suggesting coupling or 'potential' coupling between the two processes.

Nitrification rates in the Toolik lake water column were not measured because such experiments were not feasible at the low rates and would have required prohibitively long incubation times. Nitrate is barely detectable in the water column during the open water period; but, in early spring before phytoplankton growth, the nitrate concentration is sometimes in excess of 3 μmol l^{-1}. Inlet and outlet streams do not flow during the winter and, therefore, the source must be either the sediment or water column nitrification with accumulation during the winter period of phytoplankton inactivity. We amended Toolik Lake bottom water in carboys with $^{15}NH_4^+$ early one winter, incubated them in aphotic sediment and recovered them after several months in spring. Although the results are not quantitative, the fact that $^{15}NO_3^-$ appeared in the carboys in spring demonstrates water column nitrification at some low level, at least during the winter. Any summer activity is masked by phytoplankton assimilation of NO_3^-. In the Barrow ponds, no liquid water remains in winter and the period for winter nitrification *in situ* is minimal. Spring NO_3^- levels are elevated, but much of this probably enters through the spring runoff.

Although the rates of nitrification and denitrification are low in both environments, direct comparison is precluded because of differing techniques used in the two studies.

Phytoplankton

Information on the characteristics of the phytoplankton in the two waters is presented in Table 3. Based on chlorophyll concentrations as a measure of phytoplankton biomass, the algal population in the ponds is approximately twice that in Toolik Lake on a volume basis. Accordingly, primary production is also about two times greater in the ponds than in the lake. However, the ammonium uptake rate in the pond is five times that in the lake and the nitrate uptake rate in the lake is twice that in the pond. Average turnover time for dissolved inorganic nitrogen is twice as rapid in the lake as in the ponds. The higher ammonium uptake rates in the ponds probably are due to the greater availability of nutrients (Table 1) and the higher primary production. The higher NO_3^- uptake in the lake despite the low ambient levels may indicate nitrogen deficiency in Toolik Lake when compared with the ponds. Although NO_3^- may not be the preferred nitrogen source, it is utilized to a greater extent in the lake

Table 3. Nitrogen uptake dynamics: Comparison between Barrow ponds (1971) and Toolik Lake (1980).

Parameter	Barrow ponds	Toolik Lake
Ave. NH_4^+ uptake (nmol l^{-1} d^{-1})	269[a]	57
Ave. NO_3^- uptake (nmol l^{-1} d^{-1})	7[a]	13
Ave. NH_4^+ turnover time (d)	6.2	3.0
Ave. NO_3^- turnover time (d)	$\gg 50$	26.3
Chlorophyll *a*	0.6[b]	1.4
Primary production (nmol l^{-1} d^{-1})	2190[c]	910
Unweighted C/N uptake ratio (mol)	7.9	13.0

[a] Pond B only.
[b] Estimated from Fig. 5-1 in Alexander *et al.* (1980).
[c] Mean of data from four ponds.

than in the ponds because the total inorganic nitrogen availability is so low. The difference in turnover times also suggests greater nitrogen demand in Toolik Lake and resultant rapid recycling of nutrients. Our data show that as much as 66% of the nitrogen supporting primary production is recycled within the Toolik Lake water column. The C/N uptake ratios support this argument for greater N deficiency in phytoplankton in Toolik Lake than in the Barrow ponds. The ratio for the Barrow ponds is considerably closer to the Redfield number (6.5/1) than is that for Toolik Lake. In fact, the average C/N uptake ratio for Toolik Lake in 1980 is considerably higher at 18.0. Obviously, immediate contact between the whole water column and sediment in the Barrow ponds serves to increase nutrient supply for primary production. Additional data which support the argument for suboptimal levels of N in Toolik Lake include Michaelis-Menten kinetic analysis which shows first or mixed order kinetics at ambient nutrient levels at virtually all times of the year for mid-epilimnetic algae, whereas experiments in Barrow never resulted in increased assimilation in response to either NH_4^+ or NO_3^- addition.

Fertilization experiments and bioassays

In 1970, a whole pond P addition of 10 μmol l^{-1} resulted in an immediate increase in photosynthesis and phytoplankton biomass. Nitrogen uptake also showed a positive response. A steady infusion of 0.06 and 0.31 μmol l^{-1} d^{-1} to two subponds showed an analogous response in biomass, primary production and nitrogen uptake. Addition of phosphorus also stimulated nitrogen fixation in subponds, which supports the suggestion above that nitrogen fixation may be limited *in vivo* by suboptimal phosphorus levels. Another experiment underscored the role of sediment in nitrogen supply to the water column. Addition of 3.5 μmol l^{-1} PO_4^{3-} to water isolated from the sediment gave little response in primary production until 35.7 μmol l^{-1} NH_4^+-N was added. Details of this experiment may be found in Prentki *et al.* (1980). At no point was nitrogen alone tested as a treatment.

Bottle bioassay experiments were conducted in Toolik Lake during 1980-1981 to test response of primary production and nitrogen uptake to a suite of chemical amendments including P, N + P, vitamins, chelates and trace metals. Samples were collected from mid-epilimnion (3 m) into 20 l carboys. There were 10 treatments in total. Among them was the addition of NO_3^- (3 μM), NH_4^+ (3 μM), PO_4^{3-} (0.5 μM) and N + P (3 μM NO_3^{-1} + 3 μM NH_4^+ + 0.5 μM PO_4^{3-}). A six-hour incubation followed the addition of the various treatments to the various carboys. Samples were then drawn into triplicate bottles and ^{14}C or ^{15}N tracer was added. The tracer uptake measurements were done with a 24 hour incubation period, suspended at one meter depth in the lake. The date compared by single factor ANOVA, with $P < 0.01$. The treatment most stimulatory for primary production on all dates was the addition of 3 μmol l^{-1} NO_3^- and NH_4^+ in concert with 0.5 μmol l^{-1} PO_4^{3-}. On nearly all experimental occasions (12 total), N stimulated primary production as much as N + P. Rarely did any other treatment elicit a response significantly greater than the control. Strikingly, amendment with P enhanced $^{15}NO_3^-$ and $^{15}NH_4^+$ uptake on seven occasions each. The stimulatory effect of N + P or N alone on primary production and P on ^{15}N uptake was independent of thermal stratification. To our knowledge, nutrient release from Toolik Lake sediment is the lowest ever recorded (no soluble reactive phosphorus release,

11–22 mmol $m^{-2} yr^{-1}$ N release) so internal supply from this source is not important in sustaining planktonic productivity. Rather, water column recycling and allochthonous supply (water renewal time 0.5 yr) are important. This is in marked contrast to the situation in the Barrow ponds, where we have shown that sediment is the most important factor supplying N for primary production. Data for Toolik Lake can be cautiously interpreted to suggest close balance between N and P deficiency for phytoplankton. Although the data from the Barrow ponds suggest P deficiency, they also indicate that given a deeper water column in which sediment-agitated release of N is eliminated, a situation analogous to that in Toolik Lake N + P deficiency may develop. On the other hand, the entire tundra area surrounding the thaw ponds is highly P-deficient also (Chapin *et al.*, 1980).

Conclusions

Both the thaw ponds and lake waters in Alaska's Arctic contain extremely low concentrations of all forms of N and are unproductive as a result. However, there are differences in the nutrient regimes. Large lakes rely on recycling of N in the water column to support primary production as sediment recharge of N is very low due to the low rate of sedimentation, low water temperature and year-round oxic conditions. In the shallow ponds wind agitation can resupply N into the water and the interstitial waters are richer in N due to macrophytes. which are absent in larger lakes.

Although more detailed work is needed to settle the question, neither system appears to support much nitrogen fixation, probably due to the low P.

Due to the lack of sediment-water interaction, N is as important as P in regulating phytoplankton primary production in the lake, whereas the evidence for ponds suggests P deficiency as the major controlling factor. Denitrification accounts for a small loss in both environments, probably due to low NO_3^- supply in the sediment and a continuously oxic water column. Nitrification is probably occurring at a low rate in the sediment and possibly in the water column of both environments based on elevated early spring NO_3^- levels, although in the case of the ponds this is probably primarily imported by runoff. In Toolik Lake, the elevated concentrations are evident before the runoff begins.

Finally, the phytoplankton are able to use NO_3^- more effectively in the lake than in ponds, possibly because of greater N deficiency in lake waters.

Acknowledgements

The research on Toolik Lake described herein primarily was carried out by Ph. D. candidate S. Whalen and M. S. candidate K. Klingensmith and had been reported in their dissertations. Financial support was provided by the National Science Foundation (DPP75-129481 and DPP79-00815), a Noyes Foundation Fellowship to S. Whalen, and an Institute of Marine Science Fellowship to S. Whalen. Technical assistance by T. Chapman, C. Mildebrand and D. Pederson is greatly appreciated, as is the collaboration of M. Billington in the nitrogen fixation measurements and the assistance and advice of numerous colleagues while conducting the research, especially J. Cornwell. The pond research was also carried out with support from NSF, jointly sponsored by the Division of Polar Programs and the US International Biological Program. We appreciate the assistance of M. Billington, C. Chmielowski, R. Clasby, C. Coulon, L. Peyton and the collaboration of R. Barsdate and R. Prentki.

References

Alexander, V., M. Billington and D. M. Schell, 1978. Nitrogen fixation in arctic and alpine tundra, p. 539–558. *In*: L. L. Tieszen (ed.), *Vegetation and Production Ecology of an Alaskan Arctic Tundra*, Ecological Studies 29. Springer-Verlag.

Alexander, V., D. M. Stanley, R. J. Daley and C. P. McRoy, 1980. Primary Producers, p. 179–250. *In*: J. E. Hobbie (ed.), *Limnology of Tundra Ponds*. Dowden, Hutchinson and Ross, Inc., Stroudsburg, PA.

Barsdate, R. J. and V. Alexander. 1975. The nitrogen balance of arctic tundra: Pathways, rates, and environmental implications. *J. Environ. Qual.* 4: 111–117.

Clasby, F. S. III, P. C. Miller, W. D. Billings and P. I. Coyne, 1980. Carbon and nutrient budgets and their control in coastal tundra, p. 458–490. *In*: J. Brown, P. C. Miller, L. L. Tieszen and F. L. Bunnell (eds.), *An Arctic Ecosystem: The Coastal Tundra at Barrow*. Dowden, Hutchinson and Ross, Inc., Stroudsburg, PA.

Coyne, P., 1980. Chemistry, p. 76–178. *In*: J. E. Hobbie (ed.), *Limnology of Tundra Ponds*. Dowden, Hutchinson and Ross, Inc., Stroudsburg, PA.

Hobbie, J. E., 1973. Arctic limnology: A review, p. 127–168. *In*: M. E. Britton (ed.), *Alaskan Arctic Tundra*. Arctic Inst. N. Am. Tech. Pap. 25.

Kalff, J., 1970. Arctic Lake Ecosystems. p. 651–663. *In*: M. W. Holdgate (Ed.), *Antarctic Ecology*. Academic Press. London-New York. 998 p.

Klingensmith, K. M., 1981. Sediment Nitrification, Denitrification and Nitrous Oxide Production in an Arctic Lake. M. S. Thesis, University of Alaska-Fairbanks, 60 p.

Livingstone, D. A., 1963. Alaska, Yukon, Northwest Territories, and Greenland. p. 559–574. *In*: D. G. Frey (ed.) *Limnology in North America*. The University of Wisconsin Press, Madison, WI.

McRoy, C. P. and V. Alexander. 1975. Nitrogen kinetics in aquatic plants in arctic Alaska. *Aquat. Bot.* 1: 3–10.

Prentki, R. T., M. C. Miller, R. J. Barsdate, V. Alexander, J. Kelley and P. Coyne, 1980. Chemistry, p. 76–178. *In*: J. E. Hobbie (ed.), *Limnology of Tundra Ponds*. Dowden, Hutchinson and Ross, Inc., Stroudsburg, PA.

Schell, D. M. and V. Alexander, 1973. Nitrogen fixation in arctic coastal tundra in relation to vegetation and micro-relief. *Arctic* 26: 130–137.

Whalen, S. C., 1986. Pelagic Nitrogen Cycle in an Arctic Lake. Ph. D. Thesis, University of Alaska-Fairbanks, 169 p.

Walker, H. J., 1973. Morphology of the North Slope, P. 49–92. *In*: M. E. Britton (ed.), *Alaskan Arctic Tundra*. Arctic Inst. N. Am. Tech. Pap. 25.

Hydrobiologia 172: 173–182, 1989
W. F. Vincent and J. C. Ellis-Evans (eds) High Latitude Limnology

Photon dependence of inorganic nitrogen transport by phytoplankton in perennially ice-covered antarctic lakes

John C. Priscu
Department of Biology, Montana State, University, Bozeman, Montana 59717, USA

Key words: Antarctic, ^{15}N, nitrogen transport, photon dependence, phytoplankton

Abstract

^{15}N was used to examine the influence of Photosynthetic Photon Flux Density (PPFD) on NO_3^- and NH_4^+ transport by phytoplankton in the shallow and deep-chlorophyll layers of Lakes Fryxell and Vanda, Antarctica. The response observed in Lake Vanda could be modelled with a four parameter equation previously used to model photosynthesis. Only the 9 m NH_4^+ transport experiments in Lake Fryxell could be modelled with this equation. Other experiments in Lake Fryxell showed either no response to PPFD or a linear increase with no saturation at the PPFD levels used. Distinct trends were observed in transport parameters both between depths and between nitrogen species. Overall, the parameters indicate that phytoplankton in these lakes possess nitrogen transport affinities similar to those reported for other aquatic systems.

Introduction

Of all environmental factors influencing phytoplankton growth in polar aquatic systems, light has perhaps received the most attention. This interest stems from the fact that these systems are exposed to extremes in Photosynthetic Photon Flux Densities (PPFD) resulting from seasonal or permanent ice and snow cover, together with the variability associated with day length at high latitudes (e.g., periods of total darkness alternate with periods of continuous sunlight). Most of our current information on photoadaptation by polar algae has come from photosynthetic studies of marine phytoplankton or sea-ice microalgae (e.g. Tilzer, 1983; Sakshaug & Holm-Hansen, 1986; Palmisano *et al.*, 1985; Cota, 1985). As with marine systems, the relatively few studies which have examined the influence of PPFD on freshwater algal growth in high latitude systems have dealt primarily with inorganic carbon uptake (e.g. Priscu *et al.*, 1987; Vincent, 1981) despite the fact that inorganic nutrients (e.g. N, P) have also been shown to regulate algal growth.

Nitrogen assimilation by phytoplankton is ultimately dependent upon PPFD as an energy source. Priscu (1984) and Whalen & Alexander (1984) have shown that the dependence of inorganic nitrogen uptake on PPFD (up to PPFD saturation) in freshwater systems can be described by the Michaelis-Menten formulation, supporting results from marine work (MacIsaac & Dugdale, 1972; Terry, 1982; Nelson & Conway, 1979). Although the exact biochemical mechanism(s) by which PPFD regulates N assimilation remains unresolved (Miflin & Lea, 1979; Syrett, 1981), the dependence of inorganic nitrogen uptake on PPFD can have important ecological ramifications. Priscu (1984), studying influences of PPFD on nitrogen uptake in a subalpine lake, found that the half-saturation constants for both NH_4^+ and NO_3^- uptake by deep-living populations corresponded to the ambient PPFD to which they are exposed, allowing them to assimilate nitrogen despite low PPFD levels. Priscu's study further showed that dark NH_4^+ and NO_3^- uptake was about 50% of uptake at saturating PPFD which would allow the organisms to use nitrogen at night when photosynthesis ceased. These results indicate that the PPFD dependence of nutrient uptake must be known for a thorough understanding of cellular C : N balance and hence phytoplankton growth.

The first phytoplankton studies of Antarctic lakes were conducted by Goldman over 2 decades ago (Goldman, 1964). Goldman's studies focused on phytoplankton photosynthesis in lakes lying in the dry valley region of Antarctica. More recent studies on these lakes (e.g. Vincent & Vincent, 1982; Priscu *et al.*, 1987) have dealt with the influence of PPFD and nutrients on algal photosynthesis. These later studies have confirmed Goldman's findings that the phytoplankton in these dry valley lakes are physiologically adapted to the environmental conditions, in particular PPFD, in which they exist. The recent study of Priscu *et al.* (1987) showed that the I_k values for complete photosynthesis in lakes Fryxell and Vanda are among the lowest yet recorded for phytoplankton.

I report here the first data on the influence of PPFD on NO_3^- and NH_4^+ transport in Lakes Fryxell and Vanda, two permanently ice-covered lakes located in the dry valleys of Antarctica. Experiments were conducted on the shallow and deep-chlorophyll layer populations of each lake to compare physiological differences which may exist between these distinct habitats.

Study sites

Lake Fryxell (77° 35′ S 163° 15′ E) lies in the lower Taylor Valley in southern Victoria Land at an elevation of 22 m. The lake has an area of 7 km^2, a maximum peizometric depth of 19 m, and is permanently covered with about 4.5 m of ice. The lake is anaerobic below 9.5 m. Inorganic nitrogen levels (NH_4^+ and NO_3^-) are generally $<10\ \mu g\ N\ l^{-1}$ above 9 m; NH_4^+ exceeds $250\ \mu g\ N\ l^{-1}$ below this depth while NO_3^- remains low. Lake Vanda (77° 35′ S, 161° 40′ E) lies in the Wright Valley, southern Victoria Land, at an elevation of 143 m. The lake is permanently covered with about 3.5 m of ice, has a maximum peizometric depth of 68 m and has a surface of 5.2 km^2. Oxygen is absent below about 60 m. NO_3^- concentrations range from 0–10 $\mu g\ N\ l^{-1}$ between 0–50 m; from 15–250 $\mu g\ N\ l^{-1}$ between 50–60 m and are $<50\ \mu g\ N\ l^{-1}$ below 60 m. NH_4^+ levels range from 0–17 $\mu g\ N\ l^{-1}$, 17–600 $\mu g\ N\ l^{-1}$, and 600–1800 $\mu g\ N\ l^{-1}$ for these same depth ranges.

The latitudes at which these lakes lie receive about 4 months each of daylight, twilight, and darkness. Owing to the permanent ice cover, these lakes have extremely stable water columns and distinct microbial layering (Vincent, 1981; Vincent & Vincent, 1982; Priscu *et al.*, 1987).

Materials and methods

General. Sampling and in situ measurements and incubations were done through a 10 cm diameter hole in the permanent ice-cap made with a SIPRE coring device over the deepest portion of each lake. Samples were collected with a 1-litre non-metallic discrete-depth water sampler during December 1984. All depths reported refer to the piezometric water level.

Samples for chlorophyll *a* (phaeophytin corrected) analysis were filtered onto Whatman GF/F filters and stored frozen until analysis (within 1 month). The filters were thawed and extracted overnight at 4 °C in the dark in dimethyl sulfoxide (DMSO) which was then diluted 1 : 1 with 90% acetone. After centrifugation, the fluorescence of the supernatant (pure and acidified) was measured with an Aminco spectrofluorometer calibrated with standard amounts of pure chlorophyll *a*. PPFD was measured with a LI-COR cosine corrected quantum sensor.

Experimental. The influence of PPFD on inorganic nitrogen transport was determined on lake water samples (10 litre), collected just beneath the ice cap and from the deep-chlorophyll layer (Fryxell = 9 m; Vanda = 57.5 m) of both lakes. The samples were mixed in a 5-litre carboy and distributed (Fryxell = 400 ml; Vanda = 1000 ml) into polyethylene screw cap bottles. The bottles were then inoculated with $(^{15}NH_4)_2SO_4$ (99 atom%) or $Na^{15}NO_3$ (99 atom%) to a final enrichment of 150 μg N l^{-1}. A separate set of experiments showed that this concentration is saturating to uptake of both NH_4^+ and NO_3^- (Priscu, unpublished data). The samples were incubated either in a tent or small hut under a range of PPFD levels, obtained with neutral density screens, for approximately 8 hr. Illumination was from natural light. It was necessary to incubate the samples 'indoors' to avoid freezing. The temperature in the tent and hut ranged from ~0 °C to ~8 °C which, except for the 57.5 m sample from Lake Vanda, is within the range of ambient lake temperatures from which the samples were collected (i.e. Lake Fryxell: 4.6 m = 0.9 °C, 9 m = 3.0 °C; Lake Vanda: 3.5 m = 5.2 °C; 57.5 m = 19.2 °C). Uptake was terminated by filtration through 2.5 cm precombusted (475 °C for 24 h) Whatman GF/F filters. The filters were rinsed with about 20 ml of deionized water and (wet) frozen until analysis.

^{15}N Determination. The ^{15}N content of the seston concentrated on the filters was determined by optical emission spectrometry using the equipment and methods described by Timperly & Priscu (1986). Transport rates were calculated according to the equations of Neess *et al.* (1962) and Dugdale & Goering (1967). Corrections were not made for isotopic dilution during the incubation (e.g. Glibert *et al.*, 1982). Such a correction would probably be negligible given the large amount of ^{15}N added to each experimental flask. Specific rates of nitrogen transport were calculated by dividing volumetric uptake rates by the chlorophyll *a* concentration at the beginning of the experiment. Although chlorophyll *a* per cell may fluctuate with depth owing to differences in PPFD, I chose chlorophyll *a* because it is this constituent which absorbs the photons necessary to fuel cellular transport mechanisms. Chlorophyll *a* normalization will also facilitate comparison with the large body of published chlorophyll *a* normalized photosynthesis versus PPFD curves.

PPFD dependency models. The dependency of inorganic nitrogen transport on PPFD was modelled either by linear regression or a modification of the model of Platt *et al.* (1980). Platt *et al.*'s equation, with appropriate modification, is as follows:

$$N^B = N_s^B (1 - e^{-a}) e^{-b} + D^B , \qquad (1)$$

where N^B = nitrogen transport rate per unit chlorophyll (ng N μg Chl^{-1} h^{-1}); N_s^B = maximum nitrogen transport in the absence of inhibition (i.e. $\beta = 0$), it has the same units as N^B; $a = \alpha I/N_s^B$ where α = slope of the initial PPFD limited portion of the curve [(ng N μg Chl^{-1} h^{-1})/(μE m^{-2} s^{-1})]; I = PPFD (μE m^{-2} s^{-1}); $b = \beta I/N_s^B$, where β = slope of the curve showing light inhibition (same units as α), and D^B = chlorophyll specific dark transport (y-intercept) (ng N μg Chl^{-1} h^{-1}).

Equation (1) differs from Platt *et al.*'s in that nitrogen transport replaces photosynthesis and a term for dark uptake (D^B) is included. Platt *et al.* were able to assume that dark uptake of carbon was equal to zero for their photosynthetic curves. It is obvious from inspection of my data that dark inorganic nitrogen transport cannot be neglected; physiological evidence also supports such dark nitrogen transport (Priscu, 1984).

Equation (1) was used to fit all of the curves except where Marquardt's algorithm, used to fit the equation, failed to converge on the parameters within 200 iterations. In this latter case, linear regression was used to fit the data. Linear regression yields values of D^B (y-intercept) and α (slope) both with the same units as in equation (1).

In addition to the parameter values obtained with equation (1), the following derived values were also computed (Platt *et al.*, 1980).

$$N_m^B = N_s^B(\alpha/\alpha + \beta)(\beta/\alpha + \beta)^{\beta/\alpha} + D^B \quad (3)$$

where N_m^B is the chlorophyll specific transport rate at the optimal PPFD level. It has the same units as N_s^B.

$$I_k = N_m^B/\alpha \tag{4}$$

where I_k is the conventional index of light adaptation for photosynthesis introduced by Talling (1957), i.e. the PPFD level where extrapolations of α and N_m^B intersect.

$$I_s = N_s^B/\alpha \tag{5}$$

where I_s is a parameter similar to I_k which is based on N_s^B rather than N_m^B. It must always be equal to or greater than I_k.

$$I_b = N_s^B/\beta \tag{6}$$

where I_b is an index of photoinhibition with dimensions $\mu E\ m^{-2}\ s^{-1}$. This is a more consistent index of photoinhibition than β because β is scaled by the value of N_s^B (Platt *et al.*, 1980). A small value of I_b is indicative of strong photoinhibition. If β equals zero than I_b is undefined.

$$I_m = (N_s^B/\alpha)\ln((\alpha + \beta)/\beta) \tag{7}$$

where I_m is the PPFD at which maximum uptake occurs. If $\beta = 0$ then I_m is undefined.

Results

Vertical chlorophyll *a* profiles (Fig. 1) show distinct maxima at 57.5 m and 9 m corresponding to about 1% and 0.5% of PPFD incident upon the upper ice surface in Lakes Vanda and Fryxell, respectively. The overall chlorophyll levels were more than an order of magnitude higher in Lake Fryxell; chlorophyll concentration at 9 m reached 21.7 $\mu g\ l^{-1}$. Chlorophyll in Lake Vanda reached a maximum of 0.34 $\mu g\ l^{-1}$ at 57.5 m but was usually less than 0.1 $\mu g\ l^{-1}$. The ice cap attenuated more than 90% of the incident PPFD in both lakes.

Using radiant energy data (langleys) measured continuously by the New Zealand Meteorological

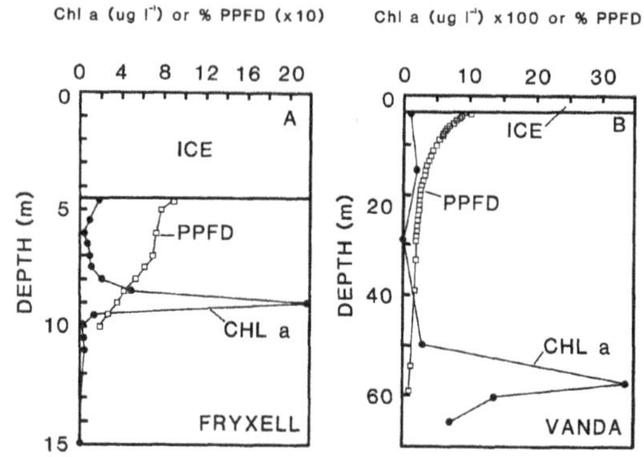

Fig. 1 A,B. Vertical profiles of PPFD (% of that incident upon the surface of the ice) and chlorophyll *a* ($\mu g\ l^{-1}$) in Lake Fryxell (A) and Lake Vanda (B). Note that % PPFD in Lake Fryxell must be divided by 10 and the chlorophyll concentrations in Lake Vanda must be divided by 100 to obtain the actual values.

Service at Lake Vanda during the study period, and assuming 550 nm as the mean wavelength of sunlight, the average PPFD was computed using the conversion factors of Wetzel (1975). The average 24 h PPFD level between 16 December 1984 – 28 December was 484 $\mu E\ m^{-2}\ s^{-1}$. The highest values were generally observed between 1200–1300 h local time; the lowest usually occurred between 0000–0200 h local time. Based on results from these calculations, the daily mean PPFD reaching the deep-chlorophyll layers during the study period in Lakes Vanda and Fryxell was 4.8 and 1.9 $\mu E\ m^{-2}\ s^{-1}$, respectively.

The response of N transport to PPFD could be modeled with equation (1) in all cases for Lake Vanda (Fig. 2). However, Marquardt's algorithm, used to fit the data to equation (1), converged only on the data for NH_4^+ at 9 m in Lake Fryxell. Visual inspection of all curves for which the algorithm converged showed characteristics on which the model was based (see Platt *et al.*, 1980). The data sets not described by equation (1) were fitted with a line of least-squares fit. Only the 4.6 m NH_4^+ experiment (Fig. 3c) showed a significant correlation between PPFD and transport ($r = 0.94$; $p < 0.01$); the slope of the line was also statistically different from zero ($p < 0.001$). The associations between NO_3^- uptake and

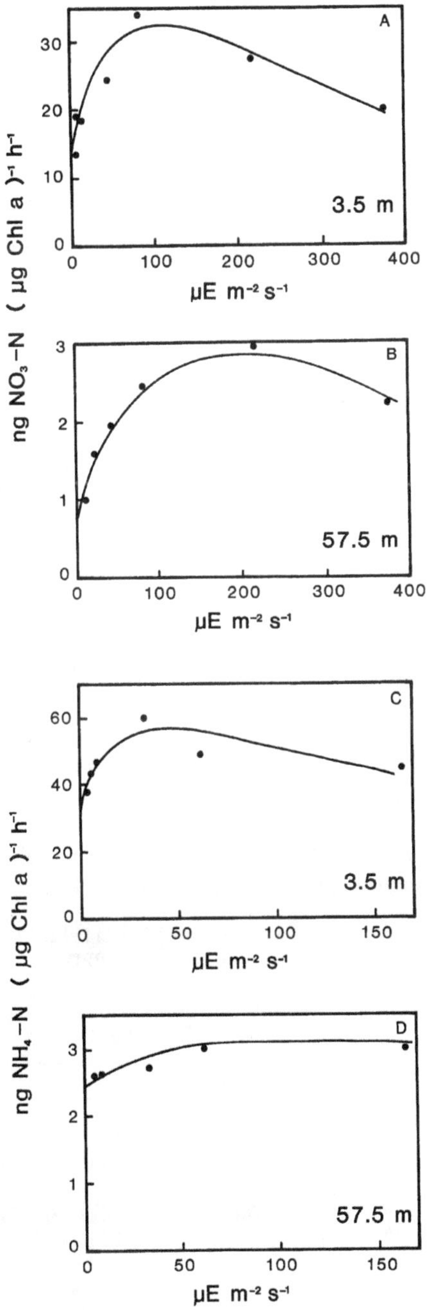

Fig. 2 A–D. Response of inorganic nitrogen transport (ng N μg Chla^{-1} h^{-1}) to PPFD (μE m^{-2} s^{-1}) for the 3.5 and 57.5 m populations in Lake Vanda.

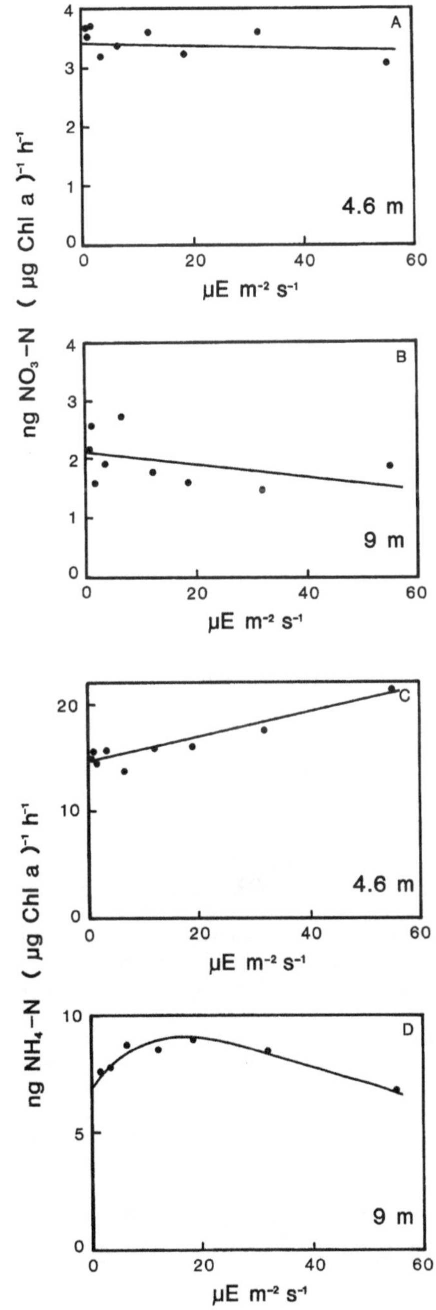

Fig. 3 A–D. Response of inorganic nitrogen transport (ng N μg Chla^{-1} h^{-1}) to PPFD (μg m^{-2} s^{-1}) for the 4.6 and 9.0 m populations in Lake Fryxell.

PPFD at both depths in Lake Fryxell were not statistically significant ($p > 0.05$); the slopes were also not significantly different from zero ($p > 0.20$).

The parameters computed for all of the curves are given in Table 1. Several trends are evident in Lake Vanda. Firstly, N_s^B and β are greater for NO_3^-, regardless of depth. This trend is also reflected in I_b which is lower for NO_3^-. Together these results show that NO_3^- transport is inhibited by PPFD more strongly than NH_4^+ transport. Secondly, N_s^B, N_m^B, α and D^B are all greatest at 3.5 m. Higher N_s^B, N_m^B and D^B values at 3.5 m indicate that the organisms at this depth possess higher specific transport rates for both NO_3^- and NH_4^+ with respect to the deep-water populations at both optimal (i.e. saturating) and zero PPFD levels. By virtue of the higher α values at 3.5 m, nitrogen transport by phytoplankton at this depth has a higher affinity for PPFD than transport in deep-water organisms. Finally, the PPFD level for optimal nitrogen transport, I_m, appears to be greater for NO_3^-. It should be noted that, owing to logistic constraints, the density of samples along the PPFD gradient is relatively low. Consequently, some of these apparent trends should be considered tentative.

The transport parameters obtained at 9 m in Lake Fryxell for NH_4^+ generally fall within the range observed in Lake Vanda. The exceptions to this occur in the I_b and I_m values, both of which are lower than those computed for Lake Vanda. The lower I_b values suggests that higher PPFD inhibition occurs in Lake Fryxell, although lack of a complete data set makes this comparison tentative. The smaller I_m value in Lake Fryxell indicates that lower PPFD levels are needed for optimal nitrogen transport, but again lack of a complete data set makes this comparison tentative.

The I_m values for nearly all experiments were greater than the mean daily PPFD reaching the respective populations. The exception occurred in the 3.5 m Lake Vanda NH_4^+ experiment in which I_m was slightly lower than the average PPFD level. This suggests that phytoplankton transport of inorganic nitrogen was below the optimal PPFD level for transport over a large portion of the day. The optimal level is however, reached or exceeded during midday in the 3.5 m community of Lake Vanda when PPFD at this depth is maximum (approx. 112 $\mu E\ m^{-2}\ s^{-1}$). Optimal PPFD for nitrogen transport is apparently never reached at 9 m for NH_4^+ in Lake Fryxell.

The y-intercepts (D^B) from least-squares fits through the Lake Fryxell 9 m, NH_4^+ experiment and both NO_3^- experiments are comparable with responses fitted with equation (1). It can be as-

Table 1. Parameters describing the characteristics of inorganic nitrogen transport, as a function of PPFD, in Lakes Vanda and Fryxell. Definitions and units are given in the text. The last two columns give the percent of ice-surface PPFD and the mean (and range) of daily $\mu E\ m^{-2}\ s^{-1}$ at each respective depth.

Lake	N_s^B	N_m^B	α	β	D^B	I_k	I_s	I_b	I_m	%	PPFD
VANDA											
NO_3^-											
3.5 m	4,906.0	32.8	0.47	42.9	13.3	69.8	10,411.6	114.1	113.4	10.2	49.4 (4.3–111.7)
57.5 m	565.9	3.0	0.04	3.3	0.7	73.8	14,142.5	171.4	170.4	1.0	4.8 (0.4– 11.0)
NH_4^+											
3.5 m	61.5	52.9	4.40	0.2	25.1	12.0	14.0	412.8	47.8	10.2	49.4 (4.3–111.7)
57.5 m	3.0	3.0	0.01	0.0	2.5	336.7	336.7	–	–	1.0	4.8 (0.4– 11.0)
FRYXELL											
NO_3^-											
4.6 m	–	–	–0.01	–	3.4	–	–	–	–	1.0	4.8 (0.4–11.0)
9.0 m	–	–	–0.01	–	2.1	–	–	–	–	0.4	1.9 (0.2– 4.4)
NH_4^+											
4.6 m	–	–	0.11	–	14.6	–	–	–	–	1.0	4.8 (0.4–11.0)
9.0 m	581.7	9.0	0.45	41.2	6.7	20.0	1,292.6	14.1	14.1	0.4	1.9 (0.2– 4.4)

sumed that, because the slopes are not significantly different from zero, D^B for the Lake Fryxell NO_3^- experiments represents N_m^B. If this assumption is valid, then the estimate of N_m^B is within the range of the experiments fitted with equation (1). The slope of the line fitted to the 4.6 m NH_4^+ experiment is statistically significant from zero and equates with α. Alpha computed in this manner is comparable to α from the initial slope of the nonlinear curves. The lack of an asymptote in this experiment indicates that PPFD saturation was not reached within the range of levels used in the experiment. That the slopes of the lines in the Lake Fryxell NO_3^- experiments were not different from zero indicates that NO_3^- transport has no response to PPFD over the levels used in the experiment.

Discussion

I have shown that the PPFD response of NO_3^- and NH_4^+ transport by phytoplankton can often be modeled successfully with equations developed by Platt *et al.* (1980) for photosynthetic response. Lewis & Levine (1984), working with tropical cyanobacteria, have also shown that the empirical model of Platt *et al.* can describe successfully the PPFD response of atmospheric nitrogen fixation. Previous work on inorganic nitrogen uptake has generally established that there is a correlation between inorganic nitrogen transport and PPFD (e.g. MacIsaac & Dugdale, 1972; Priscu, 1984; Whalen & Alexander, 1984). These studies used the Michaelis-Menten equation to fit the PPFD response up to the point of saturation; the portion of the curve showing inhibition at high PPFD was modeled either by a separate mathematical function (e.g. linear regression) or by visual inspection. This approach is unacceptable in practice because of the subjectivity involved in deciding exactly which data points belong to the photoinhibition regime. It is obvious from the data I present here that selection of points associated with photoinhibition would lack reference to any objective criterion.

Another problem with using the Michaelis-Menten equation to fit PPFD response curves for nitrogen transport is that it assumes that the curve passes through the origin (i.e. transport ceases at zero PPFD). Therefore, the Michaelis-Menten model can be used only when dark transport is a negligible proportion ($< 10\%$ of PPFD saturated transport, Nelson & Conway, 1974). MacIsaac & Dugdale (1972) and MacIsaac *et al.* (1974) suggested that dark nitrogen transport is not necessarily a constant proportion of light transport, but rather a distinct process that occurs in the dark or under low PPFD conditions. Consequently, these authors chose to ignore dark transport and forced their response curve to pass through zero. Currently no direct physiological evidence exists supporting MacIsaac and Dugdale's suggestion that dark transport is distinct from PPFD mediated transport. It is apparent that the Michaelis-Menten model is not appropriate for describing the PPFD response of inorganic nitrogen transport by phytoplankton. Not only does the photosynthetic model of Platt *et al.* (1980) appear to fit nitrogen transport data well, but it yields many useful parameters which can be used to characterize certain portions of the curve, or the physiological mechanisms responsible for these portions. Moreover, it allows one to compare directly the PPFD response of inorganic nitrogen transport to the wealth of information which exists for inorganic carbon transport (i.e. photosynthesis) by phytoplankton.

When compared to the ambient PPFD levels which exist at specific depths in Lake Vanda, the transport parameters (Table 1) do not indicate clearly physiological adaptations to ambient PPFD. For example, α is higher at 3.5 m implying that the deep community is adapted, in relative terms, to a higher PPFD environment than organisms at 57.5 m (i.e. the deep-community has a lower affinity for PPFD). This fact is supported by the lower I_k, higher β and lower I_b values expressed by the shallower water communities. In concert, these parameters indicate that the shallow-water populations in Lake Vanda are adapted to transport inorganic nitrogen in a lower PPFD environment than the populations forming the deep-chlorophyll layer.

Dark transport for NO_3^- at 3.5 m and 57.5 m was 41% and 24%, respectively, of N_m^B whereas that for NH_4^+ was 47% and 83% of N_m^B for these same depths. The relatively high dark transport of NO_3^- at the shallow depth supports other parameters suggesting that this community is low PPFD adapted with respect to the deep-living community. Dark transport of NH_4^+ shows an opposite trend. Despite these inconsistencies, the fact that the PPFD level for optimal transport (I_m) is generally equal to or greater than the average daily PPFD level at depth, implies that inorganic nitrogen transport at saturating concentrations is regulated over most of the day by ambient PPFD levels.

It is important to realize that transport rates were measured at saturating nitrogen levels ($> 150\ \mu g\ N\ l^{-1}$). Based on NO_3^- and NH_4^+ concentrations measured in these lakes (Vincent, 1981; Vincent *et al.*, 1981; Priscu, unpublished data) saturating levels of NO_3^- and NH_4^+ occurred at 57.5 m in Lake Vanda. Concentrations of NO_3^- and NH_4^+ in Lake Fryxell and at 3.5 m in Lake Vanda were always $< 50\ \mu g\ N\ l^{-1}$. This brings up the question as to whether PPFD is the major factor limiting transport of nitrogen and potential algal growth in the regions of these lakes where ambient nitrogen concentrations are not saturating. It appears that nitrogen transport in the shallow populations of these lakes where inorganic nitrogen levels are lowest and PPFD levels highest, would be regulated to a greater extent by nitrogen concentration than PPFD. The converse would be true for the deep-living algal communities. Consequently, the transport parameters in Table 1 must be used cautiously to assess the influence of PPFD on vertical patterns of in situ algal growth. The interrelationships between PPFD and nitrogen concentration in regulating in situ algal growth is definitely an area requiring further research.

As discussed previously, virtually all earlier studies which have examined the influence of PPFD on inorganic nitrogen transport modeled their responses with the Michaelis-Menten equation. In these experiments the half-saturation constant for transport with respect to PPFD (K_{LT}) was the parameter used to examine photoadaptation. For the sake of comparison, I have estimated K_{LT} for Lakes Vanda and Fryxell from the following relationship,

$$K_{LT} = K_{LT}/N_m^{\beta} \times N_m^{\beta} \qquad (8)$$

K_{LT}/N_m^{β} was estimated by multiplying the y-intercept of a linear regression of (PPFD/N-transport) on PPFD by N_m^{β} according to the Hanes-Woolf transformation of the Michaelis-Menten equation (Segel, 1976). Only values showing no photoinhibition were used in this analysis.

Values computed with this relationship are compared with K_{LT} from other systems in Table 2. Values for Antarctic lakes are within the range of those from other aquatic systems indicating that planktonic microorganisms growing under the permanent ice caps of Antarctic lakes during Austral summer possess nitrogen transport affinities similar to other systems. This result does not support completely the response of photosynthesis to PPFD in these lakes. Priscu *et al.* (1987) showed that phytoplankton photosynthesis in Lakes Vanda and Fryxell is photoadapted to lower PPFD levels than any other population yet reported. Priscu *et al.* (1987) argued that hydraulic stability induced by the permanent ice-caps in the Antarctic desert lakes allows microorganisms to adapt precisely their photosynthetic mechanisms to ambient PPFD levels. This argument does not seem to apply to PPFD mediated inorganic nitrogen transport. However, until the crossover point between PPFD and nitrogen limitation is determined in these lakes, definite conclusions regarding the importance of PPFD on *in situ* nitrogen transport and subsequent microorganism growth remain tentative.

Acknowledgments

I thank W. F. Vincent and C. Howard-Williams for allowing me to participate in New Zealand Antarctic Research event KO81 and for their

Table 2. Comparison of half-saturation constants (K_{LT}) for inorganic nitrogen transport (including both NO_3^- and NH_4^+) in various aquatic systems. K_{LT} is given as % of surface PPFD and as $\mu E\ m^{-2}\ s^{-1}$.

Region	K_{LT}		Reference
	%	$\mu E\ m^{-2}\ s^{-1}$	
East Tropical Pacific Ocean	14	–	MacIsaac and Dugdale (1972)
Peru Upwelling	0.9–13.5	–	MacIsaac and Dugdale (1972)
Northwest Africa	1.5–7.0	–	MacIsaac *et al.* (1974)
Northwest Africa	5.5–6.2	–	Nelson and Conway (1979)
Antarctic Ocean	1.1–2.6	2.3–4.4	Slawyk (1979)
Toolik Lake	6.0–3.1	7.0–29.0	Whalen and Alexander (1984)
Castle Lake, California			
surface populations	1.1–2.7	5.8–16.1	Priscu (1984)
deep-chl populations	0.4–3.7	2.3–25.5	Priscu (1984)
Lake Vanda			
surface populations	0.5–2.0	2.4–9.5	This Study
deep-chl populations	0.4–5.1	1.7–24.9	This Study
Lake Fryxell			
deep-chl populations	0.08	0.04	This Study

support in both the field and laboratory. E. White provided bench space at the Taupo Research Laboratory. G. Spearpoint supplied expert field assistance and L. Priscu, M. Downes, S. Dryden, M. Gibbs, S. Pickmere, V. Reid, M. Timperley, P. Wade and P. Woods assisted with data processing. The personnel at Scott Base and Vanda Station, and the US Navy furnished the necessary logistic support to conduct this research. Fieldwork was performed during the 1984–1985 New Zealand Antarctic Field season. This research was supported by the Antarctic Division, New Zealand Department of Scientific and Industrial Research and an NSF grant (INT 8412682) to the author.

References

Cota, G., 1985. Photoadaptation of high Arctic ice algae. Nature 219: 222–224.

Dugdale, R. C. & J. J. Goering, 1967. Uptake of new and regenerated forms of nitrogen in primary productivity. Limnol. Oceanogr. 12: 196–206.

Glibert, P. M., F. Lipschultz, J. J. McCarthy & M. A. Altabet, 1982. Isotope dilution models of uptake and remineralization of ammonium by marine plankton. Limnol. Oceanogr. 27: 639–650.

Goldman, C. R., 1964. Primary productivity studies in Antarctic Lakes. Primary Symp. S.C.A.R. Paris, 2–8 Sep. 1962. Hermann Ltd., England. 651 p.

Lewis, W. M. & S. N. Levine, 1984. The light response of nitrogen fixation in Lake Valencia, Venezuela. Limnol. Oceanogr. 29: 894–900.

MacIsaac, J. J. & R. C. Dugdale, 1972. Interactions of light and inorganic nitrogen in controlling nitrogen uptake in the sea. Deep-Sea Res. 19: 209–232.

MacIsaac, J. J., R. C. Dugdale & G. Slawyk, 1974. Nitrogen uptake in the northwest Africa upwelling area: results from Cinceca-Charcot II cruise. Tethys 6: 69–76.

Miflin, B. J. & P. J. Lea, 1977. Amino acid metabolism. Ann. Rev. Plant Physiol. 28: 299–329.

Neess, J. C., R. C. Dugdale, V. A. Dugdale & J. J. Goering, 1962. Nitrogen metabolism in lakes. I. Measurement of nitrogen fixation with ^{15}N. Limnol. Oceanogr. 7: 163–169.

Nelson, D. M. & H. L. Conway, 1979. Effects of the light regime on nutrient assimilation by phytoplankton in the Baja California and northwest Africa upwelling systems. J. Mar. Res. 37: 301–318.

Palmisano, A. C., J. B. Soohoo & C. W. Sullivan, 1985. Photosynthesis-irradiance relationships in sea ice microalgae from McMurdo Sound. Antarctica. J. Phycol. 21: 341–346.

Platt, T., C. L. Gallegos & W. G. Harrison, 1980. Photoinhibition of photosynthesis in natural assemblage of marine phytoplankton. J. Mar. Res. 38: 687–701.

Priscu, J. C., 1984. A comparison of nitrogen and carbon metabolism in the shallow and deep-water phytoplankton populations of a subalpine lake: Response to photosynthetic photon flux density. J. Plankton Res. 6: 733–749.

Priscu, J. C., L. R. Priscu, W. F. Vincent & C. Howard-Williams, 1987. Photosynthate distribution by microplankton in permanently ice-covered Antarctic desert lakes. Limnol. Oceanogr. 32: 260–270.

Sakshaug, E. & O. Holm-Hansen, 1986. Photoadaptation in Antarctic phytoplankton: Variation in growth rate, chemical composition and P versus I curves. J. Plankton Res. 8: 459–473.

Segel, I. H., 1976. Biochemical calculations, 2nd Ed. Wiley. 441 p.

Slawyk, G., 1979. ^{13}C and ^{15}N uptake by phytoplankton in the Antarctic upwelling area: results from the Antipod I cruise in the Indian Ocean sector. Aust. J. mar. Freshwat. Res. 30: 431–448.

Syrett, P. J., 1981. Nitrogen metabolism of microalgae, p. 182–210. Int: T. Platt (ed.) Physiological Basis of Phytoplankton Ecology. Can. Bull. Fish. aquat. Sci. 210: 346 p.

Talling, J. F., 1957. The phytoplankton population as a compound photosynthetic system. New Phytol. 56: 133–149.

Terry, K. L., 1982. Nitrate uptake and assimilation in *Thalassiosira weissfloggi* and *Phaedoctylum tricornutum*: Interactions with photosynthesis and with the uptake of other ions. Mar. Biol. 69: 21–30.

Tilzer, M. M., B. von Bodungen & V. Smetacek, 1985. Light dependence of phytoplankton photosynthesis in the Antarctic Ocean; Implications for regulating productivity, In: W. R. Siegfried, P. R. Condy & R. M. Laws (eds.), Proceedings of the 4th SCAR Symposium on Antarctic Biology, Wilderness, Sept. 1983, Springer-Verlag, Berlin, pp. 60–69.

Timperley, M. H. & J. C. Priscu, 1986. Nitrogen-15 analysis by optical emission spectrometry using an atomic absorption spectrometer. Analyst 3: 23–28.

Vincent, W. F., 1981. Production strategies in Antarctic inland waters: Phytoplankton eco-physiology in a permanently ice-covered lake. Ecology 62: 1215–1224.

Vincent, W. F. & C. L. Vincent, 1982. Factors controlling phytoplankton production in Lake Vanda (77° S). Can. J. Fish aquat. Sci. 39: 1602–1609.

Wetzel, R. G., 1975. Limnology. Saunders, 743 p.

Whalen, S. C. & V. Alexander, 1984. Influence of temperature and light on rate of inorganic nitrogen transport by algae in an arctic lake. Can. J. Fish. aquat. Sci. 41: 1310–1318.

Hydrobiologia 172: 183–191, 1989
W. F. Vincent and J. C. Ellis-Evans (eds) High Latitude Limnology

Patterns of energy storage in *Pseudoboeckella poppei* (Crustacea, Copepoda) from two contrasting lakes on Signy Island, Antarctica.

Andrew Clarke, J. Cynan Ellis-Evans, Mark W. Sanders & Lesley J. Holmes
British Antarctic Survey, NERC, High Cross, Madingley Road, Cambridge, CB3 OET, England

Key words: Antarctic lakes, copepod, lipid, phytoplankton, triacylglycerol, wax ester

Abstract

The copepod *Pseudoboeckella poppei* (Daday) (Calanoida, Centropagidae) was sampled from Sombre and Heywood Lakes on Signy Island, Antarctica (60° S, 45° W) between January 1984 and March 1985. Sombre Lake is clear and oligotrophic with little phytoplankton and a bottom sediment low in organic content. By contrast Heywood Lake is turbid and mesotrophic; a substantial phytoplankton develops in summer and the bottom sediments are comparatively rich in organics. Both lakes freeze over for much of the year, forcing the copepods to adopt a benthic feeding strategy over winter. Adult *Pseudoboeckella* feed on phytoplankton when this is available, but also on detritus, diatoms and short algal filaments stirred up from the sediment. In Heywood Lake, male copepods show a smooth seasonal trend in lipid content with lipid being synthesised in early summer and utilised in late summer and winter. The summer increase in lipid content is associated with an increase in dry weight. Female lipid contents show evidence of two peaks of egg production. In Sombre Lake both male and female copepods increase in size during summer and show a wider range of lipid contents than in Heywood Lake; it is likely that this is due to the poorer winter feeding conditions which necessitate the synthesis of a much larger store of reserves during the summer. In contrast to marine calanoid copepods, lipid stores are exclusively triacylglycerol with no trace of wax ester.

Introduction

The main energy store in marine and freshwater crustaceans is lipid (Båmstedt, 1986). The pattern of storage and utilisation of this lipid is dictated by:

1) The availability of food
2) The varying demands of energy for growth, reproduction and maintenance (Båmstedt, 1986, Sargent & Henderson, 1986).

In this study we have examined the seasonal cycle of lipid storage in the herbivorous copepod *Pseudoboeckella poppei* (Daday) (Calanoida, Centropagidae) from two contrasting lakes on Signy Island, South Orkney Islands, Antarctica.

Study sites

Signy Island (60° 43′ S, 45° 36′ W) is situated in the maritime Antarctic zone (Antarctic Peninsula and offlying islands) which is characteristically somewhat warmer and more humid than continental Antarctica. There are sixteen lakes on Signy Island, ranging from permanently ice-covered, extremely oligotrophic lakes to highly enriched systems which are ice-free for three-four months each year. Ice cover generally forms in

April-May and may attain a thickness of 1 m. Phytoplankton proliferate each spring (October–December) as the disappearance of snow cover permits increased light penetration of the ice. In autumn, as the ice cover reforms, rapidly decreasing solar input and the absence of wind-generated turbulence results in the phytoplankton sedimenting from the water column. In the enriched lakes phytoplankton dominate but in more oligotrophic lakes an extensive benthic community of cyanobacteria is present. Unlike the phytoplankton, these cyanobacterial mats have been shown to be capable of net photosynthesis virtually year round by developing an extreme shade adaptation (Priddle, 1980).

The two lakes examined in this study were Sombre Lake and Heywood Lake. Sombre Lake is a clear oligotrophic system with a maximum depth of 10.2 m. Heywood Lake is a turbid mesotrophic system comprising two basins. The North Basin studied here has a maximum depth of 6 m. In both lakes the annual temperature range is 0.2–6.0 °C and during the open water period in summer the entire water column is homogeneous due to wind mixing.

Materials and Methods

Copepods were collected between January 1984 and March 1985 by vertical net hauls. In winter, nets were deployed and recovered through holes cut in the overlying ice. Copepods were removed gently from the net and transported in a dewar containing lake water to the laboratory at Signy Station within one hour. In the laboratory copepods were sorted under a stereomicroscope, counted and frozen on dry ice (– 70 °C) for shipment to UK. Subsamples were also counted and preserved in 4% neutralised formaldehyde. During sorting copepods were held in lake water at environmental temperature.

In UK lipid was extracted from frozen copepods by homogenisation in methanol/chloroform (2 : 1, v/v) according to Bligh & Dyer (1959). The purified lipid was then assayed by reduction of dichromate ion. Lipid class composition was determined by high performance thin-layer chromatography (HPTLC) coupled with scanning densitometry. Full details of analytical procedures are given in Clarke *et al.* (1985). Since the number of copepods in each pooled sample was known, lipid content has been expressed in absolute terms (that is, μg lipid per copepod). Because each sample of copepods was pooled for analysis, this lipid content is equivalent to a mean; however we can say nothing about individual to individual variation in lipid content from these samples.

Copepods preserved in formalin were carefully drained of preservative, washed briefly and dried at 60 °C. Mean dry weights were obtained by weighing the entire sample on a Cahn microbalance and dividing by the number of copepods in the sample.

Results

Lipid content: males

In Heywood Lake the lipid content of male copepods showed a smooth seasonal trend. Lipid contents were lowest in mid-winter (2.7 μg per individual in June) rising during spring to a maximum in January of 19.6 μg per individual (Fig. 1). In Sombre Lake seasonal variations in lipid content were more extensive. Male lipid content declined throughout the winter to a minimum value of about 7 μg per individual in October/November and then increased dramatically during summer to a peak of over 50 μg per copepod in January/February. The fall in lipid content during winter in Sombre Lake was much steeper in 1985 than in 1984 (Fig. 1).

In both lakes the increase in lipid content was associated with the spring phytoplankton bloom, and was accompanied by an increase in dry weight (Fig. 2). In Heywood Lake the major phytoplankton bloom starts once superficial snow is reduced, but before the ice cover melts. This is usually followed by a second, less intense bloom during the open water season (Hawes, 1985). In Sombre Lake there is only a slight seasonal variation in water column phytoplankton; chlorophyll contents rarely reaching 8 mg m^{-3}, com-

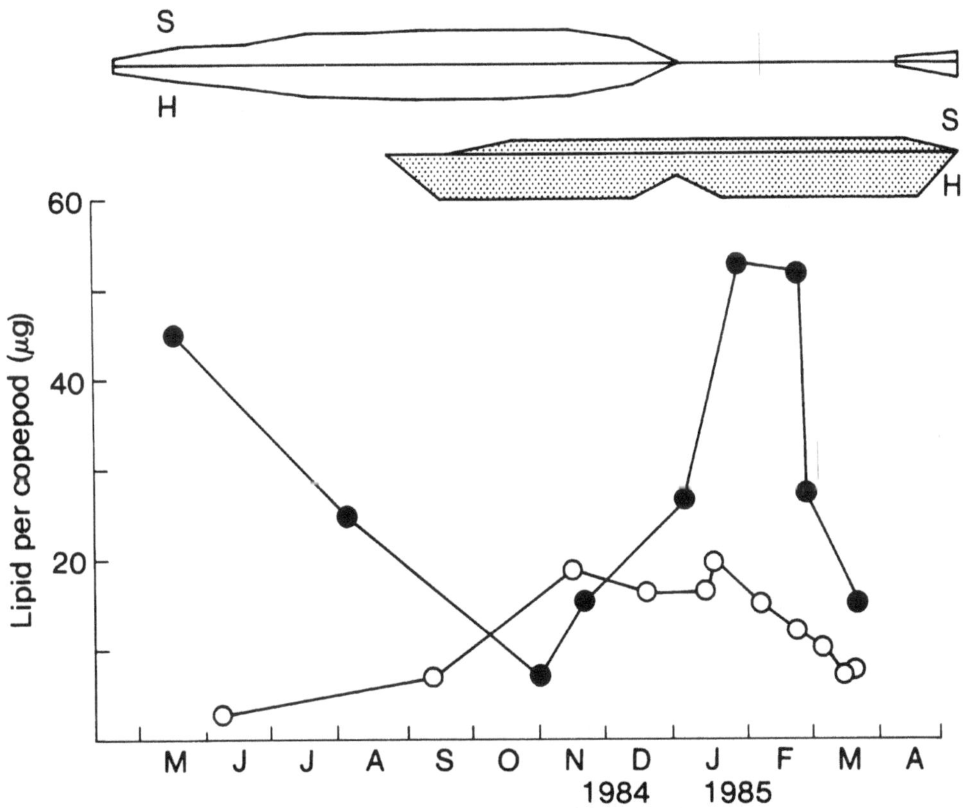

Fig. 1. Seasonal variation in mean lipid content of male *Pseudoboeckella poppei* from Heywood Lake (○) and Sombre Lake (●), Signy Island. Each data point represents the mean lipid content of an individual copepod, determined from a single extraction of a pooled sample containing a known number of individuals. The clear polygon shows the duration of ice cover during 1984 and 1985, the stippled areas the approximate duration of the phytoplankton bloom. In both cases data from Sombre Lake are above, and data for Heywood Lake are below, the horizontal line.

pared with peak levels of 50 mg m^{-3} in Heywood Lake (Hawes, 1985).

These differing patterns of food availability clearly influence the timing and extent of weight increase and lipid synthesis in male *Pseudoboeckella* from the two lakes. In Heywood Lake copepods increase in weight and lipid content once the under-ice bloom starts in October/November. Both lipid content and dry weight decrease during the (smaller) open-water bloom period. In Sombre Lake both lipid content and dry weight increase most dramatically during the open-water period, although the increase in lipid content starts as soon as the slight bloom gets under way.

Lipid content: mature females

Mature (non-ovigerous) females were those classified as adults (CVI) but which were not carrying eggs. Mature females from Heywood Lake contained between 6.1 and 21.5 µg lipid (Table 1). There was little apparent seasonal trend although mature females with lipid contents greater than 15 µg tended to be taken in summer.

Mean lipid contents of mature females from Sombre Lake were very variable (Fig. 3) and the

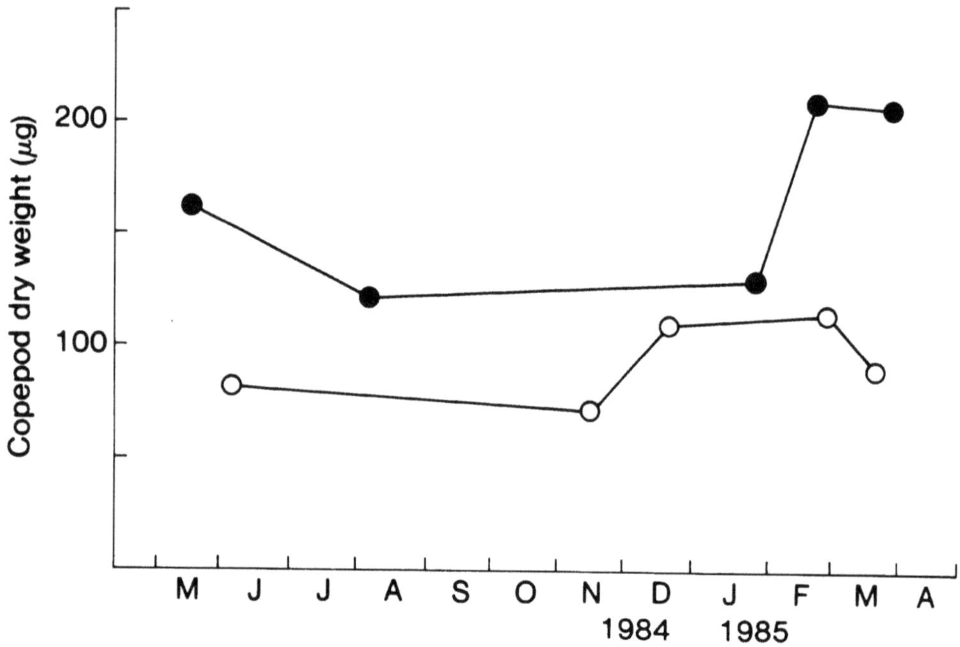

Fig. 2. Seasonal variation in mean dry weight of male *Pseudoboeckella poppei* from Heywood Lake (○) and Sombre Lake (●). Each data point represents the mean dry weight of an individual copepod, determined from the weight of a pooled sample containing a known number of individuals.

range of mean lipid contents observed (4.6 to 64.2 μg: Table 1) was much greater than in Heywood Lake. As in Heywood Lake, the seasonal pattern was obscured by large sample to sample variation, but peak contents always occurred in late summer.

Although the lipid content data are very variable, in both lakes high lipid contents tended to occur towards the end of summer. This could be due either to a deposition of storage lipid or an increase in copepod size. In both Sombre and Heywood Lakes there was an increase in the mean dry weight of mature (non-ovigerous) females during mid summer (Fig. 4). In Heywood

Table 1. Lipid content of *Pseudoboeckella poppei* from Heywood and Sombre Lakes, Signy Island. Data are presented as median values, with the overall range and total number of determinations in brackets. Each determination is a single analysis of a pooled sample containing a known number of individuals. For each class of copepod, the overall data for Heywood and Sombre Lakes were compared by calculating the Mann-Whitney U-statistic. U_o is the observed value. U_c is the critical value of U at the level P = 0.05. For a significant result $U_o < U_c$; for no class of copepod were the samples significantly different.

	Lipid content (μg per copepod)		U_o	U_c
	Heywood Lake	Sombre Lake		
Males	16.6 (2.7 to 27.1; n = 11)	24.8 (6.9 to 52.6; n = 9)	30	23
Mature Females (non-ovigerous)	13.9 (6.1 to 21.5; n = 12)	13.3 (4.6 to 64.2; n = 11)	49	33
Ovigerous Females	25.9 (4.2 to 64.6; n = 8)	39.4 (7.6 to 57.5; n = 7)	23	10

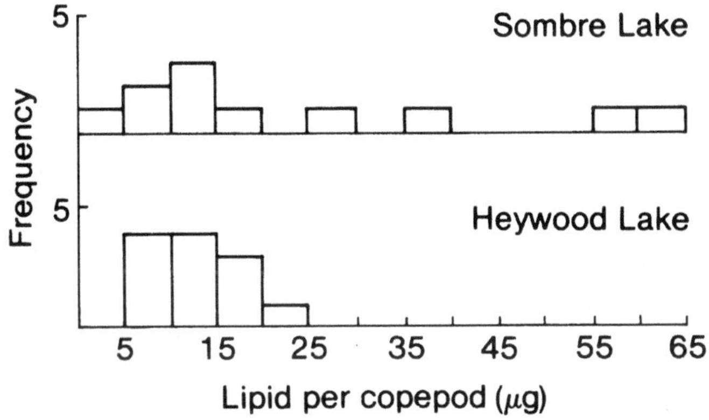

Fig. 3. Frequency histograms of mean lipid content of mature (non-ovigerous) CVI female *Pseudoboeckella poppei* from Sombre and Heywood Lakes.

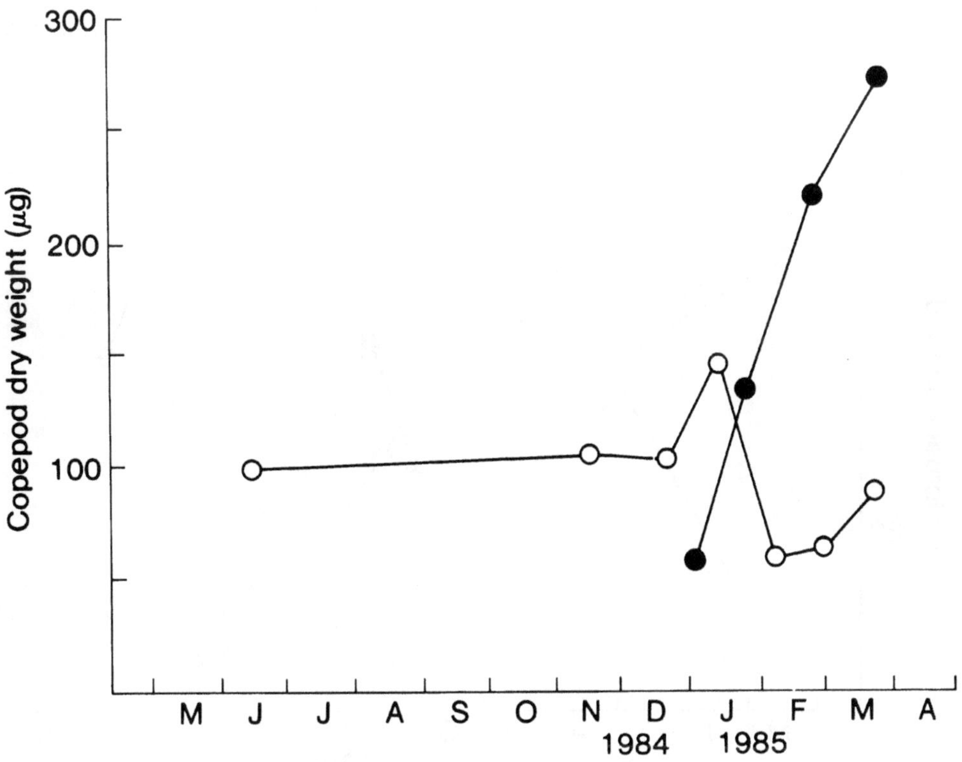

Fig. 4. Seasonal variation in the mean dry weight of mature (non-ovigerous) female *Pseudoboeckella poppei* from Heywood Lake (○) and Sombre Lake (●). Presentation as for Fig. 2.

Lake this increase was small, and followed by a sharp decrease in size. In Sombre Lake, however, there was a dramatic increase in mean dry weight, at a average rate of 2.4 $\mu g\, d^{-1}$. During the previous season (1984, data not shown) mature females in Sombre Lake showed a similar increase, but at a slower rate of 1.2 $\mu g\, d^{-1}$. The maximum mean dry weights recorded were 170 μg in May 1984 and 245 μg in March 1985. These data indicate that there are significant year to year variations in feeding conditions for *Pseudoboeckella* in Sombre Lake.

Lipid content: ovigerous females

The lipid contents of ovigerous females were similar in the two lakes. Overall range varied from 4.2 to 64.6 μg per individual and the frequency distributions for the two lakes did not differ significantly (Table 1). In Heywood Lake there were two peaks in lipid content (March and September), implying two peaks in breeding activity (Fig. 5). In Sombre Lake ovigerous female lipid contents were greatest during the summer open water period. Lipid contents then fell more or less steadily throughout the following winter. Ovigerous females can be found in Sombre Lake in most months of the year, but peak numbers occur in late winter (June to October: Weller, 1977). Breeding activity may thus explain the winter decline in lipid content, with lipid reserves being deposited in summer when feeding conditions are better.

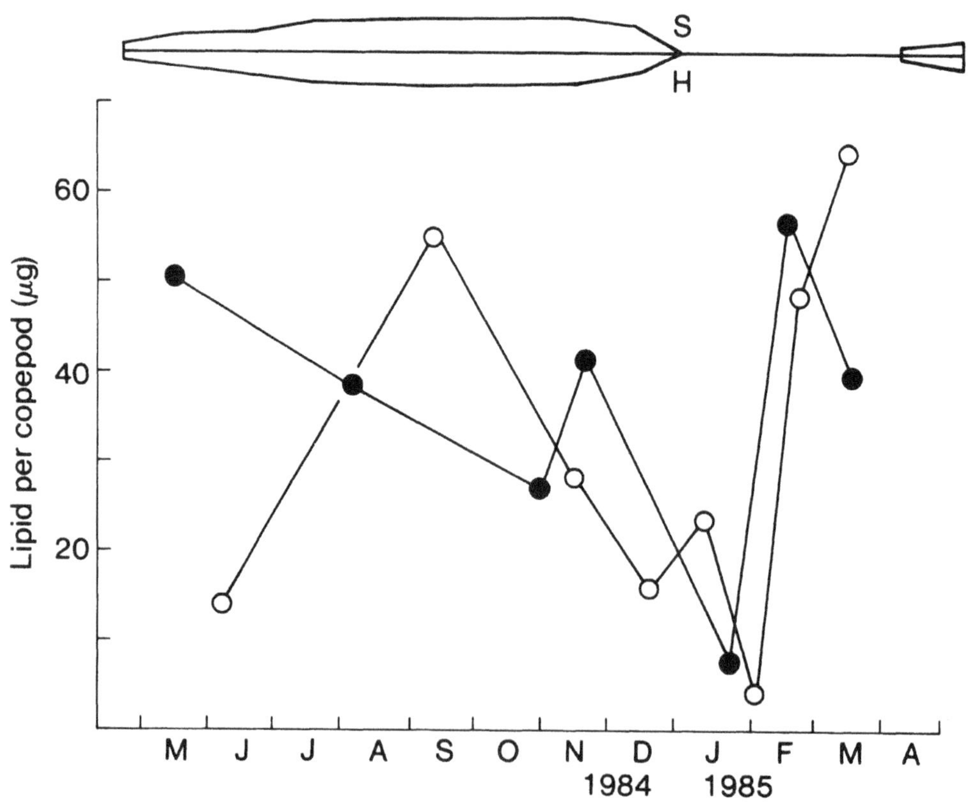

Fig. 5. Seasonal variation in mean lipid content of ovigerous female *Pseudoboeckella poppei* from Heywood Lake (○) and Sombre Lake (●). Data and presentation as for Fig. 1.

Table 2. Major features of the lipid class composition of *Pseudoboeckella poppei* from Heywood (H) and Sombre (S) Lakes, Signy Island. All data are % total lipid. PL, phospholipid; FS, free sterol; FFA, free fatty acid; TAG, triacylglycerol. n is the number of samples analysed. F is the variance ratio for a linear regression (least squares) of TAG (% total lipid) on lipid content (μg); NS, not significant (P > 0.05).

		% total lipid				n	F
		PL	FS	FFA	TAG		
Males	H	13–72	2–11	4–25	6–70	12	12.0 (P < 0.05)
	S	2–42	1–10	1–14	33–92	9	0.3 (NS)
Feamles	H	10–73	1–8	3–28	3–44	11	0.2 (NS)
	S	4–57	1–14	2–13	19–93	11	6.9 (NS)
Ovigerous Females	H	4–45	1–8	1–19	14–79	8	0.9 (NS)
	S	5–27	1–2	2–7	2–77	6	0.1 (NS)

Lipid class composition

The major lipid classes detected in *Pseudoboeckella poppei* were triacylglycerol, free fatty acid, free sterol and polar lipids (the latter will mostly comprise phospholipids). There were no traces of wax ester apart from two samples from Sombre Lake which contained 5–10% of a component with a high TLC mobility, similar but not identical to that of standard wax ester. The identity of this component is unknown.

The lipid composition of aquatic invertebrates varies to such an extent with feeding, season and reproductive state, that there can be no such thing as a definitive lipid composition for a given species. The range of compositions found in this study is shown in Table 2. In male copepods from Heywood Lake there was a significant positive relationship between lipid content and the proportion of triacylglycerol (TAG) in the lipid. This indicates that male copepods tend to store lipid as triacylglycerol. In Sombre Lake, however, there was no sign of such a relationship (Table 2). This suggests that, in Sombre Lake, copepods store both triacylglycerol and phospholipid and confirms that there are significant differences in the lipid metabolism of *Pseudoboeckella poppei* between the two lakes.

Discussion

Feeding ecology of Pseudoboeckella

The early copepodid stages of *Pseudoboeckella poppei* are filter feeders and essentially planktonic, but later on the mouthparts become modified for detritus feeding (Heywood, 1970). Copepodid stages CV and CVI are essentially nektobenthic and in summer they swarm over the algal mats which grow in the shallow shelf regions of the lakes, though adult copepods may also occur in the water column. In both lakes, the algal mats comprise a loose assemblage of filaments (green or cyanophyte) and unicells all of which can be utilized by *P. poppei* (Weller, 1977). In summer, a substantial phytoplankton develops in Heywood Lake and it is clear from gut content analysis that CV and CVI copepods make use of this plentiful food source in addition to benthic material. Phytoplankton are comparatively scarce in Sombre Lake, and here copepods are mainly benthic feeders.

In winter virtually all the shallow region of the lakes are frozen solid and the algal assemblages are therefore unavailable. In addition the deteriorating light regime and lack of wind-induced turbulence causes the water column phytoplankton to sediment, so in both lakes food is limited to benthic material on the inner fringes of the shallows and in the trough. In Sombre Lake this material largely comprises a consolidated mat of slow-growing, perennial cyanophytes and a variety of diatoms intermixed with significant quantities of inorganic material. In Heywood Lake, perennial mats are absent and the sediment has a far higher organic content due to substantial inputs of sedimented plant material. *P. poppei* has been shown to be an unselective feeder but unable to feed effectively off consolidated mats (Weller, 1977). Rather the maxillipeds of these animals are used to stir up debris, diatoms and short algal filament lengths for subsequent filtration. The nutritional potential of Heywood Lake benthic material ingested by the copepods would therefore appear to be far higher than in Sombre Lake where much of the loose material is inorganic. Life cycle studies of copepods in both lakes indicate year-round recruitment in Heywood Lake but only one recruitment, in late winter, in Sombre Lake (Heywood, 1970; Weller, 1977).

Lipid content and composition

The differing patterns of food availability between Sombre and Heywood Lakes can explain some, but not all, of the observed seasonal patterns of lipid content and reproduction in the two lakes.

Although only the early copepodid stages feed directly on phytoplankton, the organic-rich sediments and associated benthic phytoplankton in Heywood Lake suggest that food is generally available year round. This correlates with the year-round reproduction and generally low seasonal variations in lipid content in copepods from Heywood Lake. Where food is generally available there is no need to synthesise a large store of reserves.

In Sombre Lake, however, food is seasonally limiting and both growth and lipid deposition in *Pseudoboeckella* are limited to the summer. If newly hatched naupliar stages are to be able to feed, they must hatch in early summer. This in turn dictates that eggs must be produced and incubated in late winter and must be synthesised from lipid reserves laid down the previous summer. This is precisely what is observed, with striking increases in copepod dry weight and lipid content during the summer open-water season, and a steady utilisation of female lipid over winter. Male lipid decreases more rapidly in late summer and early winter. The reason for this is unclear, but it may be due to metabolic activity associated with sperm production and reproductive behaviour.

Taken together, these features indicate that, as has previously been shown (Heywood, 1970; Weller, 1977), *Pseudoboeckella poppei* is not directly dependent on the phytoplankton bloom. Rather it is the availability of benthic material, and particularly organic-rich detritus, which is important. Thus in Heywood Lake, where food appears to be generally available, reproduction is year-round, the observed range in clutch size is 18 to 200 (Heywood, 1970) and there is only a slight seasonal variation in dry weight and lipid content. In Sombre Lake where food is believed to be seasonally limiting and the detritus poor in organics, reproduction is seasonal and clutch size only 31 to 47 (Weller, 1977). In addition there is a severe seasonality in both lipid content and dry weight. It is not clear why males are bigger in Sombre Lake than Heywood Lake.

Copepods from freshwater lakes are clearly distinct from marine copepods, in that energy is stored largely as wax ester in the sea, but as triacylglycerol in freshwater. This distinction has been previously reported by Sargent (1976) and Clarke (1984), both based on analyses of copepods from Signy Island. Morris (1983, 1984) has also shown that triacylglycerol and not wax ester is the major storage lipid in *Epischura* from Lake Baikal. The reason for this dichotomy is not clear. It is not purely taxonomic, for several groups of organisms show a similar pattern (including amphipods). Neither does the suggestion that the enzyme machinery for wax synthesis simply did not evolve in freshwaters (Morris, 1983, 1984) make any sense given the wide range of marine organisms which store wax (Sargent, 1976; Clarke, 1984). Perhaps the explanation may lie in the timing of periods of synthesis and utilisation of supplies. In Heywood Lake food is converted immediately into eggs, rather than reserves. The same situation obtains in the Antarctic krill, *Euphausia superba*, which also stores triacylglycerol rather than wax ester.

In many marine copepods euphausiids, energy

is stored one season for utilisation many months later. In these cases the lipid reserve is predominantly wax ester. However *Pseudoboeckella poppei* in Sombre Lake synthesise a lipid store in one summer, to be used over the following winter for egg production. This tends towards the marine situation, and yet the storage lipid is still triacylglycerol. Clearly the reason for the wax-marine/triacylglycerol-freshwater dichotomy is not simple.

Acknowledgements

The authors would like to thank the members of Signy Station 1983-1985 for help with the fieldwork.

References

Båmstedt, U., 1986. Chemical composition and energy content. In E. D. S. Corner and S. C. M. O'Hara (eds), The Biological Chemistry of Marine Copepods. Oxford University Press, Oxford: 1–58.

Bligh, E. G. & W. J. Dyer, 1959. A rapid method of total lipid extraction and purification. Can. J. Biochem. Physiol., 37: 911–917.

Clarke, A., 1984. The lipid content and composition of some Antarctic macrozooplankton. Bull. Br. Antarc. Surv., 63: 57–70.

Clarke, A., A. Skadsheim & L. J. Holmes, 1985. Lipid biochemistry and reproductive biology in two species of Gammaridae (Crustacea: Amphipoda). Mar. Biol., 88: 247–263.

Hawes, I., 1985. Light climate and phytoplankton photosynthesis in maritime Antarctic lakes. Hydrobiologia, 123: 69–79.

Heywood, R. B., 1970. Ecology of the freshwater lakes of Signy Island, South Orkney Islands: III. Biology of the copepod *Pseudoboeckella silvestri* Daday (Calanoida, Centropagidae). Bull. Br. Antarc. Surv., 23: 1–17.

Morris, R. J., 1983. Absence of wax esters in pelagic Lake Baikal fauna. Lipids, 18: 149–150.

Morris, R. J., 1984. The endemic faunae of Lake Baikal: their general biochemistry and detailed lipid composition. Proc. R. Soc. Lond. B, 222: 51–78.

Priddle, J., 1980. The production ecology of benthic plants in some Antarctic lakes. I. *In situ* production studies. J. Ecol., 68: 141–153.

Priddle, J., I. Hawes, J. C. Ellis-Evans & T. J. Smith, 1986 Antarctic aquatic ecosystems as habitats for phytoplankton. Biol. Rev., 61: 199–238.

Sargent, J. R., 1976. The structure, metabolism and function of lipids in marine organisms. In D. Malino and J. R. Sargent (eds), Biochemical and Biophysical Perspectives in Marine Biology, Vol. 3. Academic Press, London and New York: 149–212.

Sargent, J. R. & R. J. Henderson, 1986. Lipids. In E. D. S. Corner and S. C. M. O'Hara (eds), The Biological Chemistry of Marine Copepods. Oxford University Press, Oxford: 59–108.

Weller, D. L. M., 1977. Observations on the diet and development of *Pseudoboeckella poppei* (Calanoida, Centropagidae) from an Antarctic lake. Bull. Br. Antarct. Surv. 45: 77–92.

Hydrobiologia 172: 193–206, 1989
W. F. Vincent and J. C. Ellis-Evans (eds) High Latitude Limnology

Tundra ponds of the Yukon Delta, Alaska, and their macroinvertebrate communities

John A. Maciolek[1]
National Fishery Research Center, Bldg. 204, Naval Station, Seattle, WA 98115, USA; [1] *present address: P.O. Box 7117, Mammoth Lakes, CA 93546, USA*

Key words: macroinvertebrates, pond character, thaw-basins, taxon richness

Abstract

The Yukon Delta, a low alluvial tundra in western Alaska, has more than 10^5 thaw-basin ponds within its 70000 km^2 area. In 1984 and 1985, 68 ponds in three interior areas of the Delta were surveyed to determine limnological features, macroinvertebrate fauna, and trophic character. Ponds ranged up to 90 ha in area, 2 m in depth, and 17 m in elevation, and occurred in various temporal stages of growth and senescence. Among the 18 major invertebrate taxa collected, in order of decreasing frequency of occurrence, Trichoptera, Hemiptera, Diptera, Pelecypoda, Isopoda, Coleoptera, Gastropoda, and Oligochaeta were found in over 50% of the ponds. Trichoptera, the only taxon occurring in all ponds, was represented by 22 species of 6 families. The average Delta pond had 6.6 of the nine more common taxa. This measure of faunal richness was similar among study areas but was higher in low-tundra (sea level) ponds and in older ponds on raised tundra. In comparison, lentic invertebrate communities in five other areas of Alaskan and Canadian tundra had fewer taxa and also lower average richness based on occurrence of the same nine taxa.

Introduction

North America's largest delta occupies much of Alaska's coastline along the Bering Sea above latitude 60°N. Most of its treeless, roadless expanse of alluvial tundra is wetland noted for enormous breeding populations of water birds. The Yukon Delta is the dominant feature of a national wildlife refuge having an area of 11×10^4 km^2. Within refuge boundaries are over 40 eskimo villages whose residents depend on intrinsic food resources, among which fishes are highly important. Some of the more desirable fish species feed extensively in shallow ponds during the 3-month ice-free summer. Because of their great combined area, lakes and ponds are critical to the gross aquatic productivity of the Delta.

Declining catches of subsistence fishes in recent years raised suspicion that fish feeding migrations were being obstructed by activities of beaver, a mammal that had suddenly appeared in abundance on the Delta. Concern for the beaver-fish relationship led to a study of the ponds, objectives in part being to describe environments and invertebrate communities upon which so many fishes and migratory birds feed. Lack of published information on Delta limnology was surprising for an area so large and so wet. Initially absent was the impetus for aquatic study, such as environmental concern over oil exploitation that led to surveys on north slope tundra (e.g. Bergman *et al.*, 1977; Brunskill *et al.*, 1973). Such impetus arrived with the establishment of the Yukon Delta National Wildlife Refuge with management responsibility for natural resources. Refuge sponsorship of recent and ongoing studies, of which this is a small part, should bring a large measure of understanding to a geologically simple but biologically complex region of low arctic tundra.

Study sites

The Yukon Delta (Fig. 1) is a 70 000 km^2 alluvial fan formed from the sediments of the Yukon and Kuskokwin Rivers and their tributaries. The Delta is unglaciated tundra underlain by permafrost about 200 m thick (Johnson & Hartman, 1969). Its low-lying expanse consists of Quaternary deposits of aeolian, estuarine, and floodplain silt intermixed with organic detritus (Hoare & Condon, 1971). Lava (basalt) outcroppings up to a few hundred meters high occur in some localities, but not in the areas studied. Water in the form of rivers, sinuous streams, lakes, ponds, and marshes is a dominant feature (cf. Fig. 2). Much of the land surface is low tundra at or near sea level that allows tidal fluctuations along larger streams throughout most of the Delta. Raised tundra, for the most part, is elevated a few to several meters in a gently rolling terrain.

Climate is only slightly less harsh than that of the Alaskan north slope (cf. Hobbie, 1984). Annual precipitation averages about 40 cm, much of it falling as rain in late summer. Winter snowfall varies from 100 to 200 cm yr^{-1} on different parts of the Delta. Ice develops to a thickness of a meter or more and begins to break up in late May on the major rivers. Tundra ponds, however, are generally ice free only between about mid June and late September, allowing a 3-month growing season for aquatic invertebrates, and for fishes and water birds that feed on them. Typical tundra surface is blanketed by a mat of lichens and mosses with a scattering of grasses, sedges, annual forbs, and dwarf shrubs such as birch. The living mat is underlain by a fibrous peaty layer of variable thickness. At maximum thaw observed during this study, permafrost lay less than a half meter below the vegetational surface. Deeper thaw is found along watercourses and margins of some older ponds where large willow and alder shrubs grow. Ice-wedge polygons, so prominent a feature along the north coast of Alaska (cf. Bergman *et al.*, 1977; Hobbie, 1984), occur more sparingly on the Delta and were not observed in study areas.

The three areas surveyed (Fig. 1) were selected primarily because of evidence of beaver constructional activity. Area A was the most remote, being near the geometric center of the Delta. It was studied between 14 August and 11 September 1984 during what was considered a summer drought. The latter date corresponded to the first hard frost. All of the 35 ponds sampled in Area A were on raised tundra. Areas B and C were surveyed between 22 July and 17 September 1985 during an unusually wet summer in which rainfall persisted from early August until sampling was concluded. Thirteen ponds were samples in Area B, all but one of them occurring on raised tundra. Study Area C (Fig. 2) shows the distribution of the ponds (20) sampled between both low (tidally influenced) and raised tundra a few to several meters high.

Methods

Pond sampling was done by teams of two or four persons operating from field camps that were established and supplied by amphibious aircraft operating out of Refuge headquarters in Bethel (Fig. 1). Ponds were reached by hiking across the tundra, usually after gaining proximity by boat (canoe or inflatable skiff). Because invertebrate sampling was done in conjunction with other work (e.g. fish capture and beaver observation), ponds were selected by operational proximity, time available, and size. Only those ponds large enough in area (ca. 0.2 ha) to appear on quadrangle maps were chosen; very large ponds (100 ha or more) were ignored. At each pond, note was made of the amount of surface area

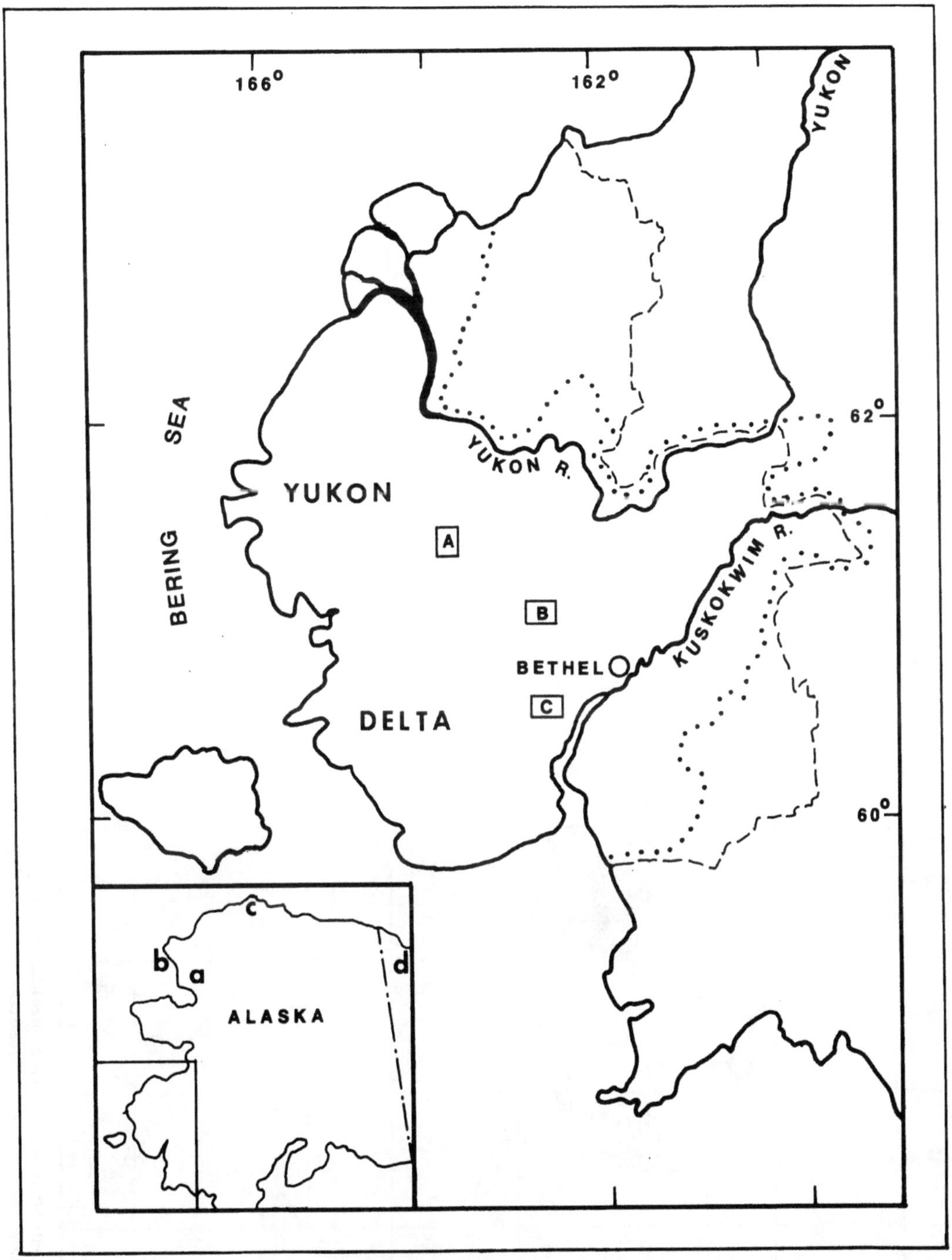

Fig. 1. Yukon Delta, Alaska, is a 70 000 km² tundra floodplain formed by the Kuskokwim and Yukon Rivers and their lower tributaries. Areas studied are shown as lettered rectangles. The dotted line is the approximate limit of alluvial deposits; dashed line is the interior boundary of the Yukon Delta National Wildlife Refuge, headquartered in the village of Bethel. Inset (lower left) shows map location and tundra areas studied by other investigators (cf. Table 2): a = Noatak River Basin, b = Cape Thompson – Ogotoruk Creek, c = Point Barrow, and d = Mackenzie Delta.

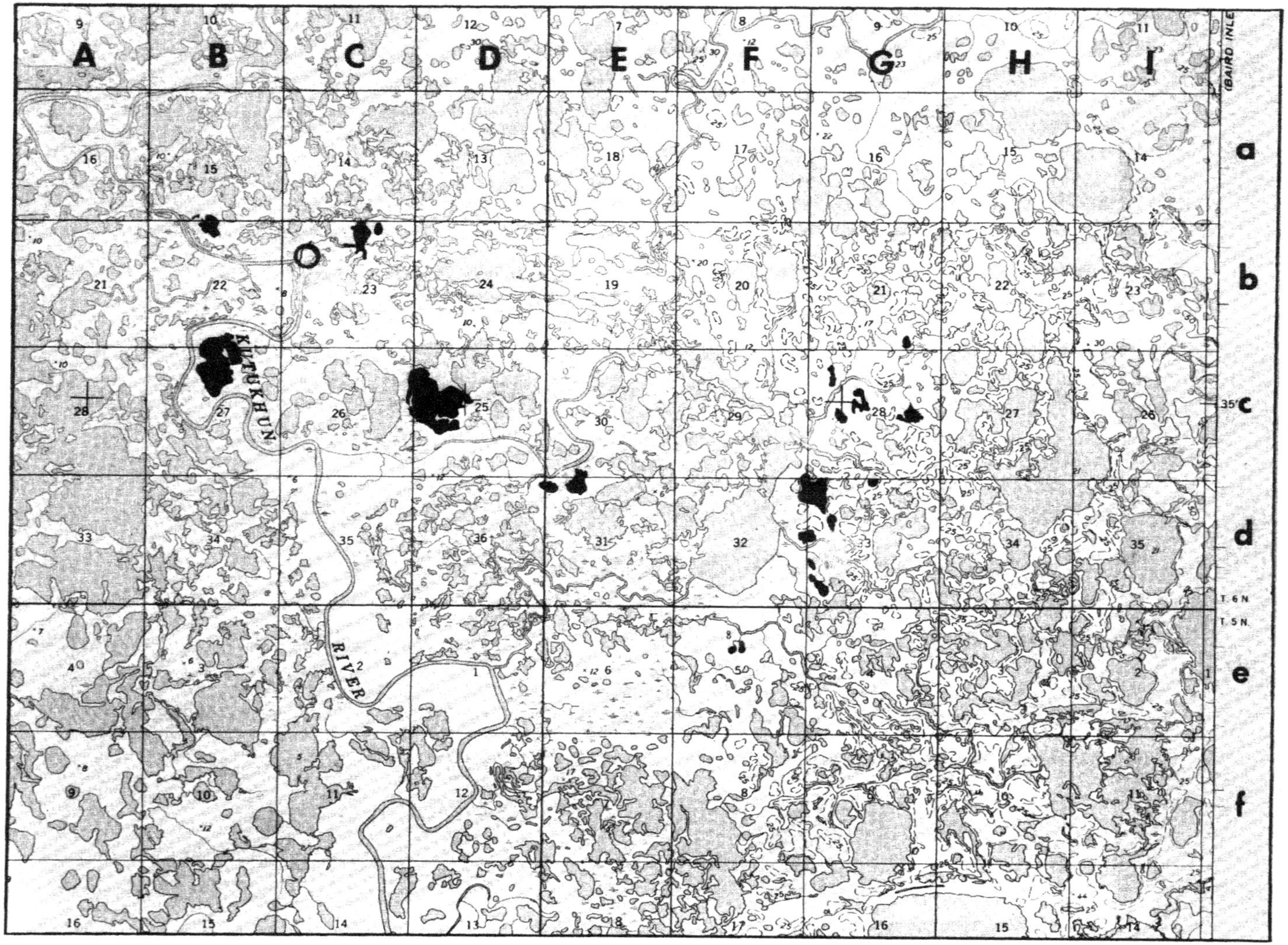

Fig. 2. Yukon Delta Study Area C is one of three locations surveyed (cf. Fig. 1). Sectional grid in mi^2 (= 2.59 km^2) is letter-coded along axes for site location reference. Circle in section Cb indicates operational base camp from which 20 ponds (blackened) were sampled. Low tundra (tide influenced) occupied left side of map (columns A–E); rolling, raised tundra is on the right (columns F–I). Highest land elevation, about 15 m, occurs along right side of figure. Total surface area of ponds in this part of the Delta exceeds that of land and is proportionally greater than in other study areas. Map is adapted from US Geological Survey Baird Inlet C–2 quadrangle, scale 1 : 63 360.

covered by emergent vegetation as an index of trophic stage (see later). All ponds sampled had at least one-fourth of the total basin as open water. Occasional probes were made with a steel rod to ascertain the depth of the permafrost surface.

Invertebrates were captured by long-handled dip nets with 0.4 m hoops and 3 mm nylon mesh. In each pond, four to six multiple sweeps of the net were made at different points along the shore to include areas of vegetational differences and bottom sediments. The intent was to sample each pond with similar effort, and to capture all forms of macroinvertebrates regardless of habitat type. Each multiple sweep was picked directly into 10% formalin, later to be sorted and transferred to alcohol. Because large numbers of individuals were caught in some taxa, only representative specimens of them were retained. Net samples were examined carefully to secure all individuals of taxa that were poorly represented. Upon completion of sampling in a given pond, the major taxa present were listed in order of decreasing numbers of component individuals.

Results

Tundra ponds

A rough estimate indicates that about 2×10^5 ponds dot the surface of the Yukon Delta. The difference between a 'lake' and a 'pond' there is subjectively based on surface area. Locally, lentic waters larger than about 1 km^2 are referred to as lakes and very large ones, although shallow, are formally named. All standing waters observed and sampled in this study were less than 2 m deep, most of them averaging near 1 m. The shallowest pond sampled, also the largest with an area of 90 ha, was only 0.2 m deep. All ponds appeared to have flat bottoms without significant littoral slope. Bottom material was mainly black ooze and sometimes fibrous peat, or a combination of the two. Two ponds, one of them the large one noted above, had firm substrates consisting of compacted light-colored silt. No mineral particles larger than silt were encountered in the sediments.

Surface elevation of the ponds ranged from 0 to 17 m; only 6 of the 68 ponds sampled were elevated more than 12 m. Water temperature maxima ranged from 12° to 14 °C without stratification, much the same as in more northerly ponds (Brewer, 1958; Livingstone *et al.*, 1958). Water was transparent but often stained with humic solutes. Mr. Rae Baxter (pers. commun.) of Bethel, who has sampled pondwater at various seasons, indicated that typical water has low mineral content, tends toward neutrality except when humic acids are high, and ponds that do not freeze to the bottom in winter can develop anoxia sometimes accompanied by sulfide generation.

All ponds had macrophytic vegetation. Besides frequently encountered filamentous green algae and *Nostoc* blobs, common submersed flora included bladderwort (*Utricularia* sp.) and pondweed (*Potamogeton* spp.). Characteristic emergent vegetation included marsh fivefinger (*Potentilla palustris*), buckbean (*Menyanthes trifoliata*), horsetail (*Equisetum* spp.), and various sedges (Cyperaceae) and grasses (Gramineae). Some ponds had fringing mats of sphagnum moss.

Delta ponds on raised tundra fall within that high-latitude class of lentic waters often referred to as 'thaw lakes' or 'cave-in lakes'. The origin and cyclic nature of such waters has been described for other locations in Alaska by Hobbie (1984, Point Barrow), Hopkins (1949, Seward Peninsula), Livingstone *et al.* (1958, Point Barrow), and Wallace (1948, eastern interior). Black (1968) presented a concise review of that phenomenon. Most descriptions emphasize the role of ice wedges but that phenomenon does not seem to apply to Delta ponds where Wallace's (1948) scenario is more applicable. An ideal situation occurs in summer in flat tundra when the insulating mat has been disrupted, allowing the underlying permafrost to thaw. With sufficient heat income, abetted by stained water and dark sediments, a pool develops that gradually deepens and expands laterally in a circular fashion (Fig. 3, top). Because most of the upper several meters of permafrost is ice (Black, 1968), the forming pond sinks below the surrounding surface as denser

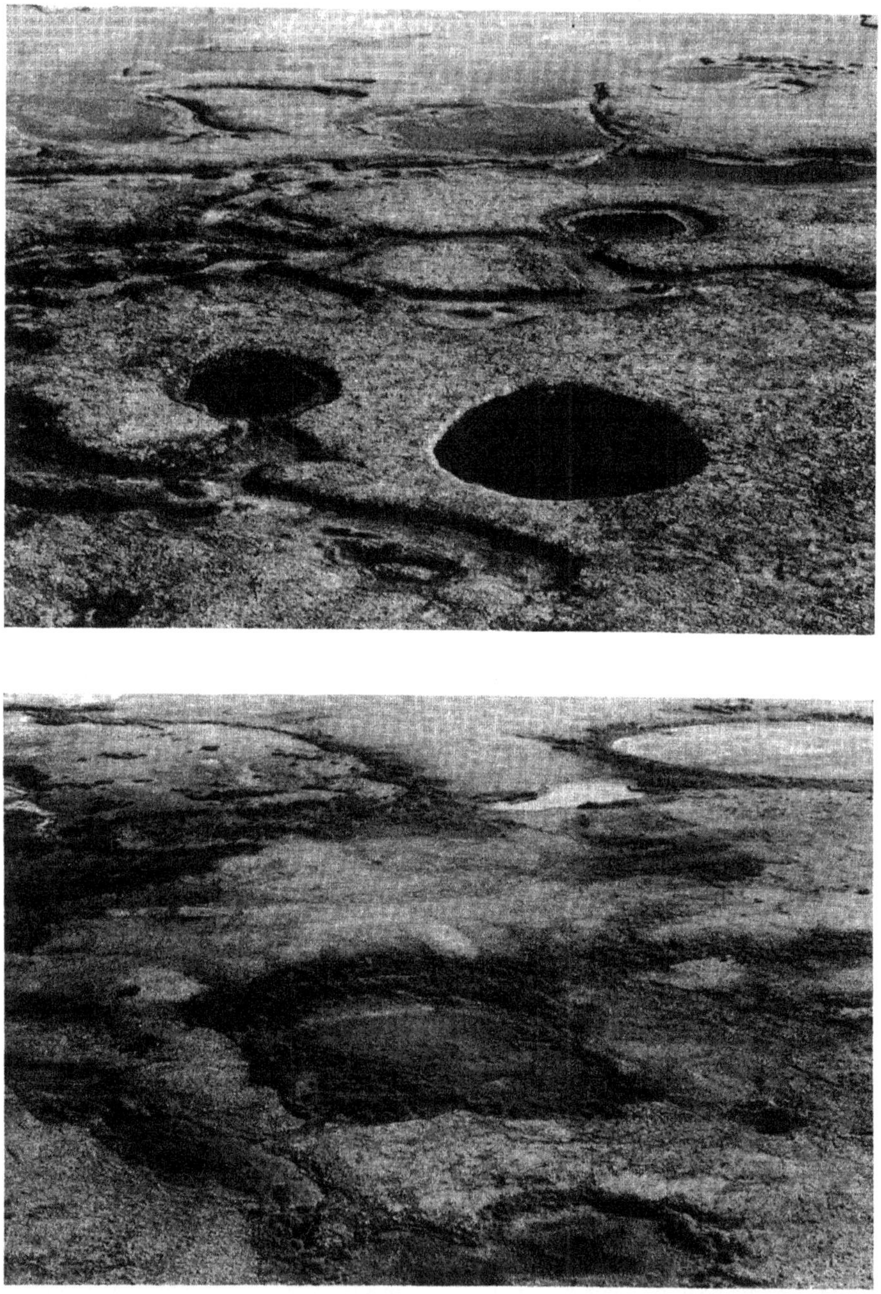

Fig. 3. Examples of Yukon Delta thaw-basin ponds at various stages of growth and senescence. Top: circular ponds in raised flat tundra. Pond without outlet in the lower right of center is growing as shown by steep banks and scant littoral vegetation; pond to the left has developed a weak outlet and is in an early stage of senescence; and pond at upper right is further senesced. Bottom: ponds in undulating tundra. Center pond has senesced slowly into a marsh bordered by shrub growth; basin in the extreme upper right has bare sediments, the probable result of rapid decantation. Permafrost reforms in basins of both types of former ponds.

thaw water replaces bulkier ice, creating steep-sided banks a meter or more high. Depth of permafrost thaw under Delta ponds exceeded 3 m where probed. Black (1968) and Brewer (1958) noted that large basins with water deeper than 2 m thaw to a depth of 60 m or more. Wallace (1948) estimated bank erosion rates from about 6 to 20 cm yr^{-1}. If such rates prevail on the Delta, a growing pond 100 m in diameter would be between 500 and 1 700 yrs old. Irregularly shaped ponds develop when two or more subcirular ponds coalesce, or more commonly on the Delta, where topography affects lateral growth rates.

Pond water is derived from permafrost melt, direct precipitation, and local runoff; it is lost by evaporation and by drainage where outlets occur. As long as a neutral water balance is maintained, a pond can continue to grow. Cessation of growth usually occurs when an outlet develops, and is followed by a period of slow senescence accompanied by an expanding band of littoral vegetation (Fig. 3, top). Eventually a marsh evolves in the basin (Fig. 3, bottom), and with the growth of sufficient insulating vegetation, permafrost is re-established to complete the cycle. Some ponds appear to have decanted rapidly, as by breaching a point of lowering topography, leaving a basin of unvegetated sediments (Fig. 3, bottom). Other ponds appear to have senesced without having developed outlets, presumably by expanding into the surrounding watershed to a point where runoff could no longer sustain a water balance. One such pond examined, a small acidic type in a late stage of senescence, had permafrost forming under a fringing sphagnum mat within a couple of meters of the water's edge and more than 10 m from the original bank. That Delta ponds senesce more slowly than they grow is suggested by the low proportion (ca. 10%) of developing ponds among the group surveyed.

In considering the cyclic life of tundra ponds, Wallace (1948) distinguished four temporal stages based on size and morphometry: youthful, early mature, late mature, and old age. Bergman *et al.* (1977) classified north slope ponds according to type and extent of emergent vegetation (see also Hobbie, 1984). Such classifications are similar to the sequence used in this study to identify age-trophic stages of the ponds sampled: stage 1 included ponds that were still growing or had reached a static size without having developed much emergent vegetation; stage 2 were ponds in early senescent phase with emergent vegetation occluding up to 20% of the water surface; and stage 3 indicated ponds with the proportion of open water diminishing to about 25% of the original basin. Ponds further senesced (stage 4) were not included in the sampling. These stages were associated with the relative abundances of invertebrates observed in sampling (see later).

Pond fauna

Interest in invertebrates resulted from common knowledge of Delta residents that certain fishes enter shallow ponds at the onset of thaw in June to feed, and leave them before freezeup in September. It is the only period during the year that some of the fishes feed (Baxter, pers. commun.). Pond productivity is therefore highly important to fish trophic ecology. In Area A (Fig. 1) northern pike (*Esox lucius*) up to 90 cm long were present in deeper drainages but not in ponds. Most ponds there contained Alaska blackfish (*Dallia pectoralis*), a forage species attaining a maximum length of 20 cm but averaging only half that long. Ponds in Area B had blackfish and a more diminutive species, the ninespine stickleback (*Pungitius pungitius*), but no larger fishes. In addition to the above forage fishes, many Area C ponds had broad whitefish (*Coregonus nasus*) to 56 cm long, humpback whitefish (*C. pidschian*) to 50 cm long, least cisco (*C. sardinella*) to 39 cm long, and northern pike. Distributional patterns indicted that the larger fishes readily entered all ponds in low tundra and some ponds in adjacent higher tundra, but avoided most elevated ponds, even those apparently accessible to them.

Invertebrates found in the ponds are listed in the Appendix to the extent that they were identified; estimates of the minimum number of species present are given for some taxonomic groups. Small invertebrates, including plankton, were not sampled because of the coarse size of net mesh employed. Two of the more noteworthy taxonomic 'finds' were the isopod *Asellus alaskensis*, previously known only from a few localities far to the north (Bowman & Holmquist, 1975), and the diversity of caddisflies (Trichoptera) comprising 22 species in 10 genera and 6 families. Larvae of these insects constructed cases only of living and detrital organic matter in the absence of mineral particles. Among lower taxa, Porifera and Turbellaria were uncommon in all study areas, Hydrozoa was found in one pond in Area A, and Bryozoa occurred in two ponds in Area C.

Distributions of the more prominent taxa by frequency of pond occurrence are given by study area in Table 1. Ephemeroptera and Odonata were uncommon (3–20% and 9–19% of the ponds, respectively) and low in abundance. Seven other orders and one class (Oligochaeta) occurred in more than half the ponds. Trichoptera was the most widely distributed taxon (all ponds, all areas) followed by Hemiptera, Diptera, Pelecypoda, Isopoda, Coleoptera, Gastropoda, and Oligochaeta. Considering the small size of some larval Diptera, it is possible that this taxon (especially Chironomidae) occurred in all ponds. Less common taxa such as Amphipoda, Ostracoda (large species only) Conchostraca, and Hydracarina (large species only) were each found in fewer than 10% of the ponds. While some orders and families (e.g. Pelecypoda, Trichoptera, Chironomidae, and Gerridae) occurred with similar fre-

Table 1. Frequency of occurrence of macroinvertebrates in 68 ponds on three study areas of the Yukon Delta, Alaska (cf. Fig. 1).

Invertebrate taxon	Percentage of ponds with taxon by area			
	Area A 35 ponds	Area B 13 ponds	Area C 20 ponds	All areas
Insecta				
Ephemeroptera	3	8	20	9
Odonata	9	15	15	12
Hemiptera				97
Corixidae	91	100	50	81
Gerridae	80	92	90	85
Trichoptera	100	100	100	100
Coleoptera				69
Dytiscidae	66	77	10	51
Gyrinidae	86	54	20	60
Diptera				96
Chironomidae	94	92	90	93
Other[1]	20	62	40	34
Annelida				
Oligochaeta	49	92	75	65
Hirudinia	11	23	25	18
Crustacea				
Isopoda	91	54	55	74
Other[2]	40	8	15	26
Mollusca				
Gastropoda	74	46	60	65
Pelecypoda	80	92	80	82
Miscellaneous[3]	17	15	35	22

[1] Ceratopogonidae, Dixidae, and Tipulidae.
[2] Amphipoda, Conchostraca, and Ostracoda.
[3] Bryozoa, Hydracarina, and Turbellaria.

quency in all areas, there were notable distributional exceptions. Low frequencies were found for the Corixidae and Coleoptera in Area C, Oligochaeta in Area A, and Gastropoda in Area B. On the other hand, Isopoda was widely distributed only in Area A. Representative specimens of two common taxa, Trichoptera and Mollusca, are shown in Fig. 4.

Generally, those invertebrate groups occurring more frequently also had the greater numbers of individuals in a pond community. This was particularly true of Hemiptera, Trichoptera, Chironomidae, and Mollusca, Field estimates indicated that in a typical pond Trichoptera (Fig. 4., top) comprised the greatest biomass, followed by Mollusca (Fig. 4., bottom), Hemiptera, and Coleoptera. Large dytiscid beetles (larvae 4 cm and adults 3 cm long) abounded in a few ponds where they exceeded other taxa in biomass. Analysis of individual pond records demonstrated, as would be expected, that both diversity and abundance of invertebrates were low in stage 1 and darkly stained ponds, but were high in stage 3 ponds.

Formal treatment of species diversity in Delta ponds is not possible with the data available, but a simple tally of the numbers of selected taxa present per pond, here referred to as taxon richness, permits comparisons of within-community diversity. Taxa selected were the nine more obtrusive ones common to all study areas (Table 1): Hemiptera, Trichoptera, Coleoptera, Diptera, Oligochaeta, Hirudinia, Isopoda, Gastropoda, and Pelecypoda. Taxon richness for all 68 Delta ponds averaged 6.6 (range, 4–9), and was similar for each study area: A = 6.7, B = 6.9, and C = 6.1. In Area C, however, there was a striking difference between low-tundra ponds (mean richness, 7.1) and raised-tundra ponds (mean richness, 5.5); approximately the same mean values of taxon richness were observed for stage 3 and stage 1 ponds, respectively.

Taxon richness and relative abundances of invertebrates observed in this study are probably representative of most ponds elsewhere on the Delta despite the proportionately small number of communities sampled. But these findings also suggest that other taxa and many more species may occur there. For example, the 35 ponds sampled in 1984 yielded 14 species of Trichoptera. Eight additional species were found the following year. More extensive sampling elsewhere, as on volcanic and fresh alluvial terranes or in oxbow ponds along the major rivers, would surely add to the regional invertebrate taxonomic record.

Discussion

Rawson (1953) was among the first to recognize the deficiency of information on the biology of American arctic lakes and ponds. Only more recently have significant reports appeared on macroinvertebrates in lentic tundra waters of northwest America. Foremost are the reports by Holmquist (1967, 1973a, 1973b, 1973c; 1974, and 1975) involving 103 lakes and ponds distributed over a broad area from the Seward Peninsula eastward into the Mackenzie Delta. Others by Butler *et al.* (1980) at Point Barrow, Watson *et al.* (1966) at Cape Thompson, Weins *et al.* (1975) on the Mackenzie Delta, and Young (1974) in the Noatak Basin involve fewer, more localized waters (cf. Fig. 1). Occurrence of 20 invertebrate taxa at order level or higher, as reported in the literature above, are given in Table 2 for comparison with Yukon Delta findings. A factor of note is that the referenced works concern some waters either deeper or much larger than those of this study (e.g. only a few of Holmquist's 103 lakes were small shallow ponds), and therefore had additional environmental complexity. A notable feature shown by this listing is the large number of species within some taxa in some areas: turbellarians and oligochaetes in northern Alaska; hydrachnids, odonates, and mollusks on the Mackenzie Delta; and coleopterans in the Noatak Basin. Dipterans comprised the greatest number of species in the two locations where this difficult taxon was analyzed. Chironomids appear to have the greatest number of species of any insect family or of any higher taxon in tundra ponds.

Comparing total taxa for each area (Table 2), the Yukon Delta has the greatest number with 18.

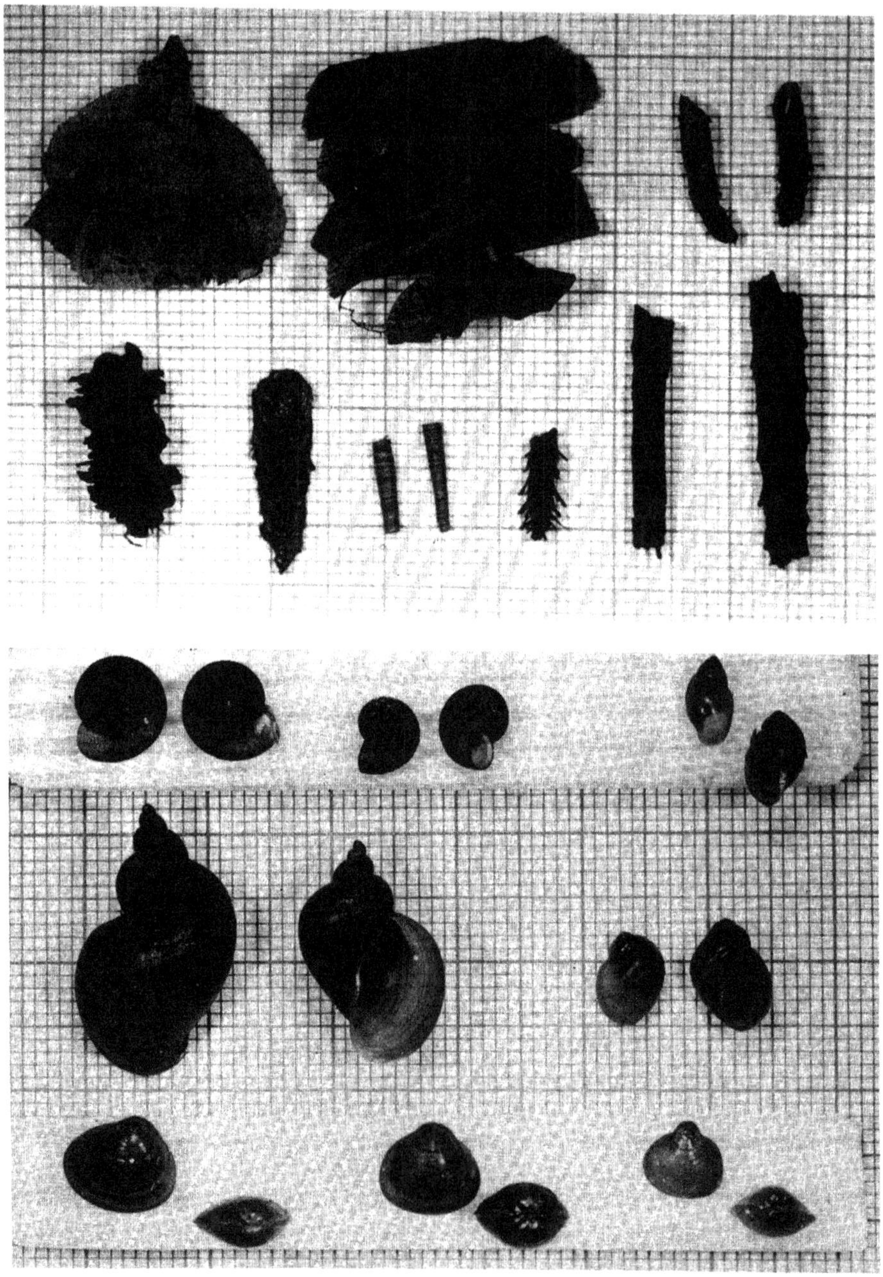

Fig. 4. Representatives of the more widely distributed taxa found in tundra ponds of the Yukon Delta. *Top*: larval cases of some of the 22 species of Caddisflies (Trichoptera, cf. Appendix 1), the only taxon found in all 68 ponds sampled. Upper row from left: 1 and 2, *Nemotaulius hostilis*; 3 and 4, *Grensia praeterita*. Lower row from left: 1, *Limnephilus* sp.; 2, *Molannodes tinctus*; 3 and 4, *Micrasema scissum*; 5–7, *Agrypnia* spp. *Bottom*: gastropod and pelecypod mollusks. Upper row from left: *Gyraulus deflectus*, *Valvata sincera*, and *Physa* sp. Center: *Lymnaea atkaensis* and *Stagnicola* sp. Lower row: *Pisidium* sp., *Sphaerium* sp., and *S. lacustre*. Background grid is metric (mm).

Table 2. Occurrence of aquatic macroinvertebrates in lakes and ponds of arctic Alaska and the deltas of the Mackenzie and Yukon Rivers (cf. Fig. 1).[1] Numerals in column headings are the numbers of waters sampled.

Invertebrate taxon[2]	No. of species or presence (x) by survey area					
	Yukon Delta 68	M'kenzie Delta 13	North Alaska 103	Noatak Basin 47	Cape Thompson 3	Point Barrow 8
Porifera	1		1			
Hydrozoa	1		3			
Turbellaria	x		10		x	1
Bryozoa	1		3			
Oligochaeta	3+	11	20+		x	2
Hirudinia	4+		5			
Polychaeta			3		x	
Branchiopoda	x	3	3	2	x	1
Isopoda	1		2	1	x	
Amphipoda	1	2	2	1	x	
Hydracarina	3+	12	x		x	1
Plecoptera				2	x	1
Ephemeroptera	1	3		7	x	
Odonata	3	15		3		
Hemiptera	3	8	x	3		
Trichoptera	22	14	6	7	x	2
Coleoptera	5+	15	x	31	x	2
Diptera	x	73	x	x	x	37+
Gastropoda	7	18	4	4		1
Pelecypoda	6	15	6	1	x	
Total taxons	18	12	17	12	13	8

[1] Data sources: North Alaska, Holmquist (1975); Mackenzie Delta, Wiens *et al.* (1975) and Oliver *et al.* (1978); Noatak Basin, Young (1974); Cape Thompson, Watson *et al.* (1966); and Point Barrow, Butler *et al.* (1980).

[2] Freshwater representatives only; Hirudinia are non-parasitic forms.

Nearly equaling it is northern Alaska whose 17 taxa are not surprising in view of the large number of waters and broad geographic coverage. Lesser numbers of taxa (8 to 13) found in the more discrete localities may represent limited distributions, lower environmental diversities, or perhaps, sampling techniques employed. Taxon richness (see preceding section) in other areas is below the 6.6 of the Yukon Delta. Based on the same nine taxa, averages were 4.5 per pond at three places: the Mackenzie Delta with 11 ponds, the Alaska north slope (Holmquist, 1973c) with 6 ponds, and the Noatak Basin with 8 ponds. Slightly higher richness (5 taxa per pond) ocurred at Cape Thompson. Greater average richness of Yukon Delta ponds probably is not due to climate alone because one or more of the ponds in each of the other localities had a richness of 6 or higher, while some of the Yukon Delta ponds had only 4. Rather, the difference seems to lie in the greater proportion of organically rich (stage 3) ponds on the Yukon Delta. Our observations concur with those of Holmquist (1975) who concluded that the '...much spoken of "biological simplicity of arctic lakes" has proved to be a chimera.'

The diversity and abundance of invertebrates observed on the Yukon Delta seem contrary to the harsh winter conditions and short growing season in tundra ponds, and apparently results from the accumulation of biomass from many

Appendix 1. Taxonomic list of macroinvertebrates in tundra ponds on the Yukon Delta, Alaska. Provisional identifications are followed by '?'.

Porifera
- *Spongilla lacustris*?

Coelenterata; Hydrozoa
- Hydridae
  - *Chlorohydra* sp.?

Turbellaria: Tricladida
- unident. planaria

Bryozoa
- Cristatellidae
  - *Cristatella* sp.?

Annelida; Oligochaeta
- 3+ unident. spp.

Annelida; Hirudinia
- 4+ unident. spp.

Arthropoda; Crustacea
- Conchostraca
  - *Caenestheria* sp.?
- Ostracoda?
  - unident. large sp.
- Isopoda
  - *Asellus alaskensis*
- Amphipoda
  - *Hyallela azteca?*

Arthropoda; Hydracarina
- 3+ unident. spp.

Arthropoda; Insecta
- Ephemeroptera
  - Baetidae
    - *Cloeon* sp.
- Odonata
  - Aeschnidae
    - *Aeschna* sp.
  - Libellulidae
    - *Sympetrum* sp.
  - Coenagrionidae
    - *Coenagrion resolutum*
- Hemiptera
  - Corixidae
    - *Arctocorisa chanceae*
    - *Callicorixa alaskensis*
  - Gerridae
    - *Gerris pingreensis*
- Trichoptera
  - Branchycentridae
    - *Micrasema scissum*
  - Leptoceridae
    - *Ceraclea* sp.
    - *Oecetis immobilis*
    - *Oecetis ochracea*
    - *Oecetis* sp.
  - Limnephilidae
    - *Grensia praeterita*
    - *Limnephilus infernalis*
    - *Limnephilus nigriceps*
    - *Limnephilus partitus*
    - *Limnephilus picturatus*
    - *Limnephilus stigma*
    - *Limnephilus* (*sublunatus* gp.) 3 spp.
    - *Nemotaulius hostilis*
  - Molannidae
    - *Molannodes tinctus*
  - Phryganeidae
    - *Agrypnia straminea*
    - *Agrypnia 2* spp.
    - *Oligotricha lapponica*
  - Psychomyiidae
    - *Polycentropus* (*crassicornis* gp.) sp.
    - *Polycentropus* sp.
- Coleoptera
  - Dytiscidae
    - *Agabus* sp.
    - *Colymbetes* sp.
    - *Dytiscus* sp.
  - Gyrinidae
    - *Gyrinus* sp.
- Diptera
  - Chironomidae
    - several unident. spp.
  - Tipulidae
    - *Limonia* sp.?
    - *Pedicia* sp.?
    - *Prionocera* sp.?
  - Ceratopogonidae
    - unident. sp.
  - Dixidae
    - unident. sp.

Mollusca; Gastropoda
- Lymnaeidae
  - *Lymnaea* cf. *atkaensis*
  - *Lymnaea* sp.
- Physidae
  - *Physa gyrina gyrina*
  - *Physa jennessi jennessi*
- Planorbidae
  - *Gyraulus deflectus*
  - *Planorbula campestris*
- Valvatidae
  - *Valvata sincera helicoidea*

Mollusca; Pelecypoda
- Sphaeriidae
  - *Pisidium casertanum*?
  - *Pisidium* cf. *compressum*
  - *Pisidium idahoense*?
  - *Sphaerium lacustre*
  - *Sphaerium nitidum*?
  - *Sphaerium* sp.

slow growing organism. Some large invertebrates such as branchiopod crustaceans have obligate annual cycles (Butler *et al.*, 1980). But two species of chironomids were found to have 7 yr cycles in tundra ponds (Butler, 1982), and similar longevity has been noted among terrestrial arctic invertebrates (Downes, 1965). It seems likely that many other aquatic insects and some of the other invertebrates such as mollusks have equally slow growth in the arctic. Multi-year existence of invertebrates in habitats that are frozed solidly during most of the year is a situation that has interested several investigators as reviewed by Downes (1965) and Holmquist (1973c). It has been demonstrated that chironomid larvae, and apparently also certain other insects, oligochaetes, and mollusks, can withstand freezing for prolonged periods at temperatures far below the freezing point. Downes (1965) presented an intriguing observation on the metabolism of freeze-adapted chironomids: '...nutritive reserves sufficient for 10 days at 0° would enable an insect to survive for 1 000 years at −23 °C.' Thus there is the theoretical possibility of finding viable 'fossil' invertebrates within former pond permafrost.

Acknowledgements

This study was conducted by personnel of the Biology Department, Montana State University, coordinated by the Montana Cooperative Fishery Research Unit in collaboration with agencies of the US Fish and Wildlife Service: the Yukon Delta National Wildlife Refuge; the Divisions of Fishery Resources and Fishery Research in Anchorage, Alaska; and the National Fishery Research Center in Seattle, Washington. Valuable research literature was provided by the Alaska Cooperative Fishery Research Unit in Fairbanks, Alaska. Mike Finley, Todd Haymaker, Mark Novak, and Martin Bridgham assisted in field surveys. Various invertebrates were identified by staff members of the US National Museum: T. E. Bowman (Isopoda), O. S. Flint (Ephemeroptera and Trichoptera) R. C. Froeschner (Hemiptera), R. Herschler and P. R. Greenhall (Mollusca), and J. A. Louton (Odonata).

References

Bergman, R. D., R. L. Howard, K. F. Abraham & M. W. Weller, 1977. Water birds and their wetland resources in relation to oil development at Storkersen Point, Alaska. US Fish Wildl. Serv., Resour. Publ. 129. 38 pp.

Black, R. F., 1968. Thaw depressions and thaw lakes, a review. Soc. Scient. Lodziensis, Biul. Perygla. 19: 131–150.

Bowman, T. E. & C. Holmquist, 1975. *Asellus (Asellus) alaskensis*, n. sp., the first Alaskan *Asellus*, with remarks on its Asian affinities (Crustacea; Isopoda: Asellidae). Proc. biol. Soc. Wash. 88: 59–72.

Brewer, M. C., 1958. The thermal regime of an arctic lake. Trans. am. geophys. Union 39: 278–284.

Brunskill, G. J., D. M. Rosenberg, N. B. Snow, G. L. Vascotto & R. Wagemann, 1973. Ecological studies of aquatic systems in the Mackenzie-Porcupine Drainages in relation to proposed pipeline and highway developments. Eviron.-Soc. Comm. N. Pipelines, Task Force north. oil Devel. Vol. 1, Rep. 73–40. Information Canada, Ottawa. 131 pp.

Butler, M. G., 1982. A 7-year life cycle for two *Chironomus* species in arctic Alaskan tundra ponds (Diptera: Chironomidae). Can. J. Zool. 60: 58–70.

Butler, M., M. C. Miller & S. Mozley, 1980. 7. Macrobenthos. In J. E. Hobbie (ed.), Limnology of Tundra Ponds. US/IBP Synthesis Ser.: 13 Dowden, Hutchinson & Ross, Inc. Stroudsburg, PA: 297–339.

Downes, J. A., 1965. Adaptations of insects in the arctic. Ann. Rev. Ent. 10: 257–274.

Hoare, J. M. & W. H. Condon, 1971. Geologic map of the Marshall Quadrangle, western Alaska. US Geol. Surv., Misc. geol. Invest., Map I-668.

Hobbie, J. E., 1984. The ecology of tundra ponds of the arctic coastal plain: a community profile. US Fish Wildl. Serv. FWS/OBS – 83/25. 52 pp.

Holmquist, C., 1967. Turbellaria of northern Alaska and northwest Canada. Int. Rev. ges. Hydrobiol. 52: 123–139.

Holmquist, C., 1973a. Fresh-water polychaete worms of Alaska, with notes on the anatomy of *Manayunkia speciosa* Leidy. Zool. Jb. Syst. Okol. Geogr. Tiere 100: 497–516.

Holmquist, C., 1973b. *Spongilla lacustris* (L.) (Porifera) from northern Alaska and northwest Canada. Zool. Anz. 191: 300–309.

Holmquist, C., 1973c. Some arctic limnology and the hibernation of invertebrates and some fishes in sub-zero temperatures. Arch. Hydrobiol. 72: 49–70.

Holmquist, C., 1974. On *Alexandrovina onegensis* Hrabé from Alaska, with a revision of the Telmatodrilinae (Oligochaeta, Tubificidae). Zool. Jb. Syst. Okol. Geogr. Tiere 101: 249–268.

Holmquist, C., 1975. Lakes of northern Alaska and northwest Canada and their invertebrate fauna. Zool. Jb. Syst. Okol. Geogr. Tiere 102: 333–484.

Hopkins, D. M., 1949. Thaw lakes and thaw sinks of the Imuruk Lake area, Seward Peninsula, Alaska, J. Geol. 57: 119–131.

Johnson, P. R. & C. W. Hartman, 1969. Environmental atlas of Alaska. Inst. arctic environ. Eng., Univ. Alaska, College, AK. 111 pp.

Livingstone, D. A., K. Bryan & R. G. Leahy, 1958. Effects of an arctic environment on the origin and development of freshwater lakes. Limnol. Oceanogr. 3: 192–214.

Oliver, D. R., D. McClymont & M. E. Roussel, 1978. A key to some larvae of Chironomidae (Diptera) form the Mackenzie and Porcupine River watersheds. Can. Fish. Mar. Serv., Tech. Rep. 791. 73 pp.

Rawson, D. S., 1953. Limnology in the North American arctic and subarctic. Arctic 6: 198–204.

Wallace, R. E., 1948. Cave-in lakes in the Nabesna, Chisana, and Tanana River valleys, eastern Alaska. J. Geol. 56: 171–181.

Watson, D. G., W. C. Hanson, J. I. Davis & C. E. Cushing, 1966. Chap. 17 Limnology of tundra ponds and Ogotoruk Creek. In N. J. Wilimovsky & J. W. Wolfe (eds.), Environment of the Cape Thompson Region, Alaska. US Atomic Energy Comm., Oak Ridge, TN: 415–435.

Wiens, A. P., D. M. Rosenberg & N. E. Snow, 1975. Species list of aquatic plants and animals collected from the Mackenzie and Porcupine River watersheds from 1971–1973. Can. Fish. Mar. Serv., Tech. Rep. 557. 39 pp.

Young, S. B. (ed.), 1974. The environment of the Noatak River Basin, Alaska. Contrib. Ctr north. Stud. No. 1. 584 pp.

Hydrobiologia 172: 207–214, 1989
W. F. Vincent and J. C. Ellis-Evans (eds) High Latitude Limnology

Vertical distributions of a planktonic harpacticoid and a calanoid (Copepoda) in a meromictic Antarctic lake

I. A. E. Bayly[1] & D. Eslake[2]
[1] *Department of Zoology, Monash University, Clayton, Vic. 3168, Australia*; [2] *Antarctic Division, Department of Science, Kingston, Tas. 7150, Australia*

Key words: Antarctic zooplankton, planktonic harpacticoid, Copepoda, meromictic

Abstract

Lake Abraxas (68° 29′ S., 78° 17′ E.) has been cut off from the sea for c. 6000 yr. In December, 1985, it had an ice cover 1.38 m thick, a thermocline at 5.2 m, and zero oxygen at 17.7 m and below. The salinity of the oxylimnion was in the range 14–19 g l^{-1}. The calanoid, *Paralabidocera antarctica* (I. C. Thompson), had a maximum density just above a thermocline at 5.0 m, at which depth adult males outnumbered females almost 3 : 1. Immediately beneath the ice adult females outnumbered males 4 : 1. Immature copepodite stages were more abundant relative to adults near the oxycline than at higher levels. The planktonic harpacticoid, *Amphiascoides* sp., had a maximum density close to the oxycline at 17.2 m.

Introduction

There are at least 13 meromictic lakes in the Vestfold Hills region of Antarctica. The ecology of the zooplankton of two of these – Burton Lake and Ace Lake – has already been studied (Bayly, 1986; Bayly & Burton, 1987).

Burton Lake retains a tenuous connection with the sea, and probably for this reason contains a more diverse zooplankton (there are four metazoan species – two calanoids, an anthomedusa and a ctenophore) than those lakes which are cut off from the sea. Generally the latter contain no more than two species of planktonic Crustacea. However, Lake Abraxas (68° 29′ S., 78° 17′ E.) is unique amongst the cut off lakes of this region in two respects: it contains truly planktonic harpacticoid copepods (Burton & Hamond, 1981) and a total of three species of micro-Crustacea, two of which are planktonic and the third essentially benthic. The present investigation was undertaken to establish the nature of the vertical distributions of the two planktonic species concerning which nothing was previously known.

Lake Abraxas was cut off from the sea by isostatic uplift after retreat of the continental icescheet c. 6000 B. P. It is 150 m from the present Antarctic shoreline at the nearest point, the surface is about 10 m above sea-level (Burton & Hamond, 1981). The maximum depth is 23.6 m, and the salinity of the oxylimnion is in the range 14–19 g l^{-1}.

Field work for this paper was carried out between 21 December 1985 and 12 January 1986. The thickness of the ice cover on 21 December was 1.38 m.

Methods

Oxygen and temperature were determined *in situ* with a 'Phox' meter Model No. 62TE. Redox potentials were measured *in situ* with a 'Phox' meter Model No. 42E fitted with a special electrode (No. 47, Mark II, for high pressure) suitable for operation to depths of up to 50 m. Salinity was determined with an 'Atago' hand refractometer model No. 2441. Light measurements were carried out with a 'Li-Cor' 188SB integrating quantum photometer fitted with a 1925B underwater sensor.

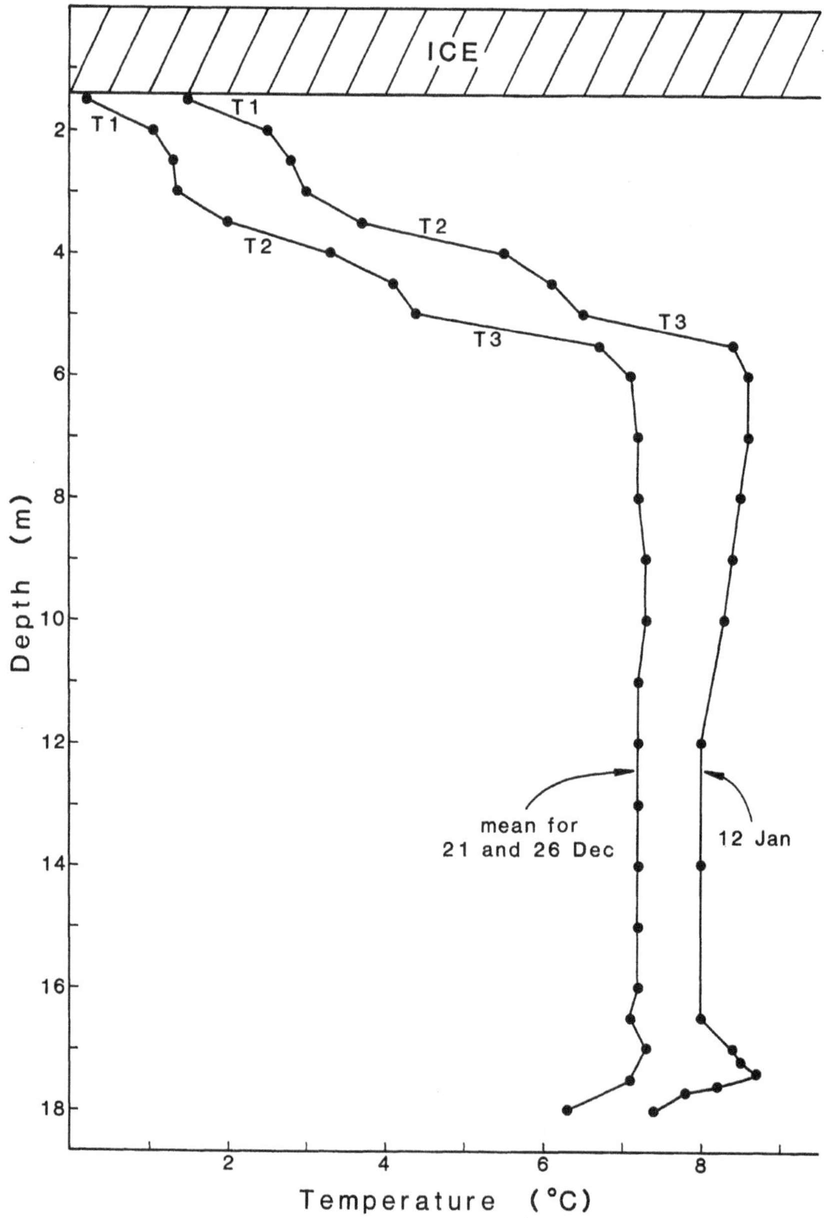

Fig. 1. Temperature-depth graphs for Lake Abraxas for the summer of 1985–86. T1–T3; three successive thermoclines.

Zooplankton was collected by the method described in detail by Bayly (1986) except that at each new depth the pump was run for a short time before the net was fitted within the suction line. This *modus operandi* avoided any contamination by plankton being carried over from one nominal level of sampling to the next.

All plankton sampling was carried out from 24–26 December, 1985. A total of 62 200-l samples were pumped from various levels in the depth range 1.5–19.0 m. Of these, eight samples were taken below the upper limit of the anoxylimnion (located at a depth of 17.7 m) and contained no living copepods. Within the oxylimnion 18 different levels were sampled, two replicates being pumped from each of nine of these levels and four replicates from each of the remaining nine levels. For those levels within the oxylimnion from which four replicate samples were taken, 95% confidence limits for the mean density were calculated by Elliott's (1977:90-1) logarithmic transformation method for small samples from a contagious distribution.

Results

Physico-chemical limnology

Temperature profiles are presented in Fig. 1. Although the January graph is displaced to the right, it is almost identical in shape to that for December. Both graphs indicate the existence of three thermoclines within the upper 6 m: at c. 1.75 m, at c. 3.75 m, and the third (and strongest) at c. 5.25 m.

Values for redox potential, salinity and light are given in Table 1. Redox potential changed from a small positive to a large negative value between 17 and 18 m, and closer investigation in this region fixed the cline at 17.7 m. Salinity values within the oxylimnion varied from 14.0 g l^{-1} at 2.0 m to 19.5 g l^{-1} at 17.0 m with a halocline between 5 and 6 m depth. Transmission of light was high with 7% of the surface value arriving at 17 m.

Oxygen concentrations and saturation levels were abnormally high throughout the whole of the depth range 2.0–16.5, but especially from 6–16 m (Table 2). The oxycline *sensu stricto* (i.e. the plane where the rate of change of oxygen with depth is maximal) was located at a depth of 17.2 m (Fig. 4).

Zooplankton

The major features of the vertical distribution of immature copepodite stages (CI-V), adult males (CVI ♂) and adult females (CVI ♀) of *Paralabidocera antarctica* (I. C. Thompson) are shown in Fig. 2 and Table 3. Data on the abundance of nauplii (calanoid and harpacticoid combined)

Table 1. Redox potential, salinity and light values for Lake Abraxas

Position[a] or depth (m)	Redox potential (mV)	Salinity (g l^{-1})	Photon flux density[b] (μmol m^{-2} s^{-1})
	21.xii.85	24.xii.85	4.i.86
A			1650
B			1650
C			1643
1			658
2	+ 70	14.0	486
3	+ 40	14.0	453
4	+ 40	15.0	429
5	+ 40	16.0	402
6	+ 40	19.0	360
7	+ 40	–	323
8	+ 40	19.0	293
9	+ 40	–	268
10	+ 40	–	245
11	+ 40	19.0	223
12	+ 40	–	206
13	+ 40	–	190
14	+ 40	19.5	175
15	+ 40	–	163
16	+ 40	19.5	150
17	+ 40	19.5	120
18	– 390	20.0	89
19	– 390	20.0	–
20	– 400	21.0	–
21	– 400	22.0	–
22	– 400	22.5	–

[a] A, in air over ice; B, in ice hole 5 mm above the air/water interface; C, in ice hole 5 mm below the air/water interface.

[b] Mean values for two series: one descending, one ascending, 13 : 30–14 : 00 hr.

Table 2. Oxygen levels in Lake Abraxas.

Depth (m)	Absolute concentration (mg l^{-1})				Saturation percentage			
	21.xii.85	26.xii.85	4.i.86	12.i.86	21.xii.85	26.xii.85	4.i.86	12.i.86
2.0	16.9		16.9	15.5	110		119	120
4.0	17.8		19.2	16.8	129		148	138
6.0	17.6		19.8	17.2	140		165	148
8.0	17.7		>20.0	17.0	148		186	150
10.0	18.0		>20.0	16.9	151		>200	148
12.0	18.1		>20.0	16.8	151		>200	146
14.0	18.1		>20.0	16.8	151		>200	143
16.0	18.5	17.8	>20.0	16.3	151	151	198	142
16.5	–	–	13.7	10.0	–	–	115	85
17.0	9.8	9.2	11.7	8.8	84	76	97	77
17.1		–	9.7	8.0		–	82	68
17.2		–	5.6	5.7		–	47	48
17.3		–	1.9	2.1		–	17	18
17.4		–	1.4	0.6		–	12	6
17.5		0.75	1.0	0.3		6	8	3
17.6			0.6	0.2			5	2
17.7			0.3	0.2			3	2
18.0			0.0	0.0			0	0

and of the entire copepodite population (CI-VI) of *Amphiascoides* sp. are presented in Fig. 3 and Table 4.

There is a great expansion in the population density of all three post-naupliar components of the *P. antarctica* population at a depth of 5.0 m (Fig. 2). The reality of this peak is established by the 95% confidence limits given in Table 3. This level is 1.0 m above the warm isothermal stratum extending from c. 6–16 m, and just above the third (T3) of a series of thermoclines located in depth range 1.5–5.5 m (Fig. 1). At this level of maxi-

Table 3. Aspects of the vertical distribution of *Paralabidocera antarctica* in Lake Abraxas, 24–26 December 1985. (Data restricted to depths at which population mean density of stages CI–VI exceeded 100 m^{-3}.)

Depth (m)	Number of samples	CI–VI density (no. m^{-3})		CVI ♀ / CVI ♂	CVI ♀ / CI-V (%)	CVI ♂ / CI-V (%)	CI-V / CVI (%)
		Mean	95% confidence limits				
1.5	4	438	133–1442	3.9	183	47	43
2.0	2	123		3.5	215	62	36
3.0	4	116	49–276	2.2	152	69	45
4.0	2	138		1.0	200	200	25
5.0	4	4043	2540–6430	0.38	118	310	23
5.25	2	420		0.92	189	206	26
5.50	2	420		0.42	175	425	17
6.0	4	160	48–532	0.77	136	177	32
8.5	2	103		1.0	136	136	37
17.2	2	155		0.59	59	100	64
17.4	4	206	84–505	0.49	19	39	170

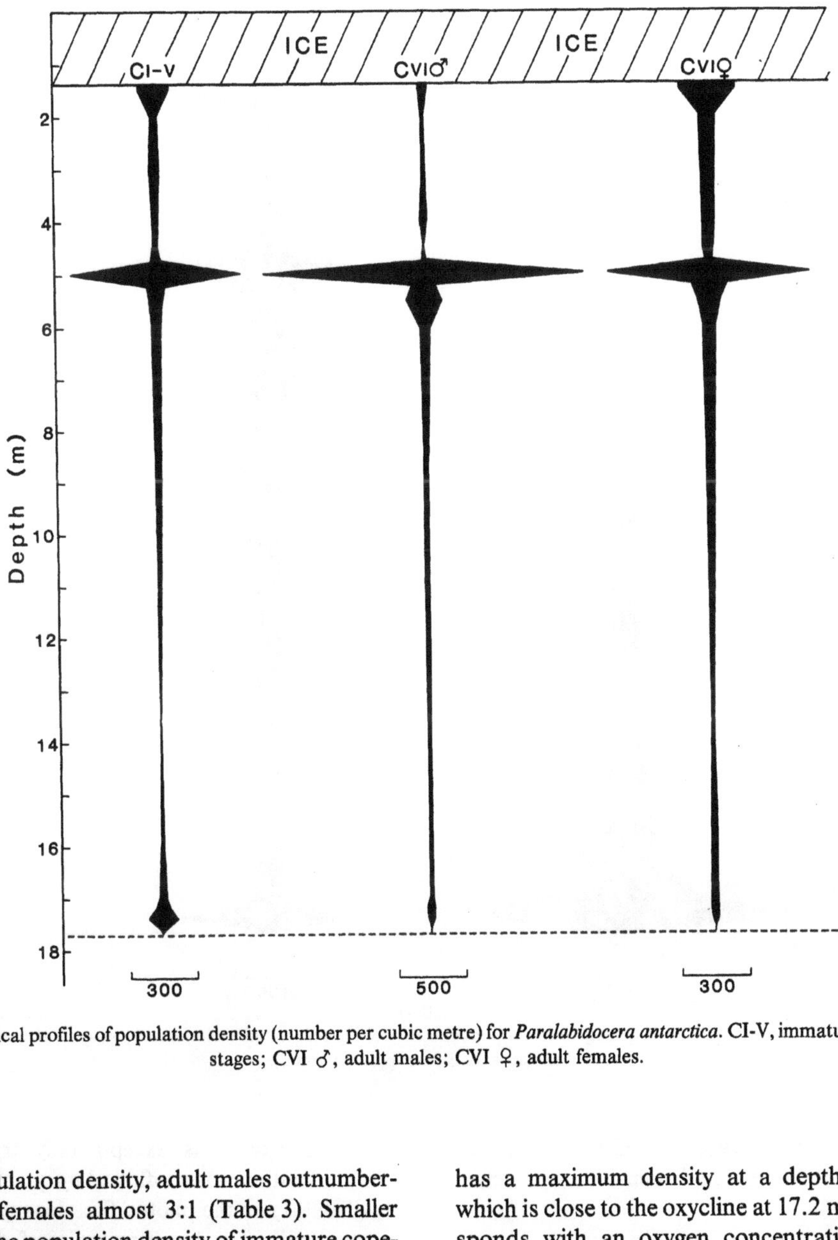

Fig. 2. Vertical profiles of population density (number per cubic metre) for *Paralabidocera antarctica*. CI-V, immature copepodite stages; CVI ♂, adult males; CVI ♀, adult females.

mum population density, adult males outnumbered adult females almost 3:1 (Table 3). Smaller peaks in the population density of immature copepodites and adult females occur immediately beneath the ice in which region adult females outnumber adult males almost 4 : 1 (Table 3). A third density mode for the immature copepodites occurs at 17.4 m.

The post-naupliar population of *Amphiascoides* has a maximum density at a depth of 17.3 m which is close to the oxycline at 17.2 m and corresponds with an oxygen concentration of only 2.2 mg/l (Figs. 3 & 4).

The density profile for nauplii (Fig. 3) exhibits three modes: one immediately beneath the ice, at 5.0 m, and at 17.3 m. It is reasonable to suppose that the first two of these peaks consists mainly of *P. antarctica* nauplii, and the third mainly of the

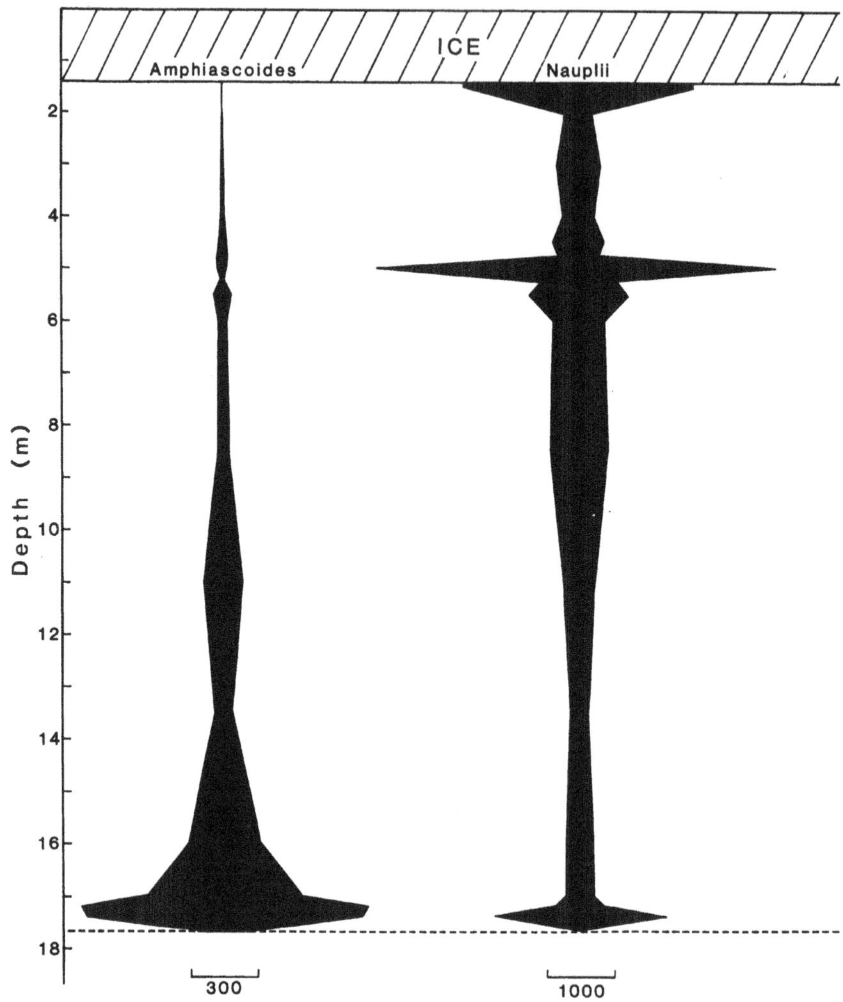

Fig. 3. Vertical profiles of population density (number per cubic metre) for post-naupliar *Amphiascoides* sp. and nauplii (calanoid and harpacticoid nauplii combined).

nauplii of *Amphiascoides*. Confidence limits for the harpacticoid and nauplii densities are given in Table 4.

Discussion

Qualitative observations on white interval-markers on black instrument cables lowered through an ice hole indicated that the water of Lake Abraxas was exceptionally transparent. This indication was confirmed by quantitative data (Table 1) on the photon flux density. The proportion of light at the air-water interface reaching down to a depth of 17 m was approximately the same as that reaching only 5–7 m in Ace Lake (Bayly & Burton, 1987) or <2 m in Burton Lake (Bayly, 1986). As a consequence, sufficient primary production is possible in the immediate vicinity of the oxycline to sustain the relatively high abun-

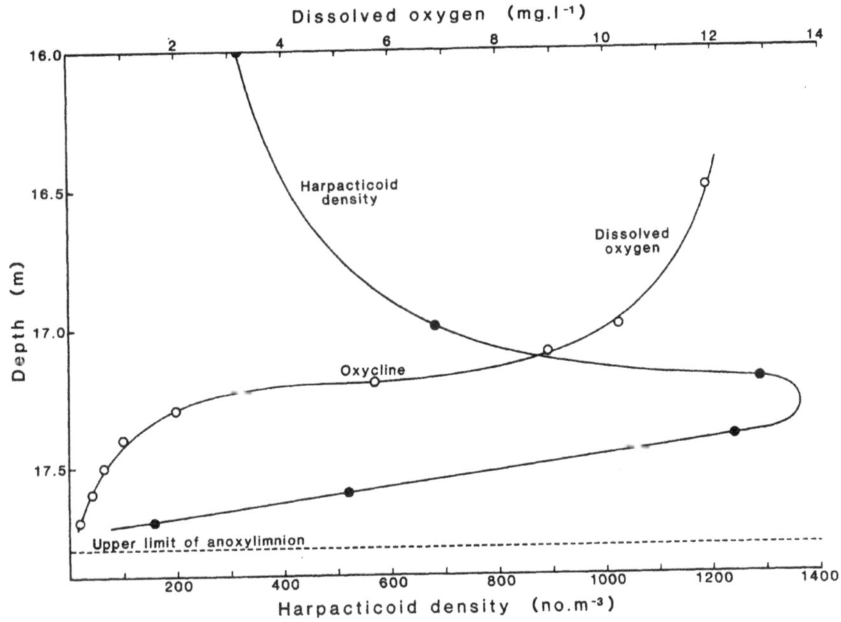

Fig. 4. Relationship between density of post-naupliar harpacticoids and dissolved oxygen concentration in the region of the oxycline.

dance of the presumptively herbivorous harpacticoids located there (Fig. 3, Table 4).

The strongest thermocline (Fig. 1), and halocline (Table 1), were located slightly below the 5.0 m level, and one or other (or both) of these apparently set lower limit to the high density stratum of *P. antarctica* located at a depth of 5.0 m (Figs. 2 & 3).

One of the most striking chemical features of Abraxas Lake is the very high levels of dissolved

Table 4. Vertical variation in the population density of *Amphiascoides* sp. and in copepod nauplii (harpacticoid and calanoid) in Lake Abraxas, 24–26 December 1985. (Data restricted to depths at which four replicate samples were taken.)

Depth (m)	*Amphiascoides* density[a] (no. m^{-3})		Nauplius density[b] (no. m^{-3})	
	Mean	95% confidence limits	Mean	95% confidence limits
1.5	0		3380	697–16395
3.0	0		625	261–1499
5.0	40	6–260	5926	2136–16446
6.0	36	6–218	733	320–1677
11.0	161	19–1375	461	78–2723
16.0	316	120–840	381	133–1094
17.0	684	200–2370	385	136–1092
17.4	1241	335–4585	2560	714–9180
17.6	518	60–4511	540	82–3572

[a] The mean density for two replicates at a depth of 17.2 m was 1290 m^{-3}.
[b] The mean density for two replicates at a depth of 17.2 m was 720 m^{-3}.

oxygen (Table 2). Considering the depth range 6.0–16.0 m, the percentage saturation was 140–151 on 21 December 1985, 142–150 on 12 January 1986, and on 4 January 1986 values in excess of 200 (the upper limit of the meter) were recorded from 10.0–14.0 m. Comparable and even greater degrees of supersaturation have been recorded for other Antarctic lakes such as Lake Hoare (Mikell *et al.*, 1984; Wharton *et al.*, 1986).

Parallels exist between some of the data presented above for *P. antarctica* and those reported by Bayly & Burton (1987) for the same species in Ace Lake. In the latter population, like that in Abraxas Lake, adult females outnumbered adult males in near-surface waters but at deeper levels the reverse was true. Bayly & Burton (1987) suggested that a portion of the adult female population may have a trophic relationship with the under surface of the ice by the consumption of epontic algae.

There is a striking spatial separation of the maximum density of the calanoid and that of the harpacticoid. The exact trophic basis for these two maxima is unknown. The harpacticoid peak corresponds with an oxygen concentration (c. $2\ mg\ l^{-1}$) that may well exclude or stress adult calanoids. Some harpacticoids are highly tolerant of low oxygen concentrations. *Apodosyllus africanus* (Kunz), for example, has been shown by Wieser *et al.* (1974) to be a facultative anaerobe. A large part of the population of this benthic species flourishes where the Eh is less than − 100 mV and little or no molecular oxygen is available. It would seem that the species of *Amphiascoides* in Lake Abraxas has carried over into the open water an adaptation to low levels of oxygen of a benthic ancestor.

Planktonic harpacticoids are something of a rarity. Hicks and Coull (1983) noted that of the approximately 2 800 described species of Harpacticoida, all are free-living and benthic except for fewer than 35 species which are symbiotic or planktonic. They further stated that of the truly planktonic harpacticoids, most have elongated setae or body shapes and some are (for the group) unusually large. However, the species of *Amphiascoides* in Abraxas Lake has neither elongated setae nor a filamentous body, and furthermore is not usually large. The mature females of this species have a mean length (for n = 5) of 0.44 mm. In contrast, the mature females of the second harpacticoid species in Abraxas Lake, which is benthic, have a mean length (for n = 7) of 0.78 mm.

References

Bayly, I. A. E., 1986. Ecology of the zooplankton of a meromictic Antarctic lagoon with special reference to *Drepanopus bispinosus* (Copepoda: Calanoida). Hydrobiologia 140: 199–231.

Bayly, I. A. E. & H. R. Burton, 1987. Vertical distribution of *Paralabidocera antarctica* (Copepoda: Calanoida) in Ace Lake, Antarctica, in summer. Aust. J. mar. Freshwater Res. 38: 537–43.

Burton, H. R. & R. Hamond, 1981. Harpacticoid copepods from a saline lake in the Vestfold Hills, Antarctica. Aust. J. mar. Freshwater Res. 32: 465–467.

Elliot, J. M., 1977. Some methods for the statistical analysis of samples of benthic invertebrates. Freshwat. Biol. Assoc. UK Sci. Publ. No. 25, 156 pp.

Hicks, G. R. F. & B. C. Coull, 1983. The ecology of marine meiobenthic harpacticoid copepods. Oceanogr. Mar. Bio. Ann. Rev. 20: 68–175.

Mikell, A. T., B. C. Parker & G. M. Simmons, 1984. Response of an Antarctic lake heterotrophic community to high dissolved oxygen. Appl. Envir. Microbiol. 47: 1062–1066.

Wharton, R. A., C. P. McKay, G. M. Simmons & B. C. Parker, 1986. Oxygen budget of a perennially ice-covered Antarctic lake. Limnol. Oceanogr. 31: 437–443.

Wieser, W., J. Ott, F. Schiemer & E. Graiger, 1974. An ecophysiological study of some meiofauna species inhabiting a sandy beach at Bermuda. Mar. Biol. 26: 235–248.

Hydrobiologia 172: 215–234, 1989
W. F. Vincent and J. C. Ellis-Evans (eds) High Latitude Limnology

Physico-chemical characteristics and origin of hypersaline meromictic Lake Garrow in the Canadian High Arctic

M. Ouellet[1], M. Dickman[2], M. Bisson[3] & P. Pagé[4]
[1] *Institut national de la recherche scientifique (INRS-Eau) Université du Québec, Sainte-Foy (Québec), G1V 4C7 Canada*; [2] *Biological Science Department, Brock University, St-Catharines (Ontario), L2S 3A1 Canada*; [3] *Ministère de l'Environment du Québec, Québec, G1V 4H2 Canada*; [4] *Département des Sciences de la Terre (GEOTOP), Université du Québec à Montréal, Montréal, H3P 3P8 Canada*

Key words: Arctic, chemistry, Cornwallis Island, heliothermic, hypersaline water, ice, lake, meromixis, mesothermic, permafrost

Abstract

Garrow Lake (75° 23′ N; 96° 50′ W), located 3 km from the southern tip of Little Cornwallis Island and 6.7 m above mean sea level, is a meromictic ecto-creno-cryogenic lake with an area of 418 ha and a maximum water depth of 49 m. The thermal stratification of this lake is mesothermic (heliothermic). Some of the solar energy that penetrates through the 2 m ice cover is stored for a long period of time in the upper level of the monimolimnion, under a greenhouse effect due the water density gradient. The energy transfer (0.06 °C m^{-1}) by conduction toward the bottom sediments is very constant from one year to the next and is likely to prevent the presence of permafrost under this water body.

In its chemical composition, this meromictic lake is quite comparable to the world's saltiest water bodies and is the first lake, with a salinity greater than sea water to be reported for the Canadian Arctic. Its anoxic monimolimnion is nearly three times (90‰) as salty as normal sea water.

This hypersaline water seems to have been derived from isostatic trapped marine waters within the present lacustrine basin as well as from underground during deglaciation of the area. The subsequent freezing-out of salt from the underground waters and the migration and accumulation of these waters in the bottom of Garrow Lake through a talik within the permafrost were the main contributing factors. The speed of formation and migration of the underground brine was a function of the postglacial isostatic uplift rate as well as the permafrost growth rate.

Introduction

Garrow Lake is one of the most saline (90‰) meromictic deep water lakes of Canada. Although some saline and permanently density-stratified deep lakes had been reported from some parts of Canada (McLaren, 1961; Williams *et al.*, 1961; Hattersley-Smith & Serson, 1964; Carter, 1965; Northcote & Halsey, 1969; Hattersley-Smith *et al.*, 1970; Jeffries *et al.*, 1984), none exceeded the salinity of the sea. High salinity lakes such as Garrow occur in several regions of the Antarctic, and limnological reviews of these are given by Heywood, 1972, 1977, 1984; Williams, 1981; Wilson, 1981 and Vincent, 1987. Salty and meromictic lakes display an unusually wide variety of environmental conditions where ecological, physiological, evolutional and paleolimnological hypotheses might be tested.

The main characteristic of a meromictic lake is its incomplete waters circulation during homeothermal (isothermal) conditions of its mixolimnion (upper most water layer). This results in a permanent anoxic condition in its monimolimnion (the bottom non-mixing water layer in the lake: Findenegg, 1935; Hutchinson, 1937). Walker & Likens (1975) have reviewed the typology of meromixis, principally as a function of water circulation patterns. The permanent chemical stratification of the water column generally tends to become more pronounced with time, and the constant accumulation of electrolytes in the monimolimnion results in a steep chemical gradient, which has a profound effect on the biotic communities. The absence of mixing in the whole lake water gives rise to a continuous permanent loss of nutrients from the mixolimnion, which explains the low productivity (oligotrophy) of the trophogenic (photosynthetic) zone in many meromictic lakes.

Garrow Lake is located at 75° 23′ N, 96° 50′ W, on Little Cornwallis Island, N.W.T., in the High Arctic approximately 96 km northwest of Resolute Bay (Fig. 1). The lake is mesothermic (heliothermic), with a midwater maximum temperature of 9 °C and deep water salinity nearly three times that of sea. This ecto-crenocryogenic meromictic lake is surrounded by gently rolling tundra desert. In November 1981, Cominco Limited's Polaris Project, a lead and zinc mining development, began discharging mine wastes into the monimolimnion of the lake. We observed the lake when it was ice-free in late August 1980 and early September 1981, as well as under ice cover in January 1983 and June 1984.

The main objectives of the present study were to broaden our knowledge of the limnological nature and origin of this unusual Arctic lacustrine environment and to determine its characteristics prior to the discharge of lead and zinc mine tailings.

Study site

From a topographic map (Fig. 2), the Garrow Lake watershed looks like a small fjord but on the ground there is little or no sign of glacial erosion. The landscape of this area of the Canadian Arctic is essentially fluvial with smooth slopes and a considerable layer of regolith. The area is dissected by an intricate network of small streams (mostly intermittent) and valleys. The presence of broken marine fossil shells within the entire catchment area of Garrow Lake is evidence of postglacial uplift and implies that Little Cornwallis Island was covered, at least in part, by ice during the last Wisconsinian glaciation. Within the Garrow Lake watershed there is no established drainage system. Periglacial structures such as polygonal networks, frost cracks and solifluction structures are abundant. The gently rolling topography of the Garrow Lake watershed looks like an immense amphitheatre with a restricted opening to the south end. This type of topography acts as a wind shelter for the lake surface water. The origin of the lake basin depression is probably tectonic. Thornsteinsson (1958) and Thornsteinsson & Kerr (1968) mapped a geosyncline with the main axis running along the east shore of the lake. Kerr (1977a, b) described the structural and tectonic aspects of the geology as well as the galena and sphalerite mineralization origin of the area. The underlying bedrock of the Innuitian orogen is composed of several thousand metres of Ordovician and Silurian calcareous shale, shaly limestone, cherty and dolomitic limestone. A thick, carbonate-rich glacial drift largely derived from the Cape Phillips formation covers the area and outcrops of bedrock are not observed within the watershed.

The sparsely scattered vegetation ($\approx 10\%$ total cover) consisted mainly of mosses, lichens, Arctic poppies and some creeping willows, all of which occur mostly along seepage plates and in the depressions originating from periglacial processes. The vegetation and the surficial materials were recently studied by Edlund (1982). The mean annual air temperature of the area is − 16 °C. In July, the mean reaches + 4.3 °C while in January

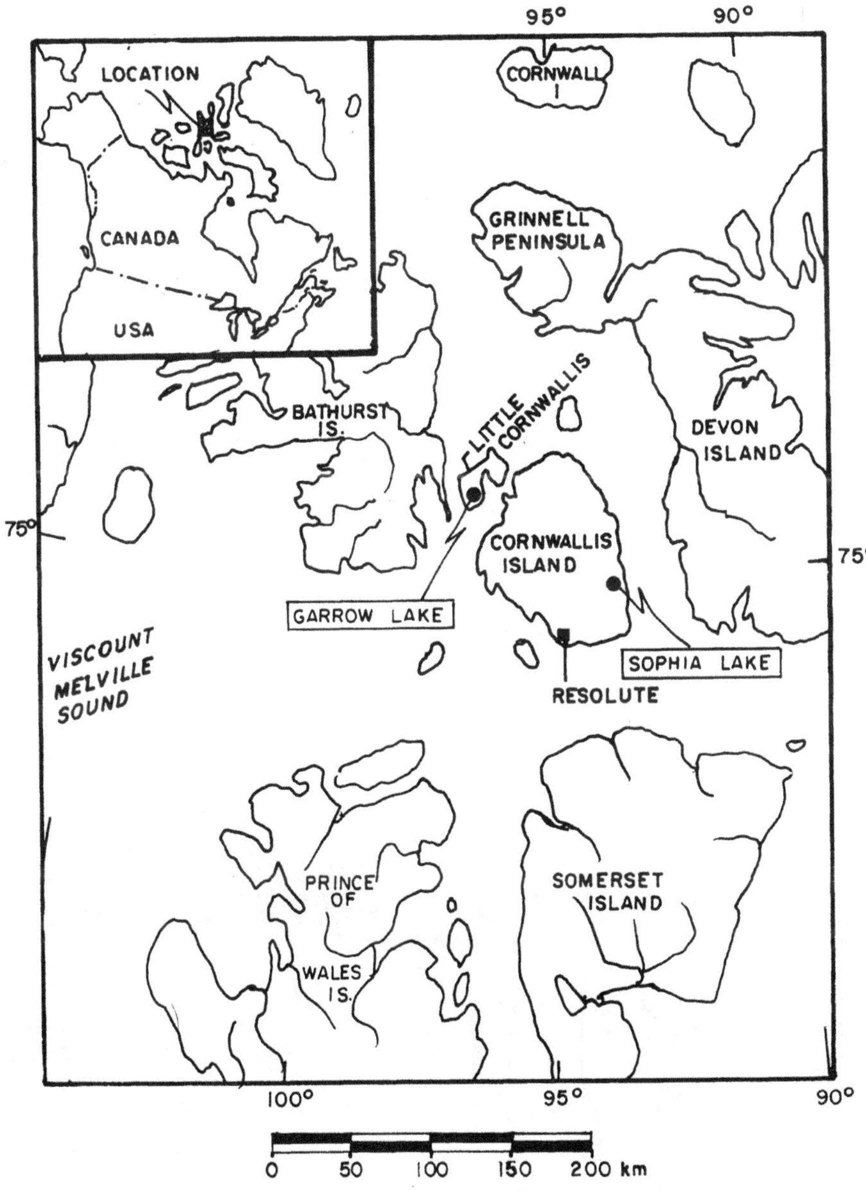

Fig. 1. Location of Garrow Lake in the Canadian Arctic.

it falls to −32 °C. The average frost-free season (>0 °C) lasts only from the 10th to the 20th of July. The total mean annual precipitation is of the order of 13.6 cm (Cook, 1960). This climatic region (Bathurst – Prince of Wales Islands; Maxwell, 1981) receives an annual net radiation of between 5 to 10 k cal cm^{-2} as compared to values of less than 5 k cal cm^{-2} over the Arctic Ocean. Sea-ice conditions are a mixture of multi-year and first-year types with the degree of summer clearing variable from year to year.

The lake volume was estimated by Rigler

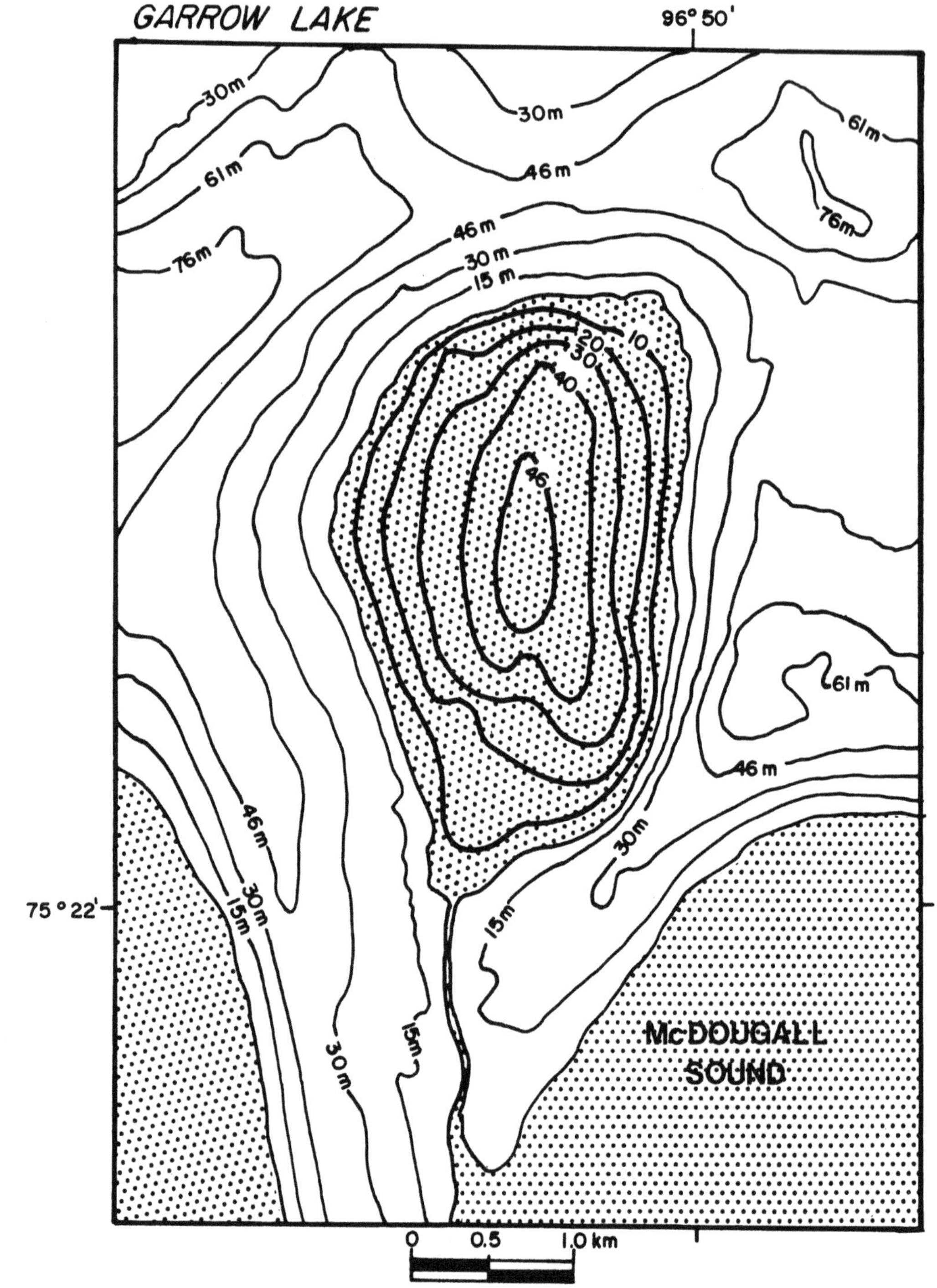

Fig. 2. Morphometric map of Garrow Lake watershed.

(1974) and later by B. C. Research (1978) to be $10.26 \times 10^6 m^3$. Lake length (3293 m) and width (1830 m) as well as mean depth (24.5 m) and maximum depth (49 m) were also estimated by the above authors. The lake's catchment area (573 ha) was only 1.4 times larger than the area (418 ha) of the lake itself. A small catchment in relation to lake size is typical of a number of meromictic basins (Walker & Likens, 1975). Garrow Lake is normally ice free for about one month toward the end of the summer (August-September).

Thornsteinsson & Kerr (1968) and more recently Washburn & Stuiver (1984) established the age of the successive raised beaches around Cornwallis Island, and from these dates it was possible to estimate Garrow Lake's age at roughly 3000 years before present. The lake is presently 6.7 m above sea level, which could mean that its basin has been rising by approximately 2.2 mm per year. Its age is of the same order as those recently established from the isotopic composition of the upper monimolimnic water (2580 years; Pagé *et al.*, 1984).

Methods

Physico-chemistry

Garrow Lake was visited in August 1980, September 1981, January 1983 and June 1984 in order to establish some of the aqueous physical, chemical and biological characteristics. All water samples were taken using a 5 litre Van Dorn bottle and transferred to new half litre plastic bottles rinsed with the water with which they were filled. Samples intended for heavy metal analysis as well as Fe and Mn analysis were acidified in the field.

Dissolved oxygen, temperature, H_2S and turbidity were measured in the field. In 1980, dissolved oxygen (YSI, model-51A) and temperature (YSI, model-33) were measured *in situ* while in 1981, Winkler dissolved oxygen analysis was performed less than six hours after sampling. For H_2S and turbidity a Hach portable chemical water quality analyzer was used according to the methods described in A.P.H.A. (1981).

The analytical methods used in the laboratory are described at length in Jones & Bisson (1980). Part of the unacidified sample was centrifuged for chloride and sulphate analyses. For the quantification of metals (major and trace), 100 ml of each sample was digested by boiling with 3 ml of concentrated HNO_3. The residue was dissolved in 25 ml of HNO_3 (0.5%), and this solution was used to measure the major cations (Na, K, Ca, Mg, Fe, Mn) and heavy metals (Cd, Co, Cu, Ni, Pb, Zn) by extraction with ammonium pyrrolidine dithiocarbonate (A.P.D.C.) and methyl isobutyl ketone (M.I.B.K.) (Tessier *et al.*, 1979). They were quantified by atomic absorption spectrophotometry using a Varian Techtron, model AA-575 or Model AA-5. The major metallic ions were quantified by flame photometry, while trace metals (Cd, Co, Cu, Ni, Pb, Zn) were measured by flameless technique on a Varian Techtron, model CRA-90, in order to lower the detection limit.

Laboratory pH was measured at 22 °C by means of a Radiometer model PHM-53 pH meter, and, was then calibrated with two buffering solutions. Laboratory conductivity was estimated at 25 °C with a Radiometer model CDM conductivity meter. Phosphate and total phosphorus were analyzed by following the method described by A.P.H.A. (1981). After specific digestion (H_2SO_4) they were estimated by colorimetry on a Technicon II auto-analyzer for low concentrations ($\leq 10\ \mu g\ l^{-1}$). The complex (phospho-ammonium molybdate) was extracted by iso-butanol and quantified by colorimetry (Stephens, 1963). The colorimetric quantification of chloride is based on the dissociation of thiocyanate ions from mercuric thiocyanate during the formation of mercuric chloride. In the presence of ferric ions, thiocyanate gives rise to colored ferric thiocyanate the intensity of which is proportional to the concentration of the original chloride ions (Technicon, 1973). The sulphates were quantified by the auto-analyzer using the methylthymol blue method (Technicon, 1972). The total carbon (TOC + TIC) and total inorganic carbon (TIC) were estimated using an infra-red Beckman carbon analyzer, model 915A. For total carbon the

Table 1. Garrow Lake water quality data in relation to depth for 24 August 1980

Depth m	pH	O_2 1980	O_2 1981	H_2S	Temp. 1980	Temp. 1981	Salinity		Turb. JTU	$PO_4^=$	Pt	Cl^-	$SO_4^=$	TIC	TOC + TIC
		mg l^{-1}			°C		‰			µg l^{-1}		mg l^{-1}			
0	6.8	14.6	14.8	0	0.8	2.0	0.1[a]	0.1(14..3)[b]	0.2	–	16	87	12.0	2	3
2	7.6	17.0	–	0	1.4	2.0	2.0	2.5(14.3)	–	1	12	1383	142	15	24
3	7.8	17.5	16.4	0	4,3	2.0	5.0	5.7(9.5)	0.2	2	16	3160	350	32	52
5	8.0	16.0	17.1	0	4.4	2.0	6.5	6.9(9.2)	–	1	11	3830	575	35	49
10	7.9	16.0	16.1	0	3.9	3.5	9.4	10.1(9.4)	0.2	1	20	5960	350	35	57
13	7.6	10.9	15.9	0	5.3	6.9	30.2	34.6(9.6)	0.3	3	21	19152	3400	52	88
17	7.6	7.0	7.6	0	8.0	8.5	65.4	74.9(9.5)	0.4	1	24	41496	5750	59	103
19	7.2	0	0	0	9.1	8.9	78.8	90.2(9.6)	1.1	1	17	–	6125	64	130
20	7.1	0	0	0.5	9.1	9.0	84.1	92.2(6.5)	4.2	6	37	51072	6125	72	130
25	7.1	0	0	45	8.8	9.2	87.7	92.8(7.8)	2.1	450	710	54264	6400	140	200
30	7.3	0	0	–	8.5	–	90.1	102.7(9.3)	–	–	890	56924	6875	140	200
35	7.2	0	0	110	8.0	–	93.4	103.7(7.4)	2.0	350	780	57456	6775	190	260
40	7.6	0	0	150	7.6	–	84.9	97.9(10.1)	2.8	410	920	54264	6125	190	270

[a] calculated from Cl^- concentration
[b] calculated from $\sum^+ + \sum^-$ (mg l^{-1}) 10^{-3} and coefficient of variation in % of the two methods
[c] coefficient of variation in % of the ionic balance

Table 1. Garrow Lake water quality data in relation to depth for 24 August 1980 (continued)

Depth m	Alk. $CaCO_3$	Na^+	K^+	Ca^{++}	Mg^{++}	$\sum^+$	$\sum^-$	Fe	Mn	Cd	Co	Cu	Ni	Pb	Zn
	mg l^{-1}					meq l^{-1}		µg l^{-1}							
0	88	17	1	3	5	1.2	2.7(52.3)[c]	≤ 50	≤ 30	≤0.5	≤2	≤1	2	≤10	9
2	–	380	7	43	77	27.4	41.7(29.3)	–	–	–	–	–	–	–	–
3	–	1200	22	90	160	70.4	95.8(21.6)	–	–	–	–	–	–	–	–
5	–	1400	26	93	180	81.0	115.0(24.5)	≤ 50	≤ 30	≤0.5	≤2	1	12	≤10	12
10	125	2800	25	97	180	142.1	174.2(14.3)	–	–	–	–	–	–	–	–
13	160	5800	70	430	1300	382.5	606.9(22.1)	≤ 50	≤ 30	1.0	≤2	≤1	10	≤10	9
17	230	15000	230	760	2200	887.3	1281.5(26.5)	90	130	≤1.2	≤5	2.4	5	≤24	24
19	150	19000	360	830	2500	1082.9	1537.4(24.5)	110	700	≤1.2	≤5	≤2.4	10	≤24	24
20	300	23000	430	830	2600	1266.8	1557.4(15.5)	340	470	≤1.2	≤5	≤2.4	11	≤24	7
25	560	23000	560	880	2600	1272.6	1652.5(18.4)	170	≤ 30	≤1.2	≤5	≤2.4	11	≤24	7
30	700	22000	690	920	2700	1242.2	1736.9(23.5)	140	≤ 30	≤1.2	≤5	≤2.4	≤ 5	≤24	10
35	850	25000	650	870	2600	1361.4	1749.7(17.6)	140	≤ 30	≤1.2	≤5	≤2.4	10	≤24	10
40	900	21000	560	760	2200	1146.7	1646.7(25.3)	200	≤ 30	≤1.2	≤5	2.4	11	≤24	34

sample was heated at 950 °C and the CO_2 recuperated in a cobalt oxide column while for TIC this conversion was done by heating at 150 °C in the presence of phosphoric acid (85%).

Results

The physico-chemical composition of the Garrow Lake water column is shown in Tables 1 to 3 respectively for our 1980, 1983 and 1984 samplings. The greater (≈25%) coefficient of variation for the 1980 ionic balances in relation to those of 1983 and 1984 (≈3%) is due to the longer wait before the water samples reached our laboratory for analysis. Some of these samples were reprocessed several times by different technicians and the results were found to be correct. The salinity values presented here were calculated from the chloride concentrations using the Wallace (1974) formula (S‰ = 1.807 Cl‰). They are also in good agreement (≈1%) with the sum in $g\,l^{-1}$ (‰) of the main anions and cations. The specific gravity of the water was estimated from the 1984 calculated salinity by using the linear increment of 0.00085 $g\,l^{-1}$ for every unit of salinity (Wetzel, 1975).

The vertical distributions of the majority of the chemical parameters studied are illustrated in Fig. 3 as a function of depth. Due to the low levels of some heavy metals such as Cd, Co, Cu, Ni and Pb, the vertical depth distribution of these elements is only reported and no interpretation is attempted.

As is typical of meromictic lakes, the vertical physico-chemical characteristics of Garrow Lake water can be described in terms of a three-part pattern: 1) a 0 to 12 m water layer (mixolimnion) of less than 10‰ salinity; 2) a 12 to 20 m layer of water (chemocline, halocline, pycnocline) which is characterized by a rapid increase in concentrations of the great majority of variables (except O_2, H_2S, temp., pH, TIC, TOC); 3) a 20 to 49 m hypersaline (90‰) layer (monimolimnion) in which the parameters reach their maximum concentrations. The increase in alkalinity, TOC and total carbon within the chemocline and the monimolimnion is more gradual and does not follow the three step pattern. These parameters increase from top to bottom and do not display any strong inflection immediately below the 20 m water stratum.

Temperature and transparency

The temperature profiles for Garrow Lake are characteristic of lakes displaying mesothermic (heliothermic) heating (Hutchinson, 1957). Temperatures rise with water density until maximum light intensity falls below a critical level. A significant drop in light intensity occurs just below 20 m as indicated by the turbidity curve. At that depth, the temperature levels off at about 9 °C and then falls gradually to reach 7 °C at the lake bottom.

To compare the interannual thermal stability of the lake water, we have also included the 1976 Fallis & Harbicht (1980) data along with the 1977 B.C. Research (1981) temperature observations. Within the chemocline and the monimolimnion, all temperature curves are largely overlapping while within the mixolimnion they are highly dependent on the sampling season.

pH, alkalinity, carbon

The depth profile for pH was more complex than we anticipated. The pH rose with increasing salinity from the surface to about 10 m (pH of 8.0) depending on the sampling season. Then it declined with depth to a value of 7.0 at 20 m, and increased again slightly below the chemocline (7.2). This small range of pH is a good indication of the highly buffered nature of the water, which is evident from the high alkalinity (800 $mg\,l^{-1}$) and bicarbonate content (1000 $mg\,l^{-1}$) of the anoxic zone.

The absence of a sharp gradient within the chemocline for alkalinity as well as total carbon and TIC is probably an indication of the role played by the heterotropic and photosynthetic bacteria, described within that stratum by Dickman & Ouellet (1983), in eroding the carbonate and bicarbonates from the top layer of the monimolimnion (overlapping stratum of O_2 and H_2S). The anaerobic photosynthetic bacteria oxidize thiosulphates or elemental sulphur, in

Table 2. Garrow Lake water quality data in relation to depth for January 1983

Depth m	Cond. mmho cm^{-1}	Salinity		Cl^-	$SO_4^=$	HCO_3^-	Alk. $CaCO_3$	Hardness	Na^+	K^+	Ca^{++}	Mg^{++}	$\sum^+$	$\sum^-$
		‰		mg l^{-1}									meq l^{-1}	
3	8.4	4.8[a]	5.0(0.14)[b]	2666	350	223[c]	183	993	1440	44	101	180	83.6	86.1(2.05)[d]
5	8.4	4.8	4.5(0.14)	2666	300	202	166	990	1460	46	100	180	84.5	84.7(0.22)
10	9.7	5.6	5.7(0.07)	3150	375	152	125	1177	1680	52	109	220	97.9	97.7(0.16)
15	63.0	44.2	44.8(0.42)	27000	3750	316	260	9608	13300	338	650	1940	779.2	774.2(0.46)
18	89.0	67.3	65.5(1.27)	37333	5800	426	350	13614	17800	540	870	2780	1060.1	1180.5(7.60)
22	103.0	81.5	81.4(0.07)	45166	6100	585	480	16256	24700	718	1000	3100	1397.7	1410.3(0.64)
24	104.0	82.7	80.6(1.48)	45833	6200	609	500	15000	23300	718	980	3050	1331.6	1431.5(5.11)
30	98.5	77.3	77.6(0.21)	46000	5700	609	500	13924	24100	493	920	2825	1344.3	1336.5(0.41)
35	112.0	89.4	89.5(0.07)	49500	6300	950	780	16697	27500	815	1000	3450	1550.7	1542.6(0.37)
40	112.5	90.2	87.4(1.98)	50000	6500	1023	840	14485	24600	880	1080	3350	1422.0	1542.1(6.64)
45	113.5	91.2	90.5(0.49)	50550	6400	987	810	16635	27300	780	1140	3350	1539.9	1574.9(1.59)

[a] = calculated from Cl^- concentration
[b] = calculated from $\sum^+ + \sum^-$ (mg l^{-1}) 10^{-3} and standard deviation of the two methods
[c] = calculated from total alkalinity
[d] = coefficient of variation in % of the ionic balance

Table 3. Garrow Lake water quality data in relation to depth for the 14 June 1984.

Depth m	pH	Temp. °C	Cond. mmho cm^{-1}	Salinity ‰		Cl^- $g\,l^{-1}$	$SO_4^=$ $mg\,l^{-1}$	HCO_3^-	TIC	Alk. $CaCO_3$	Hard.	Na^+	K^+	Ca^{++}	Mg^{++}	$\sum^-$ $meq\,l^{-1}$	$\sum^+$
Ice*	7.3	0	1.0	0.5[a]	0.5(0.0)[b]	0.27	36	13[c]	3	11	129	146	4	11	25	9.0	8.5 (4.21)[d]
0	7.3	0	0.9	0.4	0.5(0.07)	0.24	50	21	3	17	96	120	4	9	18	7.2	7.9 (4.08)
2	7.3	0	3.1	1.5	1.6(0.07)	0.88	90	45	9	37	331	444	13	30	62	26.3	26.8 (1.51)
5	8.0	0	9.5	5.4	–	2.97	373	–	32	–	–	–	–	–	–	–	91.8 (–)
7	8.0	0	10.0	5.6	5.6(0.00)	3.13	380	146	33	120	1140	1676	33	99	217	96.6	96.5 (1.09)
12	8.0	2	10.1	5.7	5.8(0.07)	3.18	382	158	32	130	1166	1724	33	101	222	99.1	97.6 (1.09)
17	7.5	8	79.8	35.2	33.1(1.48)	33.80	3900	293	58	2450	12179	5878	206	800	2474	907.2	1098.0(15.77)
19	7.3	9.1	89.8	65.8	66.6(0.56)	36.50	5900	366	65	300	13700	19550	568	904	2780	1138.7	1152.0 (0.82)
21	7.1	9	101.3	76.7	76.8(0.07)	42.50	6580	366	80	300	15791	22390	693	1031	3211	1307.2	1335.4 (1.51)
27	7.2	8.5	112.5	83.5	85.3(1.27)	46.30	7000	731	158	600	16741	25890	832	1120	3394	1482.0	1451.3 (1.48)
32	7.2	8.1	115.6	90.8	90.3(0.35)	50.30	6790	853	186	700	–	26840	902	1118	3518	1536.6	1559.7 (1.05)
42	7.2	7.6	115.3	90.2	90.6(0.28)	50.00	6580	975	212	800	–	27620	878	1108	3474	1569.4	1546.9 (1.02)
46	7.1	7.2	115.1	83.6	88.4(3.39)	46.30	7850	975	170	800	17112	27790	880	1110	3484	1573.5	1468.9 (4.86)

* = water from ice surface pool = white ice
[a] = calculated from Cl^-
[b] = calculated from $\sum^+$ and $\sum^-$ $(mg\,l^{-1})\,10^{-3}$ and standard deviation of the two methods
[c] = calculated from alkalinity
[d] = coefficient of variation in % of the balance

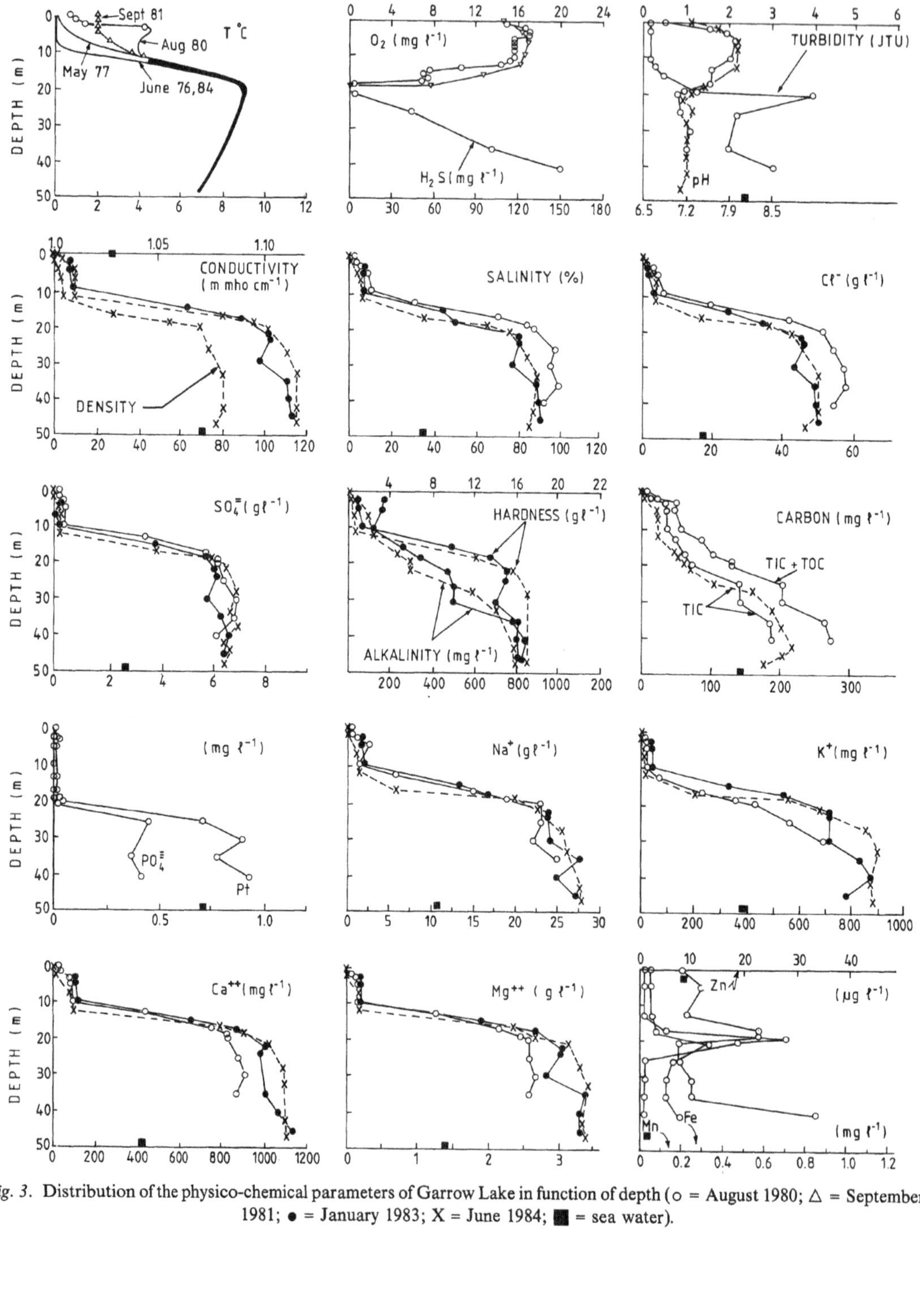

Fig. 3. Distribution of the physico-chemical parameters of Garrow Lake in function of depth (○ = August 1980; △ = September 1981; ● = January 1983; X = June 1984; ■ = sea water).

neutral or alkaline environments, by reducing the nitrate to N_2 (cf. Wetzel, 1975), giving rise also to the production of CO_2, which unfortunately we did not measure. Moreover, under such a reducing environment the precipitation of divalent metallic carbonates such as Fe^{2+} and Mn^{2+} would also occur, contributing to the smoothing of total carbon depth curve (cf. Postgate, 1979).

Dissolved O_2, H_2S and trace metals

At the bottom of the chemocline, the water becomes anoxic and H_2S is detected in low concentrations (0.5 mg l^{-1}). It is at this stratum (boundary layer between dissolved O_2 and H_2S) that the minor elements such as Fe, Mn and Zn reach their peak. These peaks are the results of several complex biological and chemical factors which are largely interdependent. At this oxic/anoxic interface photosynthetic and heterotrophic bacteria as well as adsorption and coprecipitation are likely to emphasize the particulate phases of these three metallic ions. On the immediate upper side of the O_2/H_2S interface, oxidation of reduced Mn^{2+}, Fe^{2+} and Zn^{2+} will make these more mobile, and upward diffusion will enhance their bacterial particulate conversion as well as contribute to some losses through the chemocline. On the lower side of the O_2/H_2S interface the metallic sulphide and carbonate ions are likely to precipitate in the presence of H_2S, which reaches a bottom concentration of 150 mg l^{-1}. This might serve to amplify the ionic peaks of these three elements. If so, the high value of Zn (34 μg l^{-1}) at the bottom would then have to be explained by its remobilization from the sediments due to the depletion of Fe^{2+} and Mn^{2+}. Phosphates trapped within the bottom anoxic water would also precipitate in various forms with iron. The role of bacteria as well as the chemical dynamics of O_2/H_2S interfaces of meromictic lakes have been studied in detail by several authors (Wetzel, 1975; Jörgensen *et al.*, 1979; Postgate, 1979; Cloern *et al.*, 1983; Sanchez *et al.*, 1986) and is beyond the scope of the present study.

Conductivity, salinity, density

The specific conductivity varies from about 1 to 10 mmho cm^{-1} within the mixolimnion, increased up to 100 mmho cm^{-1} in the chemocline and levels out at a maximum of 110 mmho cm^{-1} through the monimolimnion. The salinity and conductivity vertical profiles follow much the same pattern, since both parameters reflect the total soluble salts. Salinity averages about 7‰ within the 0–12 m stratum, reaches the salinity of seawater at 15 m, and keeps increasing to a concentration of 90‰ within the greater part of the monimolimnion. The specific gravity calculated from the 1984 salinity data follow much the same depth trend. The greater density (1.076) at the 32 and 42 m levels in relation to the bottom density of 1.073 is not considered to be significant. The calculation of this variable from the 1983 salinity results or from the average would not show such an inversion. The correction for density difference as a function of the water temperature would be negligible. Except for Sophia Lake (75° 06′ N; 93° 31′ W; Ouellet *et al.*, 1987), such high conductivity, salinity and specific gravity in deep water are unknown to the present authors for Canadian lakes but have been studied extensively over the past two decades in several Antarctic lakes (cf. Williams, 1981; Heywood, 1984; Vincent, 1987).

Major ions

All the major ions also exhibit the three-step water depth profile already described. From the surface to the bottom they all increase in absolute quantities (meq l^{-1}) by more than two orders of magnitude. The dominant ions in the surficial as well as the bottom water are Na^+ and Cl^-, which contribute over 80% of the total. These ions are followed by $SO_4 > Mg > Ca > K$ and HCO_3. Moreover their relative abundance, more specifically within the monimolimnion, is similar to seawater (Table 4, Fig. 4).

Table 4. Ionic ratios of Garrow Lake water compare to seawater

Ratio / Water	Na + K / Ca + Mg	Na / Mg	Na / Ca	Mg / K	Mg / Ca	Ca / K	Cl / CA	SO_4 / Ca	HCO_3 / Na	HCO_3 / K
Ice*	4.1	5.8	13.2	6.2	2.2	2.7	24.5	3.3	0.1	3.3
Mixolimnion	4.7	7.2	14.8	4.8	2.1	2.3	29.6	3.0	0.1	3.5
Monimolimnion	6.2	8.0	25.0	3.9	3.1	1.3	41.7	7.1	0.04	1.1
Sea	6.5	8.3	26.2	3.2	3.1	1.0	47.2	6.6	0.04	0.4

* water from ice surface pool = white ice

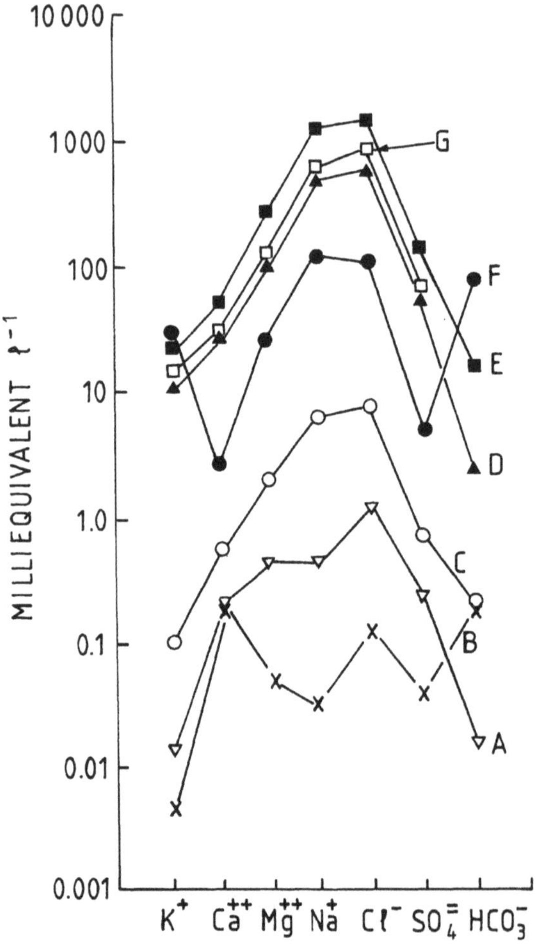

Fig. 4. Comparative major ionic composition of various waters (A = Sophia Lake black ice, Ouellet *et al.*, 1987; B = Char Lake area precipitation, Schindler *et al.*, 1974; C = Garrow Lake white ice; D = sea water; E = Garrow Lake monimolimnion; F = Fryxell Lake, Lawrence, 1982; G = Nurume Lake, Tominaga & Fukui, 1981).

Discussion

Temperature stratification

Snow cover on the smooth wind-blown ice of Garrow Lake is minimal and the clear ice and water of this ultraoligotrophic lake permits light to penetrate to its monimolimnion. The bacterial plate on top of this monimolimnion absorbs nearly all of the light energy penetrating to this depth ($\approx 2\%$). A small part of this light energy is used by the bacteria for photosynthesis, but most is transformed into heat as it is absorbed by the water. Because the water below 12 m does not circulate to the surface, this heat is trapped in the chemocline, resulting in a greenhouse effect. Such greenhouse effects have been observed in lakes at all latitudes. Anderson (1958) described a mesotherm meromictic lake that reached 56 °C in summer. Hattersley-Smith *et al.* (1970) reported a similar phenomenon for three meromictic lakes on northern Ellesmere Island. According to the recent review by Kirkland *et al.* (1983), there are only 30 lakes of this type in the world.

Studies by Ragotzkie & Likens (1964) and Hoare *et al.* (1964) on permanently frozen Antarctic lakes which were covered with about 4 m of snow and ice have demonstrated that in summer, lake surface waters gain heat by net radiation and lose it by direct evaporation and sublimation. For the same lakes, Wilson & Wellman (1962) estimated that between 5 to 8 percent of the incident light energy penetrates the ice to maintain Lake Vanda water temperatures around 7 °C at the chemocline despite a mean annual air temperature of about − 20 °C. Moreover, Goldman *et al.* (1967) have estimated for the same lake that 20% of the surface light penetrates the 4 m ice cover while close to 2% reaches the 40 m water stratum. Garrow Lake's chemocline is heated in the same fashion. In time, a steady state is reached, resulting in a maximum temperature of roughly 9 °C (year round) at the 22 m in level.

The most important heat loss pathway for the 9 °C chemocline stratum is via the mixolimnion and the atmosphere. The transfer of heat within the chemocline (12–20 m) is largely by convection through an upward mean thermal gradient of 0.6 $°C\,m^{-1}$. Heat lost by conduction is of lesser importance. Conduction losses will be primarily through the bottom sediments of the lake. This downward gradient of heat flow is 0.06 $°C\,m^{-1}$ and is remarkably constant from year to year according to our observations and those of Fallis & Harbicht (1980) and of B. C. Research (1981).

Evidence for a permanent ice-free zone below Garrow Lake

The presence of $5 \times 10^6 m^3$ of warm water (7–9 °C) in the bottom of Garrow Lake greatly influences the permafrost characteristics around and beneath the lake. Studies carried out by Livingstone *et al.* (1958) on the origin and development of Alaskan freshwater lakes revealed a net flow of heat from the water to the sediments in lakes too deep to freeze solid in winter. Johnson & Brown (1964) demonstrated that the presence of a small lake less than 300 m in diameter and with a maximal water depth of 2 m was sufficient to maintain a deep thaw zone extending from the bottom to at least 70 m below the lake's bottom. Bedrock was encountered at that depth and drilling was stopped. Permafrost occurred at all depths for holes drilled outside the periphery of the lake. Similar findings were reported by Lackenbruch & Marshall (1969) and Mackay (1979) for lakes having a bottom water temperature of about 4 °C and by geophysical and hydrological studies of the dry valley region of southern Victoria Land, Antarctica (McGinnis *et al.*, 1973, Cartwright & Harris, 1981).

We may deduce from these authors that even if the Cornwallis Island permafrost reaches a depth of 390 m (Misener, 1955), the presence of Garrow Lake water with a mean bottom temperature of 7 °C will prevent the formation of permafrost under this large deep lake. The water acts as an insulating agent and shuts off the penetration of sub-zero temperatures from the surface downwards into the ground, and the energy flow from the water body to the surrounding surface will be sufficient to maintain a thawed ground structure (chimney or talik). From heat conduction theory, Lackenbruch (1957, 1958) concluded that this type of chimney will be present within the permafrost if the minimal horizontal distance of a lake is more than twice as great as the local undisturbed permafrost depth. This convenient rule of thumb applies very well to Garrow Lake.

Geothermal heat

Garrow Lake water is not significantly heated through geothermal heat flow. Heat flow for Cornwallis Island has been estimated to be between 1 and 1.5×10^{-6} cal $cm^{-2}\,s^{-1}$ (Lackenbruch & Marshall, 1969). This amount of energy is equivalent to about 30 cal cm^{-2} $year^{-1}$ and is therefore just sufficient to melt a sheet of ice about 0.3 cm thick. The 6000 cal $cm^{-2}\,y^{-1}$ of solar energy that the Cornwallis Island lakes receive is more than enough to account for the observed heating of the water body (Rigler, 1978). From theoretical consideration of radiation heating for various water depths, Fallis & Harbicht (1980) similarly also concluded that the solar radiation was sufficient to account for the observed thermal profile in Garrow Lake without invoking geothermal heating. Mackay (1979) also concluded that the convective heat transport by groundwater flow has negligible impact in a study dealing with the pingos of the Tuktoyaktuk Peninsula, Western Canadian Arctic. The Antarctic perennially frozen lakes of southern Victoria Land have been described by McKay *et al.* (1985) to be heated the same way; for example the annual mean solar flux incident on the surface of Lake Vanda was estimated to be 1300 times greater than the geothermal heat flux.

The age of Garrow Lake
The hypersaline waters of Garrow Lake's monimolimnion may be due to several mechanisms which must have been effective since the lake became isolated from the open sea during the postglacial uplift. In order to estimate this time interval, surficial loose broken fossil sea shells were sampled from different altitudes (0–57 m) within the Garrow Lake's watershed and dated by the ^{14}C method (Table 5). These dates range from 3 540 to 8 050 ^{14}C years B. P. It is evident, particularly from the zero m level date (3 540 years), based on a surficial shell sample from within the intertidal zone at the mouth of Garrow Creek, that there was a high degree of surface displacement. By assuming that the zero m altitude date should give an approximate ^{14}C age of zero years B.P., then by subtracting the 3 540 years obtained for the actual sea shore limit from the other dates obtained at higher altituded we might estimate the time at which Garrow Lake basin became isolated from the sea. Further, by averaging the three corrected ^{14}C dates (1990, 4490, 3548) for samples from altitudes (5–7 m) close to that one of the Garrow Lake water surface (6.7 m), we obtained a ^{14}C date of 3350 B. P., which in terms of the above hypothesis is quite acceptable. This estimated date is somewhat older than that of 2580 ± 260 years derived by Pagé *et al.* (1984) from the water carbonates of the 25 m stratum but younger than the recently dated 29 and 48 m strata which gave ^{14}C ages of 3850 ± 150 and 4150 ± 150 years B. P. respectively (Pagé unpub.). On account of our ^{14}C dates and the estimated 2600 years B.P. of Fallis & Harbicht (1980), it is worth noting that the bottom waters are older than the time emergence of Garrow Lake shore line (6.7 m A.S.L.).

Table 5. Carbon-14 dating of marine molluskan shells from the Garrow Lake watershed

Field No	Lab. No	Age B.P.	Corrected[a] age B.P.	Altitude (m)
1	QU-1393	3540 ± 90	0	0
2	QU-1390	5530 ± 110	1990	5
3	QU-1385	8030 ± 110	4490	7
4	QU-1392	7220 ± 110	3568	7
5	QU-1386	7370 ± 100	3830	40
6	QU-1391	7530 ± 130	3990	41
7	QU-1387	8050 ± 110	4510	54
8	QU-1388	7540 ± 110	4000	57

[a] = The corrected dates are obtained by subtracting the age of the 0 m level sample

The origin of Garrow Lake water
Several processes might have contributed to the high salinity of Garrow Lake water over the last 4 000 years: trapped seawater, atmospheric origin, physico-chemical and biological precipitation, freezing-out of salts and groundwater influx are among the most probable.

Trapped seawater. In order to elucidate the origin of Garrow Lake, the cationic and anionic composition of its monimolimnic and surficial waters are compared in Fig. 4 to other pertinent water quality data. These include the average (n = 13) summer atmospheric precipitation (rain and snow) of the Char Lake area (Schindler *et al.*, 1974). This lake is located 95 km S.E. of Garrow within a similar physical and geological environment, and its precipitation quality is assumed to be quite representative of the southern part of Cornwallis Island. Moreover, we have also plotted the ionic water composition of a dry black ice sample taken at the one metre level within the two m Sophia Lake (75° 06′ N; 93° 31′ W, Ouellet *et al.*, 1987) ice cover, together with values for typical seawater composition (Riley & Skirrow, 1975) and those of lakes Nurume (69° 14′ S; 39° 40′ E, Tominaga & Fukui, 1981) and Fryxell (75° 35′ S; 143° 35′ E, Lawrence, 1982). According to these latter authors Lake Nurume has been shown to be a relict lake of marine origin while Lake Fryxell has been precipitating calcite and aragonite for more than 20000 years.

The main origin of the bottom of Garrow Lake is clearly indicated by the close correlation of its ionic curve (Fig. 4, curve E) with that of typical marine water (Fig. 4, D; r = 0,9946, P ≤ .001) as well as that of Lake Nurume (Fig. 4, G; r = 0.9993, P ≤ .001). The relative composition of its surficial water (curve C) is also similar ($r = 0.9365$, $P \leq .001$) to sea water but the concentration for each individual ion is lower by more than two orders of magnitude. On the other hand, the relative ionic composition of the mixolimnion (not shown) is also the same ($r = 0.9365$, $P \leq .001$) but the absolute concentrations are intermediate.

Atmospheric origin. Although sea spray has been advocated by several authors (Thomas, 1965; Wilson, 1979) to explain the origin of salty Antarctic lakes, this mechanism could not apply to Garrow Lake. Probably because of the absence of an extensive snow cover inland, the atmospheric ionic inputs of the southern edges of the Cornwallis islands (Fig. 4, B) are dominated by Ca^{2+} and HCO_3^- and moreover are deficient in Na^+. The long period of arctic sea ice cover and the presence of summer packed ice as well as the light winds typical of the short summer period are likely to minimize the generation of sea spray. If this mechanism were important, the ionic composition of the white ice of Garrow Lake would be highly correlated with that atmospheric precipitation, but the correlation coefficient is only 0.0066 ($P > .001$).

The high salt content (> 5‰) that might have resulted from atmospheric precipition and/or sea spray as well as from evaporation and sublimation (Yoshida *et al.*, 1970) was not found at the surface of the Garrow Lake ice cover, which lasts nearly 11 months per year. In order to contribute to the origin of the hypersaline bottom water, the salinity generated by these processes in the white ice would have to be at least 10 times greater. A longer residence time, favored by a greater specific density within the mixolimnion, might then have allowed the transfer of some of these ions to the monimolimnion via physico-chemical or biological precipitation. From the present discussion we have to conclude that because of the existence of a strong density gradient, the meteoritic water entering the watershed is rapidly flushed out of the lake surface with the spring melt. This was conclusively shown by Ouellet *et al.* (1987) for Lake Sophia (Fig. 1).

Physico-chemical and biological precipitation. From our present data, which unfortunately do not deal with the mineralogy of the water particulates or the sediments, it is rather difficult to quantify the role played by the precipitation-dissolution mechanism in the origin of the high saline water of the monimolimnion. On the other hand, it is also evident that any important precipitation of major ions from the water column would result in a significant alteration of the ionic ratios of the solute. Significant precipitation in the mixolimnion followed by sedimentation through the chemocline and a subsequent dissolution, wholly or in part, in the monimolimnion or at the sediment-water interface should amplify the alteration of the major ionic ratios of the water constituents. The great similarity between the ionic composition of Garrow Lake water with those of the sea and Lake Nurume has already been discussed. But when these are compared to the curve of Lake Fryxell (Fig. 4, F), their ionic ratios differ markedly. Although Fryxell is of marine origin, it has been conclusively demonstrated by mineralogical studies (Lawrence, 1982) to have precipitated calcite and aragonite for the last 20000 years, which is, according to the above discussion, not the case for Garrow Lake.

Freezing out of salts and groundwater influx. The freezing-out of salts from below the lake's basin is probably the major factor responsible for its hypersalinity. Pore sea water trapped, before and during the marine submergence, within the sediments below Garrow Lake's watershed would have been progressively frozen during isostatic rebound. As the island uplifted and the sea receded, permafrost would have encroached into the lake's watershed, freezing the pore water as it went (Fig. 5). As discussed above, the presence of a chimney (talik) below the insulating water layer would favor the displacement of the underground frozen-out water salts into the bottom of Garrow Lake.

On a small scale, the propagation of a permafrost front can be regarded as a progressive freezing of the pore water during isostatic rebound. The rate of advancement will be roughly

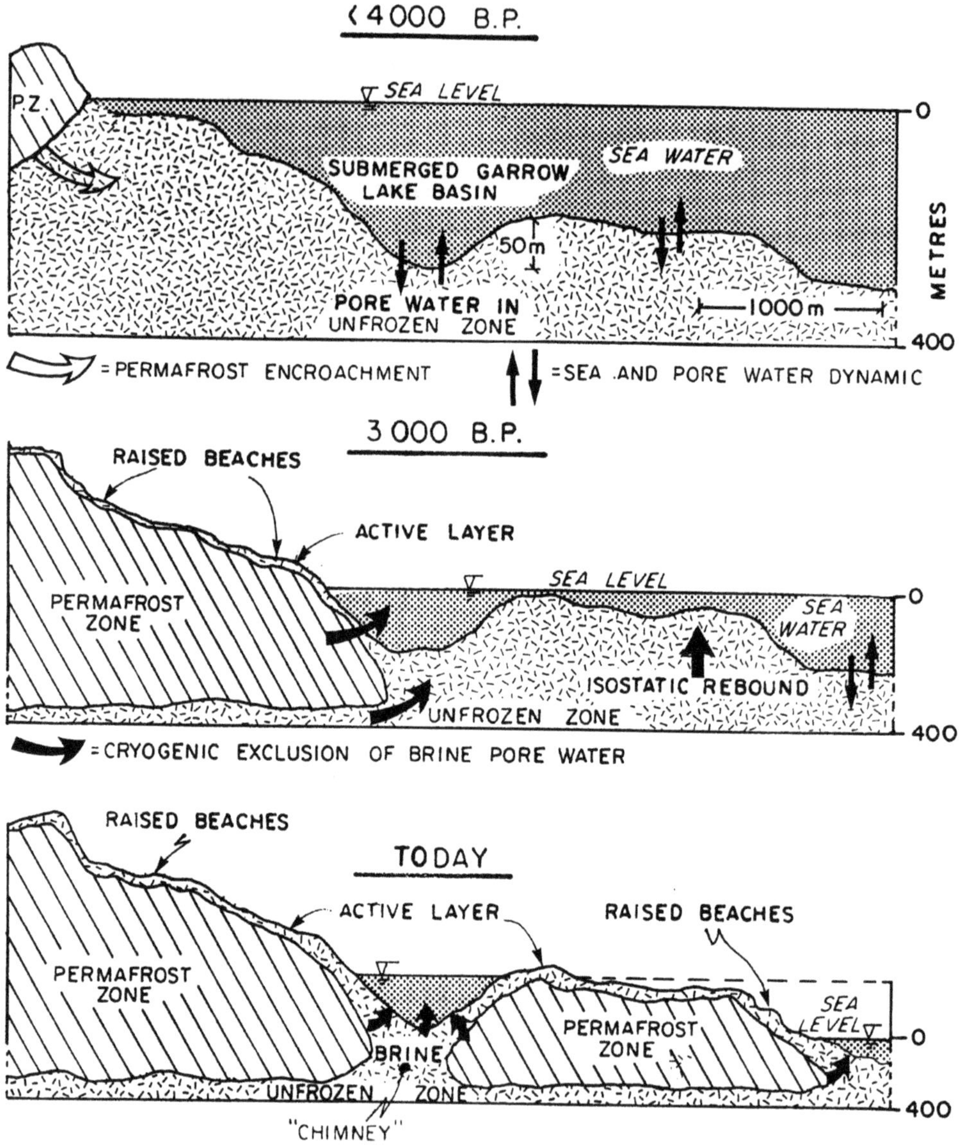

Fig. 5. Hypothesis for underground frozen-out pore water contributions to the hypersaline waters of Garrow Lake (Modified from Dickman & Ouellet, 1983.

proportional to the rate of uplift, the air temperature, the porosity of the glacial debris, the bedrock and the availability of pore water (Menzies, 1981). As the freezing-out of salts occurs at the 'front', there would follow a volumetric expansion from the advance of the permafrost wedges (Khakimov, 1957; Tsytovich, 1981). This volumetric expansion would act to force the accumulated brines into the unfrozen chimney surrounding the basin of Garrow Lake. Because of underground pressures, salinity and the insulating capacity of the ground, the Garrow Lake freezing-out process, taking place within the ground, has most probably occurred at only a few degrees below 0 °C (O to − 8). This is unlike the much lower temperatures advocated by several authors (cf. Burton, 1981) to explain the origin of Antarctic dry oasis saline lake waters whose ion ratios, because of cryogenic fractionation taking place at air-water interfaces, are very different from those in the sea. This Garrow Lake melting-freezing temperature range corresponds to the temperate ice of Liboutry (1971). Thus, according to the work of Thompson and Nelson (1956), the original ionic ratio of the resulting brine would be maintained, which is not the case when freezing-out takes place at much lower temperatures (− 20 to − 45 °C). The under-representation of Na^+ in the surface black ice of Sophia Lake (Fig. 4, curve A) is a good example of this ion segregation freezing-out mechanism at low sub-zero temperatures.

The suggested origin of Garrow Lake salt water is also supported by the work of Howard & Prescott (1973) on Alaskan lakes and Mackay (1979) on the origin of the pingos of the Tuktoyaktuk Peninsula, Western Canadian Arctic. The concentration of the unfrozen brines under two shallow (2–3 m) lakes and under the pingos was 10 to 20 times higher than the unfrozen lake waters but with little change to their ionic ratios. Studies done on similar antarctic lakes by Harris *et al.* (1979) Cartwright & Harris (1981) and Harris & Cartwright (1981) are also supporting our salt groundwater influx origin. Based on these facts we believe that the same freezing-out process is largely responsible for the origin of Garrow Lake monimolimnion hypersaline water.

The complete isolation of the lake basin from the paleo-intertidal zone happened over several hundred years and surficial runoff diluted by meteoritic water would be preferentially flushed out of the 'lake' surface by tidal currents. Moreover, according to the work of Sadler & Serson (1981), who described freshwater anchor ice, below the low tide level along the S.W. beach of Cornwallis Island, the water flowing within the surface active layer is of very low salinity ($\leq 0.2‰$). This influx of salts to the monimolimnion from surface runoff or from within the surficial permafrost active layer (Wilson, 1981), as well as from the cryogenic concentration of ions from the multi-annual lake icecover formation (Goldman *et al.*, 1967) were probably of minor importance. Such surficial phenomena (Rigler, 1978; Welch & Bergmann, 1985) would then have given rise to bottom water more depleted in ^{18}O (Pagé *et al.*, 1984, 1987; Mitchel, 1986) and the ion ratio signatures of the watershed (Ca^{2+}, Mg^{2+}, HCO_3^-) would consequently be traceable within the deep stratum. No evidence of such results were found at all (Table 4).

Along the same line, Jones & Ouellet (1983a, 1983b) have shown for several large and deep lacustrine systems that the physico-chemical dynamic of lake surface cryogenic processes induced chemical changes in a very thin ice-water stratum of little duration and of minor importance. They also conclusively established that ions of these chemical reservoirs (ice covers) are even migrating upward under the influence of various meteorological episodes as well as by capillary action and intercrystalline transfer. These authors and others (Barica & Armstrong, 1971) have also clearly pointed out that for the same system the ionic exclusion factors are quite variable in time and space and their averages are much more important for SiO_2, Ca^{2+} and Na^+ relative to K^+ and Mg^{2+}. If this surficial mechanism was the origin of the bottom brine our typical sea-water monimolimnion ionic ratios ($r = 0.999$, $P < 0.001$) should then be also strongly upset.

Moreover, and as discussed earlier, the greater ^{14}C age (> 1 000 years) of the bottom water of Garrow Lake compared to the estimated age of its postglacial isostatic separation of the sea is not in contradiction with our groundwater salt influx hypothesis. Trapping of seawater did not take place only within the lacustrine depression but also at an earlier time within the still unfrozen submerged underground of the region (pore seawater). These above facts invalidate all surficial salt sources such as were recently speculated by Stewart & Platford (1986), (cf. Ouellet & Pagé, 1987).

The influx of salts by deep groundwater to

Garrow Lake from geologically old evaporite deposits is also probably negligible. In relation to the Polaris mine exploration, several deep boreholes have been drilled and no salt formations have been encountered (Cominco, pers. commun.). As a matter of fact, in 1984, ground ice obtained from the mine gallery (250 m below ground level) was shown by Pagé *et al.* (1987) to be less concentrated in chloride (19 mg l^{-1}) by a factor of 2500 times relative to the monimolimnion (46300 mg l^{-1}). This is interpreted as being the residual cryogenic byproduct of the brine ionic exclusion generated by the permafrost growth. On the other hand, according to Kerr (1977a, b) the base of the Bay Fjord Formation (Lower Middle Ordovician) contains some gypsum and halite located several thousand meters below sea level. If the aquifers were chemically strongly influenced by old evaporites and/or sedimentary rocks before its finds its way to the unfrozen chimney beneath the lake's monimolimnion, the water ionic ratios would then be expected to deviate substantially from those of seawater. On the other hand, its $\delta^{18}O$ values, which have been shown by Pagé *et al.* (1984, 1987) to vary between − 21 to − 8‰, would cluster around + 10‰ like those of deep brines (Fritz & Frape, 1982) and would appear closer to the global meteoric water line.

These saline meromictic lakes offer a unique set of opportunities to investigate several polar limnological factors. The simplicity of their food chains as well as their unusual environmental conditions rank these among the best field laboratories which should continue to yield important new insights in order to understand aquatic systems. It is the hope of the present authors that such information will guide future generations of decision makers and water quality managers regarding the wise development of our Polar resources.

Acknowledgments

We wish to thank Dr. P. Lasalle of the Quebec Ministry of Natural Resources for C-14 dates. The manuscript was finalized by the first author while on sabbatical leave at Waikato University, New Zealand, where facilities were provided by the School of Science through Drs. J. D. Green and M. A. Chapman, to whom we are grateful.

Financial support and technical field assistance were provided by the Natural Science and Engineering Research Council of Canada, by the Quebec and Ontaria Ministries of Education as well as by the Canadian Department of Energy, Mines and Resources through its Polar Continental Shelf Programme and the COMINCO.

References

Anderson, G. C., 1958. Some limnological features of a shallow meromictic lake. Oceanogr. 3: 57–68.

A. P. H. A., 1981. Standard methods for the examination of water and wastewater, 15th edn. Am. Pub. Health Ass., Washington, DC, 1134 pp.

Barica, L. & F. A. J. Armstrong, 1971. Contribution by snow to the nutrient budget of some small northwest Ontario lakes. Limnol. Oceanogr. 16: 891–899.

B. C. Research, 1978. Polaris Mine aquatic environmental studies 1977. Unpub. rep. prepared for Cominco by B. C. Research, 3650 Westbrook Mall, Vancouver, Canada, V6S 2L2.

B. C. Research, 1981. Polaris Mine aquatic studies 1980, Garrow Lake, Pullen Strait. Unpub. rep. prepared for Cominco by B. C. Research, 3650 Westbrook Mall, Vancouver, Canada, V6S 2L2.

Burton, H. R., 1981. Chemistry, physics and evaluation of Antarctic saline lakes. Hydrobiologia 82: 339–342.

Cartwright, K. & H. J. H. Harris, 1981. Hydrology of the Dry Valley Region, Antarctica. In L. D. McGinnis (ed.), Dry Valley Drilling Project. Antarct. Res. Ser. 33: 197–214. Am. geophys. U., Washington, DC, 455 pp.

Cloern, J. E., Cole, B. E. & R. S. Oremland, 1983. Autotrophic processes in meromictic Big Soda Lake, Nevada. Limnol. Oceanogr. 28(6): 1049–1061.

Carter, J. C. H., 1965. The ecology of the calanoid copepod *Pseudocalanus minutus* Kroyer in Tessiarsuk, a coastal meromictic lake of north Labrador. Limnol. Oceanogr. 10(3): 345–353.

Cook, F. A., 1960. Periglacial geomorphological investigation at Resolute, 1959. Arctic 13(2): 132–135.

Dickman, M. & M. Ouellet, 1983. Limnological aspects of meromictic Lake Garrow. Québec Univ (INRS-Eau), Rep. 77: 85 pp.

Edlund, S. H., 1982. Vegetations of Cornwallis and adjacent islands, district of Franklin: relationships between vegetation and surficial material. Geol. Surv. Can., Ottawa, file 857.

Fallis, B. W. & S. M. Harbicht, 1980. A preliminary study of the limnology and biology of Garrow Lake, N. W. T. Tech. Rep., Western Region, Dep. fish. Oceans, Winnipeg, Canada.

Findenegg, I., 1935. Limnologische Unersuchungen im Karntner Seengebiete Ein Beitrag zur Kenntnis des Stoffhaushaltes in Alpenseen. Int. Revue ges. Hydrobiol. Hydrogr. 32: 369–423.

Fritz, P. & K. S. Frape, 1982. Comments on the ^{18}O, ^{2}H, and chemical composition of saline groundwaters on the Canadian Shield. In E. C. Perry, Jr. and C. W. Montgomery (eds), Isotopes studies of hydrologic processes. Univ. Illinois Press, DeKalb, USA: 57–63.

Goldman, C. R., D. T. Mason & J. E. Hobbie, 1967. Two Antarctic desert lakes. Limnol. Oceanogr. 12: 295–310.

Harris, H. J. H., K. Cartwright & T. Torii, 1979. Dynamic chemical equilibrium in a polar desert pond: A sensitive index of meteorological cycles. Sciences 204: 301–303.

Harris, H. G. H., K. Cartwright, 1981. Hydrology of the Don Juan Basin, Wright Valey, Antarctica. In L. D. McGinnis (ed.), Dry Valley Drilling Project. Antarct. Res. Ser. 33: 161–184. Am. geophys. U., Washington, DC, 465 pp.

Hattersley-Smith, G. & H. Serson, 1964. Stratified water of a glacial lake in Northern Ellesmere Island. Arctic 7: 108–111.

Hattersley-Smith, G., J. E. Keys, H. Serso & J. E. Mielke, 1970. Density stratified lakes in Northern Ellesmere Island. Nature 225: 55–56.

Heywood, R. B., 1972. Antarctic limnology: A review. Brit. Antarct. Sur. Bull. 29: 35–45.

Heywood, R. B., 1977. Antarctic freshwater ecosystems: a review and synthesis. pp. 801–808. In G. A. Leano (ed.), Adaptation within Antarctic ecosystems. This Symposium on Antarctic Research, Washington, Scientific Commettee for Antarctic Research Washington, DC.

Heywood, R. B., 1984. Antarctic inland waters. In R. M. Laws (ed.), Antarctic ecology. Academic Press, Lond.: 279–344.

Hoare, R. A., K. D. Popplewell, D. A. House, R. A. Henderson, W. M. Pebble & A. T. Wilson, 1964. Lake Bonney, Taylor Valley, Antarctica: a natural solar energy trap. Nature 202: 886–888.

Howard, H. H. & G. W. Prescott, 1973. Seasonal variation of chemical parameters in Alaskan Tundra lakes. Am. Midl. Nat. 90: 154–164.

Hutchinson, G. E., 1937. A contribution to the limnology of arid regions. Trans. Commun. Acad. Arts. Sc. 33: 1–132.

Hutchinson, G. E., 1957. A treatise on Limnology, 1. J. Wiley & Sons, NY, 1115 pp.

Jeffries, M. O., H. R. Krouse, M. A. Shakur & S. A. Harres, 1984. Isotopes geochemistry of stratified lake 'A', Ellesmere Island, N. W. T., Canada. Can. J. Earth Sci. 2(9): 1008–1017.

Johnson, G. H. & J. E. Brown, 1964. Some observations on permafrost distribution at a lake in the Mackenzie Delta, N. W. T., Canada. Arctic 17: 163–175.

Jones, H. G. & M. Bisson, 1980. Le Lac Laflamme: qualité de l'eau et de la neige (hiver 1979–1980). Sci. Rep. 123, INRS-Eau, Québec Univ., Sainte-Foy, Québec. 91 pp.

Jones, H. G. & M. Ouellet, 1983a. Mécaniques de translocation de matière chimique et microbiologique dans la couverture de glace de quelques lacs du Québec. Eau du Québec 16(1): 71–80.

Jones, H. G. & M. Ouellet, 1983b. Caractérsitiques physicochimiques et microbiologiques de la couverture de glace de quelques lacs du Québec. Eau du Québec 16(2): 174–183.

Jörgensen, B. B., J. G. Kuenen & Y. Cohen, 1979. Microbial transformation of sulfur compounds in a stratified lake (Solar Lake, Sinai). Limnol. Oceanogr. 24(5): 799–822.

Kerr, J. W., 1977a. Cornwallis fold belt and the mechanism of basement uplift. Can. J. Earth Sci. 14: 1374–1401.

Kerr, J. W., 1977b. Cornwallis lead-zinc district, Mississippi Valley-type deposits controlled by stratigraphy and tectonics. Can. J. Earth Sci. 14: 1402–1426.

Khakimov, K. R., 1957. Voprosy teorii i praktiki iskustvennogo zamorzhivaniya gruntov Moscow, Izdatel'stvo Akademii Nauk SSSR (English translation) Artificial freezing of soils; theory and practice. Jerusalem, Israël Program for Scientific Translations, 1966.

Kirkland, D. W., J. I. Bradbury & N. E. Dean, 1983. The heliothermic lake – a direct method of collecting and storing solar energy. Arch. Hydrobiol./Suppl. 65(1): 1–40.

Lackenbruch, A. H., 1957. Thermal effects of the ocean on permafrost. Geol. Soc. Amer. Bull. 68: 1515–1529.

Lackenbruch, A. H., 1958. Permafrost. In R. W. Fairbridge (ed.), Encyclopedia of geomorphology, Reinhold Book, NY: 833–839.

Lackenbruch, A. H. & B. V. Marshall, 1969. Heat flow in the Arctic. Arctic 22: 300–311.

Lawrence, M. J. F., 1982. Origin and occurrence of Antarctic lacustrine carbonates, with special reference to Lake Fryxell, Taylor Valley. M. Sc. thesis, Waikato Univ., Hamilton, New Zealand, 249 pp.

Liboutry, L., 1971. Permeability, brine content and temperature of temperate ice. J. Glaciol. 10: 15–29.

Livingstone, D. A., K. Jr Bryan & R. G. Leahy, 1958. Effects of an arctic environment on the origin and development of freshwater lakes. Limnol. Oceanogr. 3(2): 192–214.

Mackay, J. R., 1979. Pingos of the Tuktoyaktuk Peninsula Area, Northwest Territories. Geogr. Phys. Quaternaire 33(1): 3–61.

McKay, C. P., G. P. Clow, R. A. Wharton & S. W. Squyres, 1985. Thickness of ice on perennially frozen lakes. Nature 313: 561-562.

Mcginnis, L. D., K. Nakao & C. C. Clark, 1973. Geophysical identification of frozen and unfrozen ground, Antarctica. Proc. 2nd int. Conf. Permafrost. Nat. Acad. Sci., Washington, DC: 136–146.

Maxwell, J. B., 1981. Climatic regions of the Canadian Arctic Islands. Arctic 34(3): 225–240.

Mclaren I. A., 1961. The hydrography and zooplankton of Ogac Lake, a landlocked fiord on Baffin Island. Ph. D. thesis, Yale University 171 pp.

Menzies, J., 1981. Freezing fronts and their possible influence upon processes of subglacial erosion and deposition. Ann. Glaciol. 2: 52–56.

Michel, F. A., 1986. Isotopic geochemistry of frost-blister ice, North Fork Pass, Yukon, Canada. Can. J. Earth Sci. 23: 543–549.

Misener, A. D., 1955. Heat flow and depth of permafrost at Resolute, Cornwallis Island, N. W. T., Canada. Trans. am. geophys. U. 36: 1055–1060.

Northcote, T. C. & T. O. Halsly, 1969. Seasonal changes in the limnology of some meromictic lakes in southern British Columbia. J. Fish Res. Bd Can. 21: 1703–1787.

Ouellet, M., M. Bisson, P. Pagé & M. Dickman, 1987. Physico-chemical limnology of the meromictic saline Lake Sophia Canadian Arctic Archipelago. Arct. Alp. Res. 19(3): 305–312.

Ouellet, M. & P. Pagé, 1987. Comments on 'Hypersaline gradients in two Canadian High Lakes' by K. M. Stewart & R. F. Platford. Can. J. Earth Sci. 44(9): 545–553.

Pagé, P., J. Michaud, M. Ouellet & M. Dickman, 1987. Composition and origin of lacustrine brines in the Arctic. Can. J. Earth Sci. 24: 210–216.

Pagé, P., M. Ouellet, C. Hillaire-Marcel & M. Dickman, 1984. Isotopic analyses (^{18}O, ^{13}C, ^{14}C) of two meromictic lakes in the Canadian Arctic Archipelago. Limnol. Oceanogr. 79(3): 554-573.

Postgate, J. R., 1979. The sulfate reducing bacteria. Cambridge Univ. Press, 247 pp.

Ragotzkie, R. A. & G. E. Likens, 1964. The heat balance of two antarctic lakes. Limnol. Oceanogr. 9: 412–425.

Rigler, F. H., 1974. Report of the feasibility of using Frustration Lake as a water supply for Arvik Mine. Unpub. Rep. Cominco, Vancouver, 87 pp.

Rigler, F. H., 1978. Limnology in the High Arctic: A close study of Char Lake. Verh. int. Ver. Limnol. 20: 127–140.

Riley, J. P. & G. Skirrow, 1975. Chemical oceanography Vol.3, 2nd edn, Academic Press, NY, 564 pp.

Sadler, H. E. & H. V. Serson, 1981. Fresh water anchor ice along an Arctic beach. Arctic 34(1): 62–63.

Sanchez, A. L., J. T. Murray, W. R. Schell & L. G. Miller, 1986. Fallout plutonium in two oxic-anoxic environments. Limnol. Oceanogr. 31(5): 1110–1121.

Schindler, D. W., H. E. Welch, J. K. Kalff, G. J. Brunskill & N. Kritsch, 1974. Physical and chemical limnology of Char Lake, Cornwallis Island (75°N lat.). J. Fish. Res. Bd Can. 31: 585–607.

Stephens, K., 1963. Determination of low phosphate concentrations in lake and marine waters. Limnol. Oceanogr. 8(3): 361–362.

Stewart, K. M. & R. F. Platford, 1986. Hypersaline gradients in two canadian High Arctic Lakes. Can. J. Fish aquat. Sci. 43: 1795–1803.

Technicon, 1972. Sulfate in water and wastewater. Industrial method no. 118–71W. Technicon Industrial Systems, Tarrytown, NY 10591, USA.

Technicon, 1973. Chloride in water and wastewater. Industrial method no. 99–70W. Technicon Industrial Systems, Tarrytown, NY 10591, USA.

Tessier, A., P. G. C. Campbell & M. Bisson, 1979. Sequential extraction procedure for the speciation of particulate trace metals. Analyt. Chem. 51(7): 844–851.

Thomas, C. W., 1965. On populations in Antarctic melt water pools. Pacif. Sci. 19(4): 515–541.

Thompson, T. G. & K. H. Nelson, 1956. Concentration of brines and deposition of salts from sea water under frigid conditions. Am. J. Sci. 254(4): 227–238.

Thornsteinsson, R., 1958. Cornwallis and Little Cornwallis Island, District of Franklin, North West Territories. Geol. Surv. Can. Mem. 294: 1–134.

Thornsteinsson, R. & J. W. Kerr, 1968. Cornwallis Island and adjacent smaller Islands, Canadian Arctic Archipelago. Geol. Surv. Can. pap. 67–64: 1–16.

Tominaga, H. & F. Fukui, 1981. Saline lakes at Syowa Oasis, Antarctica. Hydrobiologia 82: 375–389.

Tsytovich, N. A., 1981. The mechanism of frozen ground. McGraw-Hill serials in modern structure. G. K. Swinson (ed.), McGraw-Hill, NY, 416 pp.

Vincent, W. F., 1987. Extreme aquatic environments II. Antarctic ecosystems. In Viner A. B. (ed.) Inland Waters of New Zealand. DSIR, New Zealand. 379–412.

Walker, K. F. & G. E. Likens, 1975. Meromixis and a reconsidered typology of lake circulation patterns. Verh. int. Ver. Limnol. 19(1): 442–458.

Wallace, W. J., 1974. The development of the chlorinity-salinity concept in oceanography. Elsevier sci. Pub. Co., NY, 227 pp.

Washburn, A. L. & M. Stuiver, 1984. Radiocarbon dates Cornwallis Island area, Arctic Canada-an interim repart. Can. J. Earth Sci. 22(4): 630–637.

Welch, H. E. & M. A. Bergmann, 1985. Water circulation in small Arctic lakes in winter. Can. J. Fish aquat. Sci. 42: 506–520.

Wetzel, R., 1975. Limnology. Saunders, Philadelphia, 743 pp.

Williams, P. M., W. H. Mathews & G. L. Pickard, 1961. A lake in British Columbia containing old sea-water. Nature 191: 830–832.

Williams, W. D. (ed.) 1981. Salt lakes. Proceedings of the International Symposium on Athalassic (Inland) Salt Lakes, held at Adelaide, Australia, Oct. 1979. 444 pp. (Reprinted from Hydrobiologia 81/82).

Wilson, T. A., 1979. Geochemical problems of the Antarctic dry areas. Nature 280: 205–208.

Wilson, A. T., 1981. A review of the geochemistry and lake physics of the Antarctic Dry Areas. In L. D. McGinnis (ed.), Dry Valley Drilling Project. Antarct. Res. Ser., 33: 185–192. Am. geophys. U., Washington, DC, 465 pp.

Wilson, A. T. & H. W. Wellman, 1962. Lake Vanda: an Antarctic lake. Nature 196: 1171–1173.

Yoshida, Y., T. Torii & N. Yamogata, 1970. Antarctic saline lakes. In E. Ingerson (ed.), Proceedings of Symposium on Hydrogeochemistry and Biogeochemistry, The Clark Co., Washington, DC: 652–660.

Hydrobiologia 172: 235–254, 1989
W. F. Vincent and J. C. Ellis-Evans (eds) High Altitude Limnology

Meromixis in an antarctic fjord: a precursor to meromictic lakes on an isostatically rising coastline

J. B. Gallagher[1], H. R. Burton[1] & G. E. Calf[2]
[1]*Antarctic Division, Department of Science, Channel Highway, Kingston, Tasmania 7150, Australia*;
[2]*Australian Atomic Energy Commission, Lucas Heights, N.S.W., Australia, 2234*

Key words: antarctic, fjords, hypersaline, lake evolution, meromixis, polar lakes, Vestfold Hills

Abstract

Physico-chemical data and isotopic studies (utilising $H^{14}CO_3$, $H_2{}^{18}O$, ^3H_2O) suggest that hypersaline meromixis in Ellis Fjord (Vestfold Hills, Antarctica) was initiated during the middle Holocene period, when hypersaline brine, excluded during the annual formation of sea-ice, gravitated in a density current to the bottom. The application of this contemporary information to the genesis of the meromictic lakes found today in the Vestfold Hills, suggest that their meromixis may have developed prior to isolation from the sea. Comparison of physico-chemical data from the meromictic basins of Ellis Fjord with that of the Vestfold Hills saline lake allows some determination of their evolutionary pathways initiated before, during and after isolation from the sea. Further evolution of each lake can be explained through the individual interaction between climate, the catchment size and basin morphology.

Introduction

In the ice free oasis of the Vestfold Hills, Antarctica, there are at least 13 saline meromictic lakes in an area of about 400 km^2 (Fig. 1). The presence of such a large proportion of the world's total number of known meromictic lakes (about 150, but still increasing; Walker & Likens, 1975) in such a small area is a result of the geomorphological history and climatic conditions characteristic of the Vestfold Hills (Burton, 1981). These lakes are relict bodies of seawater which occupy early holocene fjord basins now isolated by isostatic uplift of the land after the most recent deglaciation (8 000–10 000 years BP; Adamson and Pickard, 1986a,b). They maintain their meromixis by a large halocline (which protects the lake from wind overturn during the summer) and by ice cover during the winter (Burton, 1981). The ages of these lakes depend on their distances from the sea which controlled the time of exposure from the retreating ice (Adamson & Pickard, 1986a) and range from the early Holocene to the present day.

Meromictic Burton Lake (or lagoon; Fig. 1) is an example of a fjord basin in the process of isolation from the sea. The fjord basin is isolated from the rest of Crooked Fjord for up to 8 months each year by an ice dam formed across the entrance sill (Burke & Burton, in press). It is clear that a further rise of the land in excess of one metre relative to the sea will isolate Burton Lake completely as the entrance sill will then be above high water mark. The salinities in the monimolimnia of Burton Lake and most of the other meromictic lakes are significantly greater than seawater (Burton, 1981). This cannot be due to evaporation as Burton Lake maintains a seasonal connection to the sea; but hypersaline bottom

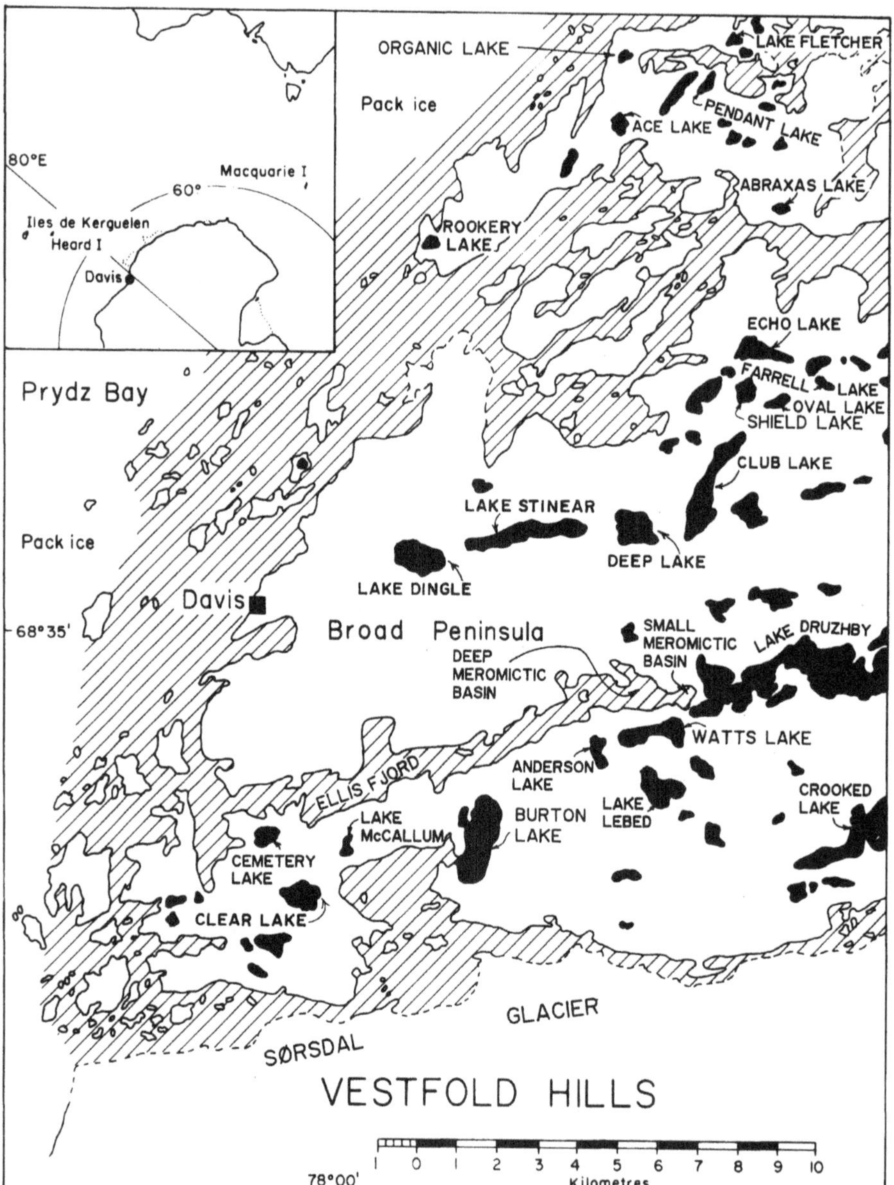

Fig. 1. Map of the Vestfold Hills showing Ellis Fjord, the position of the fjord basins and the saline lakes described in the text.

waters can also be produced by salt freeze-out from a growing ice sheet or from permafrost in high latitude situations. This applies to both lakes and bays (Pagé *et. al.*, 1984; Newbury, 1979; Matsubaya *et al.*, 1979). Hypersaline water has been measured in the bottom of one basin in Ellis Fjord on one occasion (Deep Meromictic Basin, Fig. 1; Burton, 1981) and this suggests that present day lake meromixis could have started in fjords of the middle Holocene before the land has risen sufficiently to isolate the fjord waters from the sea.

The ages and processes of formation of this early stage in the evolution of certain lakes within the Vestfold Hills are obscure. Knowledge of such details is necessary for the understanding, and hence dating, of past climatic changes recorded within the lakes' sediments. This paper provides details of the probable origin and age of the hypersaline waters found in Ellis Fjord.

Study site

Ellis Fjord is situated in the ice-free oasis of the Vestfold Hills, Antarctica (68° 35′ S, 78° 00′ E; Fig. 1). It fills a weakness zone extending east-north-east, which follows the closely-spaced parallel boundaries of several narrow bands of metamorphic rocks (Mossel gneiss and Chelnok supracrustals) (Adamson & Pickard, 1986a). It is this weakness zone which presumably allowed the ice to scour selectively out the fjord as the ice advanced approximately west-south-west across the area. Since then, there has been isostatic uplift of the land relative to the sea (Adamson & Pickard, 1986b).

Figure 2 shows the bathymetry and an idealised longitudinal section through the basins of Ellis Fjord. The fjord is up to 117 m deep and has 6 major basins, two of which to the east of sill D are meromictic and hypersaline with permanently anoxic monimolimnia. The fjord is ice covered for 11 to 12 months of the year and has a freshwater stream at its head (Ellis Rapids), which flows during the summer months for up to 6 weeks. Like the meromictic lakes in the Vestfold Hills, the Deep Meromictic Basin has a very prominent terrace on the north shore (Figs. 2, 5) and it is this basin on which this work is focused. Plate 1 shows the terrace and the general terrain of the Vestfold Hills.

A new term, hypoanachysis, is defined from the Greek, hypo (below), anachysis (estuary), which describes a perennially isolated body of water at the bottom of an estuary. This term is analogous to the word monimolimnion used to describe the lower water mass in permanently stratified lakes. A monimolimnion is a permanently stratified hypolimnion and is the lacustrine equivalent of a hypoanachysis.

Methods

Temperature and salinity profiles were obtained, through ice holes, for seven stations within the Deep Meromictic Basin (Fig. 1) every 1 to 2 days from the beginning to the end of June 1985. Four of these stations (E2, E3, E4, E5, see Fig. 2, Table 1) were further sampled every 1 to 3 weeks until mid January 1986.

Water for ^{3}H and $\delta^{18}O$ analysis (24 November 1985) was collected at the deepest site in the Deep meromictic basin (E4) with a 5 l Niskin bottle and sealed in 3.8 l glass Winchesters. These samples were then stored remotely from luminous sources. The ^{3}H analysis were performed at the Australian Atomic Energy Commission, Lucas Heights, Australia. Water was analysed for ^{18}O at the Glaciology Section, Antarctic Division, Melbourne University, Australia.

Samples for the analysis of alkalinity, pH, sulphate, hydrogen sulphide, potassium, calcium, magnesium and sodium (14 October 1985) were taken with a 5 l Niskin bottle. The pH and alkalinity were measured, immediately after collection from the Niskin bottle, in the field, at 5 °C by titration with 0.1 to 1.0 M hydrochloric acid using a glass combination electrode (Grasshoff, 1983) and a Gran function to calculate the end point (Almgren *et al.*, 1983). Hydrogen sulphide was immediately fixed with zinc acetate gel and determined colorimetrically in the field using a Hitachi model 100 U.V. visible spectrophotometer (Fonselius, 1983). Water for sulphate analysis was fixed by precipitating the hydrogen sulphide with zinc acetate gel and then frozen. The sulphate was analyzed one week later in the laboratory on a Dionex model 10 ion-chromatograph (Franzmann *et al.*, in press). The water for calcium, potassium, sodium and magnesium deter-

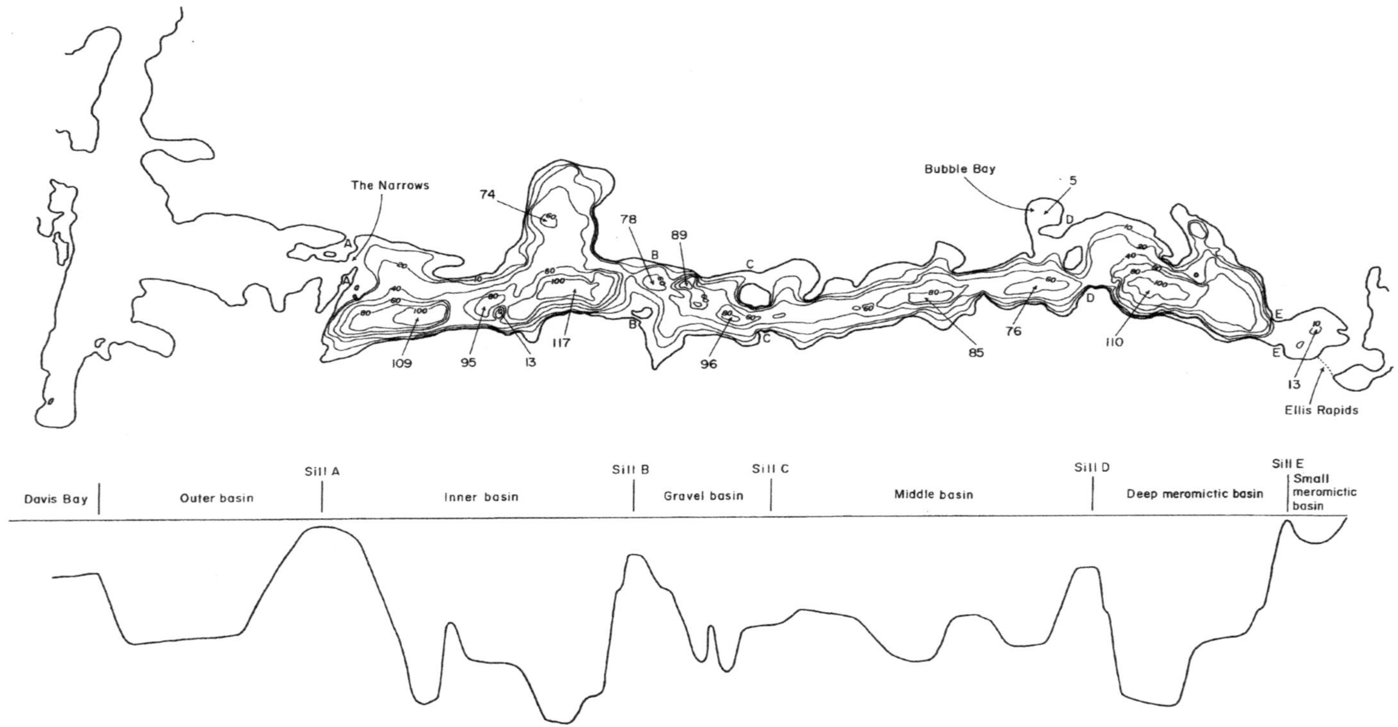

Fig. 2. Bathymetric map and idealised cross-section of Ellis Fjord, showing the position of the basins and sills. Outside contour is 10 m, further contours are then every 20 m.

minations was stored at 4 °C for 6 months and analysed at the Australian Government Analytical Laboratories, Tasmania, by flame atomic absorption after first acidifying the samples with 10% nitric acid.

Water (40 to 80 l) for $^{14}CO_2$ measurements were collected from 60, 80 and 100 m and immediately transferred under nitrogen to an all-glass purging apparatus which had been previously flushed with nitrogen. The water was acidified to pH 1 with 50% v/v sulphuric acid and warmed slowly to 60 °C. The carbon dioxide was purged for 4 hours with nitrogen gas and trapped in carbon dioxide-free concentrated sodium hydroxide solution. The precipitate of sodium carbonate was then filtered and dried under nitrogen and stored in sealed high density polyethylene containers. The ^{14}C activity was measured at the Australian Atomic Energy Commission, Sydney, Australia. The results are given with respect to 95% of the activity of NBS oxalic acid. Water for $\delta^{13}C$ measurements was collected in Niskin bottles, sealed in 500 cm^3 bottles and analysed at the Australian Atomic Energy Commission, N.S.W., Australia. Similarly, live *Laternula* elliptica King and Broderip shells were collected (by divers) in the Deep meromictic basin (10 m) and the ^{14}C activity and $\delta^{13}C$ ratio measured at the Australian Atomic Energy Commission.

Tidal measurements at Davis Bay and Ellis Fjord were measured throughout 1985 with an Applied Microsystems Model TG-12A pressure transducer tide gauge deployed in about 6 m of water. Tidal measurements in Burton Lake and immediately outside the lake in Crooked Fjord were surveyed against a common bench mark.

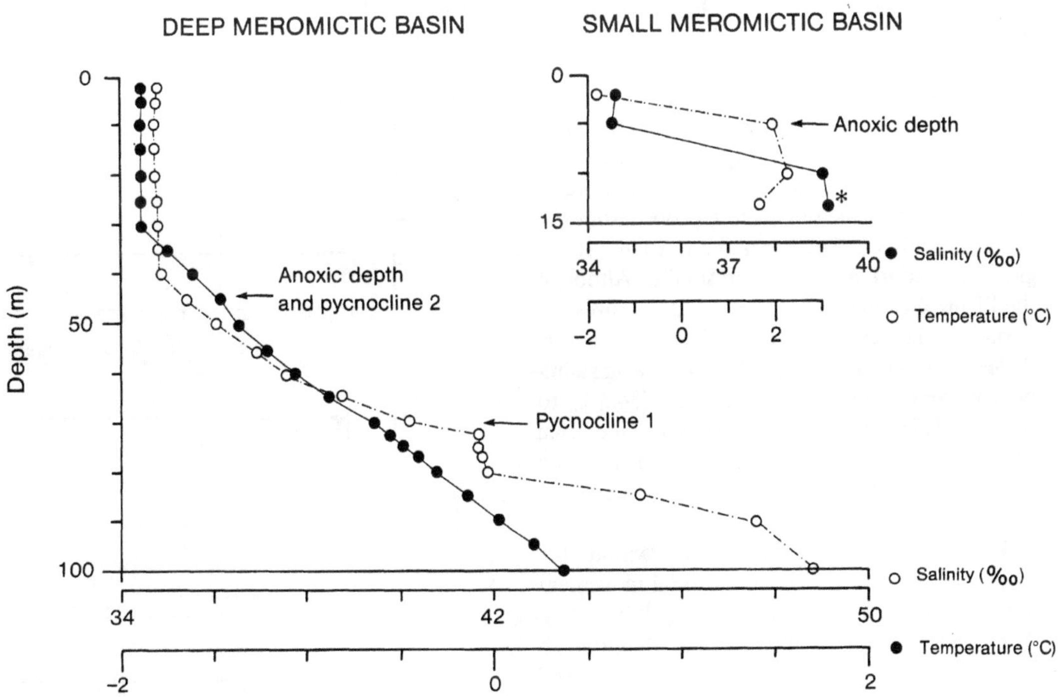

Fig. 3. Salinity and temperature curves for sampling sites E4 and E1 in the Deep and Small Meromictic Basins respectively. * denotes sample from January 1986.

Results and discussion

Meromixis in Ellis Fjord

Figure 3 shows depth profiles of salinity and temperature (mid June) at stations E4 and E1 within the Deep Meromictic Basin (DMB) and the Small Meromictic Basin (SMB) respectively. The large salinity gradients and the relatively small temperature gradients resulted in the density profiles of the water being controlled by the salinity (Fig. 3). The depth of pycnocline 2 bordering the hypoanachysis was between 45 m to 50 m in the DMB and was found to be variable in the SMB (about 5 to 10 m) over an annual cycle. The water at 45 m in the DMB also corresponded with the depth of the anoxic/oxic interface, measured by the appearance of hydrogen sulphide (Table 2). The bottom water temperature and salinity of the DMB did not significantly change throughout the 1985/86 winter to mid-summer season (Fig. 4). The observed variations in salinity from 50 m to the bottom were attributable to inaccuracy caused by tidal movement of the sampling platform relative to the basin bottom while operating in relatively large salinity gradients (0.5‰ m^{-1}). Further error could be attributed to precipitation of sulphur when samples were brought to the surface for measurement. The salinities of the oxic water at 40 m were measured with a probe, from the surface, and in these relatively small saline gradients variations were much smaller. Although the Figure 4 salinity scale somewhat obscures the variations in the oxic waters there was nevertheless still a significant trend of increasing salinity apparent over the whole year (34.3‰ to 34.9‰). This is typical of polar seas, since brine rejection during ice cover formation results in an increase in surface water salinity (Gade *et al.*, 1974).

The variations in temperature within the hypoanachysis were also due to tidal movement within a temperature gradient (0.034 °C m^{-1}). The seasonal water temperatures at 40 m (which is above the pycnocline) illustrates increased variation with respect to the hypoanachysis. Light measurements during mid summer indicated no measurable input below a depth of 50 m (Walker, pers. comm.). This illustrates that there was little advective or radiative input of heat from the surface into the hypoanachysis.

Although, there was no seasonal change of salinity in the hypoanachysis, the 'noise' in the salinity time series of the hypoanachysis may have masked any small variation caused by advective inputs of surface waters into it. The low levels of tritium (at the limit of detection) found in the surface and deep waters (Table 2) neither confirm nor deny advective input of surface waters to the hypoanachysis of today. Tritium has a half life of 12.43 years and thus should not be detected in systems older than 50 years even if there is a significant initial input to the surface waters. The very low levels of tritium in the surface water (Table 2) show that this was not so and therefore tritum cannot be used as a marker in Ellis Fjord.

Origin of the hypoanachysis

Hypersaline water (up to 37‰; Table 1) was measured near the bottom (35 m) at sample site

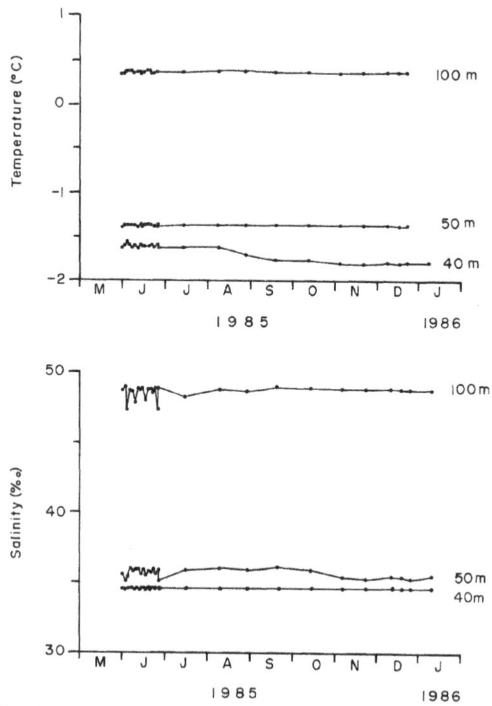

Fig. 4. Salinity and temperature time-series at selected depths in the Deep Meromictic Basin (station E4).

Table 1. Salinities at the stations in the Deep meromictic basin in June 1985

Depth (m)	Transect 1		Peripheral stations				
	E4	S2	E5	S1	E3	S3	E2
	Salinity (‰)						
2	34.4	34.4	34.4	34.4	34.4	34.4	34.5
10	34.4	34.4	34.4	–	34.4	34.4	–
15	34.4	34.4	34.4	–	34.4	34.4	–
20	34.4	34.4	34.4	–	34.4	34.4	–
25	34.4	34.4	34.4	–	34.4	34.4	–
30	34.4	34.8	–	–	34.5	–	–
35	34.5	36.9	–	–	34.7	–	–
40	34.8	–	–	–	34.8	–	–
45	34.8	–	–	–	34.9	–	–
50	34.9	–	–	–	35.4	–	–
60	37.0	–	–	–	36.0	–	–
70	38.1	–	–	–	–	–	–
80	41.9	–	–	–	–	–	–
90	46.9	–	–	–	–	–	–
100	47.4	–	–	–	–	–	–

Sea floor depth: E4 100 m; S2 35.5 m; E5 26 m; S1 2 m; E3 63 m; S3 26 m; E2 2 m.

S2 (Fig. 5) from early June to mid July. Although surface salinities continued to increase, there was no systematic change in the salinity at the bottom of site S2 ($\pm$0.2‰). This water was dense enough to penetrate into the hypoanachysis where water of an equivalent density resided at a depth of about 60 m. The formation of this hypersaline water in the DMB was authigenic as no hypersaline water was found at the entrances to the basin (E5, E2, S1; Table 1). The origin of this dense water must therefore have been brine extruded from the thickening ice sheet during the winter months. Gade *et al.*, (1974) found that such waters are formed, in the absence of large currents, in areas shallower than the well mixed layer. Brine extruded from the thickening ice sheet will tend to form a well-mixed layer, its depth governed by the previous stratification of the water column (Kantha, 1979). If the shallows are not as deep as this well-mixed layer, then brine will increase in salinity above that of the surface waters positioned over the deeper parts of the

Table 2. Physico-chemical determinands at station E4 in the Deep Meromictic Basin

Depth (m)	pH	C Alk (meq)	H_2S	SO_4^{2-}	Ca^{2+}	Na^+	K^+	Mg^{2+}	$\delta^{18}O$ SMOW (‰)	^3H (TU)	^{14}C (% modern carbon pmc)	$\delta^{13}C$ ‰
			(mmol liter^{-1})									
5	8.3	2.52	n.d.	30.0	n.d.	n.d.	n.d.	n.d.	– 1.08	n.d.	n.d.	
10	8.0	2.51	n.d.	n.d.	10.0	478	10.4	56.0	– 1.03	0.7	95.65 ± 0.72	+ 1.68
15	8.2	2.50	n.d.	27.0	n.d.	n.d.	n.d.	n.d.	n.d.	n.d.	n.d.	
20	8.0	2.52	n.d.	26.9	n.d.	n.d.	n.d.	n.d.	– 1.07	0.3	n.d.	
25	8.1	2.51	n.d.	29.3	n.d.	n.d.	n.d.	n.d.	n.d.	n.d.	n.d.	
30	8.1	2.50	n.d.	28.1	n.d.	n.d.	n.d.	n.d.	– 1.07	0.7	n.d.	
35	7.9	2.55	n.d.	31.6	n.d.	n.d.	n.d.	n.d.	n.d.	n.d.	n.d.	
40	7.6	2.41	b.d.	27.1	n.d.	n.d.	n.d.	n.d.	– 0.96	0.8	n.d.	
45	7.6	3.10	0.002	27.8	n.d.	n.d.	n.d.	n.d.	n.d.	n.d.	n.d.	
50	7.4	8.1	1.9	25.4	10.3	491	10.8	57.6	– 1.01	0.3	n.d.	
55	7.3	13.6	7.1	20.3	10.8	504	11.4	59.3	n.d.	n.d.	n.d.	
60	7.3	29.1	10.8	10.0	11.0	517	12.2	61.3	– 1.06	0.4	75.81 ± 0.73	– 6.9 ± 0.1
65	7.3	35.1	14.2	10.0	10.8	535	12.6	63.0	– 1.14	n.d	n.d.	
70	7.3	56.2	n.d.	1.4	10.2	565	13.6	66.3	– 1.27	0.6	n.d.	
75	7.2	58.8	23.3	0.9	9.4	578	13.9	67.1	– 1.32	n.d.	n.d.	
80	7.3	61.1	23.6	0.4	9.4	578	13.9	68.3	– 1.48	0.5	76.13 ± 0.64	– 7.2 ± 0.1
85	7.3	64.1	26.5	0.08	7.9	617	14.9	73.3	n.d.	n.d.	n.d.	
90	7.2	65.1	28.2	0.06	7.8	643	15.5	75.3	– 1.79	0.6	n.d.	
100	7.3	65.0	27.1	b.d.	7.0	643	15.5	77.4	– 1.82	0.2	74.48 ± 0.70	– 7.8 ± 0.1

n.d. = no data; b.d. = below level of detection; TU = Tritium units.

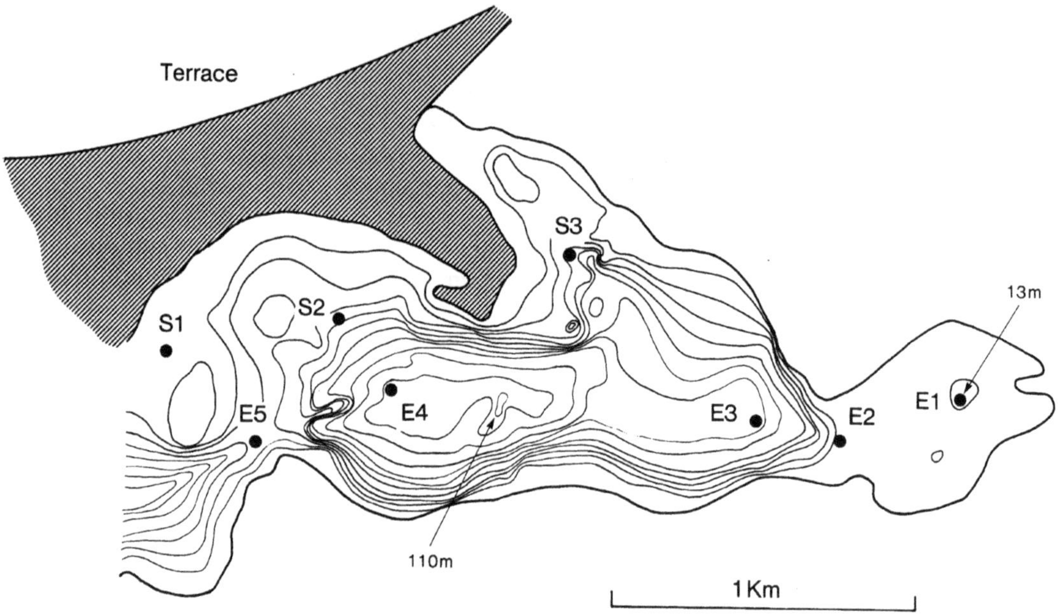

Fig. 5. Bathymetric map of the Deep Meromictic and Small Meromictic Basins in Ellis Fjord, showing the position of the sampling sites. Contours are every 10 m.

basin. In the DMB there are extensive shallows on the north shore (Fig. 2). No hypersaline water was found near the bottom of station S3. This was because of the presence of a 'freshwater' layer directly under the ice, observed during the early part of winter (diver observation). The layer was presumed to be trapped, in the bay, north of station S3, from the previous summer's flow from Ellis Rapids. Salt rejection from the growing ice sheet near station S3 would be relatively unimportant as long as the 'freshwater' layer persisted.

The hypoanachytic water has also been through a freeze stage. $\delta^{18}O$ values in the hypoanachysis paralleled a simple freezing line with salinity (Fig. 6). The fractionation constant is independent of salinity from fresh water to 35‰ sea water (O'Neil, 1968; Craig & Holm, 1968). Therefore it seems reasonable to suppose that little or any change accrued towards 48‰ salinity. The $\delta^{18}O$ values of the hypoanachysis were certainly not as depleted as waters undergoing either evaporation or equilibrium with the atmosphere of the Vestfold Hills (Matsubaya *et al.*, 1979). It seems probable, therefore, that the hypersaline waters in the hypoanachysis were formed gradually over a number of years in a manner akin to crenogenic meromixis as defined by Walker & Likens (1975). There was no significant change in temperature at 55 m in the hypoanachysis and

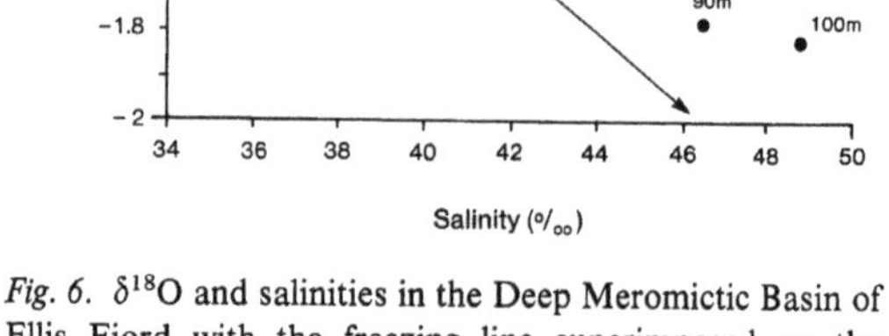

Fig. 6. $\delta^{18}O$ and salinities in the Deep Meromictic Basin of Ellis Fjord with the freezing line superimposed on the data.

this may indicate a slow passage in which the water was partly warmed from the sediment on its descent, as well as during its final equilibrium with the hypoanachytic water, as it was mixed towards the deeper stations. The temperatures at 35 m and 30 m in the shallows (S2) were about − 1.52 °C compared to about − 1.73 °C at 30 to 35 m at the deep middle stations (E4 and E3). The similar physico-chemical and isotopic data for water in the rest of the hypoanachysis below 55 m indicates that it had similar origins. The ratios of the major cations (K^+, Na^+, Mg^{2+}) were also similar to that of sea water.

Salinity of the hypoanachysis

The controlling forces which determine the formation and salinity characteristics of the hypoanachysis can be divided into two groups: those factors which form and maintain the integrity of the hypersaline water and those factors which disperse and destroy it. Factors forming and maintaining the integrity of the hypoanachysis include the large areas of shallow depths, a flat bottom topography in these shallows and a fast rate of growth of the surface ice sheet. Destructive factors are the converse of those above. An unimpeded slope does not allow the build up of high concentrations of salt in the shallows. The steeper the slope, the faster is the advection, and thus the less saline the advected water. The scale of horizontal mixing is also very relevant. The limits of the area of the shallows necessary to maintain the integrity of the hypersaline water lying on it are real, if not easily defined. If the area of mixing is in 100's km^2 (ocean scale) then the hypersaline waters are likely to be dispersed. However, in the DMB, the scale of horizontal dispersion was estimated to be on a scale of a 100 m or less (Gallagher & Burton, in press). Therefore the extent of the shallow region has to be of the same order (10,000 m^2) or greater to maintain the hypersaline waters' integrity.

As the brines flow down the sides of the basin as density currents, they may begin to be diluted with the rest of the fjord waters by breaking internal waves (Perkin & Lewis, 1980). These waves are produced at the sill and propagated along a density interface usually at sill depth. They may be countered by the convective upflow above the descending hypersaline waters. The upflow stops the hypersaline water current from mixing with the rest of the fjord as it moves down the slopes. The phenomenon has been observed in experimental tanks (Huppert *et al.*, 1986).

Tidal mixing is another destructive factor; its magnitude is controlled by net changes in the entrance sill geometry and by net changes in relative sea level (McClimans, 1978). However, in Ellis Fjord, tides are reduced in winter by the presence of ice cover (from a maximum of 1.7 m to 0.5 m; Gallagher, unpublished data).

These creative and destructive factors are to some degree self regulatory. For example, growth in ice becomes increasingly insensitive to changes in temperature when ice cover exceeds about 1.5 m (Allison, 1981). The fast rate of ice growth, and therefore of salt rejection, during the earlier colder periods of winter will also lead to faster transport of salt down towards the bottom by the density currents. The effect is to maintain a relatively constant salinity of the density current but to increase the volume of advected water. This was seen throughout the winter season in an arctic fjord, when the calculated salinity of the density current was constant throughout the period of ice growth (Gade *et al.*, 1974). In summary, the most important controlling agents of the volume and salinity of the hypoanachytic waters in the meromictic basins of Ellis fjord would appear to be the depths and areas of the shallows.

Stability of the hypoanachysis

For the hypoanachysis to maintain itself, it must be able to withstand the mixing forces of brine convection, winds and tides. The salinity of the hypoanachysis must therefore be greater than the maximum winter salinity elsewhere in the fjord (34.98‰) to prevent it being mixed with the rest of the water column through brine convection. Ice cover, while it exists, clearly protects the hypoanachysis from wind mixing. The ice cover of Ellis Fjord can last for the whole year as a total cover; but in some years there can be an open water condition for about 2 weeks during the summer months.

Opposing wind-induced mixing is the inherent stability of the density gradient of the hypoanachysis. The SMB at the head of Ellis Fjord (Fig. 2) is relatively shallow and has a lower hypoanachytic salinity (42‰) than the DMB (48‰). Its stability in the summer is also strengthened by the flow of freshwater from Ellis Rapids along the top of the two meromictic basins. This can produce a surface pycnocline in years when the rapids are flowing quickly (3 $m^3 s^{-1}$) which further protects the deeper hypoanachytic water from wind-induced mixing (Burke per. comm.).

In Burton Lake there is no significant input of freshwater and the melted ice cover is quickly replaced by the incoming seawater (Burke & Burton, in press). However, the inherent stability of its monimolimnion is enough to inhibit complete overturn. It seems that the winter input of hypersaline water replaces its small loss into the mixolimnion from wind mixing during the previous summer (Burke & Burton, in press). Tidal energy has also been shown to be important in mixing ice-covered fjord basins (Perkins & Lewis, 1980). Mixing brought about by tidal energy may indeed govern the depth of meromixis. For this reason, it is worth considering the factors which control the amount of energy which can be extracted from the tide for fjord circulation and mixing.

In short fjord inlets with a sill length/depth ratio of approximately 50 (e.g. Ellis Fjord and Burton Lake), the maximum energy flux is obtained when the inlet geometry is tailored to produce an internal tidal range 0.707 times the external tidal range (the choking coefficient; McClimans, 1978). In Ellis Fjord the internal tidal range during the summer is 0.69 times the external tide (Fig. 7), which is close to the optimum value. In Burton Lake, the internal tide is about 0.22 of the external tide (Fig. 7). It would be hard to judge the amount of energy loss in the entrance to Burton Lake where the low tide mark is beneath maximum sill depth. However, at some time in the past when the sea level was higher relative to the land, the cross-sectional area of the sill would have been larger and this would have had a choking coefficient closer to the optimum (McClimans, 1978). Tides would therefore have played a more important role in regulating the meromixis than they do today. There is, at present, not enough information to calculate the energy input from tides and wind for mixing the water column of Burton Lake.

The large range of hypolimnetic salinities in the lakes of the Vestfold Hills indicates a correspondingly large range of pycnoclines. Clear Lake (Fig. 1) has the smallest pycnocline among these meromictic systems (0.143 $g\ cm^{-3}\ m^{-1}$). The lake has lost much of its original seawater salt by the flushing effect of freshwater streams, in a manner similar to that described for Watts Lake (Pickard & Adamson, 1986).

The low-lying nature of the Vestfold Hills (plate 1), in which more than 85% of the land is below 60 m, has given most of the saline lakes extensive terracing or shallows within their basins. These conditions would be expected to produce relatively high concentrations of saline water during

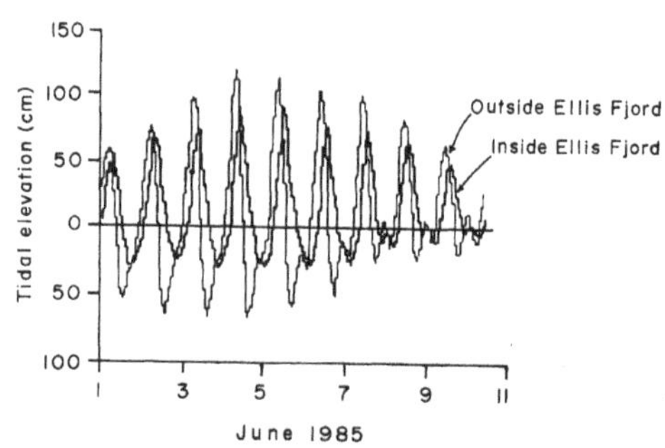

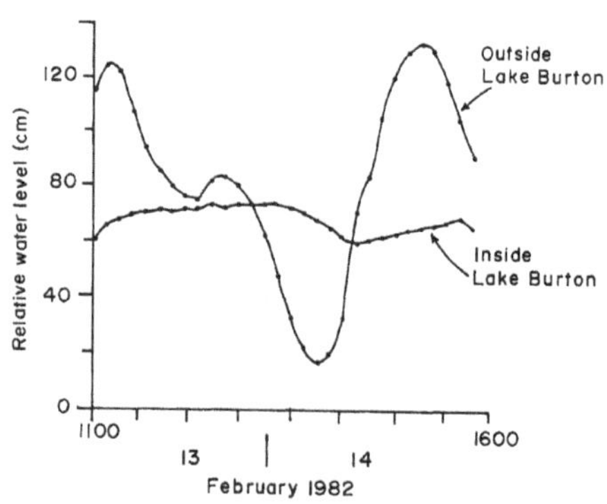

Fig. 7. Tides inside and outside of Ellis Fjord and Burton Lake.

Plate 1. Oblique aerial photograph overlooking Ellis Fjord from the east. The arrow denotes the position of the north shore terrace.

the winter. A submerged flat terrace about 3.6 m deep formed by a 1.9 m ice cover (maximum fjord ice thickness) and a tidal range of 1.7 m (Fig. 6) would produce hypersaline water of about 72‰ from summer seawater salinities (34‰). This salinity is much greater than that found in Clear Lake or in the DMB. Therefore, most of the saline lakes would be expected to have produced a stable hypoanachysis or hypolimnion during their fjord or early lake stage.

Modelling of meromictic events

Fig. 3 shows the step-like nature of the salinity profile in the hypoanachysis which is also reflected in the hydrogen sulphide profile (Table 2). There are two distinct water masses, one from about 73 m to 80 m and one from about 80 m to 110 m. The hydrogen sulphide concentrations are the highest yet recorded for natural waters (a maximum of 28.2 mmol l^{-1}). These salinity 'steps' do not seem to have formed in the same way as the large convection cells found in Lake Vanda (Hoare, 1968), a well studied system in the Victoria Land Dry Valley region. Unlike Lake Vanda, in which the salinity 'steps' reflects 'steps' of temperature, the temperature profile in Ellis Fjord crosses its salinity steps in a linear fashion (Fig. 3). This implies little or no turbulent mixing in the hypoanachysis.

It is shown below that the temperature gradient

is maintained at a steady state by balancing inputs of geothermal heat from its sediments against molecular diffusion of heat through the water column. The average flux of heat from the sediment within the hypoanachysis was calculated (10.8×10^{-6} J cm^{-2} s^{-1}) using a simple Fickian diffusion equation:

$$F = (k_z \cdot \Delta H/\Delta Z)/A \qquad (1)$$

where F is the heat flux per sediment area, H is the heat content of the water, Z is the depth, K_z is the molecular diffusivity of heat in water, and A is the sediment area. The calculated flux is of the same order as the estimated geothermal input for the area and rock type (4×10^{-6} J cm^{-2} s^{-1}; Quilty, pers. comm.) and suggests that only small vertical mixing rates operate today ($K_z = 1 \times 10^{-3}$ cm^{-2} s^{-1}). However, the hydrogen sulphide profile suggests that within each of the two hypoanachytic water bodies there has been complete mixing some time in the past, perhaps when they were formed. The sulphide concentrations are constant (Table 2) within the two water bodies, whereas their respective sediment areas change rapidly (4.4×10^3 m^2 m^{-1} and 1×10^3 m^2 m^{-1} for the 73 m to 80 m and 85 m to 110 m water bodies respectively). This is in contrast to waters above 73 m, in which the hydrogen sulphide concentrations increased with depth, and indicates different ages for the commencement of anaerobic respiration.

The step-like discontinuity in the hypoanachytic salinity suggests that after a period of stability there was a relatively rapid change in one of the factors which controls the formation and salinity of the hypersaline water. This could mean that the fjord had at least 3 stages of meromixis. The different ages of the water column within the hypoanachysis supports this. Such changes would have probably been linked to the changing nature of the shallows as isostatic uplift occurred (see 'Salinity of hypoanachysis' section) and the nature of the shallows has already been argued to be the most important factor controlling the salinity of the hypoanachysis. The other factors which affect the salinity of the hypersaline water forming in the shallows, such as horizontal mixing, are unlikely to respond dramatically enough to produce the sharp changes observed in the hypoanachytic waters (Fig. 3). Similarly, climatic changes affecting the rates of ice growth would only influence the quantity of hypersaline water formed and not the salinity (see 'Salinity of hypoanachysis' section).

Relative changes in sea level, however, may be rapid enough to effect the observed salinity changes in the hypoanachysis. The existence of flat or stepped terraces in Ellis Fjord today is witness to periods of constant sea level, relative to the land, followed by relatively rapid drops in level, even in a period of overall isostatic uplift. Therefore, it is proposed that the sharp salinity discontinuity in the DMB was a result of the terrace starting to rise from, and being lost to, the sea. Thus the first stage of meromixis was linked to the formation of the Deep Meromictic Basins' ancient terrace (Fig. 5). Figure 8a shows this process schematically.

The emergence of the terrace or shallow area from the sea would reduce the volume and perhaps the concentration of brine flowing to the bottom of the fjord. The salinities measured at the bottom of a 6 m sediment core, taken near station E4 in 1984, are greater than any in the water column (57‰; Gallagher, unpublished data). This is consistent with a shallowing of the depth of water above the terrace through the Holocene. From the ^{14}C chronologies (Table 3) it seems that the deeper water body was formed during the first stage of meromixis. It is important to note that during terrace formation the depth of the terrace does not alter (Adamson, pers. comm.). This is because the mechanism of terrace formation in this protected situation is due to ice abrasion of the shore line, a depth limited process. It follows then that the second meromictic stage (73 m to 80 m) would also have been a period of stable sea level, relative to the land. The time difference between the two meromictic stages represents the period from the emergence of the ancient terrace to either the development of a new terrace or the utilisation of deeper pre-existing extensive shallows. We favour the latter, as a new terrace is slow

Table 3. ^{14}C ages in the Deep Meromictic Basin

Depth (m)	^{14}C age (years BP)				
	Measured	Calculated by 'a' without surface correction	Calculated by 'a' with surface correction	Calculated by 'b' with surface correction	Calculated by 'c' with surface correction
10	348 ± 51				
60	2226 ± 88	5570	4890	4410	
80	2192 ± 67	5000	4240	3800	
100	2368 ± 75	5840	5030	4220	5600

'a', 'b' and 'c' calculations see Figure 9.

to form and in any case was not apparent from inspection of either surface or bathymetric contours.

The age of the water at 60 m was about the same age as that of the first meromictic stage and older than the second at 80 m. This may be entirely due to partial entrainment and subsequent dilution of the bottom hypersaline water by brine convection (Kantha, 1979) during the first meromictic stage. Hypersaline water, flowing as a consequence into the hypoanachysis during the second meromictic stage, displaced the entrained brine above it and away from its parent water body (Fig. 8b). This would protect the second water body from entrainment by penetrative convection and could explain the sharper pycnocline directly above the second meromictic water body (Fig. 3). Relative water levels appear to have fallen at a steady rate since this second meromictic stage (Adamson & Pickard, 1986b) and to have produced the observed gradual drop in salinity of the hypersaline water as the shallows become deeper (Fig. 8c).

Absolute age of the hypoanachytic waters

The absolute age of the Ellis Fjord terrace is about 5880 years BP (6225 years from Adamson & Pickard (1986b) for marine mud on the terrace less our correction for antarctic bias (reservoir age 348 years, Table 3)). The measured age of the water at 100 m in the hypoanachysis was 2368 ^{14}C years BP. Table 3 shows the apparent age for the inorganic carbon in the water calculated from the ^{14}C values (Table 2) for the bottom water body at 100 m, the middle water body at 80 m and the water above, at 60 m. How-

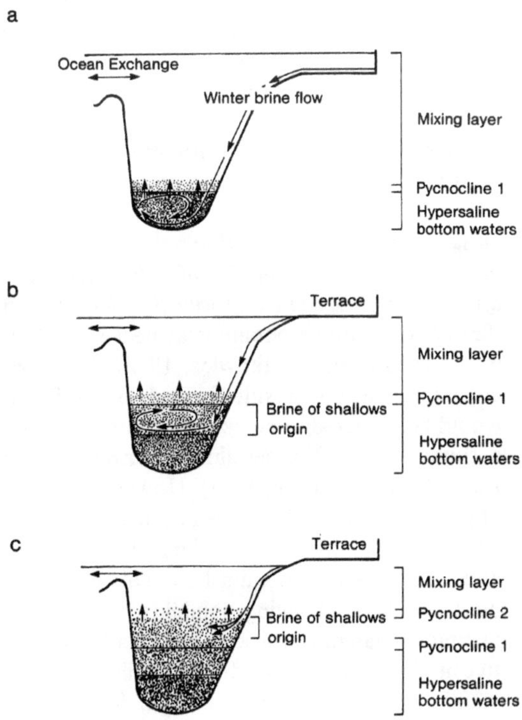

Fig. 8. Proposed model of brine production and the development of meromixis in the Deep Meromictic Basin.

ever, these dates are uncorrected for surface reservoirs effects, loss of carbon dioxide from calcium carbonate precipitation and the continued input of modern surface carbon dioxide from the fermentation of particulate organic carbon sedimented from surface waters. The importance of these three effects is illustrated by the loss of calcium in the bottom waters of the basin, the very high concentrations of carbon dioxide in the hypoanachysis and the ^{14}C pmc of the live *Laternula* shells found in the surface waters of the DMB (Table 2). Iterative models were developed which take into account the above concerns (Fig. 9). The assumptions of the models are as follows:

1) The rate of lateral mixing was instantaneous in comparison to vertical mixing. In most lake systems the lateral rate of mixing is generally 10^3 to 10^7 greater than vertical mixing (Imboden and Emerson, 1978). The fact that the temperatures in the hypoanachysis were very close to each other in stations at opposite ends of the basin (e.g. -1.08 °C at 60 m for both stations E3 and E4) supports this assumption.

2) Calcium and carbonate concentrations were at saturation at the genesis of the water body and calcium carbonate precipitation occurred from the outset at a rate governed by the production of carbon dioxide from the fermentation of sedimented algae after it had mixed laterally with the older carbon dioxide. Calcium carbonate is at near saturation in the open ocean (Pytkowicz, 1972). Therefore any more carbon dioxide formed in the DMB would be expected to quickly precipitate calcium carbonate. This has certainly occurred in Ace Lake (Hand & Burton, 1981). However, the rate of precipitation may have been equal to the calculated carbon dioxide input and may have stopped after the measured calcium loss. This may have been due to a changing solubility product adjusting to changing anoxic conditions (see 'Iterative model'). There is also the possibility that substantial carbon dioxide production from a relatively rapid rate of sulphate reduction, was largely responsible for the precipitation of the calcium carbonate. Franzmann *et al.* (in press) have found that the rate of sulphate reduction was extremely slow in the DMB in the presence of about 7 mmol l^{-1} sulphide. In the Scandinavian fjords, Hellefjord and Framvaren, sulphide concentrations were also about 7 mmol (Strom, 1936; Skei, 1983). However, the hypoanachysis in Hellefjord has been isolated for only 28 years and in Framvaren for no more than 2000 years. All this evidence suggests that end product inhibition of sulphate reduction may be significant once hydrogen sulphide concentrations reach 7 mmol l^{-1} in the water column and that this process is fast relative to the fjords' age. Assuming a respiratory quotient of about 1, the hydrogen sulphide concentration will correspond to the concentration of carbon dioxide released by fermentation, and, in Ellis Fjord, is of the same order as the calculated calcium loss to precipitation (6.98 mmol l^{-1} at 100 m). The negative $\delta^{13}C$ values (Table 2) indicate the catabolic origin of the carbon dioxide in the hypoanachysis.

3) The rate of carbon dioxide release from the sediments was constant and the rate limiting step. Microbial processes are temperature-dependent and therefore environments which have stable temperatures tend to support steady metabolic rates. Figure 4 shows the temperature profile for the 1985/86 season. The time it took to reach this steady state, considering only geothermal energy (4×10^{-6} J cm^{-2} s^{-1}, Quilty per. comm.), was in the order of 100 years. This is the time required to raise the bottom waters (110–100 m) from the average fjord surface temperature (-1.6 °C) to its present day temperature of $+0.38$ °C. This figure is small in comparison to the apparent age of these waters (Table 3); thus it seems likely that temperatures have remained stable during the life of the DMB.

The rate of diffusion of a molecule depends on its concentration gradient. The concentration gradient of carbon dioxide between the sediment and the water column certainly increased with time. However, Hesslein (1980) argued that the rate of release of dissolved substances from the sediment in a lake was constant and independent of its concentration in the water column. Any variations in the rate of release therefore depend, by this interpretation, on the rate of production in the sediments.

The rate of carbon dioxide production in the DMB sediments may or may not have been limited by the supply of algal material. Breakdown is slow (Franzmann *et al.*, in press) and sediment core examination (down to 6 m) has shown intact algal cells and high concentrations of volatile fatty acids present 30 cm into the core (Xu, pers. comm.). However, the rate of carbon dioxide production would be expected to change if the redox poise of the water altered due to changes in the dominance of microbial decomposers. For example, sulphate reducers give way to methanogens as sulphate is progressively depleted (Burton, 1981). In Burton's study, the depletion of sulphate at the bottom of meromictic Ace Lake (Fig. 1) allowed vigorous methane production. The work of Storm (1936) on Hellefjord, has shown that under fjord conditions, exceptionally high concentrations of hydrogen sulphide can build up in less than 30 years. In Ellis Fjord existing very high concentrations of hydrogen sulphide almost certainly inhibit sulphate reduction (Franzmann *et al.*, in press) and probably also inhibit any methanogenic bacteria (Mah *et al.*, 1977). Given the age of Ellis Fjord it therefore seems unlikely that there has been any significant switch to methanogenesis as the terminal step in organic mineralisation in this fjord over the majority of its lifetime.

4) The filling time of each section of the DMB hypoanachytic water was not significant with respect to the age of the fjord. This assumption would only apply in the third stage of meromixis (Fig. 8c), when the water levels continued to change, and as a consequence, the salinity of the hypersaline waters would cause them to layer on the previous years input and remain unmixed. The momentum of the incoming hypersaline water into the water body of the first and second stages of meromixis (Figs. 8a and b) was apparently enough to completely mix with the inflow of previous years. It was assumed that the input rate of hypersaline water to the hypoanachysis was constant during the filling time and the duration of the filling period for the first meromictic stage was 1500 years. This is the age difference between that of the terrace of the DMB and the age of emergence of the Watts Lake terrace near the DMB (Pickard *et al.*, 1986). A different input rate, in which the input value increased by a constant amount after each year was also used to check the sensitivity of the model. This was based on the input of hypersaline water being proportional to the terrace area, since it can be argued that the terrace area will increase by a constant amount each year. The input was in the equation form:

$$y = mx^2 + x \qquad (2)$$

where x is time and y is the volume of the hypersaline water.

Iterative model

An iterative flow diagram of the computer program is illustrated (Fig. 9). First, an estimated value of the oldest age possible was entered into the program (10000 years BP). Then, the carbon dioxide concentration at the depth in question and the carbon dioxide which was lost through calcium carbonate precipitation (calculation a; Fig. 9) were entered. The latter was calculated from the expected calcium concentration, given the water's salinity, less the measured calcium concentration.

The input of ^{14}C carbon dioxide from the fermented sedimenting algae was also corrected for photosynthetic fractionation since photosynthesis enriches the lighter carbon isotopes in particulate organic matter. The ^{14}C of particulate matter is 4% less than the ^{14}C of carbonate and bicarbonate dissolved in seawater (Craig, 1957). Finally, it was assumed that the fjord surface water ^{14}C concentration of today was constant throughout the life of the fjord and was calculated from live *Laternula* sp. shells found at 10 m in the DMB. The concentrations of ^{14}C from Antarctic surface waters were found to be less than modern wood. This gives surface water ages from about 880–2900 years, depending on the relative ratio of glacial melt water to deep upwelling seawater

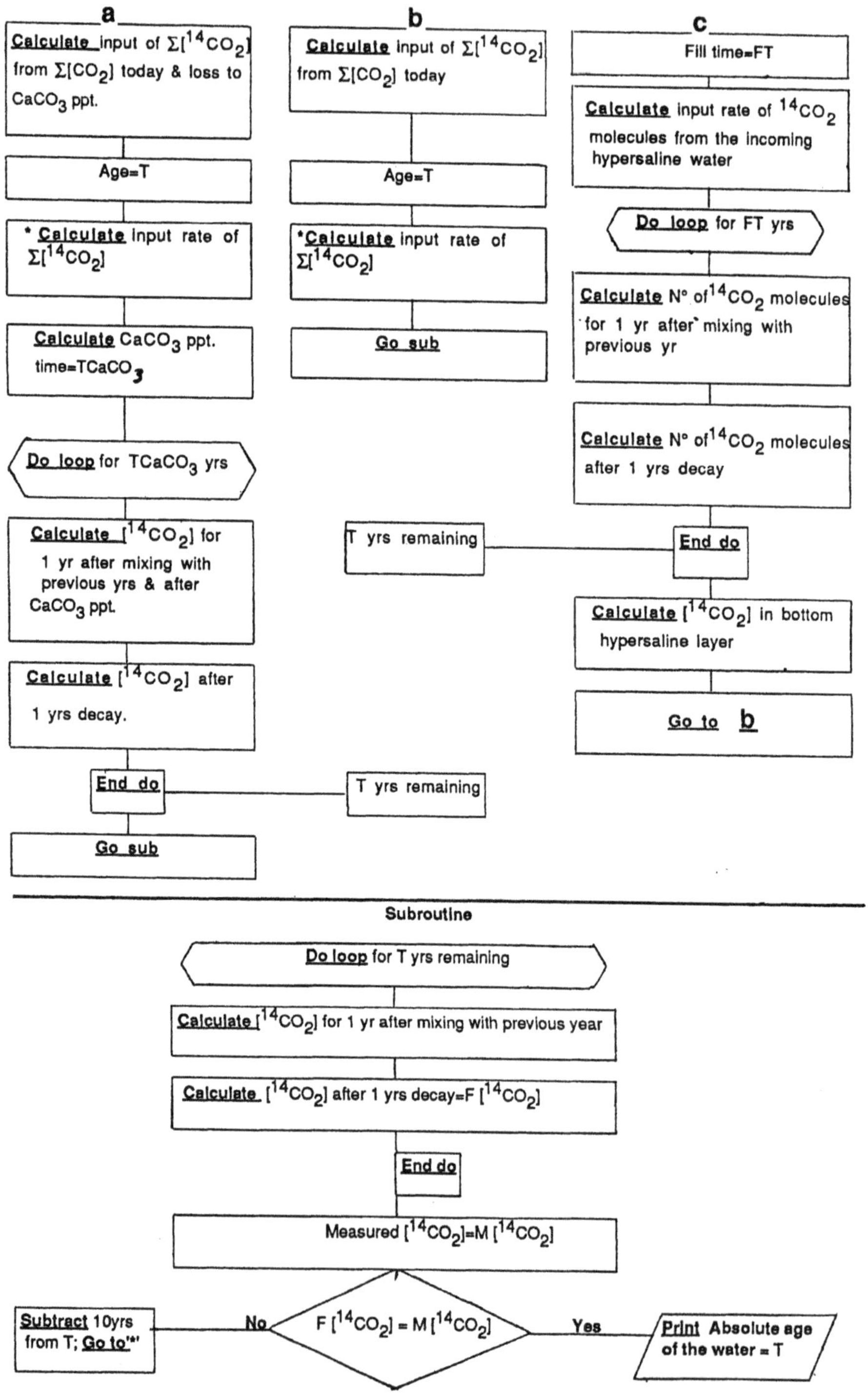

Fig. 9. Iterative schemes for calculation of absolute ^{14}C ages of the hypoanachytic waters in the Deep Meromictic Basin.

(Omoto, 1972). The input of glacial melt water from Ellis Rapids (Fig. 2) varies considerably from year to year, bringing with it waters depleted in ^{18}O relative to the sea (Matsubaya *et al.*, 1979). Therefore the age of the shells represent a more meaningful median age of the surface waters over a number of seasons, than one measurement taken from the surface water at one time in a single season. The ages were calculated from a half life of 5570 years and the absolute ^{14}C concentration from the $[CO_2]$ and $^{14}C/^{12}C$ ratio where:

$$^{14}C/^{12}C = (1.176 \times 10^{-12})(1 + 0.001\ \delta^{14}C) \quad (3)$$

Results of the models

Table 3 shows the measured and calculated ages for the hypoanachytic water bodies. Comparison of ages between the model 'a' with and without reservoir corrections (see 'Iterative model') show that the model is very sensitive to reservoir corrections. There is up to 3580 years difference at 100 m, when the age of the reservoir is just 348 years BP. Age comparisons of model 'b' with model 'a' (corrected for reservoir input) show that this model is relatively insensitive to the input rate of algal carbon dioxide to the hypoanachysis. The resultant values in 'b' only underestimate the values in 'a' by up to 810 years. However, there is a significantly larger difference from model 'b' when applying the estimated filling time (model c: 1380 years). During the filling time the concentration of carbon dioxide from the hypersaline water entering the bottom waters at 100 m was estimated at 3.4 mmol l^{-1}. this is the salinity corrected value of surface waters and no input of fermenting algal carbon dioxide was included. It was found that the input was negligible compared to the input from the hypersaline water, as the hypoanachysis was forming. The model, it seems, is very sensitive to the filling time. Comparison of this absolute age of the first meromictic stage at 100 m (5600 years) with the absolute age of the terrace (5880 years) supports the validity of the model and the estimated filling time. The age is not significantly different when using the hyperbolic input rate (equation 2). The model cannot, however, be applied to the second meromictic stage as there is no data for estimating the filling time. Nevertheless any further measurements above 60 m in which the salinity decreases continuously, would not need filling time estimates and model 'b' would suffice for calculating its age.

Evolution of fjords to meromictic lakes

It has been shown that hypersaline meromixis developed in a fjord in the middle Holocene and that some meromictic saline lakes have developed from old fjord systems. This does not necessarily mean that all saline meromictic lakes in a region such as the Vestfold Hills have derived a continuing meromixis from their parent fjord and indeed several saline lakes in these areas have clearly developed meromixis after isolation from the sea. This is illustrated by following the potential fate of the SMB at the head of Ellis Fjord. The SMB is an example of a fjord basin in the first stages of isolation with a sill depth of 1–2 m, a pycnocline at about 10 m and a hypoanachytic water salinity of about 40‰ (Fig. 3). The basin is cut off from the rest of the fjord, from the end of August for about 3 months each year, by the growing ice sheet resting on the bottom of the sill. As a consequence, during this time surface waters increase in salinity (up to 37‰) above that of seawater. When the surface waters are replaced by incoming seawater during the spring, before the Ellis Rapids start to flow, the pycnocline gradient would return to its former value. However, in the event that the mean sea level dropped below the level of the sill, as is the case in Burton Lake (Burke & Burton, in press), the basin would become isolated from the sea for at least 8 months of the year and the salinity of the surface waters would increase by further ejection of salt from the growing ice sheet. A surface mixing layer of about 7 m in depth should produce surface salinities of about 47‰, assuming 1.9 m thick ice and an initial seawater salinity of 34‰. As a consequence, when penetrative convection completely mixed the whole of the SMB's water column, the pycnocline would disappear. The highly saline waters would then be lost to the rest of the fjord when tidal exchange operated during high tides in the open summer months as is the case for Burton Lake today (Burke & Burton, in press). The development of meromictic lakes following isolation from the sea appears to be largely determined by water balance. However, isolation from the sea is itself still a relative term. The SMB of Ellis Fjord is isolated by an ice dam for about 4 months per year whilst Burton Lake is isolated totally for about 8 months and also, briefly, during low and middle tides. Lake Fletcher has occasional incursions, perhaps years apart, during unusual combinations of high tide and ice cover; and the other meromictic lakes in the Vestfold Hills appear not to have had any incursions this century. However, it is not always possible to be sure when absolute isolation of a lake has occurred, for rare contact with the sea can be made, perhaps hundreds of years apart, when extremely high tides coincide with on-shore storm winds in ice-free summers. A working definition of a lake as opposed to a lagoon or estuary might be when the mean sea level is below the level of the sill and there is no water exchange with the sea for at least 6 months of the year.

The lake catchment itself is the sole source of water once seawater contributions are removed. The arid climate of the Vestfold Hills is characterised by very low precipitation and a low mean relative humidity of 64% (Burton & Campbell, 1980), so the lake volumes of many erstwhile fjord/estuarine basins have decreased by evaporation and their epilimnia thereby increased in salinity. In those lakes where meromixis has continued, the hypoanachysis becomes a monimolimnion.

The salinities and meromictic conditions of today need not be the end results of a steady evolution however. In fact, in one lake (Ace Lake), close examination has yielded evidence from the water column distribution of sulphur isotopes to support at least two meromictic periods separated by a holomictic episode (Burton & Barker, 1979). However, disregarding the intermediate stages, it is clear that today 7 meromictic lakes are in negative water balance compared to their original seawater salinity, 5 are in positive water balance and one is still dominated by seasonal seawater fluxes (Fig. 10).

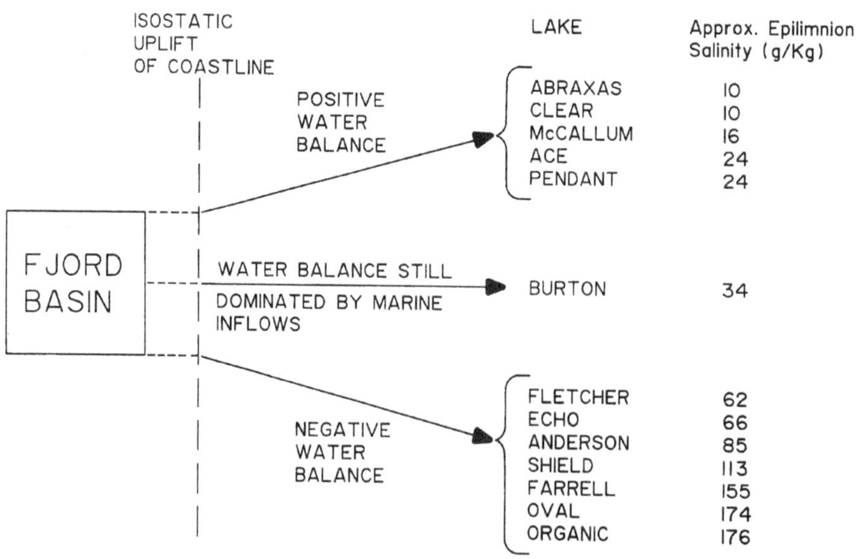

Fig. 10. The water balance status of 13 meromictic lakes in the Vestfold Hills.

It is likely that some holomictic lakes were once meromictic but are so no longer and several examples exist in the Vestfold Hills which illustrate the two incompatible paths that could bring about such a change.

1. Deep Lake, 50 m below present sea level and concentrated approximately ten times with respect to seawater, is so hypersaline that surface temperatures never fall low enough (approximately -22 °C) to form a complete ice cover (Barker, 1980). Wind-driven mixing ensures an essential holomixis once a year in later winter. Lakes Stinear, Lebed, Dingle and Club are basically similar although thin ice covers do form briefly on these lakes.
2. The other route to holomixis is to have lost virtually all the original sea salt by flushing. Lakes Crooked and Druzby, at the end of Ellis Fjord, must once have contained seawater. Today they form part of a freshwater drainage system (Tierney, 1975) that channels large volumes of surface melt water from the Sørsdal Glacier, in summer, into Ellis Fjord. Although deep (Crooked Lake is 150 m in depth) they now contain only freshwater.

These extremes of lake development from fjords and estuaries demonstrate the crucial and deciding role of water balance in the evolution of marine basin lakes in the Vestfold Hills. Many details of the evolution of these lakes are preserved in their sediments and studies of sedimentary cores from the lakes and fjords are now proceeding.

Acknowledgements

We would like to express our thanks to the Davis 1985 ANARE, and particularly R. Besso, without whose support on field trips this work would not have been possible. We also thank Dr P Franzmann and two anonymous referees for reviewing the manuscript and C. Wookey for determining $\delta^{18}O$ values.

References

Adamson, D. A. & J. Pickard, 1986a. Physiography and geomorphology of the Vestfold Hills. In: (Pickard J., ed.) Antarctic oasis. Terrestrial environments and history of the Vestfold Hills. Academic Press, Sydney: 99–139.

Adamson, D. A. & J. Pickard, 1986b. Cainozoic history of the Vestfold Hills. In: (Pickard J., ed.) Antarctic oasis. Terrestrial environments and history of the Vestfold Hills. Academic Press, Sydney: 63–67.

Allison, I., 1981. Antarctic sea ice growth and oceanic heat flux. International Association of Hydrological Sciences Publication 131 (Symposium at Canberra 1979–Sea Level, Ice and Climate Change): 161–170.

Almgren, T., D. Dryssen & S. Fonselius, 1983. Determination of alkalinity and total carbonate. In K. Grasshoff, M. Enrhart & K. Kremling (eds), Methods of Seawater Analysis, Second Revised and Extended Edition. Verlage Chemie 8: 99–123.

Barker, R. J., 1980. Physical and chemical parameters of Deep Lake, Vestfold Hills, Antarctica. Australian National Antarctic Research Expeditions (A.N.A.R.E.) Report, Publication No. 130. Antarctic Division, Kingston, Tasmania.

Burke, C. M. & H. R. Burton, in press. The ecology of photosynthetic bacteria in Burton Lake, Antarctica. Hydrobiologia: -

Burton, H. R., 1981. Chemistry, Physics and Evolution of Antarctic Saline Lakes: A Review. In W. D. Williams (ed.) Developments in Hydrobiology. Dr W. Junk. The Hague: 339–36.

Burton, H. R. & R. J. Barker, 1980. Sulfur chemistry and microbiological fractionation of sulfur isotopes in a saline Antarctic lake. Geomicrobiol. J. 1: 329–340.

Burton, H. R. & P. C. Campbell, 1980. The climate of the Vestfold Hills, Davis Station, Antarctica, with a note on its effect on the hydrology of hypersaline Deep Lake. Australian National Antarctic Research Expeditions (A.N.A.R.E.) Report, Publication No. 129. Antarctic Division, Kingston, Tasmania.

Craig, H., 1957. The natural distribution of radiocarbon and the exchange time of carbon dioxide between the atmosphere and the sea. Tellus 9: 1–17.

Craig, H. & Holm, B., 1968. Relationships of deuterium, oxygen-18, and chlorinity in the formation of sea ice. EOS Trans AGU 49: 216–217.

Fonselius, S. H., 1983. Determination of hydrogen sulphide. In K. Grasshoff, M. Enrhart & K. Kremling (eds.) Methods of Seawater Analysis, Second Revised and Extended Edition. Verlag Chemie 5: 73–80.

Franzmann, P. D., G. W. Skyring, H. R. Burton & P. P. Deprez, P. P., 1988. Sulphate reduction rates and some aspects of the limnology of four lakes and a fjord in the Vestfold Hills, Antarctica. Hydrobiologia 165: 25–33.

Gade, H. G., R. A. Lake, E. L. Lewis & E. R. Walker 1974. Oceanography of an Arctic Bay. Deep-Sea Res 21: 547–571.

Gallagher, J. B. & H. R. Burton, in press. Seasonal mixing within Ellis Fjord, Vestfold Hills, Antarctica. Estuarine, Coastal and Shelf Science.

Grasshoff, K., 1983. Determination of pH. In K. Grasshoff, M. Enrhart & K. Kremling (eds) Methods of Seawater Analysis, Second Revised and Extended Edition. Verlag Chemie 7: 85–97.

Hand, R. M. & H. R. Burton 1981. Microbial ecology of an Antarctic saline meromictic lake. In W. D. Williams (ed.) Salt Lakes: Proceedings of an International Symposium on Athalassic (Inland) Salt Lakes. Dr W. Junk, The Hague: 363–374.

Hesslein, R. H., 1980. Whole-lake model for the distribution of sediment-derived chemical species. Can. J. Fish. Aquat Sci. 37: 552–558.

Hoare, R. A., 1968. Thermocline convection in Lake Vanda, Antarctica. J. Geophys. Res. 73: 607–612.

Huppert, H. E., J. S. Turner, S. N. Carey, R. S. L. Sparks & M. A. Hallworth, 1986. Laboratory simulation of pyroclastic flows down slopes, J. Volcanol. Geotherm. Res. 30: 179–199.

Imboden, D. M. & S. Emerson 1978. Natural radon and phosphorus as a limnological tracer: horizontal and vertical eddy diffusion in Greifensee. Limnol. Oceanogr. 23:77–90.

Kantha, L. H., 1979. Turbulent entrainment at the buoyancy interface due to convective turbulence. In H. J. Freeland, D. M. Farmer & C. D. Levings (eds) Fjord Oceanography. Plenum Press, New York: 205–213.

Mah, R. A., D. M. Ward, L. Baresi & T. L. Glass, 1977. Biogenesis of methane. Ann. Rev. Microbiol. 31: 309–341.

Matsubaya, O., H. Sakai, T. Torii, H. R. Burton & K. Kerry, 1979. Antarctic saline lakes-stable isotopic ratios, chemical compositions and evolution. Geochim. Cosmochim. Acta. 43: 7–25.

McClimans, T. A., 1978. On the energetics of tidal inlets to landlocked fjords. Mar. Sci. Comm. 4: 121–137.

Newbury, T. K., 1979. Possible accumulation of heavy metals around the offshore oil production facilities in the Beaufort Sea. Arctic. 32: 42–45.

Omoto, K., 1972. A preliminary report on modern carbon datings at Syowa station and its neighbourhood, east Antarctica. Antarctic Rec. 43: 20–24.

O'Neil, J. R., 1968. Hydrogen and oxygen isotope fractionation between ice and water. J. Phys. Chem. 72: 3683–3684.

Pagé, P., M. Ouellet, C. Hillaire-Marcel & M. Dickman, 1984. Isotopic analyses (^{18}O, ^{13}C, ^{14}C) of two meromictic lakes in the Canadian Arctic Archipelago. Limnol. Oceanogr. 29: 564–573.

Perkin, R. G. & E. L. Lewis, 1980. Mixing in an Arctic fjord. J. Phys. Oceanogr. 8: 87–880.

Pickard, J., D. A. Adamson & C. W. Heath, 1986. The evolution of Watts Lake, Vestflod Hills, east Antarctica, from marine inlet to freshwater lake. Palaeogeogr. Palaeoclimatol. Palaeoecol. 53: 270–288.

Pytkowicz, R., 1972. Comments on a paper by S. Ben-Yaakov and I. R. Kaplan, 'Deep Sea In Situ Carbonate Saturometry'. J. Geophys. Res. 77: 2733–2734.

Skei, J., 1983. Geochemical and sedimentological considerations of a permanently anoxic fjord–Framvaren, South Norway. Sed. Geol. 36: 131–145.

Strom, K. M., 1936. Land-locked waters. Hydrography and bottom deposit in badly-ventilated norwegian fjords with remarks upon sedimentation under anaerobic conditions. Skr. Norske. Vidensk. Akad. Oslo 7: 85 pp.

Tierney, T. J., 1975. An externally draining freshwater system in the Vestfold Hills, Antarctica. Polar Record 17: 684–685.

Walker, K. F. & G. E. Likens, 1975. Meromixis and a reconsidered typology of lake circulation patterns. Verh. Internat. Verein. Limnol. 19: 442–458.

Hydrobiologia 172: 255–264, 1989
W. F. Vincent and J. C. Ellis-Evans (eds) High Latitude Limnology

Chemical characteristics of pond waters in the Labyrinth of southern Victoria Land, Antarctica

Tetsuya Torii[1], Shyu Nakaya[2], Osamu Matsubaya[3], Genki I. Matsumoto[4], Noriyasu Masuda[5], Tamio Kawano[6] & Haruta Murayama[7]
[1]*Chiba Institute of Technology, Tsudanuma 2-17-1, Narashino, Chiba Prefecture 275, Japan*; [2]*Department of Earth Sciences, Faculty of Science, Hirosaki University, Bunkyo-cho 3, Hirosaki, Aomori Prefecture 036, Japan*; [3]*Research Institute of Underground Resources, Mining College, Akita Prefecture 010, Japan*; [4]*Department of Chemistry, The College of Arts and Sciences, The University of Tokyo, Komaba, Meguro-ku, Tokyo 153, Japan*; [5]*Department of Chemistry, Faculty of Fisheries, Hokkaido University, Minato-machi 3-1-1, Hakodate, Hokkaido 041, Japan*; [6]*Department of Chemistry, Faculty of Education, Ōita University, Dannoharu 700, Ōita Prefecture 870-11, Japan and* [7]*Department of Chemistry, Faculty of Education, Yokohama National University, Tokiwadai 156, Hodogaya-ku, Yokohama 240, Japan*

Key words: Antarctic pond, Labyrinth, chemical element, stable isotopic ratio, water source, salt origin

Abstract

A large number of fresh water and saline ponds are located in the Labyrinth (77° 33′ S, 160° 50′ E) of the upper Wright Valley in the Dry Valleys region of southern Victoria Land, Antarctica. It is situated near the terminus of the Wright Upper Glacier between 800–1 000 m above sea level. From a limnological point of view, the most interesting problems concerning these saline ponds are the origin of their salts and their evolutional history.

Chloride ion contents vary remarkably among the ponds ranging from 0.0049 to 52.4 g kg^{-1}. Surprisingly, more than a half of the ponds are saline with the highest chloride ion content being 2.7 times greater than that of seawater. The δD and $\delta^{18}O$ values of the pond waters indicate a snow and/or glacial meltwaters origin, and that the ponds underwent subsequent alteration due to evaporation or freezing. The composition of chemical components reveal no evidence of trapped seawater. Thus the salt concentrations in the Labyrinth pond waters must be explained principally by the accumulation of atmospheric salts and subsequent repeated cycles of evaporation and freezing of the pond waters over considerable time periods.

Introduction

There are more than 16 ice-free areas, so-called 'oases', throughout Antarctica (Simonov, 1971). Their most conspicuous feature is the occurrence of many lakes and ponds, the waters of which contain chloride ions in amounts ranging from less than 10 mg kg^{-1} to about 250 g kg^{-1} (Torii *et al.*, 1977). Most of these lakes and ponds are located in depressions in ice-free areas or in drainage basins enclosed by glaciers, and are fed by surface meltwater and sub-surface flow from glaciers and adjacent snow fields during the austral summer.

Wright Valley is a major valley of the Dry Valleys (Fig. 1). Although this valley was a fjord

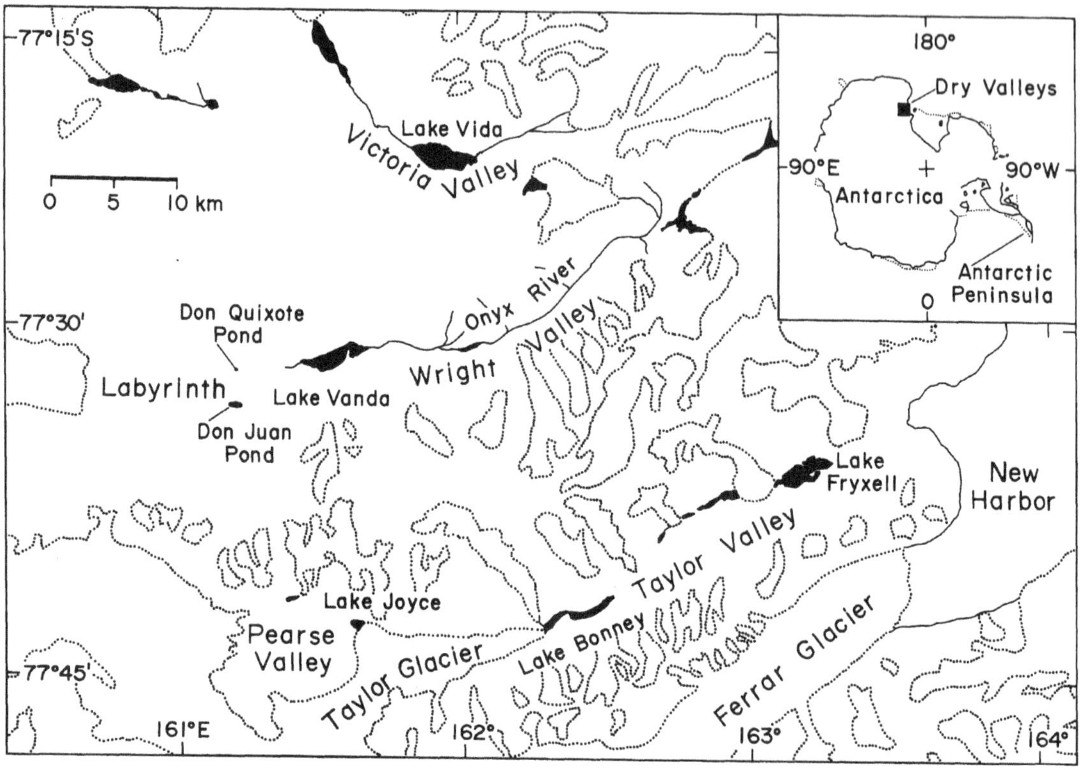

Fig. 1. The Dry Valley region, southern Victoria Land in Antarctica.

during the late Miocene (Webb, 1972), it appears that the saline waters in the valley have been derived from the accumulation of atmospheric salts rather than the alteration of trapped seawater (Torii & Yamagata, 1981; Torii *et al.*, 1981). Masuda *et al.* (1984) reported that aerosol particles also played a major role in the origin of the trace elements in certain Dry Valley lakes, such as Lakes Vanda and Canopus.

Murayama *et al.* (1983) surveyed saline ponds in the high altitude area of the upper Victoria Valley in order to examine the validity of the above theory for salt origin proposed by Torii *et al.* (1981), and found one saline pond (B-1 pond, unnamed) at the foot of Mt. Bastion (at an altitude of 1450 m) containing 17 g kg^{-1} of chloride ion.

The Labyrinth, an extensive area exposed approximately 5.5 km width and 7 km length, by glacial retreat is situated near the terminus of the Wright Upper Glacier at the west end of Wright Valley (Fig. 2).

The basement rocks in the area consist mainly of dolerite with virtually no detritius or sediment. The cross-profiles of the troughs are irregular, where the upper walls are high. They are usually composed of columns of dolerite blocks and are nearly vertical. The trough floors are mostly V-shaped with many of the lowest points being offset from the center of the troughs and, in some instances, located close under one of the walls (Selby & Wilson, 1971).

More than 60 ponds containing fresh and saline waters occur in depressions where meltwater is supplied from glacier and snow banks. The altitude of these systems ranges from 800 to 1 000 m

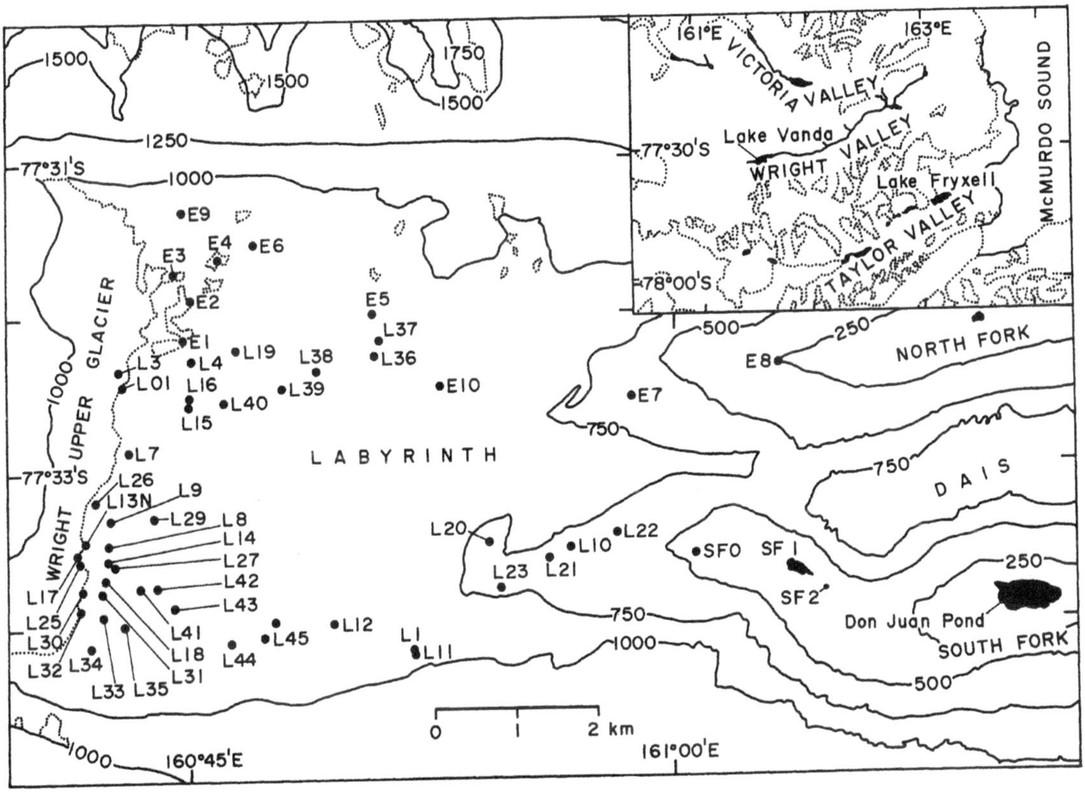

Fig. 2. Location of ponds in the Labyrinth.

above sea level where the possibility of trapped seawater cannot be easily assumed. Here we report data on chemical elements and stable isotope ratios of hydrogen and oxygen for water samples from 16 typical ponds among 61 studied in the Labyrinth to characterise the ponds and discuss the origin of pond waters and their salts.

Materials and methods

Water samples were collected during austral summers 1976–77 to 1985–86 with polyethylene bottles after drilling through the pond ice using a Sipre ice auger. Water temperature, dissolved oxygen, alkalinity, pH and electrical conductivity were measured immediately after sampling. Dissolved oxygen and alkalinity were determined by the Winkler method and by the titration method at pH 4.3 using BCG indicator, respectively. The main chemical components were analysed in the laboratory by the following methods.

Li, Na, K: Atomic absorption spectrophotometry.

Ca, Mg: Atomic absorption spectrophotometry for low salinity and the EDTA titration method for high salinity.

Cl: Mercuric thiocyanate spectrophotometry for low salinity (Iwasaki *et al.*, 1956) and mercuric nitrate titration method for high salinity (Domask & Kobe, 1952).

SO_4: Spectrophotometry for low salinity (Iwasaki *et al.*, 1957) and gravimetry for high salinity.

F: Ion selective electrode method.

Table 1. Analytical data of the typical pond waters in the Labyrinth.

Sample	Date	Sampling depth (cm)	Water Temp. (°C)	O_2 ($ml\ l^{-1}$)	Alkalinity ($meq\ l^{-1}$)	pH	Specific grafity at 25 °C
glacier ice of Wright Upper Glacier*	Dec. 26 '85	5 ~ 15	–	–	–	–	–
fresh snow	Dec. 26 '85	0 ~ 5	–	–	–	5.27	–
glacial meltwater from Wright Upper Glacier	Dec. 24 '85	0	–	–	0.8	5.65	–
L-0-1	Dec. 23 '85	95	− 2.0	14.4	11.1	7.62	1.053
L-7	Dec. 16 '84	30	+ 10.9	–	–	–	1.080
L-17	Dec. 16 '84	0	− 1.0	0.32	25.4	8.34	1.010
L-30	Dec. 27 '85	10	+ 0.8	–	0.18	9.89	1.000
L-3	Dec. 24 '85	223	0.0	25.1	0.32	9.23	1.003
L-13-N	Dec. 22 '85	74	0.0	11.5	9.93	7.98	1.002
L-1	Jan. 3 '77	0	+ 1.7	15.7	–	8.22	1.017
L-4	Dec. 14 '84	0	+ 0.3	–	–	8.11	1.080
L-9	Jan. 4 '82	15	+ 4.0	–	–	7.65	1.113
L-12	Dec. 28 '83	0	+ 2.8	14.1	0.31	10.1	1.000
L-18	Dec. 23 '83	0	+ 1.1	11.5	0.09	10.0	1.000
Inflow to L-18	Dec. 22 '83	0	+ 0.7	8.05	0.26	–	1.000
L-20	Dec. 27 '83	0	+ 0.9	9.1	0.12	7.7	1.000
E-7	Dec. 17 '84	0	− 0.8	11.5	0.12	9.71	1.000
E-10	Dec. 27 '85	12		–	2.83	9.10	1.006
			− 0.2				
B-1 Upper Victoria Valley	Dec. 21 '82	50		9.74	2.44	8.32	1.034

*: average of 12 samples.

B: Spectrophotometry (Schucker *et al.*, 1975).

D/H and $^{18}O/^{16}O$ ratios: Uranium-reduction (Friedman, 1953) and the H_2O-CO_2 equilibration method (Epstein & Mayeda, 1953), respectively. The results are reported as delta values relative to the SMOW standard (Craig, 1961).

$$X = (R_{sample}/R_{standard} - 1) \times 10^3\ (‰)$$

where X is D or ^{18}O and R is D/H or $^{18}O/^{16}O$.

Results and discussion

The general nature of the ponds

The size of ponds in Labyrinth area is typically less than 100 m in diameter with a water depth of 0.2–2.2 m. The volume of pond waters changes remarkably from year to year. During the 1977 to 1985 austral summers, pond water volumes tended to decrease, but in the 1985–86 summer season, the pond water volumes increased abruptly.

In general, most of these ponds were covered with ice even in the austral summer, and some were perennially frozen to the bottom (L-12 and L-18 ponds). On the other hand, high salinity

Na (mg kg^{-1})	K (mg kg^{-1})	Ca (mg kg^{-1})	Mg (mg kg^{-1})	Cl (mg kg^{-1})	SO_4 (mg kg^{-1})	F (mg kg^{-1})	B (mg kg^{-1})	$\delta^{18}O$ (‰)	δD (‰)
0.32	0.09	0.18	0.06	0.83	0.20	<0.01	<0.02	−38.0	−294
0.77	0.25	0.14	0.06	1.16	0.44	<0.01	<0.02	−27.1	−200
1.63	0.23	1.0	0.78	2.70	3.0	0.14	0.03	−37.2	−296
11200	52.3	990	6040	20400	8350	20	1.96	−33.3	−264
17400	354	981	9070	38100	4690	29.0	8.6	−22.3	–
2750	465	386	592	3800	2800	8.9	<0.02	−32.2	−245
6.67	0.64	1.1	1.1	8.76	3.0	0.095	<0.02	−38.7	−298
1170	11.5	270	112	392	2600	5.4	0.27	−45.4	−326
1130	12.9	118	215	1230	1110	5.4	0.59	−32.7	−261
4130	25.0	531	1920	7580	2130	8.9	0.99	−16.2	−177
14400	135	1020	10300	37200	4840	33.0	4.2	−24.5	−235
24200	243	1240	10200	52400	12100	28.0	5.7	−12.7	−174
60.3	1.3	25.8	17.5	78.2	99.6	1.4	0.03	−23.4	−212
4.8	0.54	1.9	1.1	4.9	9.5	0.21	<0.02	−37.1	−291
3.44	0.78	2.60	2.83	2.14	6.80	0.085	<0.02	−36.9	−298
42.5	1.3	4.5	10.0	93.6	15.0	0.33	0.06	−19.3	−182
24.0	1.09	2.1	5.7	48.1	24.2	0.45	0.03	−20.2	−184
1970	16.8	142	550	3030	973	5.0	0.40	−20.0	−186
9960	52.3	1070	5860	17080	1900	20.0	6.8	−25.3	−225

ponds such as L-1, L-4 and L-9 ponds were usually ice free and/or sometimes dried up. The water temperatures showed a wide range of variation from −2.0 to +10.9 °C (Table 1). All the ponds were basic, the pH values ranging from 7.62 to 10.1. Extremely high pH values were observed at L-12 and L-18 ponds and were attributed to the consumption of carbonates by intensive photosynthesis as extensive cyanobacterial mats were observed at the pond edges. Dissolved oxygen and alkalinity ranged from 0.32 to 25.1 ml l^{-1} and from 0.09 to 25.4 meq l^{-1}, respectively. The high dissolved oxygen contents can again be explained by photosynthetic activity.

Chemical composition of the pond waters

The nature of saline waters in the Dry Valleys region is variable in chemical composition and salinity, and differs considerably from valley to valley as well as from lake to lake. Dissolved salts in, for example, Lake Vanda and Don Juan Pond, in Wright Valley, are predominantly $CaCl_2$, whereas in Lake Bonney in the Taylor Valley, NaCl and $MgCl_2$ are the main salts (Fig. 3). Table 1 shows the chemical composition of representative pond waters together with those for ice of the Wright Upper Glacier, fresh snow and glacier meltwaters flowing into ponds in the Labyrinth. As with other ponds and lakes in the Antarctic, the chloride ion exceeded sulfate ion concentrations in most ponds and varied from a

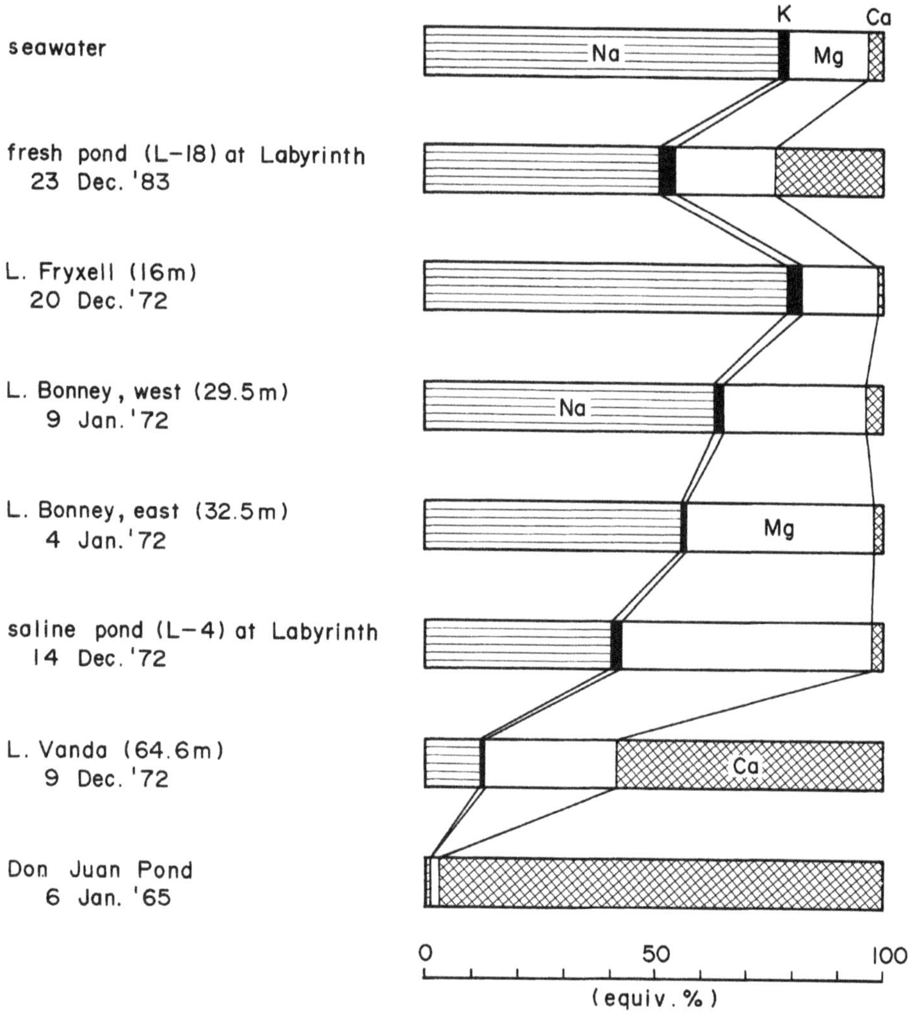

Fig. 3. Composition of cations in saline waters from the Dry Valley region in equivalent percentages.

very low concentration of 0.0049 g kg^{-1} (which is similar to those of snow and glacial meltwaters) to extremely high concentrations of 52.4 g kg^{-1} (L-9 pond). Despite proximity to the glacier, 35 of the 61 ponds were found to be saline, largely as a result of topographic barriers to meltwater input.

In general, the ratios of sodium ion to chloride ion in pond waters in the Labyrinth closely resembled that of seawater, while the ratios of Ca, Mg and sulfate ions to chloride ion were much higher than those of seawater. Figure 4 is a triangular co-ordination diagram showing relative concentrations of sodium plus potassium, magnesium, and calcium ions in equivalent percentages. The small solid triangle in the diagram represents seawater, and the solid, arrowed line shows the course of evolution of the brine under frigid conditions, as determined by Thompson and Nelson (1956).

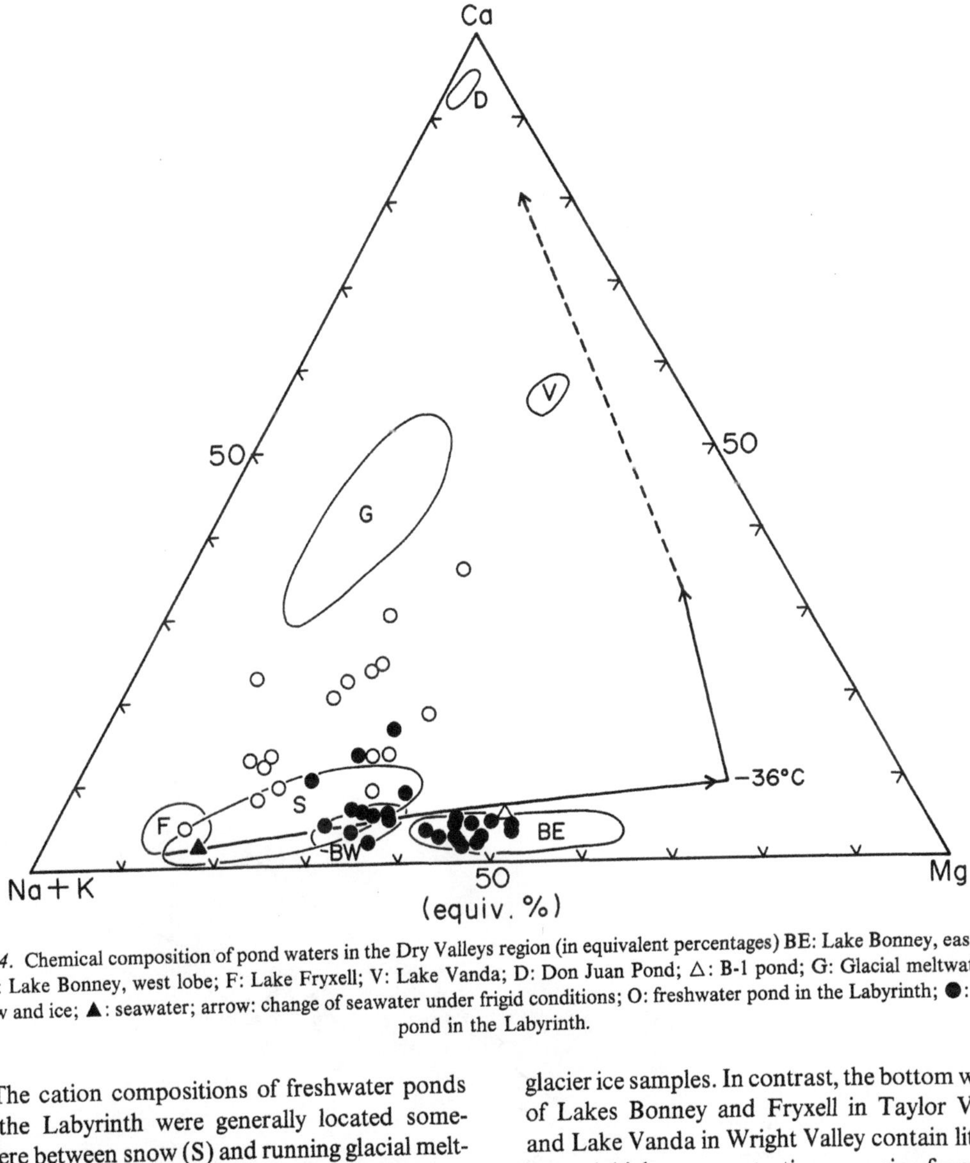

Fig. 4. Chemical composition of pond waters in the Dry Valleys region (in equivalent percentages) BE: Lake Bonney, east lobe; BW: Lake Bonney, west lobe; F: Lake Fryxell; V: Lake Vanda; D: Don Juan Pond; △: B-1 pond; G: Glacial meltwater; S; snow and ice; ▲: seawater; arrow: change of seawater under frigid conditions; O: freshwater pond in the Labyrinth; ●: saline pond in the Labyrinth.

The cation compositions of freshwater ponds in the Labyrinth were generally located somewhere between snow (S) and running glacial meltwater (G), while that of the saline waters was similar to those of Lake Bonney in Taylor Valley and B-1 pond in the upper Victoria Valley (Fig. 4), discussed in Torii & Yamagata, 1981. This shows that the salt compositions of pond waters in the Labyrinth resemble those of ponds and lakes elsewhere in the Dry Valleys region.

Lithium was found to be less than 0.02 mg l^{-1} in all the pond waters as well as in all snow and glacier ice samples. In contrast, the bottom waters of Lakes Bonney and Fryxell in Taylor Valley, and Lake Vanda in Wright Valley contain lithium in much higher concentrations, ranging from 0.13 to 24.0 mg l^{-1} (N. Takamatsu, pers comm).

The ratios of fluoride ion to chloride ion (F^-/Cl^- : 0.534×10^{-3} – 17.8×10^{-3}) in pond waters, both fresh and saline, are much larger than that of seawater (6.91×10^{-5}). On the other hand, whilst ratios of boron to chloride ion (B/Cl^- : 2.3×10^{-4} – 140×10^{-4}) in waters of freshwater ponds are much larger than that of seawater (2.4×10^{-4}), the ratios ($.36 \times 10^{-4} - 7.0 \times 10^{-4}$) in saline ponds are almost of the same order (Table 1).

The high F^-/Cl^- and B/Cl^- ratios of the freshwater ponds can be explained by the fact that meltwater (F^-/Cl^- : $4.0 \times 10^{-2} - 5.2 \times 10$-2, B/Cl^- : $93 \times 10^{-4} - 110 \times 10^{-4}$, Table 1) of the Wright Upper Glacier is high in boron and fluoride and that both snow and rain are also usually enriched in fluoride ion (Sugawara, 1964) and boron (Muto, 1956), compared with seawater. The low B/Cl^- ratios in saline ponds are probably due to the gradual volatilisation of boron during freezing and evaporation processes under the extreme frigid conditions.

Source of the pond water

Net isotopic effects due to the evaporation of lake water are dependent on the surface temperature of the water, humidity of overlying atmosphere, the isotopic ratios of atmospheric moisture, and salinity and chemical composition of lake water (Craig *et al.*, 1963; Gonfiantini, 1965; Craig & Gordon, 1965; Sofer & Gat, 1972, 1975). According to Matsubaya *et al.* (1979), the isotopic ratios of lake waters in the Dry Valleys region are controlled by the steady-state balance between evaporation and inflow. Thus, the isotopic ratio in pond waters does not explain the history of the ponds. δD and $\delta^{18}O$ values for the Wright Upper Glacier ice, snow and pond waters in the Labyrinth are

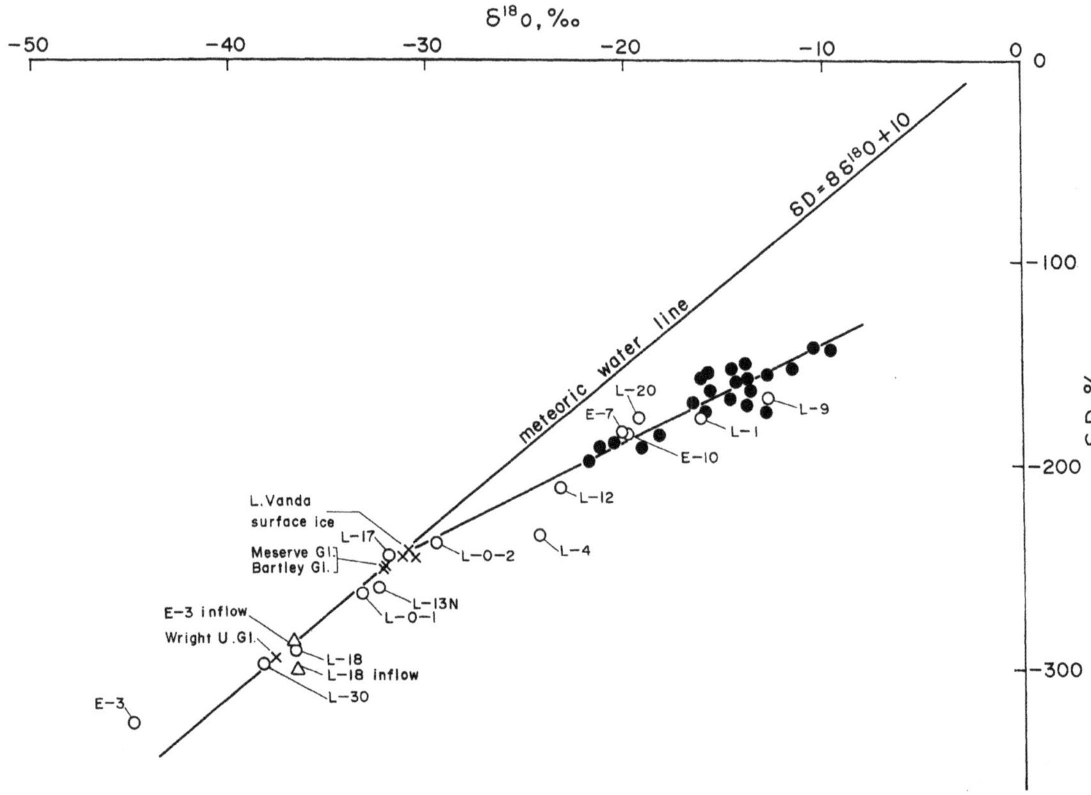

Fig. 5. Relationship between δD and $\delta^{18}O$ values of pond water, glacier ice and inflow water to pond in the Dry Valleys region. O: ponds in the Labyrinth; ●: Don Juan Pond; X: glacier ice; Δ: inflow water.

summarised in Fig. 5 together with those of other lakes and ponds in the Dry Valleys region. The isotopic ratios of pond waters in the Labyrinth mostly range from − 174 to − 326‰ in $\delta^{18}O$. The very low values found in the pond clearly indicate that the present pond water is not seawater.

The pond waters can be classified into three groups from their isotopic ratios. The first group (L-18, L-30 and E-3 ponds) is located at the terminus of the Wright Upper Glacier. The δD and $\delta^{18}O$ values in the L-18 and L-30 pond waters are similar to that of the Wright Upper Glacier, and fit closely to the meteoric water relationship of $\delta D = 8\delta^{18}O + 10$. From geomorphological observations these ponds are directly fed by surface and subsurface meltwaters from the Wright Upper Glacier. Although E-3 pond receives inflow water directly from the Wright Upper Glacier, its δD and $\delta^{18}O$ values are lower by 30 and 8.2‰, respectively, than that of the Wright Upper Glacier, and their relationship has moved away somewhat from the meteoric water line. This difference in δD and $\delta^{18}O$ can be explained by isotopic fractionation during fractional freezing of the pond water.

Although the second group (L-0-1, L-13-N and L-17 ponds) is also located at the terminus of the Wright Upper Glacier, the δD and $\delta^{18}O$ values are higher by 30–50‰ in δD and by 4–7‰ in $\delta^{18}O$ than those of the first group, and follow the meteoric water relationship. These values are similar to those of the alpine glaciers (the Meserve and Bartley Glaciers) in Wright Valley, and those of water in Lake Vanda and are also in the same range as glaciers on the side wall of Taylor Valley (δD: − 216 to − 258‰, $\delta^{18}O$: − 28.4 to − 33.6‰) (Matsubaya *et al.*, 1979). The alpine glaciers are fed by local snow and likewise these ponds may also be fed by local snowbelt.

The third group (L-1, L-4, L-7, L-9, L-12, L-20, E-7 and E-10 ponds) is distant from the terminus of the Wright Upper Glacier and seems to be fed by meltwater from surrounding snow banks. The δD and $\delta^{18}O$ values of this group vary from − 174 to − 235‰ in δD and − 12.7 to − 24.5‰ in $\delta^{18}O$, respectively, and lie on the right hand side of the meteoric water line in Fig. 5. These values are much higher than those of the second group, and similar to that of hypersaline Don Juan Pond in Wright Valley. This isotopic feature shows that significant evaporation of pond water is taking place in these ponds even though some ponds (e.g. E-7, L-20 and L-12) are far less saline than others (e.g. E-10).

Salt origin in pond water and historical events

As shown in Table 1, very high concentrations of salts were observed in some saline pond waters. The salts dissolved in lake and pond waters in the Dry Valleys region have been studied by many workers and the suggested origins include thermal springs, a marine invasion of the valley, subsurface hypersaline glacial meltwaters, atmospheric salts and rock weathering. Most workers have suggested multiple sources to account for all the salts.

The presence of lithium can be interpreted as salts having been trapped from seawater and/or reaction of the rocks with hydrothermal waters (Takamatsu *et al.*, 1983, 1986). However, none of the ponds in the Labyrinth, including those with a high salt concentration, contained any significant lithium. This also supports the argument that saline water can be formed by sources other than trapped seawater and/or hydrothermal activity. Though we cannot deny the likely contribution of salts from rock weathering, we would suggest that the evolution of saline pond waters in the Labyrinth follows the sequence outlined earlier for Dry Valley lakes by Torii & Yamagata (1981). Namely, the ponds start as glacial meltwater and snowmelt held in depressions and that the composition gradually approaches that of seawater and then moves to the right of the chemical composition diagram (Fig. 4), as does seawater, under frigid conditions, in the Thompson and Nelson model.

Torii *et al.* (1979) have calculated that the greatest possible estimated age of a saline pond (L-1) in the Labyrinth was 56 000 years, based on sodium ion concentration (4.13 g kg^{-1}), but they did not include in their calculations the salt supply from rock weathering and ion exchange from the surrounding catchment area. In the former case, sodium ion was assumed to be supplied only by meteoric water. We think this age is reasonable, but that the true age would be much younger taking the contribution of salts from rock weathering and evaporites into consideration. Thus the estimated age of the saline ponds in the Labyrinth would appear to be less than 1×10^5 years.

Acknowledgements

We are greatly indebted to the Antarctic Division, DSIR, New Zealand and US Navy for their kind support in our Antarctic researches. We thank Mr Takamatsu of Toho University for the lithium and boron analyses.

References

Craig, H., 1961. Standard for reporting concentrations of deuterium and oxygen-18 in natural waters. Science 133: 1833–1834.

Craig, H., L. I. Gordon & Y. Horibe, 1963. Isotopic exchange effects in the evaporation of water. 1. Low temperature results. J. Geophys. Res. 68: 5079–5087.

Craig, H. & L. I. Gordon, 1965. Isotopic oceanography: deuterium and oxygen-18 variations in the ocean and marine atmosphere. In D. R. Schink & J. T. Corless (eds.), Proceedings of the Symposium on Marine Geochemistry. The University of Rhode Island, Rhode Island: 277–374.

Domask, W. G. & K. A. Kobe, 1952. Mercuric determination of chlorides and water-soluble chlorohydrines. Anal. Chem. 24: 989–991.

Epstein, S. & T. Maeda, 1953. Variation of ^{18}O content of water from natural sources. Geochim. Cosmochim. Acta 4: 213–224.

Friedman, I., 1953. Deuterium content of natural waters and other substances. Geochim. Cosmochim. Acta 4: 89–103.

Gonfiantini, R., 1965. Effetti isotopici nell'evaporazione di acque salate. Atti Soc. Toscana Sci. Nat. Pisa. Mem. P.V. Ser. A. 72: 3–22.

Iwasaki, I., S. Utsumi, K. Hagino & T. Ozawa, 1956. A new spectrophotometric method for the determination of small amounts of chlorine using the mercuric thiocyanate method. Bull. Chem. Soc. Japan 29: 860–864.

Iwasaki, I., S. Utsumi, K., Hagino, T., T. Tarutani & T. Ozawa, 1957. Spectrophotometric method for the determination of small amounts of sulfate ions. Bull. Chem. Soc. Japan 30: 848–851.

Masuda, N., S. Nakaya & T. Torii, 1984. Origin of salt in Antarctic saline lake waters through trace element analysis. Mem. Natl. Inst. Polar Res., Spec. Issue 33: National Institute of Polar Research, Tokyo: 194–203.

Matsubaya, O., H. Sakai, T. Torii, H. Burton & K. Kerry, 1979. Antarctic saline lakes-stable isotopic ratios, chemical compositions and evolution. Geochim. Cosmochim. Acta 41: 7–25.

Murayama, H., Y. Yusa, G. Matsumoto & T. Torii, 1983. Activities of Japanese summer party in the Dry Valleys region during the 1981–1982 field season. Antarctic Rec. 79: 134–144 (in Japanese).

Muto, S., 1956. Distribution of bor[illegible]on in natural waters. Bull. Chem. Soc. Japan 29: 532–536.

Schucker, G. D., T. S. Magliocca & Yao-Sin Su, 1975. Spectrophotometric determination of boron in siliceous materials with azomethine H. Anal. Chim. Acta 75: 95–100.

Selby, M. J. & A. T. Wilson, 1971. The origin of the Labyrinth, Wright Valley, Antarctica. Geol. Soc. Amer. Bull. 82: 471–476.

Simonov, I. M., 1971. Oases of East Antarctica. Hydrometeorological Publishing Office, Leningrad.

Sofer, Z. & J. R. Gat, 1972. Activities and concentrations of oxygen-18 in concentrated aqueous salt solutions: Analytical and geophysical implications. Earth Planet. Sci. Lett. 15: 232–238.

Sofer, Z. & J. R. Gat, 1975. The isotope composition of evaporating brines: Effect of the isotopic activity ratio in saline solutions. Earth Planet. Sci. Lett. 26: 179–186.

Sugawara, K., 1964. Migration of elements through phases of atmosphere and hydrosphere. J. Water Waste. 6: 57–60 (in Japanese).

Takamatsu, N., M. Imahashi, K. Shimodaira & H. Kamiya, 1983. The dissolution of lithium minerals in salt solutions: Implication for the lithium content of saline waters. Geochem. J. 7: 153–160.

Takamatsu, N., M. Imahashi, K. Kamimura & M. Tsutsumi, 1986. Geochemical implication of the lithium content of saline spring waters in Japan. Geochem. J. 20: 140–151.

Thompson, T. G. & K. H. Nelson, 1956. Concentration of brines and deposition of salts from sea water under frigid conditions. Amer. J. Sci. 254: 227–238.

Torii, T., N. Yamagata, J. Ossaka & S. Murata, 1977. Salt balance in the Don Juan basin. Antarctic Rec. 58: 116–130.

Torii, T., N. Yamagata, J. Ossaka & S. Murata, 1979. A view on the formation of saline waters in the dry valleys. Mem. Natl. Inst. Polar Res., Spec. Issue 13: National Institute of Polar Research, Tokyo: 22–33.

Torii, T., S. Murta & N. Yamagata, 1981. Geochemistry of the dry valley lakes. J. Royal Soc. NZ. 11: 387–399.

Torii, T. & N. Yamagata, 1981. Limnological studies of saline lakes in the dry valleys. Antarct. Res. Ser. (Dry Valley Drilling Project) 33: 141–159.

Webb, P. N., 1972. Wright fjord, Pliocene marine invasion of an antarctic dry valley. Antarct. J.U.S. 7: 227–234.

Hydrobiologia 172: 265–289, 1989
W. F. Vincent and J. C. Ellis-Evans (eds) High Latitude Limnology

Biogeochemical study of organic substances in Antarctic lakes

Genki I. Matsumoto
The University of Tokyo, The College of Arts and Sciences, Department of Chemistry, Komaba, Meguro-ku, Tokyo 153, Japan

Key words: Antarctic lake, biogeochemistry, organic substance, stratification, hydrocarbon, fatty acid, sterol, phenolic acid, hydroxy acid, ketone

Abstract

The features of organic constituents in Antarctic lakes and ponds of the McMurdo, Syowa and Vestfold oases are summarised from a biogeochemical viewpoint. Total organic carbon or dissolved organic carbon contents in saline lakewaters are generally extremely high and much higher than those in freshwater lakes. The concentrations and/or compositions of hydrocarbons, fatty acids, sterols, phenolic acids and hydroxy acids in lake and pond waters and sediments vary markedly, probably reflecting differences in biological activity and source organisms. Long-chain alkenes, such as n-$C_{29:2}$ (carbon chain length: numbers of unsaturated bonds) are found as the major hydrocarbons in some anoxic lake sediments. Unusually, long-chain n-alkanoic acids are abundant in some Antarctic lake sediments and 24-ethylcholest-5-en-3β-ol is the most prominent sterol in most of the lakes studied. It is suggested that some bacteria, and cyanobacteria and algae are important sources of long-chain n-alkanoic acids and 24-ethylcholest-5-en-3β-ol, respectively, as previously reported from environments of the mid and lower latitudes. The dominance of p-hydroxybenzoic acid among the phenolic acids found together with the absence of syringic, p-coumaric and ferulic acids in the Antarctic lakes reflects the absence of vascular plants in the areas studied.

In three Antarctic saline lakes (Vanda, Fryxell and Ace) the kinds and amounts of organic constituents differ with depth due to the zonation of microorganisms. The maximum fatty acid contents are found at depths just above the anoxic layer, corresponding to the photosynthetic maxima in the lakes, and the depths of maximum phytoplankton populations. In the bottom sediments of the lakes, the composition of organic substances is significantly different from that in the water columns, indicating that the sinking organic substances are degraded rapidly by microorganisms on the lake bottom.

Introduction

Although most of Antarctica is covered with a thick ice sheet of approximately 2300 m, there are a number of ice-free areas, so-called 'oases', in and around the coastal regions of the continent (Fig. 1). Antarctica is an extremely harsh environment for living organisms, and is characterized by the absence of vascular plants, except in the northern part of the Antarctic Peninsula. Cryptogamic organisms, such as algae, cyanobacteria, bacteria and fungi are widely distributed in the oases, in addition to mosses and lichens. Many kinds of endemic species of organisms are distributed throughout Antarctica, which is situated farthest from the industrialized areas, so that its environment is thought to be the least polluted in the world. A large number of lakes and ponds

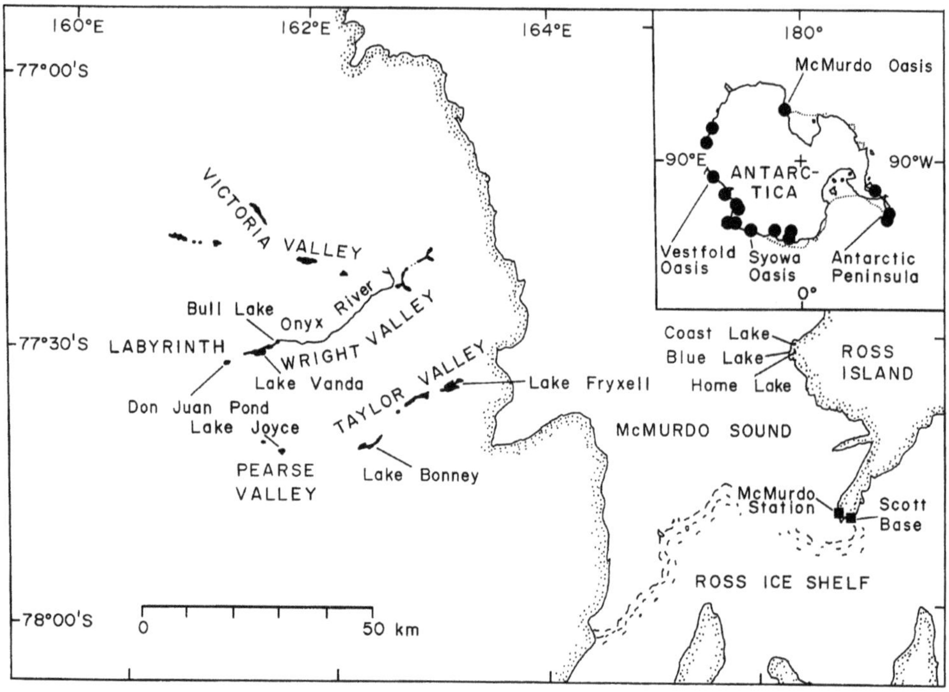

Fig. 1. Lakes and ponds in the McMurdo Oasis. ●: oasis.

with a wide variety of salt contents ranging from pure snow meltwater to 13 times that of seawater occur in the oases. The salt compositions of the lakes also vary significantly (Torii *et al.*, 1975, 1977; Torii & Yamagata, 1981). In most Antarctic lakes organic substances are largely autochthonous, since terrestrial vegetation in the catchment areas is virtually non-existent (Burton, 1981). Thus the composition and contents of organic substances in these lakes and ponds can be expected to be different not only within the lakes but also from those in the mid and lower latitudes.

Since International Geophysical Year (1957), a large number of investigators have studied inorganic constituents of Antarctic lakes (*e.g.*, Burton, 1981; Torii & Yamagata, 1981), but little attention had been paid to the study of organic substances before we started our research. Recently, organic carbon and lipid components, such as hydrocarbons, fatty acids, sterols, phenolic acids and hydroxy acids have been studied in the lakes from the McMurdo Oasis of southern Victoria Land, Syowa Oasis in Enderby Land and from Vestfold Oasis (Vestfold Hills) of Princess Elizabeth Land (Figs. 1 and 2). Matsumoto *et al.* (1984c) outlined the range of organic substances found in the oases of Antarctica. Here I characterize in detail the organic substances present in lake and pond waters as well as sediments from the Antarctic oases, discuss these in relation to source organisms, and compare results with those obtained for lacustrine environments in the mid and lower latitudes. In addition, stratification of organic substances in saline lakes is also discussed.

High abundance of organic carbon in saline waters

In the Dry Valleys of the McMurdo Oasis, Parker *et al.* (1974) reported high concentrations and rapid seasonal fluctuations of total organic

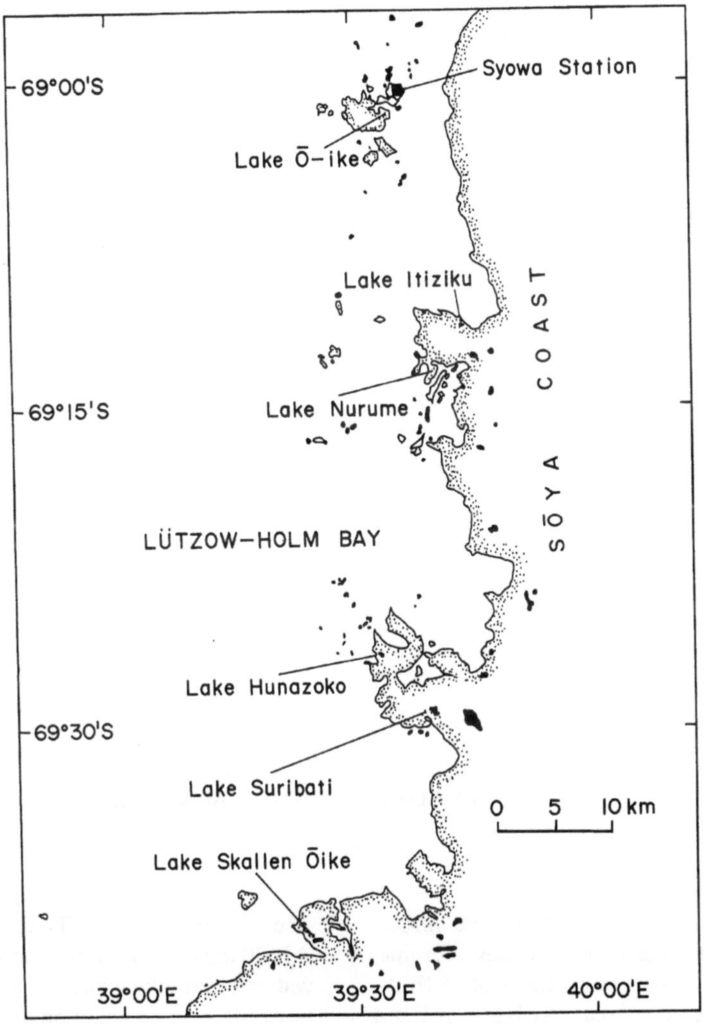

Fig. 2. Lakes in the Syowa Oasis.

carbon (TOC) in Lake Bonney, Taylor Valley and suggested an annual cycle of photosynthetic production and dark or photo-assimilation by the unique microbial communities in this permanently ice-covered and stratified polar lake (Table 1). Matsumoto and his co-workers (1977, 1979, 1984b, 1987a, 1987b, 1988) reported high concentrations of TOC in the middle and bottom waters of Lakes Bonney, Vanda and Fryxell as well as certain saline ponds of the Labyrinth in the Dry Valleys and surface water of Home Lake in Ross Island, while those of Lakes Joyce (Matsumoto *et al.*, 1979) and Miers (DOC, Downes *et al.*, 1986) were considerably lower. High TOC concentrations in saline waters have also been reported in meromictic Ace Lake (60 mg C l^{-1}, Burton, 1980) and hypersaline Deep Lake (50 mg C l^{-1}, Barker, 1980; Hand, 1980), two lakes in the Vestfold Oasis. In the Syowa Oasis, Tominaga & Fukui (1981) observed extremely

Table 1. TOC and DOC results for lake, pond and stream waters from the McMurdo, Syowa and Vestfold oases in Antarctica.

Locality (sample depth, m)	TOC, DOC ($mgC\,l^{-1}$)	References*[1]
McMurdo Oasis		
Blue Lake (surface)	0.43	a
Home Lake (surface)	110	a
Lake Vanda (5.4–68.0), L1 Pond (surface)	0.31–63.8	a,b,c,d
Ten unnamed ponds in the Labyrinth (surface–2.6)	0.78–23.1	d
Lake Bonney, east lobe (3.5–31.5)	1.2–32.2	e
Lake Bonney, east lobe (5.4–33.5)	0.52–28.0	a,b
Lake Bonney, west lobe (3.5–31.5)	0.14–20.5	e
Lake Bonney, west lobe (5.4–30.4)	0.44–18.6	a,b
Lake Fryxell (5.0–18.5)	1.4–29.1	b,f
Lake Joyce (5.4–30.4)	0.31–3.4	a
Lake Miers (4.9–15.5)	0.6–0.8*[2]	g
Stream waters (surface)	0.72–31*[2]	a,g,h
Snow packs	0.8–2.6*[2]	g
Glacier ice	1.2–2.8*[2]	g
Syowa Oasis		
Lake Nurume (0–15)	2–30*[2]	i
Lake Hunazoko (0–7)	103–186*[2]	i
Lake Suribati (0–29)	20–130*[2]	i
Thirty-five freshwater lakes	1–2*[2]	i
Vestfold Oasis		
Ace Lake (?)	> 60	j
Deep Lake (?)	50	k

*[1] a, Matsumoto *et al.* (1979); b, Matsumoto and Hanya (1977); c, Matsumoto *et al.* (1984b); d, Matsumoto *et al.* (1987b); e, Parker *et al.* (1974); f, Matsumoto *et al.* (1988); g, Downes *et al.* (1986); h, Allnut *et al.* (1981); i, Tominaga and Fukui (1981); j, Burton (1980); k, Barker (1980).
*[2] DOC.

high concentrations of dissolved organic carbon (DOC) in saline waters, from Lakes Nurume, Hunazoko and Suribati, while those of 35 freshwater lakes in the same area were between 1 and 2 $mg\,C\,l^{-1}$ (Table 1). The highest DOC concentration was found in Lake Hunazoko (186 $mg\,C\,l^{-1}$).

In general, these TOC or DOC concentrations are very high compared with those of ordinary lakewaters in Japan. For example, the TOC values of an eutrophic lake, Lake Suwa, range from 1.8 to 3.7 $mg\,C\,l^{-1}$ (Sakamoto *et al.*, 1975).

For these high TOC or DOC concentrations, a number of explanations are proposed. Parker *et al.* (1974) suggested that while much organic substances in Lake Bonney may be of long-term origin, a significant fraction of the total is produced and recycled annually in the ecosystem. The changes in TOC contents (e.g., 7.5–30 $mg\,C\,l^{-1}$) in the bottom highly saline waters of the lake which show very low temperatures (0– –4 °C) could be due to seepage input. This hypothesis is interesting, but would not seem to fully explain such large TOC variations, so further study is clearly needed. Barker (1980) suggested for Deep Lake that TOC, derived both from meltstreams and the lake itself, are accumulated due to the absence of significant microbial decomposition.

At Lake Vanda, Matsumoto *et al.* (1984b) suggested that the phytoplankton was prevented from sinking by the high density of bottom water (specific gravity, 1.1 $g\,cm^{-3}$). Moribund plankton material undergoes microbial attack in the warm anoxic bottom water (about 24 °C) and thus refractory organic substances are accumulated there over a long period after the degradation of labile organic constituents. The high abundance of ammonium-N and orthophosphate-P in the bottom waters (Vincent *et al.*, 1981; Matsumoto *et al.*, 1985b) supports the contention that degradation of large amounts of organic substances has occurred. The low TOC content (1.9 mg C l^{-1} dry wt.) of the bottom sediment from Lake Vanda also supports this explanation (Matsumoto *et al.*, 1984c). The high TOC concentration of the anoxic bottom layers in Lake Fryxell may also be due to accumulation of refractory organic substances over a long period following the degradation of more labile organic constituents (Matsumoto *et al.*, 1988). On the other hand, high TOC concentration in Home Lake, which is bordered by an Adélie Penguin rookery, is explained by the influence of the excretions of the penguins and of skuas nesting near the lake (Matsumoto *et al.*, 1979).

Lakes Nurume, Hunazoko and Suribati are located on the Soya Coast and their salt composition is similar to that of seawater, though their concentrations are much higher (Murayama, 1977; Hirabayashi & Ossaka, 1977; Tominaga & Fukui, 1981). Lake Hunazoko is believed to be a relict lake (Torii *et al.*, 1977). The high salt concentrations in Lakes Nurume and Suribati are also probably due to concentration of sea salts. Although there are considerable differences in the DOC concentration, both among the lakes as well as with depths, a significant correlation is found between DOC and chlorinity (Fig. 3, Tominaga & Fukui, 1981). The suggestion that high DOC values in the saline lakes are attributable to concentration of organic substances in seawater

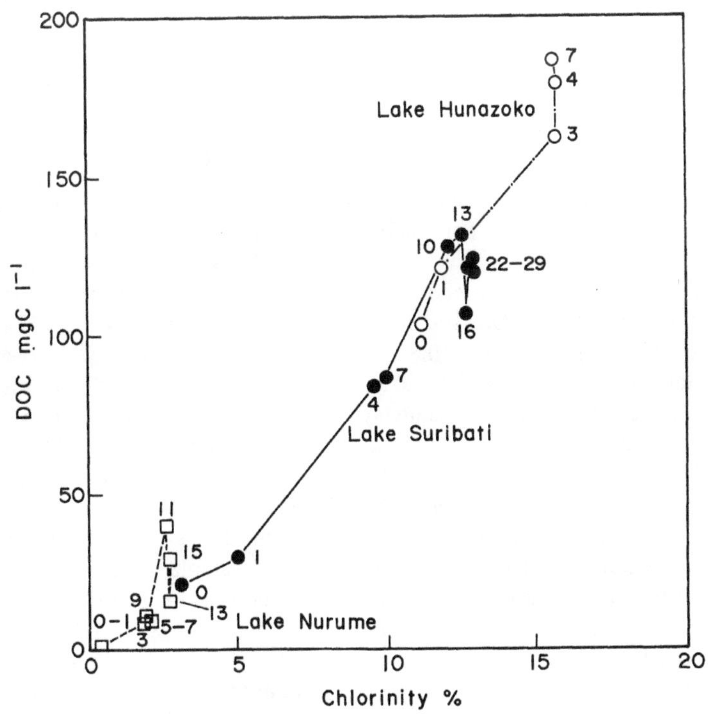

Fig. 3. Correlation between chlorinity and DOC for Lakes Nurume, Suribati and Hunazoko in the Syowa Oasis (After Tominaga & Fukui, 1981).

under freezing conditions can be excluded, because the original seawater (chlorinity, 1.9%) would have to contain an extremely high DOC content of about 10 mg C l^{-1}. Thus it is more likely that organic substances in the coastal saline lakes of the Syowa Oasis are attributable to biological activity in and around the lakes during concentration of sea salts.

A high correlation coefficient of 0.82 ($n = 12$) was obtained between TOC and electrical conductivity (at 25 °C) for the Labyrinth ponds (Matsumoto *et al.*, 1987b). The Labyrinth lies approximately 50 km from the coast between altitudes of 600 and 1 000 m above sea level. More than 60 small ponds with various salt contents in chlorinity ranging from 0.0049 to 58.3 g kg^{-1} are distributed there (Matsumoto *et al.*, 1985b; Torii *et al.*, 1988). For the Labyrinth ponds, direct influences of sea salts are unlikely, so atmospheric salts may be important sources of the pond salts. For instance, Torii & Yamagata (1981) have estimated that an unnamed saline pond (L1 pond) in the Labyrinth having a sodium ion content of 4.13 g kg^{-1} (2.6 times that of seawater) can be formed in 56 000 years simply from snow. Thus the salts supplied through meltwaters from snow and/or glaciers in the catchment areas are clearly important in the formation of saline lakes and ponds (Torii & Yamagata, 1981; Torii *et al.*, 1988).

Downes *et al.* (1986) reported that snow and glacier ice from the McMurdo Oasis contain amounts of DOC ranging from 0.8 to 2.8 mg C l^{-1} (Table 1). The TOC or DOC values in meltstream waters are also, therefore, relatively high, ranging from 0.72 to 5.6 mg C l^{-1} (Matsumoto *et al.*, 1979; Downes *et al.*, 1986). Allnut *et al.* (1981) reported an exceptionally high DOC value (31 mg C l^{-1}) in one Taylor Valley meltstream. At the same time, biological activity in and around the lakes and ponds produces organic substances, but their degradation rates will be relatively low under frigid conditions.

In general, therefore, high TOC or DOC concentrations of saline lakes and ponds (including the Labyrinth ponds) which have no outflows, can be explained as follows: organic substances, both from meltwaters and *in situ* biological activity, concentrate in waters, together with dissolved salts, by evaporation of waters over long time periods after the initial degradation of more labile organic constituents.

The contents of organic carbon extractable with ethyl acetate (EOC: lipid-carbon) may reflect an abundance of living microorganisms in water columns, since all microorganisms contain lipids (Matsumoto *et al.*, 1984b). The EOC values in lake waters from the McMurdo Oasis range from 0.05 to 9.1 mg C l^{-1}, differing markedly not only among the lakes but also with depth, as in the case of the TOC values (Matsumoto & Hanya, 1977; Matsumoto *et al.*, 1979, 1984c). Generally, the EOC values of Lakes Vanda and Bonney increase with depth, and are similar to that of the TOC results. The EOC/TOC values of the McMurdo Oasis range from 3.4 to 31%, also varying considerably both with lake and depth. The high EOC/TOC values (19–31%) for depths between 30.4 and 55.4 m in Lake Vanda suggest an abundance of fresh organic substances, including living organisms (Matsumoto *et al.*, 1984b).

Organic carbon and total nitrogen in sediments

The concentrations of TOC, EOC and total nitrogen (TN) in both lake and pond sediments from the McMurdo, Vestfold and Syowa oases vary remarkably, ranging from 0.18 to 190 mg C g^{-1} of dry sediment, 0.0015 to 11 mg C g^{-1} and from 0.021 to 21 mg N g^{-1}, respectively, reflecting probably differences in the distribution and activity of microorganisms in both lakes and ponds (Matsumoto *et al.*, 1979, 1981, 1983b, 1984c). These concentrations are generally low, but an extremely high TOC value was found in the sediment of Lake Skallen Ōike in the Syowa Oasis. This sample comprised mostly algal material (Chlorophyta, mainly *Cosmarium* spp.; Matsumoto *et al.*, 1983a). The EOC/TOC and TOC/TN values of the lake and pond sediments change remarkably among the samples, ranging from 0.83 to 29% and from 3.7 to 10, respectively, as in the case of the TOC values (Matsumoto *et al.*, 1979, 1981, 1984c).

Features of organic compounds

Hydrocarbons

Hydrocarbons, *n*-alkanes, alkenes, acyclic and cyclic isoprenoids, are ubiquitous compounds, widely distributed in geological and biological environments, and are used as biomarkers, although alkenes are usually less abundant in geological samples. Hydrocarbons in lake and pond waters in Antarctica have been studied only in the McMurdo Oasis. Generally the concentrations of hydrocarbons in lakewaters are extremely low and cannot be detected in the water columns of Lakes Vanda, Bonney, Fryxell and Joyce, except for the bottom layers of Lake Vanda (Table 2,

Table 2. Hydrocarbons found in Antarctic samples

Samples	Range	Conc. ($\mu g\ g^{-1}$)*1	Major five constituents (%)*2	Lake type
McMurdo Oasis				
Water*3				
Home Lake	C_{15}–C_{31}	4.8	$C_{17:2}$ (75), $C_{19:2}$ (9), $nC_{17:0}$ (8), $nC_{18:0}$ (1), $nC_{23:0}$ (0.8)	Saline
Lake Vanda (66 m)	C_{16}–C_{36}	1.8	$bC_{18:0}$ (67), $nC_{17:0}$ (18), $nC_{18:0}$ (4), $bC_{19:0}$ (2), $nC_{32:0}$ (2)	Saline
Lake Vanda (65.9 m)	C_{16}–C_{32}	6.5	$bC_{18:0}$ (75), $nC_{17:0}$ (14), $nC_{18:0}$ (4), $bC_{19:0}$ (4), $nC_{19:0}$ (1)	Saline
L1 Pond (surface)	C_{16}–C_{33}	0.22	$nC_{27:0}$ (13), $nC_{26:0}$ (13), $nC_{25:0}$ (12), $nC_{28:0}$ (10), $nC_{24:0}$ (10)	Saline
Sediment*3				
Home Lake	C_{14}–C_{29}	40	$nC_{17:0}$ + $C_{17:1}$ + $iC_{19:0}$ (48), $nC_{15:0}$ (29), $C_{20:0}$ (7)	Saline
Lake Vanda (edge)	C_{14}–C_{35}	0.13	$C_{19:1}$ (22), $C_{18:1}$ (13), $C_{28:1}$ (9), $C_{17:1}$ (9), $nC_{19:0}$ (8)	Saline
Lake Vanda	C_{13}–C_{30}	7.4	$bC_{18:0}$ (42), $nC_{17:0}$ (37), $C_{17:1}$ (4), $nC_{23:0}$ (3), $bC_{19:0}$ (3)	Saline
Don Juand Pond-1	C_{17}–C_{27}	0.0015	$nC_{19:0}$ (13), $nC_{17:0}$ (12), $nC_{21:0}$ (10), $nC_{23:0}$ (9), $bC_{18:0}$ (8)	Saline
Don Juand Pond-2	C_{17}–C_{29}	0.0017	$nC_{25:0}$ (13), $nC_{26:0}$ (11), $nC_{27:0}$ (11), $nC_{24:0}$ (10), $nC_{23:0}$ (10)	Saline
Don Juand Pond-3	C_{15}–C_{28}	0.0040	$C_{19:1}$ (40), $bC_{18:0}$ (12), $nC_{15:0}$ (5), $nC_{17:0}$ (5), $nC_{19:0}$ (4)	Saline
L1 Pond	C_{15}–C_{34}	1.1	$nC_{17:0}$ + $C_{17:1}$ (97), $C_{16:1}$ (0.9), $C_{19:1}$ (0.6), $nC_{21:0}$ (0.2)	Saline
L12 Pond	C_{15}–C_{30}	0.016	$C_{17:1}$ (62), $C_{19:1}$ (11), $nC_{17:0}$ (10), $C_{18:1}$ (3), $nC_{23:0}$ (2)	Saline
L13 Pond	C_{17}–C_{33}	0.0020	$nC_{17:0}$ (13), $nC_{33:0}$ (11), $nC_{19:0}$ (9), $nC_{23:0}$ (8), $nC_{25:0}$ (7)	Saline
Lake Bonney, west lobe	C_{14}–C_{34}	5.2	$C_{29:2}$ (92), $nC_{27:0}$ (2), $nC_{25:0}$ (1), $nC_{23:0}$ (0.9), $nC_{17:0}$ (0.8)	Saline
Lake Fryxell	C_{15}–C_{33}	1.3	$nC_{29:2}$ (33), $nC_{17:0}$ (21), $nC_{27:0}$ (6), $bC_{19:0}$ (5), $bC_{18:0}$ (4)	Saline
Lake Joyce	C_{14}–C_{35}	0.51	$C_{27:1}$ (28), $C_{25:1}$ (21), $C_{29:1}$ (16), $nC_{23:0}$ (9), $nC_{28:0}$ (7)	Saline
Epibenthic algae*4				
Coast Lake	C_{12}–C_{36}	140	$C_{18:1}$ (20), $C_{17:1}$ (12), $C_{19:1}$ (10), $C_{17:0}$ (8), $bC_{18:1}$ (7)	
Bull Lake	C_{14}–C_{31}	15	$nC_{17:0}$ + $C_{17:1}$ (67), $bC_{18:1}$ (11), $C_{18:1}$ (8), $C_{19:1}$ (5)	
Lake Vanda	C_{12}–C_{35}	11	$nC_{17:0}$ + $C_{17:1}$ (61), $C_{18:1}$ (24), $C_{19:1}$ (11), $nC_{16:0}$ (0.7)	
Don Juan Pond	C_{15}–C_{27}	4.6	$C_{17:1}$ (69), $nC_{17:0}$ (16), $C_{18:1}$ (10), $C_{19:1}$ (3), $C_{16:1}$ (0.5)	
Lake Fryxell	C_{15}–C_{29}	7.2	$nC_{17:0}$ (42), $C_{17:1}$ (40), $C_{18:1}$ (5), $C_{19:1}$ (2), $nC_{16:0}$ (2)	
Syowa Oasis				
Sediment*5				
Lake Ō-ike	C_{14}–C_{36}	13	$nC_{23:0}$ (39), $nC_{25:0}$ (18), $nC_{22:0}$ (9), $nC_{21:0}$ (8), $nC_{27:0}$ (7)	Freshwater
Lake Itiziku	C_{12}–C_{37}	2.9	$iC_{19:0}$ (29), $nC_{15:0}$ (20), $iC_{18:0}$ (11), $iC_{16:0}$ (11), $nC_{17:0}$ (7)	Saline
Lake Nurume	C_{15}–C_{30}	0.72	$nC_{23:0}$ (19), $nC_{21:0}$ (11), $nC_{18:0}$ (8), $nC_{30:0}$ (7), $nC_{22:0}$ (7)	Saline
Lake Hunazoko	C_{15}–C_{33}	0.38	$iC_{19:0}$ (22), $iC_{20:0}$ (10), $nC_{17:0}$ (8), $nC_{23:0}$ (8), $nC_{15:0}$ (7)	Saline
Lake Skallen Ōike	C_{14}–C_{32}	32	$nC_{17:0}$ (37), $nC_{23:0}$ (19), $nC_{25:0}$ (6), $nC_{24:0}$ (6), $nC_{20:0}$ (5)	Freshwater
Vestfold Oasis				
Sediment				
Ace Lake*6	C_{15}–C_{36}	90	$iC_{30:4}$, $iC_{20:0}$, $nC_{34:5}$; fern-7-ene, $iC_{25:0}$ (decreasing abund.)	Saline
Deep Lake*5	C_{13}–C_{36}	1.5	$iC_{19:0}$ (24), $nC_{15:0}$ (20), $iC_{18:0}$ (11), $nC_{17:0}$ (8), $iC_{16:0}$ (6)	Saline

*1 Except for water samples ($\mu g\ l^{-1}$). *2 $nC_{m:n}$, $bC_{m:n}$ and $iC_{m:n}$ are normal, branched and isoprenoid, respectively. $C_{m:n}$, carbon chain length: number of unsaturated bonds. *3 Matsumoto (1980); Matsumoto *et al.* (1979, unpublished results). *4 Mainly cyanobacteria such as *Oscillatoria* spp., *Phormidium* spp. and *Calothrix* spp. Matsumoto (1980); Matsumoto *et al.* (1979, unpublished results). *5 Matsumoto *et al.* (1981). *6 Volkman *et al.* (1986a); sediment core (0-5 cm).

Matsumoto *et al.*, 1979, 1984b, 1984c, 1988). Alkenes, $C_{17:2}$ (carbon chain length: number of unsaturated bonds) and $C_{19:2}$, are found as the prominent components in Home Lake and may be attributable to certain algae and cyanobacteria occurring in this penguin-influenced lake (Matsumoto *et al.*, 1979). Small amounts of *n*-alkanes ranging from *n*-C_{16} to *n*-C_{33} have been found in L1 pond of the Labyrinth, with a predominance of long-chain components ($\geqslant C_{20}$). They come probably from certain bacteria (Albro, 1976) and/or fungi (Weete, 1976). In the bottom waters of Lake Vanda, branched alkanes (2,6-dimethylhexadecane and 2,6,10 trimethylhexadecane, tentatively identified) and *n*-C_{17} are found with the major constituents. These branched alkanes are also found in the bottom sediment of Lake Vanda as major constituents (Fig. 4), and, in small quantity, in sediment samples from Don Juan Pond and Lake Fryxell. The occurrence of these branched alkanes is of special interest because they are not found in any environmental samples, including living microorganisms, other than in Antarctica, suggesting the presence of some unknown microorganisms in the bottom layers of Lake Vanda as well as, possibly, other Antarctic environments.

Recently Volkman *et al.* (1986a) reported hydrocarbons in a sediment core (160 cm long) from Ace Lake of the Vestfold Oasis. The range of carbon chain lengths of hydrocarbons (C_{13}-C_{36}) in lake and pond sediments from the McMurdo Oasis are similar to those of the Syowa and Vestfold oases (Table 2). Concentrations of these hydrocarbons change remarkably among the surface lake and pond sediments, ranging from 0.0015 to 90 $\mu g\, g^{-1}$ of dry sample. The highest value (125 $\mu g\, g^{-1}$), found in an Ace Lake sediment core (20–25 cm section), is amongst the highest reported for non-polluted recent sediments (Volkman *et al.*, 1986). The lowest result is observed in Don Juan Pond, which is extremely saline, and where microbial activity is very limited (Torii & Yamagata, 1981). Major constituents of hydrocarbons are different among the sediment samples, indicating differences of source organisms. The virtual absence of any contribution of allochthonous organic substances to the lakes and ponds may also be responsible for the large differences in organic substances in Antarctic lakes compared to lakes elsewhere.

Alkenes are often predominant hydrocarbons in Antarctica, and occurred as the major constituents in sediment samples from Home Lake,

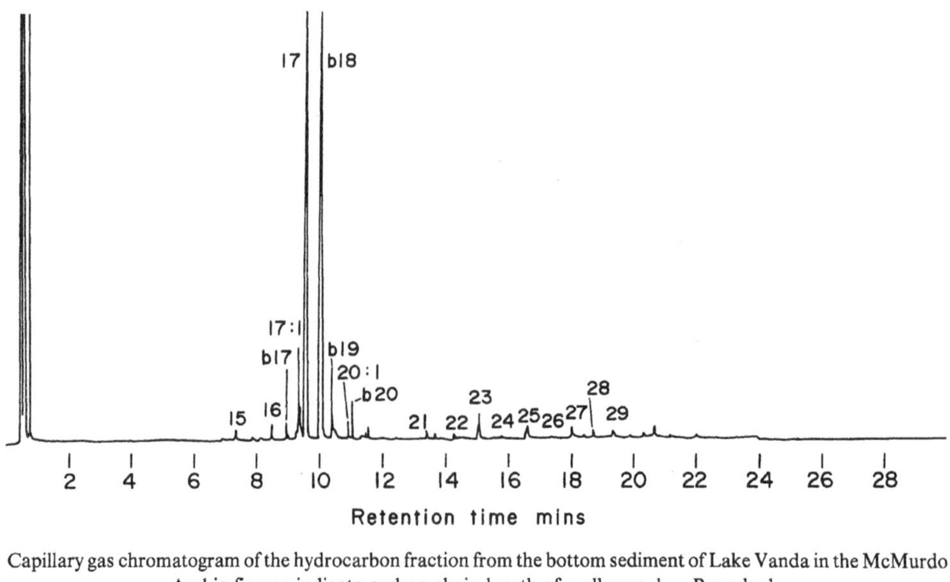

Fig. 4. Capillary gas chromatogram of the hydrocarbon fraction from the bottom sediment of Lake Vanda in the McMurdo Oasis. Arabic figures indicate carbon chain length of *n*-alkanes. b = Branched.

Lakes Vanda, Bonney, Fryxell and Joyce, as well as L1 and L12 ponds of the McMurdo Oasis (Table 2). Long-chain alkenes, such as $C_{29:2}$ and $C_{27:1}$ have been detected in Lakes Bonney, Fryxell and Joyce as the most prominent hydrocarbon (Matsumoto *et al.*, 1979, 1988). Hydrogenation of $C_{29:2}$ alkene from Lake Fryxell gave n-C_{29} alkane (Matsumoto *et al.*, 1988). Thus the former is a straight chain compound. Also Volkman *et al.* (1986a) has reported alkenes in an Ace Lake sediment core with an abundance of n-$C_{29:2}$ and n-$C_{34:3-5}$ alkenes. These long-chain alkenes were not found throughout the water columns of Lakes Bonney and Fryxell (Matsumoto *et al.*, 1979, 1988). Although it is possible that some algae and cyanobacteria produce these long-chain alkenes under heterotrophic conditions (Patterson, 1967), it is very likely that these long-chain alkenes come from microorganisms such as bacteria in the bottom anoxic sediments (Albro, 1976; Matsumoto *et al.*, 1988), while short-chain n-alkanes and alkenes will be largely due to algae and cyanobacteria as shown by their composition (Table 2).

Acyclic isoprenoid hydrocarbons having i-C_{16}, i-C_{18}, i-C_{19} (pristane), i-C_{20} (phytane), i-C_{25} (2, 6, 10, 15, 19- pentamethyleicosane) and i-C_{30} (tetrahydrosqualene) carbons are found as the major constituents in the sediments from Home Lake in McMurdo Oasis, Lakes Itiziku and Hunazoko in the Syowa Oasis and Deep Lake and Ace Lake in the Vestfold Oasis (Table 2). Pristane and phytane have been found commonly in surface lake and marine sediments from the mid and lower latitudes, although the occurrence of isoprenoid alkanes, i-C_{16} and i-C_{18}, in the modern environments is not common (Matsumoto, 1980). Isoprenoid alkanes, i-C_{16}, i-C_{18}, i-C_{19} and i-C_{20}, are believed to be products of the diagenesis of the phytyl side chain of chlorophylls, although other sources are possible (Powell & McKirdy, 1973; Didyk *et al.*, 1978). For instance, photosynthetic bacteria and zooplankton contain isoprenoid alkanes. Nissenbaum *et al.* (1972) suggest the lipids of halophilic bacteria as a further source of isoprenoid alkanes. The abundance of isoprenoid hydrocarbons, i-C_{20}, i-C_{25} and i-C_{30}, in Ace Lake sediment is clearly attributable to the dominant archaebacteria (methanogens, Volkman *et al.*, 1986).

Pristane is generated from phytol by an oxidation pathway, such as *via* phytanic and/or phytenic acids and subsequent defunctionalization, whereas i-C_{20} is formed through various reductive paths. Thus the ratio of i-C_{19}/i-C_{20} in a anoxic environment tends to give a low value, while the ratio in an oxic environment is reversed (Brooks *et al.*, 1969; Powell & McKirdy, 1973; Didyk *et al.*, 1978). In the sediments from Lakes Itiziku and Hunazoko of the Syowa Oasis and Deep Lake of the Vestfold Oasis, i-C_{19} is much more abundant than i-C_{20}, thus reflecting formation under oxic conditions which are known to occur in the bottom waters of Lakes Itiziku and Hunazoko (Murayama, 1977; Tominaga & Fukui, 1981). Deep Lake is monomictic, becoming completely mixed once each year during winter, and is fully saturated with dissolved oxygen at all times (Hand, 1980; Burton, 1981). Thus the isoprenoid alkanes found in the lake sediments will be mainly attributable to algae and cyanobacteria. In Ace Lake sediment, however, i-C_{20} and i-C_{25} are abundant, reflecting anoxic conditions (Volkman *et al.*, 1986). In the extremely highly saline waters of Lake Itiziku and Deep Lake, it is possible that halophilic bacteria are also an important source of isoprenoid alkanes in addition to algae and cyanobacteria (Matsumoto *et al.*, 1981).

Triterpenoids, fern-7-ene, hop-22(29)-ene, hop-17(21)-ene, 17β(H)-hop-21-ene and neohop-13(18)-ene and sterenes, such as stera-3,5-dienes; stera-3,5,22-trienes and ster-2-enes have been identified in an Ace Lake sediment core (Volkman *et al.*, 1986). Hop-22(29)-ene is the major hopene in the lake, and has also been found in sediment samples from Lakes Fryxell and Vanda (Matsumoto *et al.*, 1987a, 1988). These triterpenes mainly come from bacteria (Volkman *et al.*, 1986; Matsumoto *et al.*, 1988).

Fatty acids

Fatty acids are major lipid components and widely distributed in water and sedimentary environments. Fatty acids found in waters and sediments from Antarctic lakes and ponds are summarized in Table 3. Normal alkanoic acids ranging in carbon chain length from n-C_8 to n-C_{32} are found in waters and sediments of the McMurdo Oasis with a predominance of even-carbon numbers, in addition to n-alkenoic and branched (*iso* and *anteiso*) acids. They are similar to those of lake sediments from the Syowa and Vestfold oases. The concentrations of fatty acids differ remarkably among the water and sediment samples, ranging from 2.5 to 1600 $\mu g\,l^{-1}$ and 0.045 to 2000 $\mu g\,g^{-1}$, respectively, as in the case of the TOC and hydrocarbon contents, although these values are generally considerably lower than those, for instance, in a temperate site, the polluted Tama River (Matsumoto, 1980; Matsumoto *et al.*, 1979, 1981, 1984c, unpublished). The lowest value was found in Don Juan Pond sediment, whereas the highest value was detected in Home Lake sediment. The major fatty acids in water and sediment samples from the McMurdo Oasis, excepting the bottom sediments of Lakes Vanda and Joyce, were all short-chains ($<C_{20}$), *i.e.*, n-C_{16}, n-C_{18}, u-C_{16} (unsaturated), u-C_{18} n-C_{14}, n-C_{12} and b-C_{15} (*iso*- and *anteiso*- -branched), while those in the sediments from Lakes Ō-ike, Itiziku and Skallen Ō-ike of the Syowa Oasis as well as Lakes Vanda and Joyce in the McMurdo Oasis contain unusual long-chain components ($\geqslant C_{20}$), *i.e.*, n-C_{22} and n-C_{24}, in spite of the absence of vascular plants in the areas studied. Figure 5 shows an example of the presence of long-chain n-alkanoic acids in Antarctic lake sediment. Volkman *et al.* (1986b) have also reported the high abundance of long-chain n-alkanoic acids along with the absence of short-chain n-alkanoic acids in Ace Lake sediment.

It is possible that these long-chain n-alkanoic acids are derived from aeolian dusts from the mid and lower latitudes, because some aeolian dusts collected at sea near land areas contain long-chain n-alkanoic acids in considerable quantity (*e.g.*, Simoneit *et al.*, 1977). If so, long-chain n-alkanoic acids could be expected to be distributed relatively homogeneously throughout Antarctica. Long-chain n-alkanoic acids, however, were found in only 6 lake and pond sediment samples among 19 samples studied (Table 3). Also they were not abundant in the water columns of Lakes Vanda and Joyce of the McMurdo Oasis (Matsumoto *et al.*, 1979, 1984b, 1987a) and those of Ace Lake of the Vestfold Oasis (Volkman *et al.*, 1986b). Furthermore, the high concentrations of fatty acids in Lake Ō-ike (Fig. 5, 240 $\mu g\,g^{-1}$) and Lake Skallen Ōike (1900 $\mu g\,g^{-1}$) clearly demonstrate that these long-chain n-alkanoic acids are due to *in situ* microbial activity rather than aeolian dusts. Volkman *et al.* (1986b) also concluded that long-chain n-alkanoic acids found in Ace Lake are due to bacteria in the bottom sediment. Thus aeolian dusts are unlikely sources of long-chain n-alkanoic acids. These results instead indicate that microorganisms such as bacteria are significant sources of long-chain n-alkanoic acids in Antarctic lake sediments. Lakes Vanda, Joyce, Itiziku and Ace are all saline and the lake bottoms are anoxic, except for Lake Itiziku. The bottom sediments of these lakes have an unique hydrocarbon composition, as discussed above (Table 2), probably reflecting unique microbial communities. However, Lakes Ō-ike and Skallen Ōike are freshwater, and their major hydrocarbons are only long- and short-chain n-alkanes, suggesting that the microbial communities are not unusual. It is, therefore, more likely that significant amounts of long-chain n-alkanoic acids in environmental samples from the mid and lower latitudes are derived from microorganisms.

Saturated fatty acids can be classified into three groups based on their possible source materials: Short-chain n-alkanoic acids are distributed in lipids in every plant and animal; long-chain n-alkanoic acids are abundant in the waxes of vascular plants; branched (*iso* and *anteiso*) acids are dominant in bacterial lipids (Matsuda & Koyama, 1977; Matsumoto & Hanya, 1980a). In addition, fatty acids incorporated into sediments undergo significant modification in their amounts and composition during early stages of sedimen-

Table 3. Fatty acids found in Antarctic samples

Sample	Range	Conc. ($\mu g\,l^{-1}$)*1	Major five constituents (%)*2
McMurdo Oasis			
Water*3			
Blue Lake (surface)	C_{12}–C_{32}	5.9	$nC_{16:0}$ (39), $nC_{18:0}$ (16), $uC_{18:n}$ (11), $uC_{16:n}$ (10), $nC_{14:0}$ (6.4)
Home Lake (surface)	C_{8} –C_{28}	1600	$uC_{18:n}$ (32), $nC_{16:0}$ (29), $uC_{16:n}$ (28), $nC_{18:0}$ (3.3), $bC_{15:0}$ (2.4)
Onyx River (surface)	C_{14}–C_{30}	4.6	$nC_{16:0}$ (25), $uC_{16:n}$ (12), $uC_{18:n}$ (11), $nC_{14:0}$ (4.4), $nC_{18:0}$ (4.4)
Lake Vanda (55.4 m)	C_{12}–C_{30}	61	$uC_{16:n}$ (63), $uC_{18:n}$ (20), $nC_{16:0}$ (8.9), $nC_{14:0}$ (6.5), $nC_{18:0}$ (0.7)
L1 Pond (surface)	C_{12}–C_{30}	7.1	$nC_{16:0}$ (59), $nC_{18:0}$ (10), $nC_{14:0}$ (6.8), $uC_{16:n}$ (5.2), $uC_{18:n}$ (5.2)
Lake Bonney,			
east lobe (33.5 m)	C_{10}–C_{24}	14	$nC_{16:0}$ (42), $nC_{14:0}$ (28), $uC_{18:n}$ (13), $uC_{16:n}$ (11), $nC_{12:0}$ (3.3)
west lobe (18.4 m)	C_{12}–C_{18}	2.5	$nC_{16:0}$ (38), $nC_{14:0}$ (20), $uC_{18:n}$ (16), $uC_{16:n}$ (14), $nC_{18:0}$ (8.3)
Lake Fryxell (10.0 m)	C_{12}–C_{24}	66	$uC_{18:n}$ (38), $nC_{16:0}$ (35), $uC_{16:n}$ (13), $nC_{18:0}$ (5.8), $nC_{14:0}$ (3.5)
Lake Joyce (30.4 m)	C_{10}–C_{28}	5.9	$uC_{18:n}$ (40), $nC_{16:0}$ (20), $nC_{18:0}$ (16), uC_{16n} (4.4), $nC_{14:0}$ (4.1)
Sediment*4			
Home Lake	C_{8} –C_{32}	2000	$nC_{16:0}$ (33), $uC_{18:n}$ (18), $uC_{16:n}$ (14), $bC_{15:0}$ (7.9), $nC_{18:0}$ (7.6)
Lake Vanda (edge)	C_{10}–C_{32}	8.4	$uC_{18:n}$ (40), $uC_{16:n}$ (33), $nC_{16:0}$ (13), $nC_{14:0}$ (3.6), $bC_{15:0}$ (2.3)
Lake Vanda	C_{8} –C_{30}	15	$uC_{18:n}$ (46), $nC_{16:0}$ (21), $uC_{16:n}$ (12), $nC_{18:0}$ (4.4), $nC_{22:0}$ (3.6)
Don Juan Pond-1	C_{10}–C_{32}	0.071	$nC_{16:0}$ (39), $nC_{18:0}$ (17), $nC_{14:0}$ (7.0), $bC_{15:0}$ (5.0), $uC_{18:n}$ (3.5)
Don Juan Pond-2	C_{10}–C_{30}	0.045	$nC_{16:0}$ (37), $nC_{18:0}$ (20), $nC_{14:0}$ (6.1), $bC_{15:0}$ (5.3), $uC_{18:n}$ (4.1)
Don Juan Pond-3	C_{10}–C_{28}	0.13	$nC_{16:0}$ (26), $uC_{16:n}$ (25), $uC_{18:n}$ (16), $nC_{14:0}$ (6.6), $nC_{18:0}$ (5.1)
L1 Pond	C_{8} –C_{32}	64	$uC_{16:n}$ (60), $uC_{18:n}$ (29), $nC_{16:0}$ (7.9), $nC_{14:0}$ (0.6), $bC_{15:0}$ (0.4)
L12 Pond	C_{10}–C_{28}	4.0	$uC_{16:n}$ (46), $nC_{16:0}$ (21), $nC_{12:0}$ (9.0), $bC_{15:0}$ (7.8), $nC_{14:0}$ (7.6)
L13 Pond	C_{10}–C_{32}	5.3	$nC_{16:0}$ (23), $bC_{15:0}$ (15), $uC_{16:n}$ (15), $bC_{14:0}$ (11), $uC_{18:n}$ (7.6)
Lake Bonney, west lobe	C_{8} –C_{32}	140	$uC_{18:n}$ (30), $nC_{16:0}$ (27), $nC_{14:0}$ (19), $uC_{16:n}$ (15), $nC_{18:0}$ (1.6)
Lake Fryxell	C_{8} –C_{32}	350	$nC_{16:0}$ (58), $uC_{18:n}$ (16), $uC_{16:n}$ (12), $nC_{18:0}$ (5.9), $nC_{14:0}$ (3.0)
Lake Joyce	C_{8} –C_{32}	94	$nC_{16:0}$ (57), $uC_{18:n}$ (13), $nC_{18:0}$ (5.4), $nC_{28:0}$ (4.8), $uC_{16:n}$ (4.4)
Epibenthic algae*5			
Coast Lake	C_{8} –C_{32}	4100	$uC_{18:n}$ (29), $uC_{16:n}$ (24), $nC_{16:0}$ (23), $bC_{15:0}$ (6.7), $nC_{14:0}$ (4.2)
Bull Lake	C_{12}–C_{32}	180	$nC_{16:0}$ (39), $uC_{16:n}$ (23), $uC_{18:n}$ (21), $nC_{14:0}$ (5.5), $nC_{18:0}$ (3.0)
Lake Vanda	C_{8} –C_{32}	240	$uC_{18:0}$ (55), $nC_{16:0}$ (18), $uC_{16:n}$ (18), $nC_{14:0}$ (2.5), $nC_{18:0}$ (2.4)
Don Juan Pond	C_{8} –C_{28}	420	$uC_{18:n}$ (50), $uC_{16:n}$ (26), $nC_{16:0}$ (14), $bC_{15:0}$ (3.1), $nC_{14:0}$ (1.9)
Lake Fryxell	C_{8} –C_{30}	460	$nC_{16:0}$ (31), $uC_{18:n}$ (29), $uC_{16:n}$ (18), $nC_{14:0}$ (8.0), $nC_{18:0}$ (7.9)
Syowa Oasis			
Sediment*6			
Lake Ō-ike	C_{8} –C_{34}	240	$nC_{16:0}$ (23), $nC_{24:0}$ (12), $nC_{26:0}$ (9.1), $nC_{18:0}$ (7.7), $uC_{18:n}$ (6.0)
Lake Itiziku	C_{8} –C_{34}	27	$uC_{18:n}$ (34), $nC_{16:0}$ (32), $nC_{18:0}$ (7.3), $nC_{24:0}$ (3.8), $uC_{16:n}$ (2.8)
Lake Nurume	C_{8} –C_{32}	300	$nC_{16:0}$ (49), $uC_{16:n}$ (16), $nC_{14:0}$ (9.7), $uC_{18:n}$ (8.0), $nC_{18:0}$ (3.1)
Lake Hunazoko	C_{10}–C_{32}	120	$nC_{16:0}$ (38), $uC_{18:n}$ (25), $uC_{16:n}$ (14), $nC_{14:0}$ (4.7), $nC_{18:0}$ (3.7)
Lake Skallen Ōike	C_{8} –C_{34}	1900	$uC_{16:n}$ (27), $nC_{16:0}$ (22), $uC_{18:n}$ (21), $nC_{24:0}$ (5.5), $nC_{18:0}$ (3.9)
Vestfold Oasis			
Particulate matter			
Ace Lake (10 m)*7	C_{12}–C_{26}	34	$nC_{16:0}$ (31), $uC_{18:n}$ (30), $uC_{16:u}$ (18), $nC_{18:0}$ (9.2), $nC_{14:0}$ (6.0)
Sediment			
Ace Lake*7	C_{20}–C_{26}	180	$nC_{24:0}$ (76), $nC_{22:0}$ (19), $nC_{26:0}$ (5.1), nC_{20} (<0.2)
Deep Lake*6	C_{8} –C_{32}	92	$nC_{16:0}$ (64), $uC_{18:n}$ (12), $nC_{18:0}$ (4.6), $nC_{14:0}$ (4.4), $uC_{16:n}$ (2.4)

*1 Except for water samples ($\mu g\,l^{-1}$). *2 $nC_{m:0}$, $uC_{m:n}$ and $bC_{m:0}$ are n-alkanoic, unsaturated and branched (*iso* and *anteiso*) acids, respectively. *3 Typical samples; Matsumoto (1980), Matsumoto *et al.* (1979, 1987). *4 Matsumoto (1980), Matsumoto *et al.* (1979, 1981, unpublished results). *5 Mainly cyanobacteria such as *Oscillatoria* spp., *Phormidium* spp. and *Calothrix* spp.; Matsumoto *et al.* (1979, unpublished results). *6 Matsumoto *et al.* (1981). *7 Volkman *et al.* (1986b).

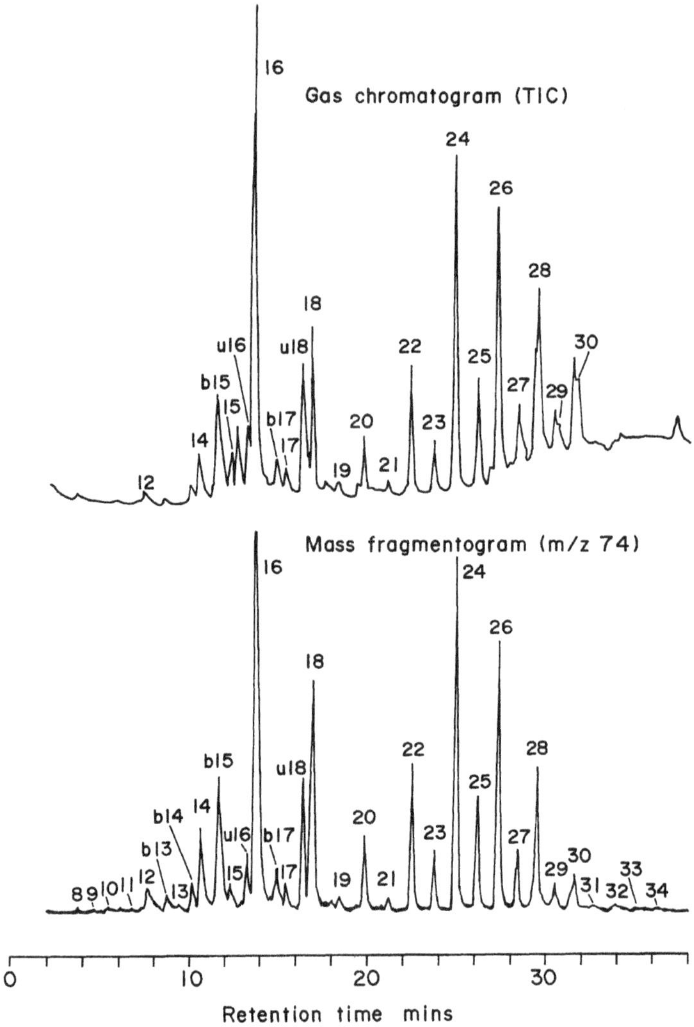

Fig. 5. Gas chromatogram (TIC) and mass fragmentogram (m/z 74) of the fatty acid fraction as methyl esters from the Lake Ō-ike sediment in the Syowa Oasis. Arabic figures on the peaks denote carbon chain length of *n*-alkanoic acids. b and u = Branched (*iso* and *anteiso*) and unsaturated acids, respectively.

tation. The decrease in the fatty acid contents is mainly due to the degradation of unsaturated fatty acids and short-chain *n*-alkanoic acids (Farrington *et al.*, 1977; Matsuda & Koyama, 1977; Matsuda, 1978). Consequently, the proportion of long-chain *n*-alkanoic acids increases after sedimentation. A triangular diagram (Fig. 6) shows the distribution of the three component fatty acids from Antarctic lakes, as compared with those from the Tokyo area (temperate zone, highly polluted, vascular plants are present) and the Ogasawara (Bonin) Islands in Japan (subtropical zone, unpolluted, vascular plants are present; Matsumoto, 1980). Short-chain *n*-alkanoic acids are most abundant in all the water samples and most of the sediment samples, although their sources are significantly different. In general, the percentages of long-chain *n*-alkanoic acids in the

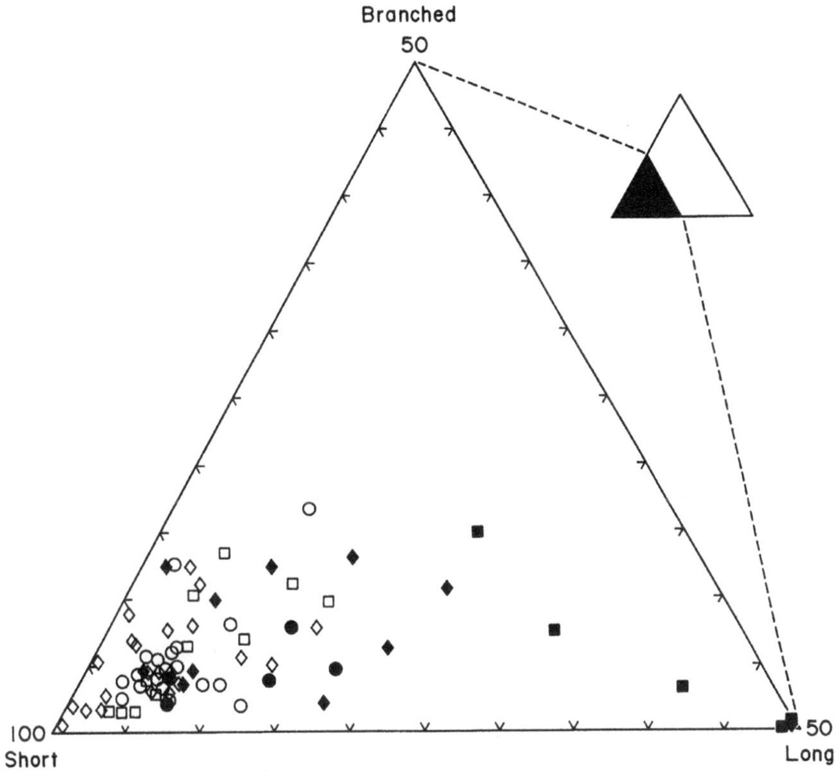

Fig. 6. The distribution of the three component fatty acids in waters and sediments from the oases of Antarctica, Tokyo area and Ogasawara (Bonin) Islands (Japan). Long: n-C_{20}–n-C_{32}. Short n-C_{12}–n-C_{19}. Branched: *Iso*- and *anteiso*-C_{13}–C_{17}. Antarctica: ◇, water; ◆, sediment. Tokyo area: ○, water; ●, sediment. Ogasawara Islands: □, water; ■, sediment.

water samples are smaller than those in the sediments of the corresponding areas. This suggests that preferential degradation of short-chain n-alkanoic acids has occurred in the sedimentary environments in all three areas. Also, it is likely that long-chain n-alkanoic acids are produced by microorganisms in the lake bottom as discussed above. Surprisingly, long-chain n-alkanoic acids in Antarctic lake and pond sediments are comparable with those in the Tokyo area and Ogasawara Islands. These results imply again that microorganisms are important sources of long-chain n-alkanoic acids in environmental samples even in the mid and lower latitudes. Furthermore, it is probable that the significant difference in environmental conditions does not markedly affect the composition of fatty acids.

Sterols

Sterols are one of the most widely distributed group of organic compounds in environments, ranging in geological age from Recent to Cretaceous, and are commonly found in soils and contemporary lake and marine sediments. C_{29} sterols (24-ethylcholesta-5,22-dien-3β-ol, 24-ethylcholest-5-en-3β-ol and 24-ethyl-5α-cholestan-3β-ol) are abundant in vascular plants, while C_{27} sterols (cholest-5-en-3β-ol and 5α-cholestan-3β-ol) are often dominant in phyto- and zooplanktons. Thus C_{29}/C_{27} sterol ratios are believed to be indicators both of allochthonous and autochthonous contributions of organic substances for lakes and of terrigenous sources of organic substances for marine environments

(Huang & Meinschein, 1976, 1979; Nishimura, 1977, 1978; Ishiwatari *et al.*, 1980; Matsumoto *et al.*, 1982). Vascular plants are absent in the areas studied. Thus the sterol composition of lake sediments could be expected to be much different from those in the mid and lower latitudes.

Matsumoto *et al.* (1982, 1983a, 1985a) reported 6 stenols (cholesta-5,22-dien-3β-ol, cholest-5-en-3β-ol, 24-methylcholesta-5,22-dien-3β-ol, 24-methylcholest-5-en-3β-ol, 24-ethylcholesta-5,22-dien-3β-ol and 24-ethylcholest-5-en-3β-ol) and 3 stanols (5α-cholestan-3β-ol, 24-methyl-5α-cholestan-3β-ol and 24-ethyl-5α-cholestan-3β-ol) in lakewaters, foams and/or sediments of the McMurdo, Syowa and Vestfold oases. The concentrations and compositions of sterols in the Antarctic samples are summarized in Table 4. The total sterol concentrations in the water column of Lake Vanda were low, ranging from 0.11 to 1.4 $\mu g\,l^{-1}$ (Matsumoto *et al.*, 1984b). The highest value was obtained in the bottom waters of the lake. The total sterol concentrations in sediment samples from the McMurdo Oasis ranged from 0.13 to 26 $\mu g\,g^{-1}$, which was somewhat higher than those from the Syowa Oasis (0.079–9.0 $\mu g\,g^{-1}$, Matsumoto *et al.*, 1982, 1983a). Although only 9 major sterols were identified in our study, using packed column gas chromatography-mass spectrometry, Volkman (1988) has recently identified 25 sterols in Ace Lake sediment by capillary gas chromatography-mass spectrometry. The total sterol concentration in the Ace Lake sediment is very high (1 170 $\mu g\,g^{-1}$) when compared with those in the lakes of the McMurdo and Syowa oases as well as those in the mid and lower latitudes.

For sediment samples from Lakes Nurume and Hunazoko and Deep Lake, the most dominant sterol was 24-methylcholest-5-en-3β-ol. It is rare for C_{28} sterols to be the most abundant (Table 4, Matsumoto *et al.*, 1983a), because in contemporary lake and marine sediments, the most abundant sterol reported is either cholest-5-en-3β-ol or 24-ethylcholest-5-en-3β-ol. Significant amounts of 24-methylcholest-5-en-3β-ol are found in a few diatoms and dinoflagellates and it is common in certain green algae which do not biosynthesize predominantly Δ^7-unsaturated sterols (Volkman, 1988). Further, 24-methylcholest-5-en-3β-ol is abundant in Antarctic mosses, *Pottia heimii*, *Sarconeurum glaciale* and *Bryum pseudotriquetrum* (Matsumoto *et al.*, 1982; Matsumoto & Kanda, 1985). Therefore, certain phytoplankton and/or mosses are possible sources of 24-methylcholest-5-en-3β-ol. It is likely that mosses are the most important source in Lake Hunazoko, where they are abundant (Matsumoto *et al.*, 1983a).

Surprisingly, 24-ethylcholest-5-en-3β-ol is the most dominant sterol in most water, foam and sediment samples from the McMurdo Oasis, and sediment samples from Lakes Ōike, Skallen Ōike and Suribati in the Syowa Oasis and from Ace Lake in the Vestfold Oasis (Table 4). Certain epibenthic organisms (mostly cyanobacteria) and green algae have been shown to contain 24-ethylcholest-5-en-3β-ol as the most dominant sterol. Thus high abundances of C_{29} sterols can be attributed to cyanobacteria and green algae in and around the lake (Matsumoto *et al.*, 1982, 1983a; Volkman, 1988). Such findings suggest that cyanobacteria and algae are also important sources of C_{29} sterols in lacustrine and coastal marine environments of the mid and lower latitudes.

The C_{29}/C_{27} sterol ratios for Antarctic waters, foams and sediments are very high, ranging from 0.49 to 5.3, 0.39 to 1.4 and from 0.33 to 7.2, respectively (Table 4). They are comparable with those in sediment samples from the mid and lower latitudes (0.48–8.5) in spite of the absence of vascular plants in the areas studied (Matsumoto *et al.*, 1982, 1983a, 1984c, 1985a). Therefore, the C_{29}/C_{27} sterol ratios cannot be applied as source indicators in the manner used for sediments from the mid and lower latitudes.

Phenolic acids

It is well known that a series of phenolic acids (*p*-hydroxybenzoic, vanillic, syringic, *p*-coumaric and ferulic acids) are widely distributed in vascular plants (Pearl *et al.*, 1961; El-Basyouni *et al.*, 1964; Higuchi *et al.*, 1967; Hartley *et al.*, 1976; Matsumoto, 1982). Thus the features of phenolic

Table 4. Sterols found in Antarctic samples

Sample	Conc.*1 (μg g⁻¹)	Composition (%)*2			C_{29}/C_{27}	References
		C_{27}	C_{28}	C_{29}		
McMurdo Oasis						
Water						
Blue Lake (surface)	0.24	26	27	47	1.8	–
Onyx River (surface)	0.041	67	0	33	0.49	a
Lake Vanda (5.4 m)	0.11	47	3	50	1.1	a
(30.4 m)	0.14	27	0	73	2.7	a
(50.4 m)	0.090	21	0	79	3.8	a
(55.4 m)	0.14	26	0	74	2.8	a
(60.4 m)	0.21	34	2	64	1.9	a
(65.9 m)	1.4	14	12	74	5.3	a
L1 Pond (surface)	0.11	23	0	77	3.3	–
Lake Joyce (30.4 m)	0.036	50	0	50	1.0	–
Foam						
Bull Lake-1	–	34.6	20.0	45.4	1.3	b
Bull Lake-2	–	35.1	16.2	48.7	1.4	b
Lake Vanda	–	59	0	41	0.69	b
L19 Pond	–	72	0	28	0.39	b
Sediment						
Lake Vanda	4.1	16.7	7.6	75.7	4.5	c
Don Juan Pond	0.13	32.0	22.5	45.5	1.4	c
L4 Pond	2.2	39.8	5.2	55.0	1.4	c
L8 Pond	1.9	37.0	7.4	55.6	1.5	c
Lake Fryxell	26	7.5	38.2	54.3	7.2	c
Lake Bonney	8.3	16.3	39.1	44.6	2.7	c
Lake Joyce	17	19.3	37.7	43.0	2.2	c
Epibenthic algae						
L8 Pond*3	6.8	42.1	11.3	46.6	1.1	c
L12 Pond*4	15	15.1	9.0	75.9	5.1	c
Lake Fryxell*5	0.56	50.1	20.7	29.2	0.60	c
Lake Bonney*6	6.0	25.2	29.1	45.7	1.8	c
Moss						
Crater Hill*7	130	3.5	51.8	44.7	13	d
Fryxell Hut*7	5.2	2.7	57.8	39.5	15	c
Miers Valley-1*8	200	0.8	62.7	36.5	46	d
Miers Valley-2*8	300	0.8	63.4	35.8	45	d
Miers Valley-3*9	160	0.8	55.4	43.8	55	d
Syowa Oasis						
Sediment						
Lake Ō-ike	0.67	39.2	12.9	47.9	1.2	e
Lake Itiziku	0.079	22.8	44.3	32.9	1.4	e
Lake Nurume	3.8	21.6	64.5	13.9	0.64	e
Lake Hunazoko	1.0	30.4	47.1	22.5	0.74	e
Lake Skallen Ōike	9.0	13.3	27.8	58.9	4.4	e
Vestfold Oasis						
Sediment						
Deep Lake	3.6	38.5	48.8	12.7	0.33	e
Ace Lake	1170	17.2	18.4	64.4	3.7	f

*1 Except for water samples (μg l⁻¹). *2 C_{27}; Cholesta-5,22-dien-3β-ol, cholest-5-en-3β-ol and 5α-cholestane-3β-ol; C_{28}, 24-methylcholesta-5,22-dien-3β-ol, 24-methylcholest-5-en-3β-ol and 24-methyl-5α-cholestane-3β-ol; C_{29}, 24-ethylcholesta-5,22-dien-3β-ol, 24-ethylcholest-5-en-3β-ol and 24-ethyl-5α-cholestane-3β-ol. *3 Cyanobacteria, mainly *Synechococcus* spp., *Phormidium* spp. and *Oscillatoria* spp. *4 Mainly *Phormidium* spp. *5 Mainly *Oscillatoria* spp. *6 Mainly *Phormidium* spp. and diatom, *Hantzschia amphioxys*. *7 *Sarconeurum glaciale*. *8 *Pottia heimii*. *9 *Bryum pseudotriquetrum*. References: a, Matsumoto *et al.* (1984b); b, Matsumoto *et al.* (1985a); c, Matsumoto *et al.* (1982); d, Matsumoto and Kanda (1985); e, Matsumoto *et al.* (1983a); f, Volkman (1988).

acids in Antarctic lakes can be expected to be much different from those in the mid and lower latitudes. A typical mass fragmentogram from an Antarctic unnamed pond sediment (L13 pond) is compared with that of a Yatsuse River sediment, Ogasawara Islands (Fig. 7). *p*-Hydroxybenzoic acid is the most prominent phenolic acid in the L13 pond sediment, whilst the Yatsuse River sediment contains a series of phenolic acids. *p*-Hydroxybenzoic acid is identified and the presence of *o*- and *m*-hydroxybenzoic, and vanillic acids is indicated for Antarctic lake waters and sediments (Table 5), but some samples contain *o*- and/or *p*-hydroxyphenylacetic acids (Matsumoto *et al.*, 1979; 1981, 1984b, 1984c). *p*-Hydroxybenzoic acid is the most prominent phenolic acids both in water and sediment samples. The concentrations of *p*-hydroxybenzoic acid in water and

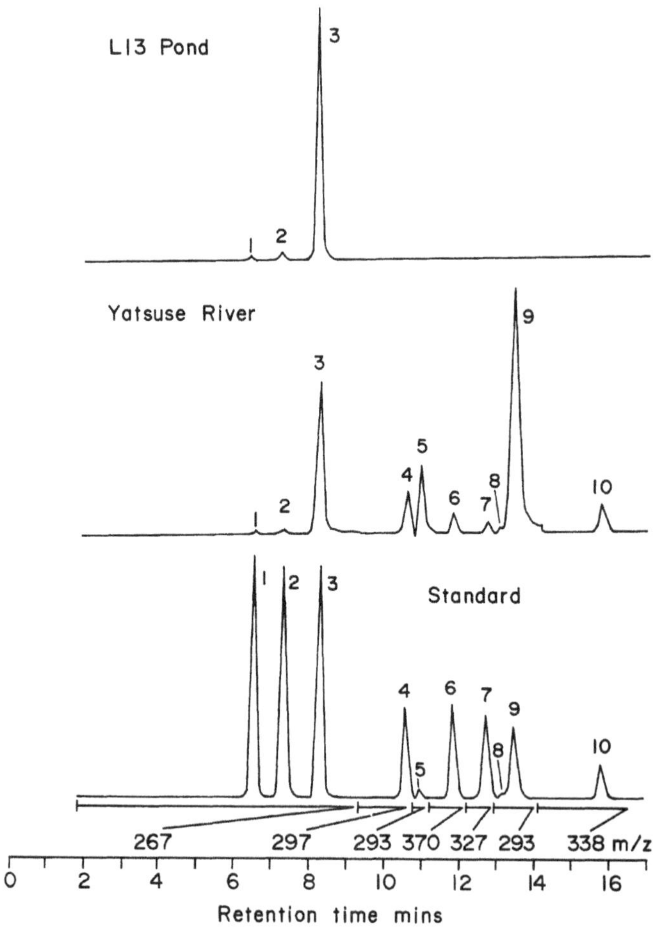

Fig. 7. Mass fragmentogram of the phenolic acid fraction from the L13 pond sediment in the McMurdo Oasis as compared with that of Yatsuse River of the Ogasawara Islands. The mass numbers at m/z 267, 297, 327, 293, 370 and 338 correspond to molecular ion (M)–15 of *o*-, *m*-, and *p*-hydroxybenzoic, vanillic, syringic and *p*-coumaric acid-, and M of protocatechuic and ferulic acid-trimethylsilyl derivatives, respectively. Identification: 1, *o*-hydroxybenzoic acid; 2, *m*-hydroxybenzoic acid; 3, *p*-hydroxybenzoic acid: 4, vanillic acid; 5, *p*-coumaric acid (*cis*); 6, protocatechuic acid; 7, syringic acid; 8, ferulic acid (*cis*); 9, *p*-coumaric acid (*trans*); 10, ferulic acid (*trans*).

Table 5. Phenolic acids found in Antarctic samples ($\mu g\ g^{-1}$)*[1]

Sample (numbers analysed)	OHA*[2]	MHA*[2]	PHA*[2]	VA*[2]	SA*[2]	PCA*[2]	FA*[2]	OPA*[2]	PPA*[2]	References
McMurdo Oasis										
Water (23)	ND–0.2	ND–0.4	ND–12	ND–0.5	ND	ND	ND	ND – +	ND – +	a
Sediment (12)	ND–0.07	ND–0.23	ND–2.0	ND–0.008	ND	ND	ND	ND	ND	a,b
Epibenthic algae (5)	ND–0.2	ND–0.90	0.16–3.9	ND–0.029	ND	ND	ND	ND	ND	a,b
Moss (1)	ND	ND	1.4	ND	ND	ND	ND	ND	ND	b
Syowa Oasis										
Sediment (5)	0.07–0.6	0.22–0.8	0.47–6.5	ND	ND	ND	ND	ND	ND	c
Vestfold Oasis										
Sediment (1)	0.002	0.002	0.006	ND	ND	ND	ND	ND	ND	c

*[1] Except for water samples ($\mu g\ l^{-1}$), *[2]OHA, *o*-hydroxybenzoic acid; MHA, *m*-hydroxybenzoic acid;PHA, *p*-hydroxybenzoic acid; VA, vanillic acid; SA, syringic acid; PCA, *p*-coumaric acid; FA, ferulic acid; OPA, *o*-hydroxyphenylacetic acid; PPA, *p*-hydroxyphenylacetic acid. ND: Not detected. + : Present but not quantified.
References: a, Matsumoto *et al.* (1979); b, Matsumoto *et al.* (unpublished results); c, Matsumoto *et al.* (1981).

sediment samples vary significantly from 0.00 to 12 $\mu g\,l^{-1}$ and from 0.00 to 6.5 $\mu g\,l^{-1}$, respectively. Generally the concentration of *p*-hydroxybenzoic acid in the bottom waters of Lakes Vanda and Bonney are considerably higher than those in the overlying waters. The highest value of *p*-hydroxybenzoic acid in water samples was found in the bottom water of the west lobe of Lake Bonney, while that in the sediment samples was observed in Lake Skallen Ōike.

Very little is known of the occurrence of phenolic acids in organisms other than vascular plants. Microorganisms, such as fungi, including *Aspergillus* spp. and *Penicillium* spp. (Shivrina *et al.*, 1959; Simonart *et al.*, 1960; Simonart & Verachtert, 1966) actinomycetes (Ratledge, 1964) and bacteria (Ratledge, 1964) may be important sources of Antarctic phenolic acids. Analytical results of cyanobacteria also reveal that they are essential sources of phenolic acids (Table 5; Matsumoto *et al.*, 1979; unpublished results).

Phenolic acid composition in Antarctic lakes and ponds are quite different from those in the Tokyo area and Ogasawara Islands in which a series of phenolic acids were found in most samples (Matsumoto *et al.*, 1977; Matsumoto, 1980, 1982; Matsumoto & Hanya, 1980b). Phenolic acids in Antarctic lakes are characterized by the predominance of *p*-hydroxybenzoic acid along with the absence of syringic, *p*-coumaric and ferulic acids. These are clearly features of phenolic acids in areas devoid of vascular plants.

Figure 8 shows the frequency distribution of the ratios of *p*-coumaric and ferulic acids-carbon (CFC) to TOC and EOC values for water and sediment samples from the three areas (Matsumoto, 1980). A wide variety of the CFC/TOC and CFC/EOC values were obtained for samples from the Tokyo area and the Ogasawara Islands samples, probably reflecting variable contributions of vascular plants for aquatic environments, whereas those in the Antarctic samples are all zero. Consequently, these ratios must be a good indicator of the contribution of vascular plants to environmental studies.

Hydroxy acids

Hydroxy acids occur in vascular plants and a wide variety of microorganisms, such as bacteria, yeasts and fungi, but little is known of their presence in cyanobacteria and algae (Pohl & Wagner, 1972; Boon *et al.*, 1977; Cardoso & Eglinton, 1983; Matsumoto & Nagashima, 1984; Matsumoto *et al.*, 1984a). 2-, 3- and ω-1-Hydroxy acids of carbon chain lengths between C_8 and C_{28} have been detected in sediment samples from Lakes Vanda, Bonney, Fryxell and Joyce and in epibenthic organisms (mostly cyanobacteria) from the Dry Valleys of the McMurdo Oasis (Matsumoto *et al.*, 1984c). In general, the major 2-hydroxy acids are short- and long-chain normal and branched (*iso*- and *anteiso*) constituents, while those of 3-hydroxy acids are short-chain normal and branched acids except for the Lake Vanda sediment. The major ω-1-hydroxy acids are all long-chains (C_{24}, C_{26} and C_{28}). Very recently, Matsumoto *et al.* (1988) reported normal 3-hydroxy acids of carbon chain length ranging from C_{10} to C_{20} with a predominance of even-carbon numbers in the water column and sediment of Lake Fryxell, together with branched and mono-unsaturated 3-hydroxy acids (Fig. 9). The major 3-hydroxy acids ($\geqslant 10\%$) were *n*-saturated with even-carbon numbers (*n*-C_{10}, *n*-C_{14}, *n*-C_{16} and *n*-C_{18}) and branched (*iso*-C_{15}, *iso*-C_{17}, *anteiso*-C_{17} and *iso*-C_{18}) mainly with odd-carbon numbers. Their total concentrations were much lower than those of fatty acids, ranging from 0.041 to 0.96 $\mu g \cdot l^{-1}$. *Iso*- and *anteiso*-branched acids were abundant in all the samples studied, indicating that they mainly came from bacteria.

Ketones

Volkman *et al.* (1988) have identified high molecular weight straight-chain unsaturated methyl and ethyl ketones and fatty acid esters in the Ace Lake sediment, including $C_{37:4}$ and $C_{37:3}$ methyl ketones $C_{38:4}$ and $C_{38:3}$ ethyl ketones, $C_{37:4}$ methyl ester. This is the first report of their occurrence in the Antarctic environment. They are similar to compounds isolated from the

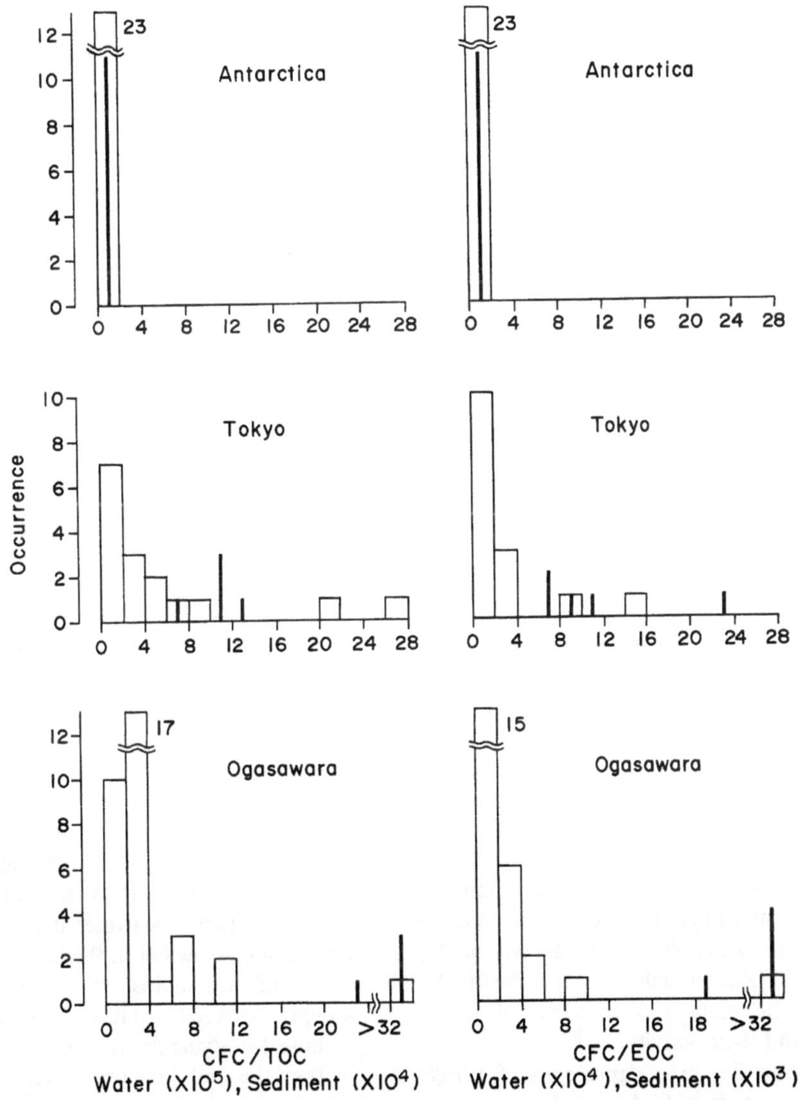

Fig. 8. The ratios of *p*-coumaric and ferulic acids-carbon to TOC (CFC/TOC) and to EOC (CFC/EOC) contents for waters and sediments from Antarctica, the Tokyo area and Ogasawara Islands. □: Water. ▌: Sediment.

marine Prymnesiophyte *Emiliania huxleyi* (Loch) Hay et Mohler. These long-chain compounds were not detectable in any of the water column particulate matter samples. Thus the green algae, *Pyramimonas gelidicola* or *Cryptomonas* sp., in the water column were unlikely sources. The occurrence of these unusual compounds in Ace Lake provides circumstantial evidence that an unidentified microflagellate in the lake may be a Prymnesiophycean alga.

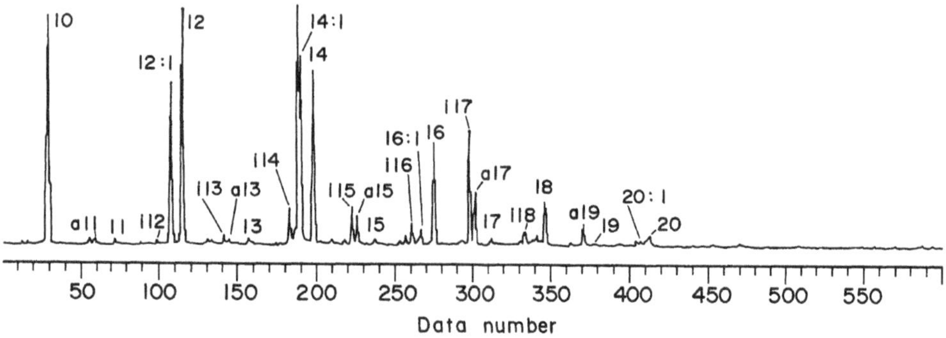

Fig. 9. Mass chromatogram of 3-hydroxy acid trimethylsilyloxy ether methyl esters (m/z 175) from the Lake Fryxell sediment in the McMurdo Oasis. Arabic figures on the peaks denote carbon chain length of 3-hydroxy acids. i = *Iso.* a = *Anteiso.* m : n = carbon chain length: numbers of unsaturation.

Stratification of organic substances

In general, Antarctic saline lakes have no outlets, are covered perennially with thick ice and thus are stratified strongly by salt water. Koob & Leister (1972) have reported significant zonation of rod-shaped organisms, unicellular coccoid algae, *Chlamydomonas subcaudata* and bacteria in meromictic, Lake Bonney. Thus it can be expected that vertical distribution of microorganisms and organic constituents in chemically stratified Antarctic lakes will be quite different with depth. The vertical distribution of organic constituents in Lakes Vanda and Fryxell of the McMurdo Oasis and Ace Lake of the Vestfold Oasis will be discussed here in relation to the stratification of lakewater and the vertical distribution of microorganisms and their activity.

Lake Vanda lies in a depression of Wright Valley at an altitude of 93.6 m, and with an area of 5.21 km^2 and maximum depth of 70 m (January, 1985) from the ice surface. The lake is perennially covered with thick ice of about 3 m. It is meromictic and the chlorinity of the bottom waters is up to 4 times greater than that of seawater, though the major salt is calcium chloride. The lake is known as a natural solar energy trap with density stratification by salt water. The bottom temperature of the lake can therefore attain about 24 °C in spite of a mean air temperature of − 20 °C (Fig. 10). Lake Fryxell is a relatively productive meromictic lake situated in the lower Taylor Valley at an altitude of 22 m, with an area of 7 km^2 and maximum depth of 19 m. The salt composition of the lakewater is similar to that of seawater, though the maximum salt content is one-fifth of seawater. Matsumoto *et al.* (1984b, 1987a, 1988) have reported the vertical distribution of organic constituents, *i.e.*, TOC, EOC, hydrocarbons, phytol, sterols, fatty acids, phenolic acids or hydroxy acids in both Lakes Vanda and Fryxell. Ace Lake is located in the Vestfold Oasis, has an area of 1.3×10^5 m^2 with a maximum depth of 24.5 m in 1984. Salinities range from about 80% that of seawater at 3 m to 30% greater than that of seawater at the bottom. The compositions of carotenoids, chlorophylls and lipids at 4 depths in Ace Lake have been determined as a means of studying the vertical zonation of species in the lake and for comparison with the lipids found in the bottom sediments (Volkman *et al.*, 1988).

The concentrations of TOC, EOC and sterols in Lake Vanda rose with depth and attained maximum values of 25 and 1.5 mg C l^{-1} and 1.4 $\mu g\,l^{-1}$ at the bottom, respectively (Fig. 10, Matsumoto *et al.*, 1984b). The DOC values also increased with depth and reached a maximum value at the bottom (Matsumoto *et al.*, 1987a). In the bottom anoxic layers of Lakes Vanda and Fryxell, refractory organic substances accumulated over long periods as discussed above.

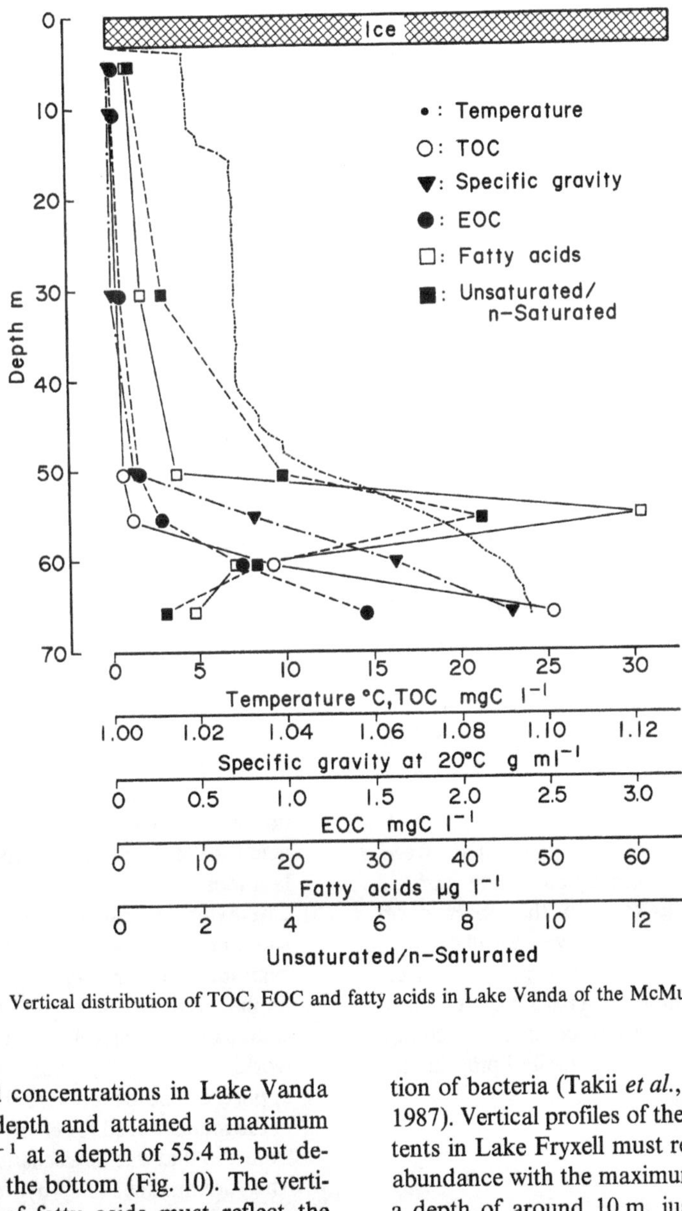

Fig. 10. Vertical distribution of TOC, EOC and fatty acids in Lake Vanda of the McMurdo Oasis.

The fatty acid concentrations in Lake Vanda increased with depth and attained a maximum value of 61 $\mu g\,l^{-1}$ at a depth of 55.4 m, but decreased towards the bottom (Fig. 10). The vertical distribution of fatty acids must reflect the abundance of phytoplankton, since Vincent *et al.* (1981) found the maxima of photosynthesis and DNA synthesis at around this depth. The proportion of branched (*iso* and *anteiso*) fatty acids in the top and bottom waters of the lake was somewhat greater than that in the middle waters. This is consistent with the pattern of vertical distribution of bacteria (Takii *et al.*, 1986; Konda *et al.*, 1987). Vertical profiles of the total fatty acid contents in Lake Fryxell must reflect phytoplankton abundance with the maximum value occurring at a depth of around 10 m, just above the anoxic layer. The proportions of branched fatty acids as well as branched 3-hydroxy acids in the anoxic layers were much higher than that in the oxic layers, again consistent with those of bacterial distribution (T. Konda *et al.*, pers. commun.). Fatty acid distributions in Ace Lake were very different at 10 and 11 m, and indicated that the algal biomass at 11 m was less than 3% of that at 10 m. The most abundant alga was *Pyramimonas gelidicola* McFadden (Chlorophyta) which occurs mainly at 10 m, the base of the oxic layer. Bacteriochlorophyll *c* was the major pigment at 11 m, the top of the anoxic layer, reflecting the abundance of green sulfur bacteria, *Chlorobium* spp. (Volkman *et al.*, 1988).

In Lake Vanda, hydrocarbons occurred only in the bottom layers (⩾60.4 m) and sediment. Matsumoto *et al.* (1984b) concluded from the features of organic constituents and biochemical data, that the vertical distribution of microorganisms and their activity were markedly different with depth. The most biologically active depths in the lake were between 55 and 60 m, which is the bottom of the oxic layer. The vertical distributions of hydrocarbons, fatty acids and 3-hydroxy acids in Lake Fryxell revealed that the distribution of lake organisms was also quite different with depth, as was also the case in Lake Vanda.

The compositions of organic substances in the water columns and sediment, both in Lakes Vanda and Fryxell, were quite different, indicating that organic substances formed in the water columns were decomposed quickly by microorganisms in the lake bottom. Volkman *et al.* (1988) also reported that non-polar lipids were minor constituents of the lipids in the surface sediments of Ace Lake, and the absence of triacylglycerols implied extensive degradation of these lipids by bacteria. The distribution of free fatty acids in the sediment was again quite different from those in the water column, indicating almost complete turnover of labile lipids in the sediment.

Conclusions

The features of organic substances in lake and pond waters and sediments from the McMurdo, Vestfold and Syowa oases in Antarctica are summarized as follows

1) Organic carbon contents vary remarkably among the lake and pond waters, ranging from 0.14 to 186 mg C l^{-1}. Generally, high organic carbon contents are found in saline waters, which can be attributed to the concentration of refractory organic substances after the degradation of labile organic constituents under frigid conditions over a long time period.
2) The kinds and amounts of organic constituents among the lake and pond waters and sediments differ considerably, and should reflect differences of the kinds and activity of lake microorganisms. The absence of significant contributions from allochthonous organic substances may also be responsible for the differences in organic compositions in the lakes.
3) Long-chain alkenes, such as n-$C_{29:2}$ and n-$C_{34:3-5}$ are the predominant hydrocarbons in anoxic sediments from Lakes Bonney, Fryxell, Joyce and Ace. These alkenes may be attributable to certain microorganisms in the lake bottom.
4) Unusually long-chain *n*-alkanoic acids are abundant in some lake sediments, in spite of the absence of vascular plants in the areas studied, and are believed to originate from microorganisms such as bacteria in the lake bottom. It suggests that microorganisms are important sources of long-chain *n*-alkanoic acids in environments of the mid and lower latitudes.
5) 24-ethylcholest-5-en-3β-ol is the dominant sterol in most of the lakes, and originates from cyanobacteria and green algae. This result implies that these organisms are essential sources of this sterol in environments of the world. The C_{29}/C_{27} sterol ratios cannot be applied as source markers for Antarctic lakes in the manner used for environmental samples from the mid and lower latitudes.
6) The dominance of *p*-hydroxybenzoic acid along with absence of syringic, *p*-coumaric and ferulic acids reflect the absence of vascular plants. Thus the CFC/TOC and CFC/EOC ratios must be important markers of the contribution of vascular plants in environmental studies.
7) Organic substances in Lakes Vanda and Fryxell, and Ace Lake are stratified with depth

due to the zonation of microorganisms. The highest fatty acid contents are found just above the anoxic layers of the lakes, probably reflecting the abundance of phytoplankton.

8) The composition of organic substances in the sediments from Lakes Vanda, Fryxell and Ace is very different from that in the water columns, indicating that organic substances formed in the water column is rapidly degraded by microorganisms in the lake bottom.

Acknowledgements

The author is greatly indebted to Professors Tetsuya Torii, Chiba Institute of Technology, and Kunihiko Watanuki, the University of Tokyo, for their encouragement. I thank editors J. C. Ellis-Evans and W. F. Vincent for their help in making it more readable.

References

Albro, P. W., 1976. Bacterial waxes. In P. E. Kolattukudy (ed.), Chemistry and Biochemistry of Natural Waxes. Elsevier, Amsterdam: 419–445.

Allnut, F. C. T., B. C. Parker, K. G. Seaburg & G. M. Simmons Jr., 1981. *In situ* nitrogen (C_2H_2) fixation in lakes of southern Victoria Land, Antarctica. Hydrological Bulletin 15: 99–109.

Barker, R. J., 1980. Physical and chemical parameters of Deep Lake, Vestfold Hills, Antarctica. ANARE Sci. Rep. 130, (cited in Burton, H. R., 1981).

Boon, J. J., F. de Lange, P. J. W. Schuyl, J. W. de Leeuw & P. A. Schenk, 1977. Organic geochemistry of Walvis Bay diatomaceous ooze. II Occurrence and significance of the hydroxy fatty acids. In R. Campos & J. Goñi (eds.), Advances in Organic Geochemistry 1975, Enadimsa, Madraid: 255–272.

Brooks, J. D., K. Gould & J. W. Smith, 1969. Isoprenoid hydrocarbons in coal and petroleum. Nature 222: 257–259.

Burton, H. R., 1980. Methane in a saline Antarctic lake. In P. A. Trudinger & M. R. Walter (eds.), Biogeochemistry of Ancient and Modern Environments Proceedings of the Forth International Symposium on Environmental Biogeochemistry (ISEB). Canberra, Australian Academy of Science (cited in Burton, H. R. 1981).

Burton, H. R., 1981. Chemistry, physics and evolution of Antarctic saline lakes: A review. Hydrobiologia 82: 339–362.

Cardoso, J. N. & G. Eglinton, 1983. The use of hydroxyacids as geochemical indicators. Geochim. Cosmochim. Acta 47: 723–730.

Didyk, B. M., B. R. T. Simoneit, S. C. Brassell & G. Eglinton, 1978. Organic geochemical indicators of palaeoenvironmental conditions of sedimentation. Nature 272: 216–222.

Downes, M. T., C. Howard-Williams & W. F. Vincent, 1986. Sources of organic nitrogen, phosphorous and carbon in antarctic streams. Hydrobiologia 134: 215–225.

El-Basyouni, S. Z., D. Chen, R. K. Ibrachim, A. C. Neish & G. H. N. Towers, 1964. The biosynthesis of hydroxybenzoic acids in higher plants. Phytochemistry 3: 485–492.

Farrington, J. W., S. M. Henrichs & R. Anderson, 1977. Fatty acids and Pb-210 geochronology of a sediment core from Buzzards Bay, Massachusetts. Geochim. Cosmochim. Acta 41: 289–296.

Hand, R. M., 1980. Bacterial populations of two saline Antarctic lakes. In P. A. Trudinger & M. R. Walter (eds.), Biogeochemistry of Ancient and Modern Environments. Proceedings of the Fourth International Symposium on Environmental Biogeochemistry (ISEB). Australian Academy of Science, Canberra: 123–129.

Hartley, R. D., E. C. Jones & T. M. Wood, 1976. Carbohydrates and carbohydrate esters of ferulic acid released from cell walls of *Lolium multifrorum* by treatment with cellulolytic enzymes. Phytochemistry 15: 305–307.

Higuchi, T., Y. Ito & I. Kawamura, 1967. *p*-Hydroxyphenylpropane component of glass lignin and role of tyrosine-ammonia lyase in its formation. Phytochemistry 6: 875–881.

Hirabayashi, J. & J. Ossaka, 1977. Seasonal variation in chemical composition and the origin of the saline lakes around Syowa Station, Antarctica. Antarct. Rec. 58: 93–107 (in Japanese).

Huang, W.-Y. & W. G. Meinschein, 1976. Sterols as source indicators of organic materials in sediments. Geochim. Cosmochim. Acta 40: 323–330.

Huang, W.-Y. & W. G. Meinschein, 1979. Sterols as ecological indicators. Geochim. Cosmochim. Acta 43: 739–745.

Ishiwatari, R., K. Ogura & S. Horie, 1980. Organic geochemistry of a lacustrine sediment (Lake Haruna, Japan). Chem. Geol. 29: 261-280.

Konda, T., S. Takii, M. Fukui, Y. Kusuoka, G. I. Matsumoto, S. Nakaya & T. Torii, 1987. Vertical distribution of bacterial cell size and biomass in Lake Vanda, an Antarctic lake. Antarct. Rec. 31: 109–117 (in Japanese).

Koob, D. D. & G. L. Leister, 1972. Primary productivity and associated physical, chemical, and biological characteristics of Lake Bonney: A perennially ice-covered lake in Antarctica. Antarctic Research Series (Amer. Geophys. Union) 20: 51–68.

Matsuda, H. & T. Koyama, 1977. Early diagenesis of fatty acids in lacustrine sediments-I. Identification and distribution of fatty acids in recent sediment from a freshwater lake. Geochim. Cosmochim. Acta 41: 777–783.

Matsuda, H., 1978. Early diagenesis of fatty acids in

lacustrine sediments-III. Changes in fatty acid composition in the sediments from a brackish water lake. Geochim. Cosmochim. Acta 42: 1027–1034.

Matsumoto, G., 1980. Comparative Geochemical Study on Organic Constituents in Aquatic Environments. D. Sc. Thesis, Tokyo Metropolitan University. Tokyo.

Matsumoto, G., 1982. Comparative study on organic constituents in polluted and unpolluted inland aquatic environments-III. Phenols and aromatic acids in polluted and unpolluted waters. Water Res. 16: 551–557.

Matsumoto, G. & T. Hanya, 1977. Organic carbons and fatty acids in Antarctic saline lakes. Antarct. Rec. 58: 81–88.

Matsumoto, G. & T. Hanya, 1980a. Organic constituents in atmospheric fallout in the Tokyo area. Atmos. Environ. 14: 1409–1419.

Matsumoto, G. & T. Hanya, 1980b. Gas chromatographic-mass spectrometric identification of phenolic acids in recent sediments. J. Chromatogr. 193: 89–94.

Matsumoto, G. I. & H. Nagashima, 1984. Occurrence of 3-hydroxy acids in microalgae and cyanobacteria and their geochemical significance. Geochim. Cosmochim. Acta 48: 1683–1687.

Matsumoto, G. I. & H. Kanda, 1985. Hydrocarbons, sterols and hydroxy acids in Antarctic mosses. Antarct. Rec. 87: 23–31.

Matsumoto, G., R. Ishiwatari & T. Hanya, 1977. Gas chromatographic-mass spectrometric identification of phenols and aromatic acids in river waters. Water Res. 11: 693–698.

Matsumoto, G., T. Torii & T. Hanya, 1979. Distribution of organic constituents in lake waters and sediments of the McMurdo Sound region in the Antarctic. Mem. Natl Inst. Polar Res., Spec. Issue 13: 103-120.

Matsumoto. G., T. Torii, F. Fukui & T. Hanya, 1981. The features of organic constituents in lake sediments from Syowa Oasis in East Antarctica. Verh. Internat. Verein. Limnol. 21: 703–707.

Matsumoto, G., T. Torii & T. Hanya, 1982. High abundance of algal 24-ethylcholesterol in Antarctic lake sediment. Nature 299: 52–54.

Matsumoto, G., T. Torii & T. Hanya, 1983a. Stenols and phytol in lake sediments from Syowa and Vestfold oases in the Antarctic. Geochem. J. 17: 1–8.

Matsumoto, G., K. Chikazawa, H. Murayama, T. Torii, H. Fukushima & T. Hanya, 1983b. Distribution and correlation of total organic carbon and mercury in Antarctic dry valley soils, sediments and organisms. Geochem. J. 17: 241-246.

Matsumoto, G. I., M. Shioya & H. Nagashima, 1984a. Occurrence of 2-hydroxy acids in microalgae. Phytochemistry 23: 1421-1423.

Matsumoto, G., T. Torii & T. Hanya, 1984b. Vertical distribution of organic constituents in an Antarctic lake: Lake Vanda. Hydrobiologia 111: 119–126.

Matsumoto, G. I., T. Torii & T. Hanya, 1984c. A review of organic geochemistry in Antarctica. Mem. Natl Inst. Polar Res. Spec. Issue 33: 204–217.

Matsumoto, G. I., T. Torii & T. Hanya, 1985a. Sterols and fatty acids in foams from Antarctic lakes of the Dry Valleys in south Victoria Land. Geochem. J. 19: 91–96.

Matsumoto, G. I., T. Torii, T. Kawano, T. Cho & Y. Kobata, 1985b. Japanese geochemical studies on the McMurdo Sound region in Antarctica during the 1983–84 austral summer. Antarct. Rec. 86: 108-118 (in Japanese).

Matsumoto, G. I.., K. Watanuki & T. Torii, 1987a. Further study on the vertical distribution of organic constituents in an Antarctic lake: Lake Vanda. Proceedings of the NIPR Symposium on Polar Biology. 1: 219–232.

Matsumoto, G. I., K. Watanuki & T. Torii, 1987b. Total organic carbon in pond waters from the Labyrinth of southern Victoria Land in the Antarctic. Antarct. Rec. 31: 171–196.

Matsumoto, G. I., K. Watanuki & T. Torii, 1988. Vertical distribution of organic constituents in an Antarctic lake: Lake Fryxell. Hydrobiologia (this volume).

Murayama, H., 1977. General characteristics of the Antarctic lakes near Syowa Station. Antarct. Rec. 58: 43–62 (in Japanese).

Nishimura, M., 1977. Origin of stanols in young lacustrine sediments. Nature 270: 711–712.

Nishimura., M., 1978. Geochemical characteristics of the high reduction zone of stenols in Suwa sediments and the environmental factors controlling the conversion of stenols into stanols. Geochim. Cosmochim. Acta 42: 349–357.

Nissenbaum, A., M. J. Baedecker & I. R. Kaplan. 1972. Organic geochemistry of Dead Sea sediments. Geochim. Cosmochim. Acta 36: 709–727.

Parker, B. C., J. T. Whitehurst & R. C. Hoehn, 1974. Observations of *in situ* concentrations and production of organic matter in an Antarctic meromictic lake. Virginia J. Sci. 25: 136–140.

Patterson, G. W., 1967. The effect of culture conditions on the hydrocarbon content of *Chlorella vulgaris*. J. Phycol. 3: 22–23.

Pearl, I. A., D. L. Beyer & D. Whitney, 1961. Alkaline hydrolysis of representative hardwoods V. Additions and corrections to previous studies. Tappi 44: 656–661.

Pohl, V. P. & H. Wagner, 1972. Fattsäuren im Pflanzen- und Tierreich (eine Übersicht) II: Trans-ungesättigte, Alkin-, Hydroxy-, Epoxy-, Oxo-, Cyclopropan- und Cyclopropen-Fettsäuren. Fette, Seifen, Anstrichmittel 74: 541–550.

Powell, T. G. & D. M. McKirdy, 1973. Relationship between ratio of pristane to phytane, crude oil composition and geological environment in Australia. Nature 243: 37–39.

Ratledge, C., 1964. Relation between the products of aromatic biosynthesis in *Mycobacterium smegmatis* and *Aerobacter aerogens*. Nature 203: 428–429.

Sakamoto, M., H. Kurasawa & T. Okino, 1975. Productivity and nutrient metabolism of communities in Lake Suwa. In S. Mori & G. Yamamoto (eds.), Productivity of Communities in Japanese Inland Waters. JIBP Synthesis, 10. University of Tokyo Press, Tokyo: 107–147.

Shivrina, A. N., O. P. Nizkovskaya, E. V. Lovyagina, E. G. Platonova & N. M. Miloba, 1959. Chemical composition of bracket fungi in various stages of development. Bot. Zh. 44: 1724-1727 (Chem. Abstr. 54: 16536h, 1960).

Simonart, P. & H. Verachtert, 1966. Phenolic compounds produced by *Aspergillus fumigatus* and *Penicillium spinulosum*. Bull. Soc. Chim. Biol. 48: 943–950.

Simonart, P., A. Wiaux & H. Verachtert, 1960. Biochemical investigations of *Penicillium griseofulvum*. Zentralbl. Bakteriol., Parasitenkd., Infektionskr. Hyg., Abt. II, 113: 209 (Chem. Abstr. 61, 4749e, 1964).

Simoneit, B. R. T., R. Chester & G. Eglinton, 1977. Biogenic lipids in particulates from the lower atmosphere over the eastern Atlantic. Nature 267: 682–685.

Takii, S., T. Konda, A. Hiraishi, G. I. Matsumoto, T. Kawano, & T. Torii, 1986. Vertical distribution in and isolation of bacteria from Lake Vanda: An Antarctic lake. Hydrobiologia 135: 15–21.

Tominaga, H. & F. Fukui, 1981. Saline lakes at Syowa Oasis, Antarctica. Hydrobiologia 82: 375-389.

Torii, T. & N. Yamagata, 1981. Limnological studies of saline lakes in the dry valleys. Antarct. Res. Ser. (Dry Valley Drilling Project) 33: 141–159.

Torii, T., N. Yamagata, S. Nakaya, S. Murata, T. Hashimoto, O. Matsubaya & H. Sakai; 1975. Geochemical aspects of the McMurdo saline lakes with special emphasis on the distribution of nutrient matters. Mem. Natl Inst. Polar Res., Spec. Issue 4: 5–29.

Torii, T., N. Yamagata, S. Nakaya & S. Murata, 1977. Chemical characteristics of Antarctic saline lakes. Antarct. Rec. 58: 9–19 (in Japanese).

Torii, T., S. Nakaya, O. Matsubaya, G. I. Matsumoto, N. Masuda. T. Kawano & H. Murayama, 1988. Chemical characteristics of pond waters in the Labyrinth of southern Victoria Land, Antarctica. Hydrobiologia (this volume).

Vincent, W. F., 1981. Production strategies in Antarctic inland waters: Phytoplankton eco-physiology in a permanently ice-covered lake. Ecology 62: 1215–1224.

Vincent, W. F., M. T. Downes & C. L. Vincent, 1981. Nitrous oxide cycling in Lake Vanda, Antarctica. Nature 292: 618–620.

Volkman, J. K., 1986. A review of sterol markers for marine and terrigenous organic matter. Org. Geochem. 9: 83–99.

Volkman, J. K., D. I. Allen, P. L. Stevenson & H. R. Burton, 1986. Bacterial and algal hydrocarbons in sediments from a saline Antarctic lake, Ace Lake. Org. Geochem. 10: 671–681.

Volkman, J. K., H. R. Burton, D. A. Everitt & D. I. Allen, 1988. Pigment and lipid compositions of algal and bacterial communities in Ace Lake, Vestfold Hills, Antarctica. Developments in Hydrobiology (in press).

Weete, J. D., 1976. Algal and fungal waxes. In P. E. Kolattukudy (ed.), Chemistry and Biochemistry of Natural Waxes. Amsterdam, Elsevier: 349-418.

Hydrobiologia 172: 291–303, 1989
W. F. Vincent and J. C. Ellis-Evans (eds) High Latitude Limnology

Vertical distribution of organic constituents in an Antarctic lake: Lake Fryxell

Genki I. Matsumoto[1], Kunihiko Watanuki[1] & Tetsuya Torrii[2]
[1]*The University of Tokyo, The College of Arts and Sciences, Department of Chemistry, Komaba, Meguro-ku, Tokyo 153, Japan and* [2]*Chiba Institute of Technology, Tsudanuma 2-17-1, Narashino, Chiba Prefecture 275, Japan*

Key words: Antarctic lake, Lake Fryxell, vertical distribution, water, sediment, organic constituents, total organic carbon, hydrocarbon, fatty acid, hydroxy acid

Abstract

Vertical distribution of organic constituents, i.e. total organic carbon (TOC), hydrocarbons, fatty acids and hydroxy acids in water and sediment samples from Lake Fryxell (77° 35′ S, 163° 15′ E) of southern Victoria Land, Antarctica were studied to elucidate their features in relation to stratification of the lake waters and likely distribution of microorganisms. The TOC content of the surface water (5.0 m; just below the ice cover of 4.50 m thickness) was 1.4 mg l^{-1}. It increased markedly with depth and attained a maximum value of 21.7 mg C l^{-1} at a depth of 17.5 m, but decreased to the bottom (13.3 mg C l^{-1}). The high TOC content of the anoxic bottom layers (> 15 m) is attributable to the concentration of refractory organic substances over long periods following the degradation of labile organic constituents. Hydrocarbons were not found in the water column, but the major constituent of the bottom sediment was n-$C_{29:2}$ alkene. Total concentrations of fatty acids in the oxic layers (⩽ 10 m) were highest at 10.0 m and much higher than those in the anoxic layers (> 10 m), probably reflecting the phytoplankton population. The content of branched (*iso* and *anteiso*) fatty acids and 3-hydroxy acids in the anoxic layers were much greater than those in the oxic layers which would seem to reflect the distribution of bacterial abundance. The differences of organic composition between the water column and sediments imply that sinking dead organisms were quickly degraded in the lake bottom. Also, the composition of microorganisms in the water column must be very different from that in the sediments.

Introduction

A large numbers of lakes and ponds, including saline water bodies, are distributed in the Dry Valleys of southern Victoria Land in Antarctica. Lake Fryxell (77° 35′ S, 163° 15′ E) is one of the largest lakes in the Dry Valleys, is relatively productive, lies in the lower Taylor Valley at an altitude of 22 m, and has an area of 7 km^2 with a maximum depth of 19 m (Vincent, 1981). The lake is perennially covered with 4-5 m of ice, which prevents direct wind-induced turbulence. The lake is density stratified by salt water. Torii *et al.* (1975) reported on the major elements and nutrients in the lake. The major salt composition of the lake is similar to that of seawater with the highest concentration being approximately one-fifth that of seawater. Vincent (1981) has studied the relative importance of light, temperature and nutrient supply for strategies of planktonic primary production in this lake.

Hydrocarbons and fatty acids are ubiquitous

organic constituents and distributed widely in living and geological environments. Hydroxy acids also occur in a wide variety of microorganisms, such as bacteria, yeasts and fungi, and higher plants, although little is known of their occurrence in cyanobacteria and other algae (Matsumoto & Nagashima, 1984; Matsumoto *et al.*, 1984a). Only microorganisms occur in Lake Fryxell, although there is scattered moss vegetation around the lake. The major organisms in the water column of the lake can be expected to be algae, cyanobacteria and bacteria. Normal-saturated and unsaturated fatty acids may reflect the abundances of algae and cyanobacteria (*e.g*, Weete, 1976), while *iso*- and *anteiso*-branched fatty acids (*e.g.*, Kaneda, 1967) and 3-hydroxy acids (*e.g.*, Wilkinson *et al.*, 1972; Matsumoto & Nagashima, 1984) may correlate with the bacterial distribution. Also short-chain ($<C_{20}$) *n*-alkanes and alkenes may be a marker of certain algae and cyanobacteria, whereas long-chain ($\geqslant C_{20}$) hydrocarbons may have arisen from some bacteria (Albro, 1976) and fungi (Weete, 1976).

Little is known of the organic constituents in Lake Fryxell. Matsumoto & Hanya (1977) reported data on total organic carbon (TOC) and fatty acids in the lake. The TOC value of the bottom layer is very high (29.1 mg C l^{-1}). The major fatty acids are *n*-C_{16}, unsaturated-C_{18} (u-C_{18}) and u-C_{16} with decreasing abundances, respectively. 24-Ethylcholest-5-en-3β-ol is the prominent sterol in the sediments of the lake (Matsumoto *et al.*, 1982), but no one has yet recorded hydrocarbons and hydroxy acids in the lake. Here we report on vertical profiles of TOC, hydrocarbons, fatty acids, and 2- and 3-hydroxy acids in Lake Fryxell, and discuss these in relation to lake stratification and microbial distribution.

Materials and methods

Samples

All samples were collected from the deepest point of the lake (19 m) on 12th December 1985. A hole was bored through lake ice 4.50 m thick with a 10-cm diameter Sipre ice auger. Water and sediment samples were collected using a Kitahara-type water sampler (1 l). Water samples for organic analysis were transferred immediately to 1 l glass bottles with teflon caps and acidified with concentrated hydrochloric acid (pH < 2). Water samples for TOC analysis were kept in 50 ml polyethylene bottles and again acidified immediately as above. Particulate matter was obtained by *in situ* filtration of 5 l of water with a glass-fiber filter having a pore size of 0.65 μm. The filters were enveloped in aluminium foil for return to the laboratory. Glass bottles were stored at about 5 °C until analysis. The 50 ml polyethylene bottles, sediment and particulate matter samples were kept frozen until analysed.

In situ measurement

Temperature and pH were measured using a Cho-type thermister-thermometer (Tama Kagaku Co. Ltd.) and a pH meter (YEW model PH51, Yokogawa Electric Co.), respectively. Dissolved oxygen was determined by Winkler titration (Japan Meteorological Agency, 1970).

Analysis of organic constituents

TOC was determined by the method of Menzel and Vaccaro (1964). The analytical procedures were essentially similar to those of previous studies (Matsumoto *et al.*, 1979, 1984a; Matsumoto & Nagashima, 1984). Organic constituents were extracted from each 1 l water sample three times (200 ml x 1, 100 ml x 2) with ethyl acetate and concentrated to dryness under reduced pressure at 30 °C. The ethyl acetate extracts were saponified with 10 ml 0.5M-potassium hydroxide/methanol (80 °C, 2 h) and extracted with ethyl acetate after acidification with concentrated hydrochloric acid. The particulate matter on the glass fiber filters and sediments were saponified with 0.5M potassium hydroxide methanol (80 °C, 2 h) and extracted with ethyl acetate after acidification. Both the ethyl acetate extracts were dried and redissolved in 50 μl benzene: ethyl acetate (1 : 1) and chromatographed through a silica gel column (160x5 mm i.d., 100 mesh, 5% water). Hydrocarbons, fatty acids and hydroxy acids were eluted with 2 column volumes of hexane, 3 column volumes of benzene: ethyl acetate (95 : 5) and 2 column volumes of ethyl acetate, respectively. Fatty acids and hydroxy acids were methylated with 14% boron trifluoride/methanol solution (80 °C, 2 h). Hydroxy acid methyl esters were further treated with 25% *N,O*-bis(trimethylsilyl)acetamide in acetonitrile solution to obtain their trimethylsilyloxy ethers. To ascertain the structure of alkenes from the sediment sample, hydrogenation was necessary. A 1 ml hexane solution of the hydrocarbon fraction, containing 10 mg platinum dioxide, was placed in a 1 ml glass ampoule and bubbled with hydrogen gas for 1 h at room temperature.

Hydrocarbons and methyl esters of fatty acids were analysed using a Shimadzu GC-8A gas chromatograph (FID) equipped with a Hewlett Packard fused silica capillary column (SE 54, 17 mx0.2 mm i.d.). The column oven temperature was programmed from 80 to 300 °C at 10 °C min^{-1}. Injector and detector temperatures were kept at 330 °C. The flow rate of helium carrier gas was 1 ml min^{-1}. The hydroxy acid methyl ester trimethylsilyloxy ethers were analysed, in splitless mode, using a Shimadzu QP-1 000 gas chromatograph/mass spectrometer equipped with the same fused silica capillary column used in the gas chromatography. Column oven temperature was programmed from 60 to 120 °C at 30 °C min^{-1} and then from 120 to 300 °C at 10 °C min^{-1}. The temperatures of injector, molecular separator and ion source were maintained at 300, 300 and 250 °C, respectively, during analysis. Mass spectra (m/z 50-600) were obtained continuously at 1.5 sec intervals at an ionization energy of 70 eV. The identification of each compound was made by comparison of retention times and mass spectra with those of authentic standards and published literature (Eglinton *et al.*, 1968; Boon *et al.*, 1977a; Matsumoto *et al.*, 1979, 1984a; Matsumoto & Nagashima, 1984).

Results and discussion

Physicochemical properties

The temperature of the surface water was 0 °C, rising sharply with depth to a maximum of 2.7 °C at a depth of 8.5 m, before decreasing gradually to the bottom (2.1 °C, Fig. 1). This profile is broadly similar to those obtained in previous studies (Torii *et al.*, 1975; Vincent, 1981; Wharton *et al.*.1982). Chlorinity increased gradually with depth and attained a maximum value of 3.71 g l^{-1} at a depth of 16 m (Torii *et al.*, 1975). pH values were near constant (8.0) throughout the water column. Dissolved oxygen in the surface layers was supersaturated, but water below 10 m was anoxic and hydrogen sulfide was observed *in situ*. Our sample from a depth of 10 m contained both dissolved oxygen and hydrogen sulfide, indicating that this depth was the boundary of the oxic and anoxic layers in the lake.

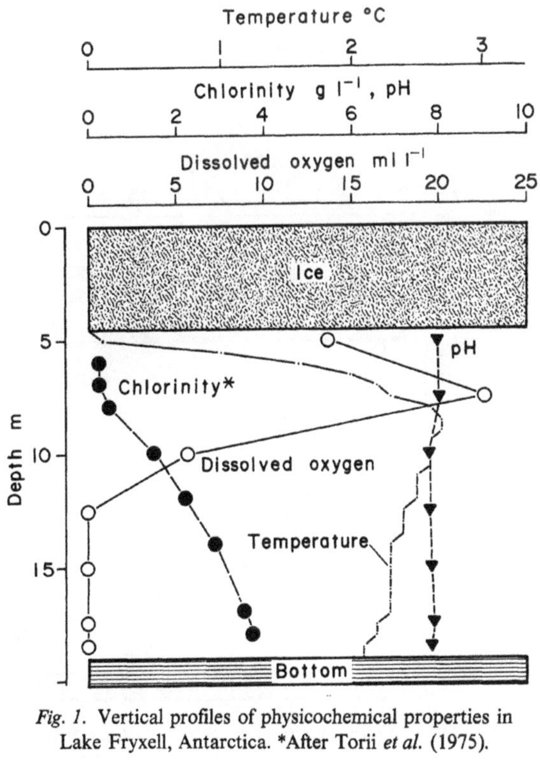

Fig. 1. Vertical profiles of physicochemical properties in Lake Fryxell, Antarctica. *After Torii *et al.* (1975).

TOC

TOC content at a depth of 5.0 m was 1.4 mg C l^{-1}, increased to a maximum value of 21.7 mg C l^{-1} at 17.5 m, but then decreased rapidly to the bottom (13.3 mg C l^{-1}, Fig. 2). The TOC values of the bottom layers ($\geq$15 m) were considerably higher than values recorded in temperate systems (*e.g.*, polluted Tama River waters at Chohfu in Tokyo, average 7.6 mg C l^{-1}; Matsumoto & Hanya, 1981), but were similar to those of other Dry Valleys lakes such as Lakes Bonney and Vanda (Parker *et al.*, 1974; Matsumoto & Hanya, 1977; Matsumoto *et al.*, 1979, 1984b). There is no outflow in this lake and no losses of organic substances or dissolved salts. This lake (including Lake Washburn which occupied the lower Taylor Valley approximately 20000 years B.P.) is a very old system (Stuiver *et al.*, 1981; Lawrence & Hendy, 1985). It seems therefore that refractory organic substances have been concentrated in the bottom anoxic layers over a long period of time after the degradation of more labile organic constituents. High concentrations of ammonium-N and orthophosphate-P in the anoxic layers (Torii *et al.*, 1975) support the suggestion that degradation of a large amounts of organic substances has occurred in the bottom waters.

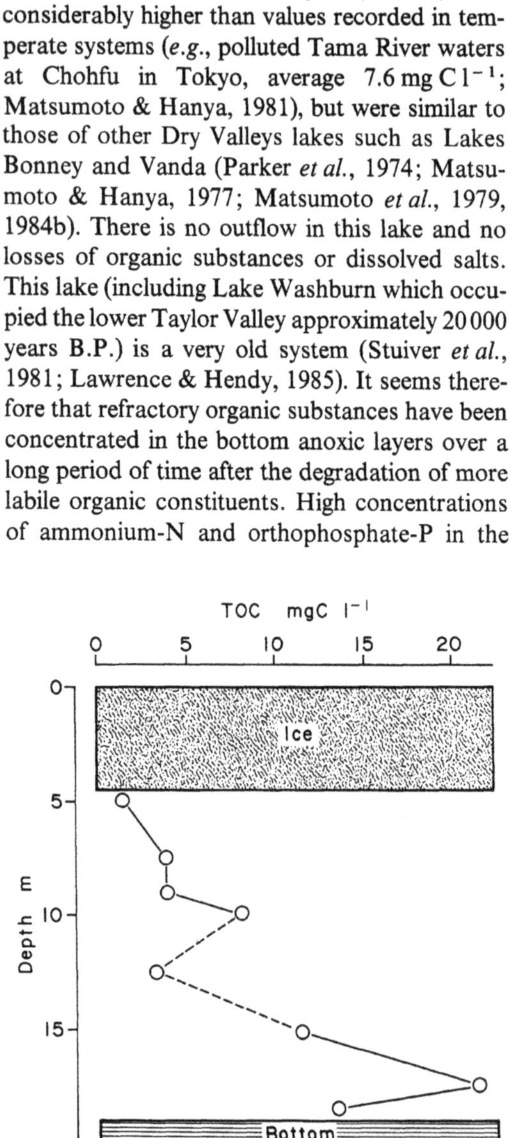

Fig. 2. Vertical distribution of TOC in Lake Fryxell, Antarctica.

Hydrocarbons

The concentrations of hydrocarbons in lake-waters were extremely low and indeed could not be found in either dissolved or particulate samples throughout the water column. Hydrocarbons were however found in the sediments (Fig. 3). Normal alkanes ranging in carbon chain length from n-C_{15} to n-C_{28} were found with a predominance of odd-carbon numbers, in addition to branched-C_{18}, $C_{29:2}$ (carbon chain length: number of unsaturated bonds) alkenes and hop-22(29)-ene (Dastillung, 1977). The major peaks were $C_{29:2}$ alkenes with a concentration of 9.9 $\mu g\ g^{-1}$ dry sediment. After hydrogenation three $C_{29:2}$ alkene peaks were reduced into a single peak corresponding to n-C_{29} alkane. Thus the three alkene peaks were attributable to the presence of *cis-trans* isomers with straight chains. Long-chain alkenes $C_{29:2}$ and $C_{29:1}$ have been found in the sediments from Lakes Bonney and Joyce, respectively, in the Dry Valleys (Matsumoto *et al.*, 1979). Recently n-$C_{29:1}$ and n-$C_{34:1-4}$ alkenes, together with the predominant hydrocarbons have also been detected in Ace Lake sediment from the Vestfold Oasis in Princess Elizabeth Land (Volkman *et al.* 1986).

Little is known of the occurrence of long-chain alkenes in living organisms for anoxic conditions. Certain bacteria contain long-chain alkenes, though they are branched (Albro, 1976). Also, Patterson (1967) reported that only heterotrophic conditions, *Chorella vulgaris* produce long-chain alkenes. Hopanoids involving hop-22(29)-ene have been shown to be widely distributed among bacteria and cyanobacteria (Dastillung, 1977; Ourisson *et al.*, 1979). Thus it is very likely that hydrocarbons in the sediments have arisen from specific bacteria and/or other microorganisms which differ from those in the water column. For the contribution of bacteria, it is possible however that the absence of hydrocarbons in the water

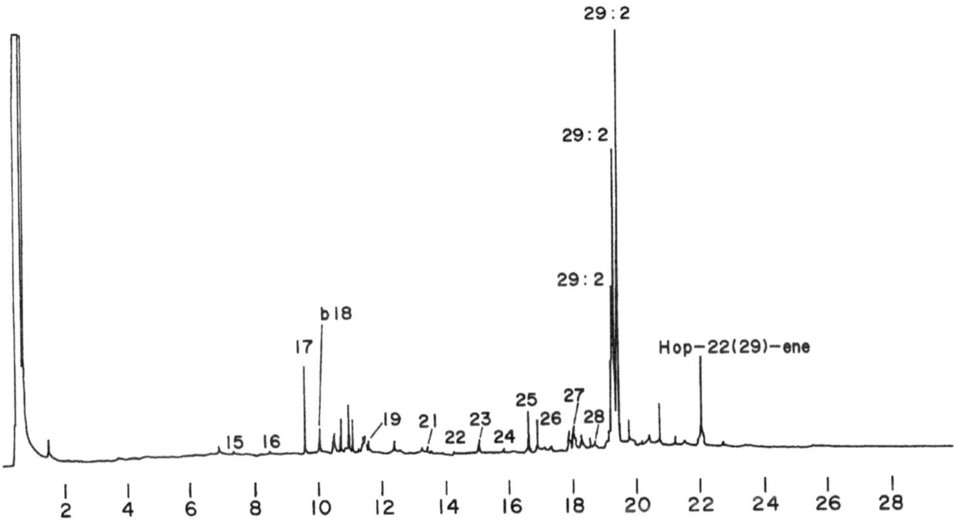

Fig. 3. Capillary gas chromatogram of the hydrocarbon fraction from the sediment of Lake Fryxell, Antarctica. Arabic figures on the peaks indicate carbon chain length. b = Branched.

column is due to the lower abundance of bacteria per unit volume in the water column than in the sediments. *Anteiso*-C_{15} fatty acid (a-C_{15}), the dominant branched fatty acids, can be considered to be a bacterial marker, as discussed below. The concentrations of a-C_{15} in the bottom water (18.5 m) and sediments were 0.56 $\mu g\, l^{-1}$ and 3.9 $\mu g\, g^{-1}$ respectively, noted below (ref. Table 1, Fig. 5). The concentration of n-$C_{29:2}$ alkenes in the sediments was 9.9 $\mu g\, g^{-1}$. If the bacterial species in the sediments are similar to those in the bottom water, then the *n*-$C_{29:2}$ and a-C_{15} concentrations and the *n*-$C_{29:2}$/a-C_{15} ratio for the sediments should give *n*-$C_{29:2}$ concentration of 1.4 $\mu g\, l^{-1}$ for the bottom water. However, it was not detected in the bottom water sample. This result indicates that the composition of bacterial species in the sediments was very different from that in the water column. It is possible that some cyanobacteria or microalgae produce long-chain alkenes under anoxic conditions.

Fatty acids

Capillary gas chromatogram of the fatty acid fraction of a water sample from a depth of 10.0 m in the lake is shown in Fig. 4. Normal-alkanoic acids in carbon chain length ranging from *n*-C_{12} to *n*-C_{20} were found in the chromatogram with a predominance of even-carbon numbers, in addition to alkenoic (u-C_{16} and u-C_{18}) and branched (*iso*- and/or *anteiso*-C_{13}-C_{15}) acids. Fatty acid compositions for the unfiltered water (total), particulate matter and bottom sediment are summarized in Table 1. Normal alkanoic, alkenoic and branched acids were found in all the samples. Mass spectra of the alkenoic acids showed that they comprised mono-, di and trienes. The major fatty acids were *n*-C_{16}, *n*-C_{18}, u-C_{16} and/or u-C_{18} at all depths.

The total fatty acid concentration (particulate and dissolved) at a depth of 5.0 m, just below the ice cover (4.50 m) was 36 $\mu g\, l^{-1}$, increased to a maximum value of 66 $\mu g\, l^{-1}$ at a depth of 10.0 m, but decreased considerably to the bottom (Fig. 5). Chlorophyll *a* and photosynthetic maxima have been observed just above (*ca.* 0.5 m) the boundary of the oxic and anoxic layers (Vincent, 1981). Thus, the total fatty acid contents would seem to reflect phytoplanktonic abundances, and indeed they are known as major lipid components of phytoplankton. Dominant algal species at this

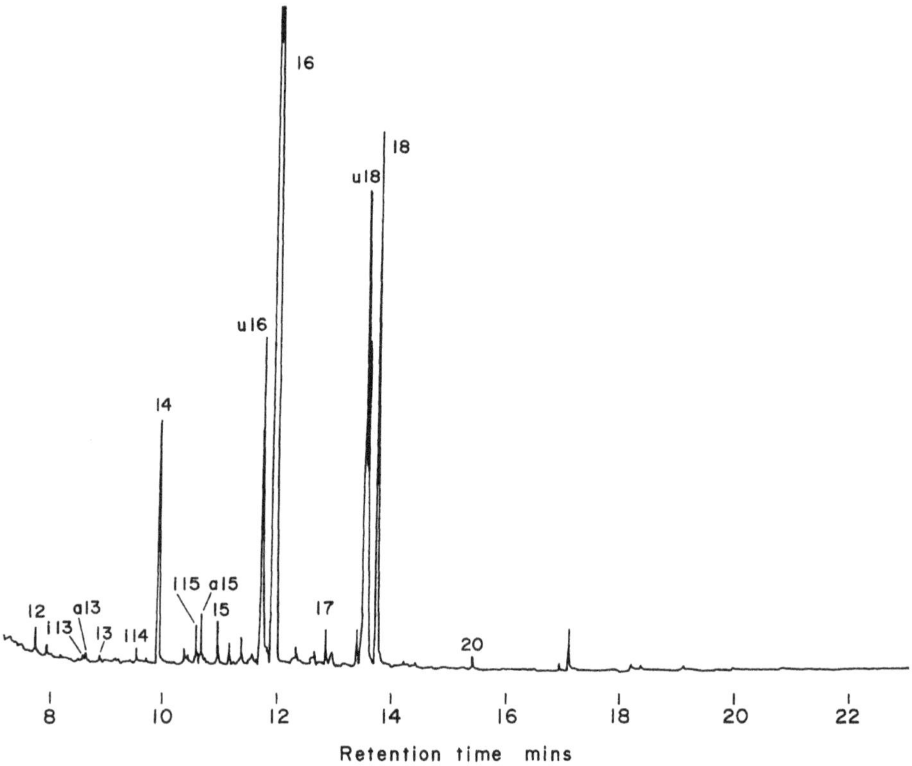

Fig. 4. Capillary gas chromatogram of the fatty acid fraction from a water sample (10.0 m) in Lake Fryxell, Antarctica. Arabic figures on the peaks denote carbon chain length. u = Unsaturated. i = *Iso*. a = *Anteiso*.

depth are *Chroomonas lacustris* and *Pyramimonas* sp. (Vincent, 1981). Therefore, our total fatty acid peak should be due to these algal species.

Fatty acid content in particulate matter (particulate fatty acids) at a depth of 5.0 m was 27 $\mu g\,l^{-1}$, rising considerably down to a depth of 7.5 m (44 $\mu g\,l^{-1}$), before decreasing to the bottom (Fig. 5). The amounts of particulate fatty acids in the anoxic layers were much lower than those in the oxic layers in spite of the increase in TOC contents, probably reflecting the decrease in contribution of the photosynthetic plankton.

The total concentration of fatty acids in the sediments was 230 $\mu g\,g^{-1}$, which is considerably higher than those in other Dry Valleys lakes such as Lakes Vanda, Bonney and Joyce (Matsumoto *et al.*, 1979, 1984b).

The particulate/total fatty acid ratios in the oxic layers (75–92%) were much higher than those in the anoxic layers (12–34%) with a maximum value at a depth of 7.5 m (Fig. 5). This result revealed, *vice versa*, that dissolved fatty acids (total particulate) in the anoxic layers were much greater than those in the oxic layers. Fatty acids in the oxic layers can be explained by the contribution of photosynthetic plankton, whereas the dissolved fatty acids in the anoxic layers represent either seston degradation products or components of 0.65 μm filterable small microorganisms, such as certain bacteria. The vertical distribution of fatty acid composition showed that branched fatty acids were much more abundant in the particulate fraction than those in the total fatty acids discussed below (Figs. 6 and 7). Hence dissolved

Table 1. Fatty acid compositions for the unfiltered water (total), particulate matter and sediment samples from Lake Fryxell, Antarctica (%).

Carbon chain length	Total						Particulate							Sediment
	5.0 m	7.5 m	10.0 m	12.5 m	15.0 m	18.5 m	5.0 m	7.5 m	10.0 m	12.5 m	15.0 m	17.5 m	18.5 m	
n-Alkanoic														
12	0.9	0.5	0.5	1.0	1.1	2.6	0.4	0.1	0.2	0.5	1.0	1.2	1.1	0.5
13	0.3	0.1	0.2	0.4	0.3	0.6	0.1	Trace	0.1	0.2	0.4	0.6	0.5	0.1
14	4.9	4.8	5.8	5.9	4.5	10.0	6.0	3.5	3.5	3.8	6.2	7.8	7.6	5.2
15	1.1	1.1	1.1	2.1	1.2	4.1	1.0	0.7	0.7	1.8	1.1	2.1	2.4	2.8
16	37.4	58.7	45.8	25.8	22.7	28.3	54.4	58.6	35.4	16.4	18.0	22.4	20.5	52.1
17	0.9	0.8	0.8	1.1	0.7	1.5	0.7	0.8	0.6	0.6	0.8	0.8	1.0	1.4
18	9.5	7.8	12.9	19.8	7.1	15.2	6.7	6.7	5.8	6.0	5.5	10.8	6.7	4.2
19	–	–	–	–	–	–	–	Trace	0.1	0.1	0.3	–	–	–
20	0.2	0.2	0.3	0.6	0.3	0.7	0.1	0.1	0.2	0.3	0.4	0.4	0.4	0.2
21	–	–	–	–	–	–	0.1	Trace	Trace	–	–	–	–	–
22	–	–	–	–	–	–	0.1	Trace	0.1	0.2	0.4	–	0.4	0.1
23	–	–	–	–	–	–	–	–	–	–	–	–	–	–
24	–	–	–	–	–	–	0.2	0.1	0.1	0.2	0.5	–	0.6	–
25	–	–	–	–	–	–	Trace	Trace	–	–	–	–	–	–
26	–	–	–	–	–	–	0.1	Trace	–	–	–	–	–	–
Branched*														
Iso-13	0.2	0.2	0.2	0.8	0.5	0.9	0.1	Trace	0.2	2.0	2.8	1.9	1.7	0.3
Anteiso-13	0.1	–	0.2	1.2	0.8	0.9	Trace	0.1	0.2	2.6	3.4	2.0	1.8	0.2
Iso-14	0.4	0.3	0.4	0.9	0.8	1.4	0.2	0.1	0.2	1.0	1.5	1.6	1.6	0.6
Iso-15	0.6	0.5	0.8	3.3	2.2	3.3	0.3	0.3	0.8	6.2	7.3	6.6	6.0	1.1
Anteiso-15	0.4	0.4	1.0	4.8	3.6	4.2	0.2	0.1	1.0	8.3	9.7	8.5	7.5	1.7
Alkenoic														
16	5.3	9.4	9.4	11.0	29.6	9.6	10.6	6.7	13.2	21.2	17.8	20.1	18.2	12.8
18	37.8	19.2	20.6	21.3	24.6	16.7	18.7	22.1	37.6	28.6	22.9	13.2	22.0	16.7

* *Iso*- and *anteiso*-C_{17} acids must be present, but not quantified because of unresolved mixtures. Trace: Less than 0.05%.

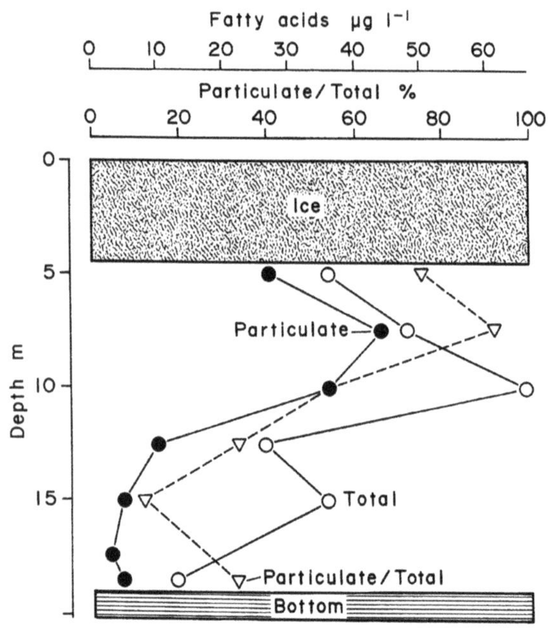

Fig. 5. Vertical distributions of total (particulate and dissolved) and particulate fatty acid contents, and the particulate/total fatty acid ratios (%) in Lake Fryxell, Antarctica.

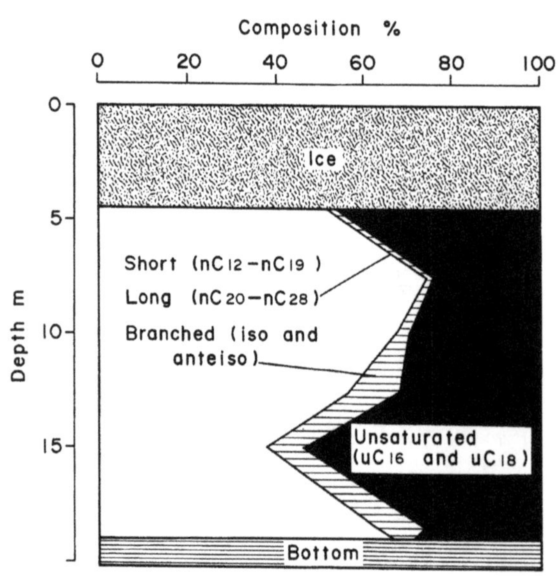

Fig. 6. Total fatty acid compositions for the water column and sediments in Lake Fryxell, Antarctica.

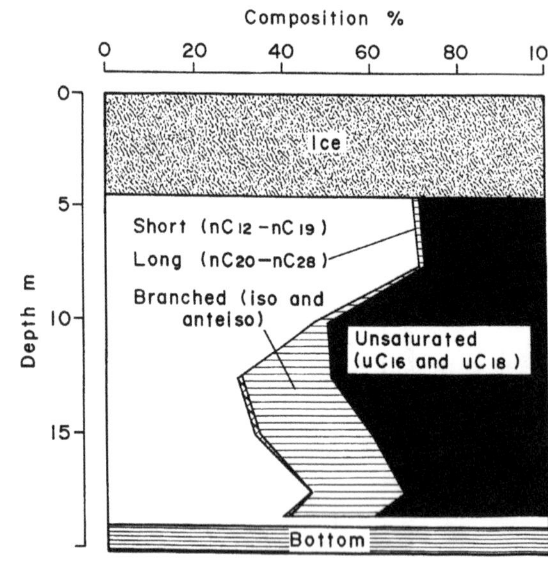

Fig. 7. Vertical distribution of particulate fatty acid composition in Lake Fryxell, Antarctica.

fatty acids in the anoxic layers occur mainly in non-biological forms derived from degrading seston rather than as components of viable microorganisms.

Vertical distribution of total and particulate fatty acid compositions showed that short-chain *n*-alkanoic acids (n-C_{12}–n-C_{10}) and alkenoic acids (u-C_{16} and u-C_{18}) comprised the major fatty acids in all the water column and sediment samples (Figs. 6 and 7). The proportions of long-chain *n*-alkanoic acids (n-C_{20}–n-C_{26}) were less than 1% at all the depths. Branched acids increased markedly from a depth of 10.0 m, probably reflecting the distribution of bacterial populations, since branched acids are abundant in bacterial lipids (Leo & Parker, 1966; Kaneda, 1967; Tornabene & Oró, 1967). The proportion of branched fatty acids in the particulate fatty acids are much greater than that in the total fatty acids, probably reflecting the fact that bacteria are largely retained on the glass fiber filter. Our branched acid distributions were consistent with those for bacterial numbers determined on the same samples by the acridine orange epifluorescence direct count method, *viz.* bacterial numbers increased dramatically from a depth of 10.0 m towards the bottom (T. Konda *et al.*, unpublished results).

Hydroxy acids

2- and 3-hydroxy acids were studied by mass chromatography, 2-hydroxy acids were not detected either in unfiltered water or particulate matter samples. The major 2-hydroxy acids in the sediments were n-C_{14}, n-C_{16} and n-C_{18} with decreasing abundances respectively, which may arise from certain bacteria (Boon *et al.*, 1977a, 1977b). Normal 3-hydroxy acids in carbon chain lengths ranging from n-C_{10} to n-C_{20} were found on the m/z 175 mass chromatogram of a water sample from a depth of 15.0 m with a predominance of even-carbon numbers, along with branched (*iso* and *anteiso*) and unsaturated hydroxy acids (Fig. 8). The analytical results for 3-hydroxy acids in the unfiltered water, particulate matter and sediments are summarised in Table 2. The major 3-hydroxy acids ($\geqslant 10\%$) were normal even-carbon numbers (n-C_{10}, n-C_{14}, n-C_{16} and n-C_{18}) and branched odd-carbon numbers (*iso*-C_{15}, *iso*-C_{17} and *anteiso*-C_{17}) except for water samples from 5.0 and 7.5 m, which contained *iso*-C_{18} as a major constituent. It is likely that the distribution of microorganisms in the upper layers of the lake differ considerably from those in the underlying water bodies.

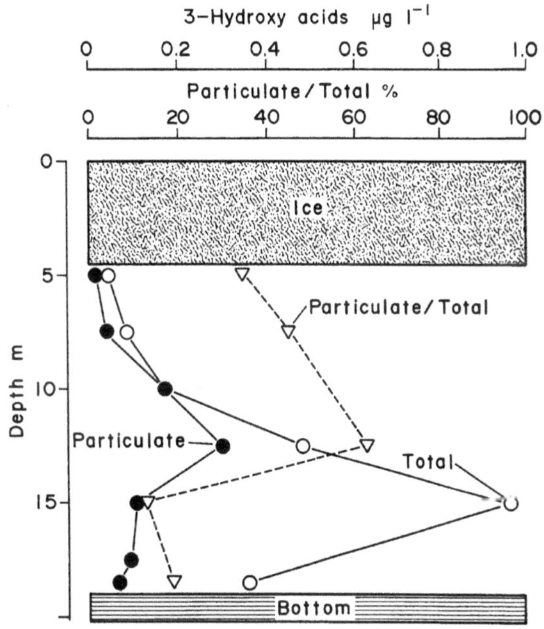

Fig. 9. Vertical distributions of total (particulate and dissolved) and particulate 3-hydroxy acid contents, and the particulate/total 3-hydroxy acid ratios (%) in Lake Fryxell, Antarctica.

The total concentration of 3-hydroxy acids at 5.0 m depth was 0.041 $\mu g\, l^{-1}$ increasing to 0.96 $\mu g\, l^{-1}$ at a depth of 15.0 m and then decreasing towards the bottom (0.36 $\mu g\, l^{-1}$, Fig. 9). Particulate 3-hydroxy acid concentrations also rose

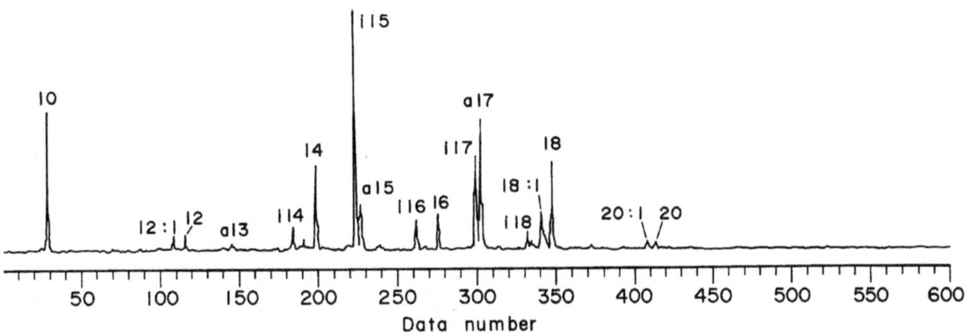

Fig. 8. Mass chromatogram (m/z 175) of 3-hydroxy acid trimethylosilyloxy ether methyl esters from a water sample (15.0 m) in Lake Fryxell, Antarctica. Arabic figures on the peaks denote carbon chain length. i = *Iso*. a = *Anteiso*.

Table 2. 3-Hydroxy acid compositions for the unfiltered water (total), particulate matter and sediment samples from Lake Fryxell, Antarctica (%).

Carbon chain length	Total						Particulate							Sediment
	5.0 m	7.5 m	10.0 m	12.5 m	15.0 m	18.5 m	5.0 m	7.5 m	10.0 m	12.5 m	15.0 m	17.5 m	18.5 m	
n-Alkanoic														
10	11.6	8.2	4.4	2.2	19.9	3.9	4.3	3.3	19.5	13.8	11.6	14.5	8.1	14.8
11	–	–	–	–	–	–	–	–	–	–	–	–	–	0.4
12	–	–	2.0	–	–	2.2	4.9	–	1.5	1.6	3.0	3.1	3.5	15.6
13	–	–	–	–	–	–	–	–	–	–	–	–	–	0.4
14	12.1	9.2	7.0	3.0	5.1	2.2	14.2	17.9	4.1	8.0	10.0	8.9	9.8	11.2
15	–	–	2.5	–	–	–	–	–	–	0.7	–	1.0	–	0.4
16	14.2	27.6	14.0	6.3	5.0	5.8	33.4	41.1	5.3	3.6	5.3	6.5	7.6	6.5
17	–	–	2.0	–	–	2.6	–	–	–	0.3	–	–	–	0.4
18	9.5	9.2	14.8	14.2	11.3	4.0	5.9	7.4	13.8	8.5	5.5	4.7	3.9	2.9
19	–	–	–	–	–	–	–	–	–	–	–	–	–	0.1
20	–	–	2.3	2.2	–	–	–	–	0.4	0.7	1.0	–	1.0	0.6
Branched*														
Anteiso-11	–	–	–	–	–	–	–	–	–	–	–	–	–	0.4
Iso-12	–	–	–	–	–	–	–	–	–	–	1.0	2.0	–	0.3
Iso-13	–	–	–	–	–	–	–	–	–	–	–	–	–	0.6
Anteiso-13	–	–	–	–	–	–	–	–	–	–	–	1.0	–	0.3
Iso-14	–	–	–	–	2.1	–	–	–	–	2.4	2.0	1.6	2.0	2.4
Iso-15	14.7	4.3	4.6	10.1	9.4	4.6	16.6	7.6	11.4	23.1	19.7	10.3	11.2	2.6
Anteiso-15	–	7.9	2.6	4.4	2.1	4.5	6.1	–	2.7	4.6	8.0	7.4	7.9	1.9
Iso-16	–	–	3.8	6.6	2.5	5.0	4.9	–	4.5	3.2	3.2	3.6	4.7	1.3
Iso-17	20.2	22.4	15.0	25.6	12.5	30.3	9.7	16.7	13.5	8.9	9.8	13.7	15.1	7.4
Anteiso-17	–	–	12.0	17.3	27.7	30.9	–	2.2	7.7	12.4	17.1	15.9	21.2	3.4
Iso-18	17.7	11.2	3.3	–	–	4.0	–	–	–	1.8	1.3	–	1.2	0.7
Anteiso-19	–	–	–	–	–	–	–	–	–	–	–	–	–	1.4
Unsaturated														
12 : 1	–	–	–	–	–	–	–	–	–	0.8	–	3.1	–	10.3
14 : 1	–	–	–	–	–	–	–	–	–	1.2	–	1.2	1.2	11.9
16 : 1	–	–	2.1	–	–	–	–	3.8	5.3	–	–	–	–	1.0
18 : 1	–	–	5.0	5.2	2.4	–	–	–	9.7	3.6	1.5	1.5	1.6	0.5
20 : 1	–	–	2.6	2.9	–	–	–	–	0.6	0.8	–	–	–	0.3

with depth, attaining a maximum value of 0.30 $\mu g\, l^{-1}$ at a depth of 12.5 m. These 3-hydroxy acid concentrations were much lower than those for fatty acids The highest particulate/total 3-hydroxy acid ratio was obtained at a depth of 10.0 m (63%). The proportion of particulate 3-hydroxy acids relative to dissolved forms in the oxic layers was much greater than that in the anoxic layers, as in the case of fatty acids. However, the vertical distribution of 3-hydroxy acids was considerably different from those for the fatty acids (Fig. 5), suggesting that the major source organisms of fatty acids and 3-hydroxy acids in this lake were different.

The vertical profiles of total 3-hydroxy acid composition showed that *n*-saturated, *iso*- and *anteiso*-3-hydroxy acids comprised the major portions at all depths (Fig. 10). Mono-unsaturated 3-hydroxy acids were found mainly at 10.0 and 12.5 m depths and in the sediments, although their proportion was less than 10% for all water samples. The abundance of mono-unsaturated 3-hydroxy acids in the sediments (24%) again suggests the presence of specific microorganisms which differ from those in the water column. *Iso*- and *anteiso*-3-hydroxy acids were major components of 3-hydroxy acids in all the water column samples. These 3-hydroxy acids are thought to be derived mainly from bacteria, notably sulfate reducing bacteria, because branched 3-hydroxy acids are abundant in bacterial lipids (Wilkinson, 1972; Moss *et al.*, 1973; Yano *et al.*, 1976; Boon *et al.*, 1977b; Mayberry, 1980; Matsumoto & Nagashima, 1984). *Anteiso*-3-hydroxy acids increased from a depth of 10 m, suggesting variations in vertical distribution of bacterial species. As in the case of hydrocarbons, the composition of 3-hydroxy acids in the sediment was very different from those in the water column. Most particularly, *iso*- and *anteiso*-3-hydroxy acids were not abundant, whilst mono-unsaturated components were dominant. These results again imply that the source organisms in the sediments were different from those in the water column. The composition of 3-hydroxy acids in the particulate matter was generally similar to those of the total 3-hydroxy acids (Fig. 11). That is, *iso*- and *anteiso*-3-hydroxy acids comprised the major pro-

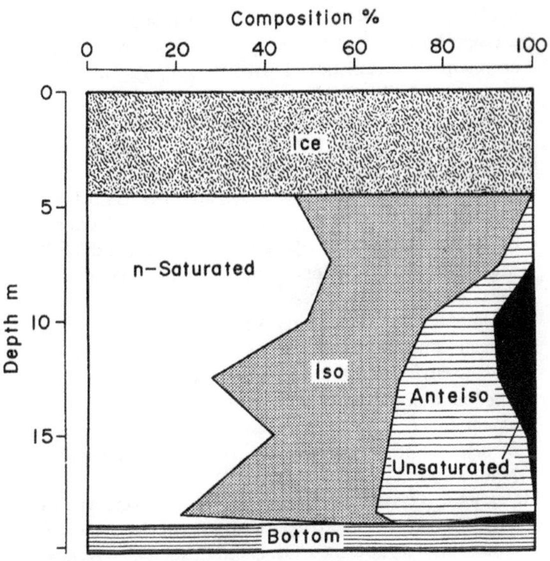

Fig. 10. Total 3-hydroxy acid compositions for the water column and sediments in Lake Fryxell, Antarctica.

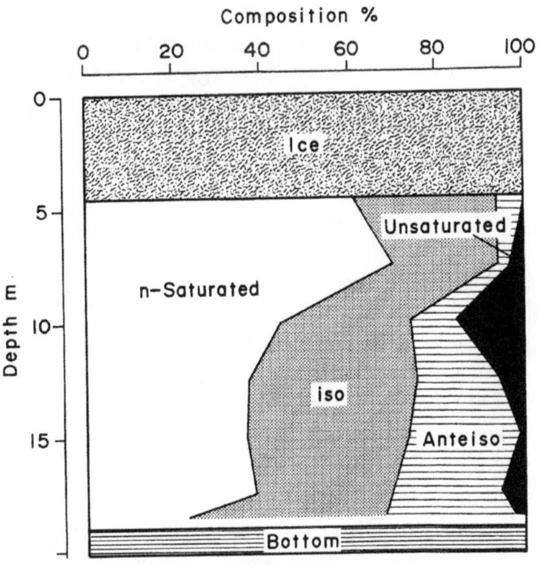

Fig. 11. Vertical distribution of particulate 3-hydroxy acid composition in Lake Fryxell, Antarctica.

portion of 3-hydroxy acids. In addition, the proportion of *anteiso*-3-hydroxy acids increased below a depth of 10.0 m, again reflecting the differences in bacterial species with depth.

Conclusions

The TOC concentration in the anoxic bottom layers (15–19 m) of Lake Fryxell was much higher than that in the oxic layers (5–10 m), implying that the concentration of refractory organic substances in the anoxic layers has been occurring over a long time period following the degradation of more labile organic constituents. The vertical distributions of hydrocarbons, fatty acids and 3-hydroxy acids have revealed that the distribution of microorganisms in the lake is very different with depth. Total fatty acid contents imply that photosynthetic plankters were most abundant at about 10 m depth, just above the anoxic zone. Most fatty acids in the oxic layers can be associated with these planktonic autotrophs, while most dissolved fatty acids in the anoxic layers may be attributable to degradation of settling plankton. Branched (*iso* and *anteiso*) fatty acids and 3-hydroxy acids indicated that bacterial numbers were more abundant in the anoxic layers than in the oxic layers. The organic composition in the water column was very different from that in the sediments, suggesting that the sinking dead organisms were quickly degraded in the lake bottom.

Acknowledgments

The authors are greatly indebted to the Antarctic Division, DSIR, New Zealand, US Navy and K211 members (Mr. K. Yoshimizu, Dr. N. Masuda and Mr. T. Takeuchi) for their kind support of our Antarctic research. We thank also editors J. C. Ellis-Evans and W. F. Vincent for their help in making it more readable, and Dr. M. Ochiai of Tokyo Metropolitan University for the opportunity to use the TOC analyzer.

References

Albro, P. W., 1976. Bacterial waxes. In P. E. Kolattukudy (ed.), Chemistry and Biochemistry of Natural Waxes. Elsevier, Amsterdam 419–445.

Boon, J. J., F. de Lange, P. J. W. Schuyl, J. W. de Leeuw & P. A. Schenk, 1977a. Organic geochemistry of Walvis Bay diatomaceous ooze. II. Occurrence and significance of the hydroxy fatty acids. In R. Campos and J. Goñi (eds.), Advances in Organic Geochemistry, 1975. Enadimsa, Madrid: 255–272.

Boon, J. J., J. W. de Leeuw, G. J. v. d. Hoek & J. H. Bosjan, 1977b. Significance and taxonomic value of *iso* and *anteiso* monoenoic fatty acids and branched β-hydroxy acids in *Desulfovibrio desulfuricans*. J. Bacteriol. 129: 1183–1191.

Dastillung, M., 1976. Lipides de sédiments récents. D. Sc. Thesis, Université Louis Pasteur, 147 pp. Strasbourg (cited in R. P. Philp, ed., Fossil Fuel Biomarkers. Applications and Spectra. Method in Geochemistry and Geophysics No. 23. Elsevier, Amsterdam: 183).

Eglinton, G., D. H. Hunneman & K. Douraghi-Zadeh, 1968. Gas chromatographic-mass spectrometric studies of long chain hydroxy acids-II. The hydroxy acids and fatty acids of a 5000-year-old lacustrine sediment. Tetrahydron 24: 5929–5941.

Japan Meteorological Agency (ed.), 1970. The Manual of Oceanographic Observation (Kaiyo Kansoku Shihin), The Oceanographical Society of Japan, Tokyo: 145–209 (in Japanese).

Kaneda, T., 1967. Fatty acids in the Genus *Bacillus*. I. *Iso*- and *anteiso*-fatty acids as characteristic constituents of lipids in 10 species. J. Bacteriol. 93: 894–903.

Lawrence, M. J. F. & C. H. Hendy, 1985. Water column and sediment characteristics of Lake Fryxell, Taylor Valley, Antarctica. N.Z.J. Geol. Geophys. 28: 543–552.

Leo, R. F. & P. L. Parker, 1966. Branched-chain fatty acids in sediments. Science 152: 649–650.

Matsumoto, G. & T. Hanya, 1977. Organic carbons and fatty acids in Antarctic saline lakes. Antarct. Rec. 58: 81–88.

Matsumoto, G., T. Torii & T. Hanya, 1979. Distribution of organic constituents in lake waters and sediments of the McMurdo Sound region. Mem. Natl Inst. Polar Res., Spec. Issue 13: 103–120.

Matsumoto, G. & T. Hanya, 1981. Comparative study on organic constituents in polluted and unpolluted inland aquatic environments-I. Features of hydrocarbons for polluted and unpolluted waters. Water Res. 15: 217–224.

Matsumoto, G., T. Torii & T. Hanya, 1982. High abundance of algal 24-ethylcholesterol in Antarctic lake sediment. Nature 299: 52–54.

Matsumoto, G. I & H. Nagashima, 1984. Occurrence of 3-hydroxy acids in microalgae and cyanobacteria and their geochemical significance. Geochim. Cosmochim. Acta 48: 1683–1687.

Matsumoto. G. I., M. Shioya & H. Nagashima, 1984a. Occurrence of 2-hydroxy acids in microalgae. Phytochemistry 23: 1421–1423.

Matsumoto, G., T. Torii & T. Hanya, 1984b. Vertical distribution of organic constituents in an Antarctic lake: Lake Vanda. Hydrobiologia 111: 119–126.

Mayberry, W. R., 1980. Hydroxy fatty acids in *Bacteroides* species: D-(-)-3-hydroxy-15-methylhexadecanoate and its homologues. J. Bacteriol. 143: 582–587.

Menzel, D. W. & R. F. Vaccaro, 1964. The measurement of dissolved organic and particulate carbon in seawater. Limnol. Oceanogr. 9: 138–142.

Moss, C. W., S. B. Samuels, J. Liddle & R. M. McKinney, 1973. Occurrence of branched-chain hydroxy fatty acids in *Pseudomonas maltophilia*. J. Bacteriol. 114: 1018–1024.

Ourisson, G., P. Albrecht & M. Rohmer, 1979. The hopanoids. Palaeochemistry and biochemistry of a group of natural products. Pure & Appl. Chem. 51: 709–729.

Parker, B. C., J. T. Whitehurst & R. C. Hoehn, 1974. Observations of *in situ* concentrations and production of organic matter in an Antarctic meromictic lake. Virginia J. Sci. 25: 136–140.

Patterson, G. W., 1967. The effect of culture conditions on the hydrocarbon content of *Chlorella vulgaris*. J. Phycol. 3: 22–23.

Stuiver, M., G. H. Denton, T. J. Hughes & J. L. Fastook, 1981. History of the marine ice sheet in West Antarctica during the last glaciation: A working hypothesis. In G. H. Denton & T. J. Hughes (eds.), The Last Great Ice Sheets. John Wiley & Sons, New York: 319–436.

Torii, T., N. Yamagata, S. Nakaya, S. Murata, T. Hashimoto, O. Matsubaya & H. Sakai, 1975. Geochemical aspects of the McMurdo saline lakes with special emphasis on the distribution of nutrient matters. Mem. Natl Inst Polar Res., Spec. Issue 4: 5–29.

Tornabene, T. G. & J. Oró, 1967. ^{14}C incorporation into the fatty acids and aliphatic hydrocarbons of *Sarcina lutea*. J. Bacteriol. 94: 349–358.

Vincent, W. F., 1981. Production strategies in Antarctic inland waters: Phytoplankton eco-physiology in a permanently ice-covered lake. Ecology 62: 1215–1224.

Volkman, J. K., D. I. Allen, P. L. Stevenson & H. R. Burton, 1986. Bacterial and algal hydrocarbons in sediments from a saline Antarctic lake, Ace Lake. Org. Geochem. 10: 671–681.

Weete, J. D., 1976. Algal and fungal waxes. In P. E Kolattukudy (ed.), Chemistry and Biochemistry of Natural Waxes. Amsterdam, Elsevier: 349–418.

Wharton, R. A., Jr., B. C. Parker, G. M. Simmons, Jr., K. G. Seaburg & F. G. Love, 1982. Biogenic calcite structures forming in Lake Fryxell, Antarctica. Nature 295: 403–405.

Wilkinson, S. G., 1972. Composition and structure of the ornithine-containing lipid from *Pseudomonas rubescens*. Biochim. Biophys. Acta 270: 1–17.

Yano, I., Y. Ohno, M. Masui, K. Kato, E. Yabuchi & A. Ohyama, 1976. Occurrence of 2- and 3-hydroxy fatty acids in high concentration in the extractable and bound lipids of *Flavobacterium meningosepticum* and *Flavobacterium* sp. IIb. Lipids 11: 685–688.

Hydrobiologia 172: 305–320, 1989
W. F. Vincent and J. C. Ellis-Evans (eds) High Latitude Limnology

Perennially ice-covered Lake Hoare, Antarctica: physical environment, biology and sedimentation

Robert A. Wharton Jr[1], George M. Simmons Jr[2] & Christopher P. McKay[3]
[1]*Desert Research Institute/Department of Biology, University of Nevada System, Reno, Nevada 89506, USA*; [2]*Department of Biology, Virginia Polytechnic Institute and State University, Blacksburg, VA 24061, USA*; [3]*Solar System Exploration Branch, Life Science Division, NASA Ames Research Center, Moffett Field, CA 94035, USA*

Key words: Antarctica, climatic changes, gas solubility, ice-covered lakes, Lake Hoare

Abstract

Lake Hoare (77° 38′ S, 162° 53′ E) is a perennially ice-covered lake at the eastern end of Taylor Valley in southern Victoria Land, Antarctica. The environment of this lake is controlled by the relatively thick ice cover (3–5 m) which eliminates wind generated currents, restricts gas exchange and sediment deposition, and reduces light penetration. The ice cover is in turn largely controlled by the extreme seasonality of Antarctica and local climate. Lake Hoare and other dry valley lakes may be sensitive indicators of short term (< 100 yr) climatic and/or anthropogenic changes in the dry valleys since the onset of intensive exploration over 30 years ago. The time constants for turnover of the water column and lake ice are 50 and 10 years, respectively. The turnover time for atmospheric gases in the lake is 30–60 years. Therefore, the lake environment responds to changes on a 10–100 year timescale. Because the ice cover has a controlling influence on the lake (*e.g.* light penetration, gas content of water, and sediment deposition), it is probable that small changes in ice ablation, sediment loading on the ice cover, or glacial meltwater (or groundwater) inflow will affect ice cover dynamics and will have a major impact on the lake environment and biota.

Introduction

The largest relatively glacier-free region on the Antarctic continent is the southern Victoria Land 'dry valleys' (~4000 km^2) near McMurdo Sound (Heywood, 1972). The dry valleys, which have also been called the southern Victoria Land 'oasis' (Parker *et al.*, 1982a) or 'Ross desert' (Friedmann & Weed, 1987), exhibit glacial and periglacial features, temperatures usually below freezing, low precipitation, cyclonic storms, high velocity winds, and four months each of continuous sunlight, twilight, and darkness (Solopov, 1967). The southern Victoria Land dry valleys contain several closed basins in which perennially ice-covered lakes are found. The ice covers on these lakes overlie liquid water which contains plankton and benthic microbial communities.

The lakes of southern Victoria Land were first discovered during the Expeditions of R. F. Scott during the early 1900's (Scott, 1905; Huxley, 1913). For half a century after their discovery, the limnology, geology, geomorphology, and climate of the southern Victoria Land dry valleys and lakes remained unknown (Parker *et al.*, 1982a). It was not until the International Geophysical Year

(1957–58) and the establishment of United States and New Zealand scientific research bases on Ross Island that studies in the dry valleys were resumed.

From the late 1950's until 1978, limnological data from these lakes were limited to samples and measurements taken through 10–23 cm drill holes in the ice. These studies focused primarily on the characterization of the water column and have been reviewed by Heywood (1984), Hobbie (1984), and Vincent & Howard-Williams (1985). Beginning in 1978, we developed a method for melting holes through the 3–5 m thick perennial ice covers for the purpose of allowing research divers to work under the ice (Simmons *et al.*, 1979; Love *et al.*, 1982). This development and the data generated by these activities opened a new dimension to limnological studies of the dry valley lakes.

Some of the more important results of this research have led to a better understanding of gas exchange mechanisms between the atmosphere, ice, and water column (Wharton *et al.*, 1986, 1987), the toxicity effects of high oxygen concentrations (Mikell *et al.*, 1984), the biological adaptations to low levels of light (Palmisano & Simmons, 1987) and temperature (Seaburg *et al.*, 1982), the relationship between sediment accumulation on the ice cover and resulting ice cover dynamics (McKay *et al.*, 1985; Nedell *et al.*, 1987b; Simmons *et al.*, 1986), and the species composition, distribution, morphology, and ecology of benthic microbial mats (Allnutt *et al.*, 1981; Love *et al.*, 1983; Parker *et al.*, 1981, 1982b; Simmons *et al.*, 1983; Wharton *et al.*, 1982, 1983).

The majority of the studies listed above lead to an important conclusion; namely, environment in a dry valley lake is to a large extent controlled by the presence of a relatively thick perennial ice cover. The ice cover eliminates wind generated currents within a lake (Ragotzkie & Likens, 1964; Hawes, 1983b) and greatly restricts exchange of gases between the water column and atmosphere (Wharton *et al.*, 1986, 1987). The ice cover also greatly reduces light penetration (Palmisano & Simmons, 1987) and restricts sediment deposition (Simmons *et al.*, 1986; Nedell *et al.*, 1987b) into the water column below.

The ice cover is in turn largely controlled by the extreme seasonality of Antarctica and local climate. Therefore, it is possible that the perennially ice-covered lakes could represent a sensitive indicator of short term (< 100 yr) climatic and/or anthropogenic changes in the dry valleys resulting from intensive exploration over the past 30 years. As shown in a subsequent section, the time constants for turnover of the water column and lake ice are ~50 and ~10 years, respectively. The turnover time for atmospheric gases in the lake is 30–60 years (Wharton *et al.*, 1987). Thus, the lake environment responds to changes on a 10–100 year timescale. Because the ice cover has a controlling influence on the lake (*e.g.* light penetration, gas content of water, and sediment deposition, etc.), it is probable that small changes in ablation, sediment loading, and glacial meltwater (or ground water) inflow will affect ice cover dynamics and will have a major impact on the lake environment and biota.

In this paper, we synthesize our recent research findings for Lake Hoare, Taylor Valley, Antarctica. Specifically, we will discuss the climate of Taylor Valley, the physics of Lake Hoare's ice cover, and the gas balance within the lake. We will also consider the effects of the ice cover on the water column and benthic environments, as well as sediment deposition. We will conclude with a discussion of future research objectives, which we believe necessary to complete our understanding of the lake environment and its potential as an indicator of short term climatic and/or anthropogenic changes in the dry valley region.

Study site

Lake Hoare (77° 38′ S, 162° 53′E) is at the eastern end of Taylor Valley in southern Victoria Land, Antarctica. The lake is 58 m above sea level, 4.1 km long, 1.0 km wide, with a surface of 1.8 km^2, a maximum depth of 34 m, and a mean depth of 14.2 m (Fig. 1). The perennial ice cover of Lake Hoare overlies water at a temperature of

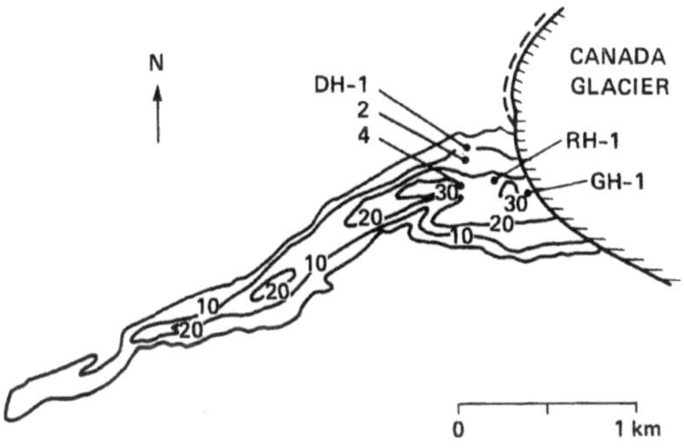

Fig. 1. Map of Lake Hoare, Taylor Valley southern Victoria land Antarctica. DH, GH, RH indicate location of research dive holes. Depth contours are in meters.

0 °C. The lake receives both water and sediment from glacial meltstreams and from nearby Lake Chad during the austral summer; lacking outflow streams, it loses water only by ablation and sublimation at the surface of the ice and by evaporation from the moat.

Results and discussion

Taylor Valley climate

The climate of the dry valleys is strongly seasonal as a result of the southern latitude. There is ~4 months of sunlight in summer and 4 months during which the sun does not cross above the horizon. The light, temperature and wind regimes follow this basic polar cycle (Clow *et al.*, in press).

In order to develop a in-depth understanding of any ecosystem it is necessary to have year-round data as to the climatic conditions. To achieve this objective without the cost or environmental impact of an over-winter team we have deployed automatic sensing and recording systems at Lake Hoare. A meteorological station was deployed in December 1985 and is situated on a small peninsular kame in the northeastern (down valley) end of Lake Hoare (about 100 m up valley from the Canada Glacier meltstream). The station consists of a cup anemometer, wind direction indicator, relative humidity measurement device, and a shielded T-type thermocouple with a electronic reference thermistor. Identical wind and temperature instruments were placed at two elevations, 1 and 3 m above ground level. A Li-Cor PAR quantum sensor, which measures photosynthetically active radiation (PAR, 400–700 nm), was attached to the surface of a flat rock approximately 10 cm above the ground, near the station. The data acquisition and storage system is based on a Campbell Scientific Data Recorder.

Two environmental variables of prime interest in understanding the ecology of Lake Hoare are temperature and light. These variables also are characterized by unique cycles associated with the polar regions. For these reasons, we focus on these variables in this paper, while a more complete description of the meteorology and instrumentation is in Clow *et al.* (in press).

Figure 2 shows the six-hour averages of temperature at 3 m height above the ground surface. Minimum temperatures during the austral winter were slightly less than − 40 °C. However, during the winter foehn winds the temperature rises sharply often to as high as − 10 °C. Summer temperatures were often above freezing and in mid January 1987 there were days in which the temperature never fell below freezing. The mean annual temperature for 1986 was − 17.3 °C, which is very close to the value reported for Lake Vanda of -20.0 °C (Thompson *et al.*, 1971).

Figure 3 shows the six-hour averages of the PAR incident on the lake surface. A diurnal cycle, even on a clear day, is evident in the summer. This is due to the ~26° difference between the zenith angle of the sun at noon and at midnight. As can be seen by comparing Figs. 2 and 3, the air temperature closely follows the incident sunlight. The summer climate is dominated by the presence of sunlight (Clow *et al.*, in press). This is true both on the scale of seasonal variations and on day by day variations in the summer. Hence, the warmest day of the year will typically be in late December or early January, close to the summer solstice. This is quite unlike lower latitudes in which there is a much longer delay between the summer light maximum and the warmest temperatures. During the summer, days with cloud cover are associated

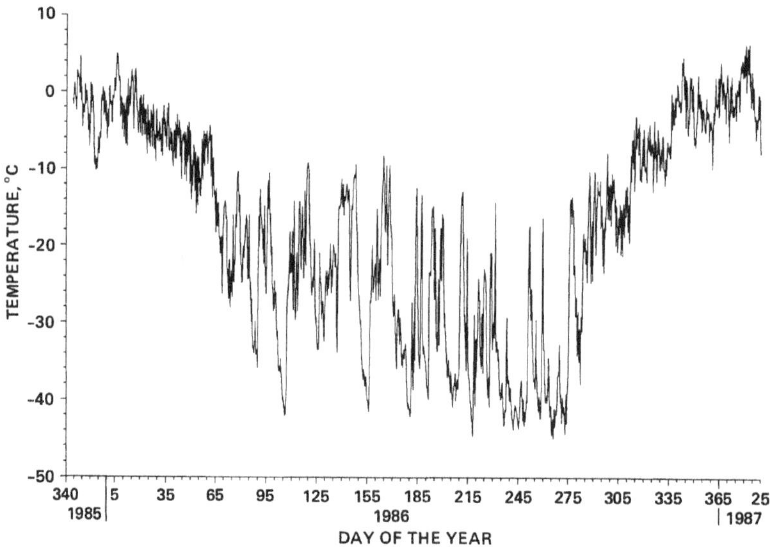

Fig. 2. Temperature at 3 meter height using shielded (against direct sunlight) T-type thermocouple with an electronic reference temperature (thermistor). Each data represents the average of data points taken every 30 seconds over six hour intervals. The data from the meteorology station is plotted versus the day of the year; zero on the x-axis corresponds to Jan. 1, 1986. The data for Dec. 1985 are plotted as well. A more detailed discussion of these results is in Clow *et al.* (in press). Errors are about ± 0.5 °C. The 1986 yearly average value is -17.3 °C.

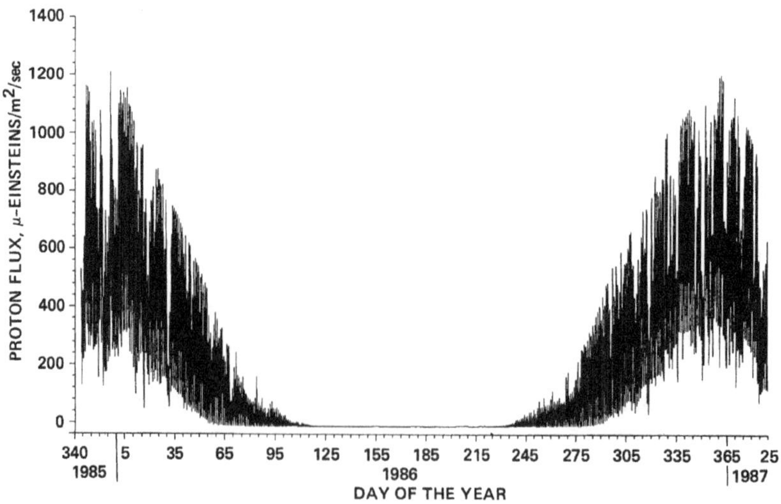

Fig. 3. Photosynthetic quantum flux at Lake Hoare (meteorology station) obtained with a Li-Cor PAR quantum sensor sensitive to light between 400–700 nm. The yearly average PAR is 188 $\mu E\, m^{-2}\, s^{-1}$ with a maximum error of 4.6%. The x-axis is as defined in Fig. 2.

with significant drops in temperature, as can be seen in Fig. 2 for the interval between days 358 and 361 of 1985 (corresponding to 25–28 Dec. 1985). The total amount of photosynthetically active sunlight incident on the lake can be determined by integrating the curve in Figure 3 which gives $5.93 \times 10^9\ \mu E\ m^{-2}$. This corresponds to a yearly average PAR of $188\ \mu E\ m^{-2}\ s^{-1}$. The maximum error on this value based upon the temperature sensitivity of the photodiode (0.15% per degree from 25 °C) is 4.6%.

Figure 4 shows the cumulative PAR as a function of time of year beginning on the first of September. To compile this figure, the data in Fig. 3 were continued by assuming that the latter portion of January and all of February and March 1987 were identical to the corresponding months in 1986. As can be seen in the figure only a small percentage of the total light (~5%) is incident before the start of typical field operations in mid October.

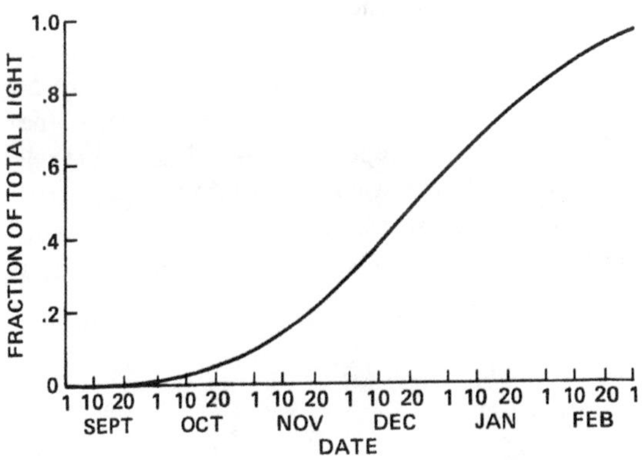

Fig. 4. Cumulative sunlight as a function of time of year beginning on the first day of sunrise. This is the fraction of the total summer sunlight that has reached the ground by the date specified on the x-axis. To compile this figure, the data in Fig. 3 for the summer season 1986–1987 were continued by assuming that the latter portion of January and all of February and March 1987 were identical to the corresponding months in 1986.

Physics of the ice cover

The dry valley lakes are perennially ice-covered because the mean annual temperatures are so low (−20 °C). The presence of liquid water beneath the ice is primarily due to the fact that for a few days in the summer the air temperatures are above freezing. Consequently, the presence of a year-round ice-water interface in a lake, unique to Antarctica, is due to this combination of very cold mean temperatures and comparatively warm summer maximums. A smaller seasonal temperature distribution would result in a lake that is either frozen completely or one that melts fully in the summer.

McKay *et al.* (1985) have developed a simple model which relates the thickness of ice on a perennially ice-covered lake to climatological variables. In this annually averaged, steady state model, the thickness of ice is determined primarily by the balance between the conduction of energy from the ice and the inputs of energy via sunlight and the transport of latent and sensible heat by the summer meltstream. The latent heat released upon freezing at the ice-water interface is the largest term in this equation. Because steady state conditions are assumed, the freezing rate of water at the ice bottom must be offset by ablation from the ice surface. Through this indirect relationship, ablation is the key variable that predicts the ice thickness. Other factors constant, higher ablation rates correspond to thinner steady state ice cover thicknesses. The ice thickness is given by (McKay *et al.*, 1985):

$$Z = \frac{b \ln(T_o/T_s) + c(T_s - T_o) - S_o(1-a)(1-r)h(1 - e^{-Z/h})}{v\rho l + F_g}$$

where T_o is the temperature of the ice-water interface, T_s is the yearly averaged temperature of the surface, both in Kelvin, Z is the equilibrium thickness of the ice cover, v is the rate of formation of new ice averaged over the entire year (equal to the ablation rate), ρ is the density of the ice, l is the heat of fusion of water, a is the albedo of the lake, r is the fraction of the lake that is covered by dark absorbing material such as sand and silt, S_o is the annual average solar radiation incident on the lake surface, h is the mean e-folding extinction path length multiplied by the cosine of the mean solar zenith angle, b and c are constants that define the thermal conductivity of the ice, $k = b/T - c$, taken to be 780 W m^{-1} and 0.615 W m^{-1} K^{-1} respectively and F_g is the geothermal heat flux.

We can use the difference in climate between Wright and Taylor Valleys to calculate the difference in steady state ice cover thickness on Lake Vanda and Lake Hoare. The mean annual temperature for 1969–70 at Lake Vanda (Wright Valley) was -20.0 °C (Thompson *et al.*, 1971), while that for Lake Hoare over 1986 was -17.3 °C. There is also a significant difference in the yearly average light reaching the surface. A value of 104 W m^{-2} has been reported for Lake Vanda (Thompson *et al.*, 1971). A value for Lake Hoare can be determined by converting the average PAR (188 μE m^{-2} s^{-1}) into total radiation (Clow *et al.* in press), which gives a value of 92 W m^{-2}. Using Eq. (1) and the nominal values of all parameters other than light (the nominal values are those from curve '1' from Fig. 2 of McKay *et al.*, 1985, with a 30 cm/yr ablation), we have computed that the ice thickness corresponding to 104 W m^{-2} is 3.36 m, and the ice thickness corresponding to 92 W m^{-2} is 4.43 m. Hence, we can predict that the ice cover on Lake Hoare would be about a metre thicker than the ice cover on Lake Vanda.

The model of McKay *et al.* (1985) assumes that the ice cover is in steady state, however, there is evidence that the thickness of the ice on Lake Hoare has undergone a significant change over the last ten years. Fig. 5 shows the thickness of the ice on Lake Hoare as determined by drill and melt holes over the past ten years. The 1980–81 point labeled with a solid circle and a set of error bars represents the results of an extensive survey of the ice cover that season. Thirty-five holes were drilled along the length and breath of the lake. The average, with standard deviation, of the ice-thickness values obtained is 4.77 ± 0.34 metres, N = 35. It is also significant that 88% of the values were within one standard deviation of the mean and no values were more than two standard deviations from the mean. The distribution is significantly tighter than a gaussian. These statistics imply that the ice cover is very uniform throughout the lake and that single measurements can be used with a fair degree of confidence. In general, any single measurement will be within ±0.34 m of the mean ice thickness. The error associated with the mean of three measurements will be significantly less. For these reasons, we conclude that the trend shown in Fig. 5, a decrease of ice thickness on Lake Hoare, is not a measurement error but represents a pronounced, and as yet unexplained, change in the lake.

The data show a clear trend; the ice thinned at a roughly uniform rate of about 28 cm/yr from 1977 to 1986. This value is comparable to the ablation rates reported for Lake Fryxell (30 cm/yr, Henderson *et al.*, 1965) and predicted for Lake Hoare from meteorological data (35 cm/yr) by Clow *et al.* (in press). This indicates

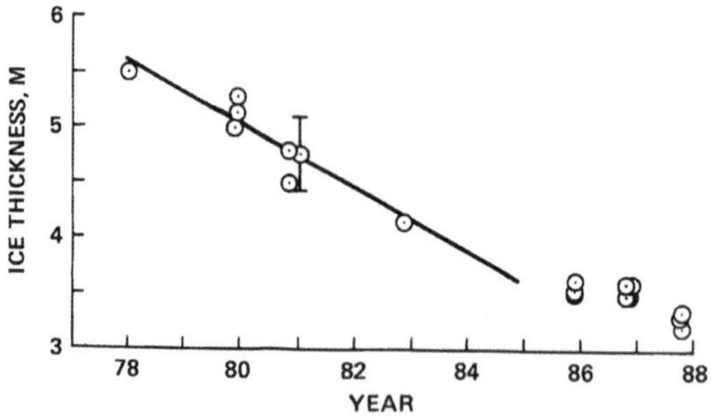

Fig. 5. Measured values of ice thickness on Lake Hoare, Taylor Valley, Antarctica over the last ten years as a function of the date. The data point plotted on 20 Jan 1981 represents the mean of 35 data points. The range indicated is ± one standard deviation. The linear fit to the data from 1977–1986 has a slope of -28.3 cm/year.

that there has been either a change in the local climate in Taylor Valley (*e.g.* mean annual temperature) or a change in the physical properties of the ice cover on Lake Hoare (*e.g.* sediment loading). To model these changes in the ice cover requires extending the results of McKay *et al.* (1985) to include time dependent effects such as seasonal changes in sunlight and ablation.

One of the significant biological implications of the ice cover is on the quantity and spectral distribution of radiation reaching the liquid phase beneath the cover. The reduction in light transmission due to the thick ice cover also affects the heat budget (Hoare, 1966; Ragotzkie & Likens, 1964; Bydder & Holdsworth, 1977; Adams & Lasenby, 1978); primary production, and plankton distribution and composition (Rodhe, 1956; Tominaga, 1977; Rigler, 1978; Priddle, 1980; Light *et al.*, 1981; Vincent, 1981; Vincent & Vincent, 1982; Cathey *et al.*, 1982; Parker *et al.*, 1982a; Hawes, 1983a,b, 1985). From an ecological perspective, this reduction in the quantity and alteration of the spectral distribution of light will exert selective pressures on the photoautotrophs (Seaburg *et al.*, 1983; Priscu *et al.*, 1987).

Palmisano & Simmons (1987) recently discussed the spectral downwelling PAR irradiance (400–700 nm) in Lake Hoare. For measurements taken near noon at summer solstice, the full waveband PAR beneath the ice was $<44\ \mu E\ m^{-2}s^{-1}$ or $\leqslant 3\%$ of surface downwelling irradiance. The ice cover absorbed longer wavelengths and maximum light transmission was in the blue region between 400–550 nm. The bulk attenuation coefficient of the water column ranged between 0.45 and 1.33 for five depths measured beneath the ice at two dive holes. Light attenuation by phytoplankton was greatest in the 400–550 nm and 656–671 nm regions. The spectral distribution of sunlight penetrating the lake ice depends upon seasonal factors including solar zenith angle, day length, cloud cover, and ice characteristics.

Goldman *et al.* (1967) first reported data on the seasonal variations in ice cover optical properties. To further document these seasonal effects, we have extended the transmission measurements of Palmisano & Simmons (1987). In Fig. 6, we show the ice cover transmission from early in the season (11 Nov. 1986), at summer solstice (23 Dec. 1982, from Palmisano & Simmons, 1987), and late in the season (11 Jan 1987). The ice cover thickness was roughly comparable in the 1982 data to the 1986–87 data (see Fig. 5). The results show an interesting pattern. Early in the season when the ice is fairly clear, the transmission of blue light greatly exceeds that of red light. This is because the transmission of blue light is dominated by scattering, which is minimal in the clear ice. However, in the red, transmission is dominated by the absorptive properties of ice which is fairly independent of whether the ice is clear or not. As the season progresses, the surface of the ice clouds due to freeze-thaw processes on the surface, and as a result, the scattering optical depth greatly increases, attenuating the blue light (Fig. 6). This scattering has little effect on the red light. The result is that later in the season the spectral

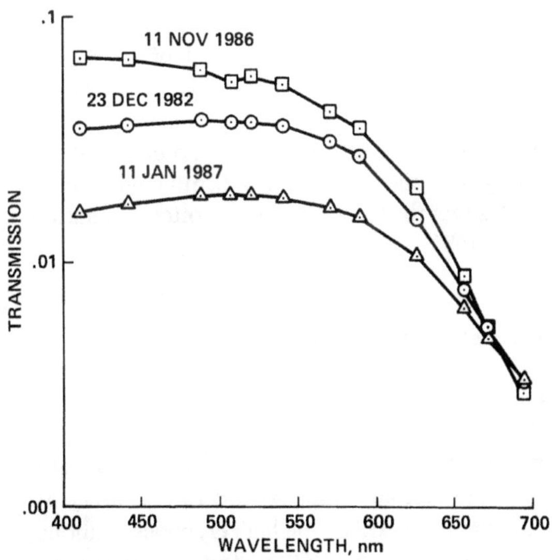

Fig. 6. Light transmission through a 3.7 m thick ice cover at various times during the season. Uppermost curve is for early in the season (11 Nov. 1986), middle curve is at summer solstice (23 Dec. 1982) from Palmisano & Simmons (1987) and lower curve is for late the season (11 Jan. 1987). The ice cover thickness was similar in the 1982 data to the 1986–87 data (see Fig. 5).

properties of the ice become more uniform with wavelength.

Roulet & Adams (1984) found spatial heterogeneity of light penetration in an ice-covered Canadian lake. They emphasized that a single point measurement could overestimate the quantity of light by 700% compared to an integrated areal measurement. During the 1986 austral summer, we measured the spatial heterogeneity of light penetration over a 10 m diameter circle immediately beneath the ice at DH4 (Fig. 1) and found the total quanta penetrating the ice surface to range between 15–51 $\mu E\ m^{-2}\ s^{-1}$ ($\bar{x} = 28.5 \pm 12.7$, N = 6) and the percent transmission to range between 0.43–1.45 ($\bar{x} = 0.81 \pm 0.36$, N = 6). The variation between our highest and lowest reading was 58.8% and most of the readings (5) were $\leqslant \pm 30\ \mu E\ m^{-2}\ s^{-1}$. This indicates that the lake is relatively uniform with respect to optical properties, which is in agreement with ice thickness measurements.

Commonly encountered conditions indicate that light penetration is ~1% of surface irradiance values. The phytoplankton and water mass then combine to absorb much of the remaining light, so that the quantity of light reaching the bottom of the lake is $\ll$1%. This aspect of light penetration and attenuation is particularly interesting given the abundance of benthic photoautotrophs that thrive on the lake's bottom (Wharton *et al.*, 1983).

Gas balance in the lake

One of the most unusual features of these lakes is the occurrence of supersaturated O_2 and N_2 in the water column ranging from slightly over saturation to over 400% for O_2 and 160% for N_2. To quantitatively explain the high O_2 concentrations, we developed a bulk O_2 budget. There are two primary net sources of O_2: a physical source resulting from gases carried into the lake by the meltstreams and forced into the water column when water freezes onto the bottom of the ice cover, and a biological source resulting from photosynthesis and from the burial of reduced carbon in the lake sediments.

While it has been known for many years that the dry valley lakes are supersaturated with O_2, quantitative budgets were not available. We have recently developed quantitative models of both oxygen and nitrogen in Lake Hoare (Wharton *et al.*, 1986, 1987). In our model of gas flow into and out of the lake, we considered both biological and nonbiological sources and sinks. Our model (Wharton *et al.*, 1986) predicted, and direct measurements have subsequently shown that there is a supersaturation of N_2 in the lake water, as well as oxygen. Dissolved N_2 levels of 145% and 163% were determined from samples taken just below the ice cover and at a depth of 12 m, respectively (Wharton *et al.*, 1987).

The two principal atmospheric gases (N_2, O_2) are both influenced by the non-biological processes affecting the gas balance of the lake water (such as freezing of water and bubble formation). Non-biological processes will act on N_2 and O_2 equally, maintaining the N_2/O_2 ratio at that value characteristic of water in equilibrium with the atmosphere (~1.8). However, biological processes (primarily photosynthesis and respiration) will affect the O_2 concentrations to a much larger extent than they will affect N_2, thus altering the N_2/O_2 ratio (Wharton *et al.*, 1987). Therefore, this ratio can be a useful 'signature' of biological and non-biological gas production. In Lake Hoare, this ratio was 1.20 at the ice/water interface and 1.O5 at 12m; considerably different from the ratio in equilibrium with air (~1.8). Based on these results, we have determined that about half of the net O_2 production in the lake is the result of biological processes.

The approach discussed above by which we can infer net biological productivity was validated by the independent publication of a study of the open ocean by Craig & Hayward (1987), in which ratios of supersaturated gases were used in a similar way.

By considering the total reservoir of gas in the lake and the inflows and outflows of O_2 and N_2, we have estimated the residence time of these gases in Lake Hoare. We obtain a value of ~30 years for oxygen and ~60 years for nitrogen (Wharton *et al.*, 1987). Thus, gases are cycled relatively rapidly through the lake.

The importance of the gas concentrating mechanisms found in perennially ice-covered Antarctic lakes includes the development of lift-off benthic mat in shallow waters, the natural selection for organisms that can adapt to perennial supersaturated oxygen levels (Mikell *et al.*, 1984), and the production of gas channels in the ice, which may serve conduits for sediment to penetrate the ice cover.

Lake biology

Because of the combination of snow cover, sediment and gas bubbles, light transmission through the ice cover is much less than would be expected with an equivalent column of pure ice or water. This reduction in light limits the plankton density, and the lack of internal currents that keep free floaters suspended (*e.g.* Langmuir spirals) also restricts the plankton population to mainly swimming forms (Parker *et al.*, 1982a). Over 200 measurements of chlorophyll *a* (Chl *a*) were made during the 1985–86 austral summer at four different locations (DH1, DH2, DH4, RH1) in Lake Hoare (Fig. 1). Chl *a* was measured fluorometrically using the technique described in Parker *et al.* (1982a). The integrated Chl *a* concentration for the season was 16.92 mg m^{-2}. However, if a hypsographic curve is used to accommodate changes in volume with depth, the value then becomes 22.26 mg m^{-2} at the lake surface. When these data are compared against 5 integrated profile measurements made during the 1979–80 season, the 1985–86 values decrease by 43.2% (22.6 vs. 39.19). The reasons for the decrease measured during the 1985–86 season may have been due to the melting of ice which would have diluted the concentrations of Chl *a* in the upper water column. Also, during the 1979–80 season the first Chl *a* collections were made nearly three weeks before we began the 1985–86 collections. In fact, the highest Chl *a* measurement of the 1979–80 season was obtained on the first profile measurement. When the 1985–86 data were examined on a seasonal basis, they showed that the maximum Chl *a* level was at 13 m piezometric depth (maximum O_2 concentration was at 10 m).

The fate of the phytoplankton community during the Antarctic winter remains to be determined. The future of the phytoplankton community under a dynamically changing ice cover is important to our understanding of this lake ecosystem. If the ice cover continues to thin and sand continues to be dumped from the ice cover, we would expect light transmission to increase, and therefore phytoplankton density, as measured by Chl *a*, to also increase.

Microbial mats composed primarily of the cyanobacteria, eukaryotic algae, and heterotrophic bacteria occur abundantly throughout the benthic region of Lake Hoare (Wharton *et al.*, 1983). These microbial mats are precipitating calcite, iron, and sulfur, and trapping and binding sediment forming alternating laminae of organic and inorganic material. An unique feature of many of these benthic mats is their development into modern, cold water stromatolites (defined by Awramik *et al.*, 1976 as organosedimentary structures produced by sediment trapping, binding, and/or precipitation as a result of the growth and metabolism of microorganisms). Wharton *et al.* (1983) have described microbial mats resulting in four types of modern Antarctic stromatolites, including lift-off, pinnacle, aerobic and anaerobic prostrate mats.

One interesting effect of the elevated gas levels is on the formation of stromatolites. Beneath the perennial ice in the shallower, more brightly lit areas of the lake is a sizeable biomass of columnar lift-off mat. In these mats, an excess of dissolved gas (primarily N_2 and O_2) accumulates as bubbles causing the mats to lift off the substrate (Parker *et al.*, 1982b; Wharton *et al.*, 1983). These pieces of mat often tear loose from the substrate and float up to the underside of the ice, thus disturbing the integrity, or prohibiting the development of a laminated structure. Therefore the formation of bubbles, which is controlled by the environment, can directly influence mat morphology and stromatolite formation.

Wharton *et al.* (1976) have developed a formula for relating O_2 and N_2 dissolved in the water column to bubble formation with depth:

$$z = 10.3[0.21(S_o - 1) + 0.78(S_n - 1) + p_w] \quad (2)$$

where z is the maximum depth of bubble formation, S_o is the supersaturation of O_2 with respect to equilibrium with the atmosphere, S_n is the corresponding number for N_2, and p_w is the vapor pressure of water. For example, in Lake Hoare the high concentration of dissolved gases in the water column results in bubble formation to a depth of ~10 m. Below 10 m bubbles cannot form in the water as a result of the increased hydrostatic pressure. Consequently, we expect bubble formation and mat lift-off to disrupt mat morphology at depths above 10 m, while at depths below 10 m laminated, prostrate mats would be expected. Indeed, the former and latter cases are observed *in situ* (Wharton *et al.*, 1983).

Another interesting aspect of the elevated gas levels in the water column is on the oxygen content of the sediment. During the 1986–87 field austral summer, oxygen profiles of the sediment were obtained with a Diamond Electro-Tech dissolved-O_2 microprobe calibrated to microwinkler determinations of water from the sediment-water interface. This method was used to determine the oxygen level of sediment in cores taken from different locations at the bottom of the lake. Contrary to what is typically observed in lake sediment, sediment collected in 10 m water depth and above remained oxic to a depth of 30 cm (maximum core depth). Sediment collected from 23 m depth and below were anoxic several mm below the sediment water interface. The sediment which remained oxic to 30 cm are overlaid with water supersaturated with O_2 (and other gases) while the anoxic sediment is overlaid by anoxic water or water with saturated levels of O_2 similar to what would be observed in a temperate lake. These results are another indication of the important role that the supersaturation of atmospheric gases play in regulating the lake environment.

Sediment/ice interactions and sedimentation

The physical environment for sedimentation in Lake Hoare is unusual because of the presence of the perennial ice cover. The ice catches and traps wind blown sediment and provides a surface for the movement (by saltation, rolling and drift on the ice) of larger sediment particles into the middle of the lake. Although the ice cover does contain large boulders right in the middle of the lake, most of the mass of the ice cover burden is in the form of sand-sized and finer particles. Based on numerous melt holes we estimate that the average sediment loading is ~0.2–2 g cm^{-2}. This range of values corresponds to less than 0.6% of the mass of a 4 m ice cover.

The sediment in the ice will have a major influence on the optical properties of the ice cover. Sediment opacity causes heating of the ice cover, which would tend to thin the ice. At the same time, sediment opacity also prevents the transmission of light through the ice, which would tend to thicken the ice (McKay *et al.*, 1985). Consequently, the amount of sediment in the ice cover is an important variable in regulating ice thickness.

Small particles of sediment will not melt their way directly through the ice cover (Simmons *et al.*, 1986). This can be shown theoretically by considering the energy balance of a radiatively heated particle. A small, dark sand-sized particle on the surface of the ice or embedded in the ice cover absorbs sunlight. If the heating rate is sufficient to raise the surface temperature of the particle above the melting point, the particle will sink into the ice cover. Because the particles are very small compared to the thickness of the ice cover, the particle surface temperature can be determined by the spherically symmetric heat equation:

$$F(1 - \omega)\pi r^2 = 4\pi r k \Delta T \quad (3)$$

where F is the radiation field in the ice cover averaged over the upward and downward directions (including scattered light), ω is the single scattering albedo of the particle ($\omega \simeq 0.2$), r is the radius of the particle, k is the thermal conductivity of the ice (at −1 °C, k ~2.3 W K^{-1} m^{-1}), and ΔT is the difference between the temperature of the particle surface and the temperature of the ice at the depth of the particle. Using this equation,

Simmons *et al.* (1986) have shown that in order to melt through ice that is only 1° below freezing requires a particle of 1.5 cm radius at the surface, 3.8 cm at a depth of 1 m, and 9.3 cm at a depth of 2 m. Melting through colder ice requires even larger particles. Hence, sand-sized particles that have radii much less than 1 cm, will not melt through the ice cover and are carried into the ice cover by surface meltwater percolation during the austral summer. Aggregates of individual particles are effectively like a particle with a larger size and are therefore able to more effectively melt the ice, but only in approximately the first meter. From this we conclude that sediment will not melt through a several meter-thick ice cover and must be carried by water through cracks or gas-bubble channels in the ice cover.

The results (Table 1) from four sets of sediment traps placed in the northeastern end of the lake have helped us unravel the unusual sedimentation processes (Simmons *et al.*, 1986; Nedell *et al.*, in prep.). Traps from DH1 (~8 m depth) showed a sedimentation rate of 4.11 mg cm^{-2} yr^{-1}. Sediment traps from DH2 (~11 m depth) and DH4 (~27 m depth) averaged 3.76 and 2.87 mg cm^{-2} yr^{-1}, respectively. It is interesting that one trap (DH2 A and DH4 C) from each of these latter two sites contained significantly more sand-sized sediment than the other traps from the same site. Traps from GH1 (~23 m depth) contained a substantial quantity of sediment and had a mean sedimentation rate of 142 mg cm^{-2} yr^{-1}. The sediment traps at GH1 were predominantly coarse sand, while farther away from the glacier at DH4, both coarse sand and finer, silty material were collected. In DH1 and DH2, which are closer to shore, the traps collected silt and clay-sized particles. The observation of different amounts of sediment from the same area supports the hypothesis that sediment enters the lake through the ice cover at distinct locations via cracks in the ice and/or gas bubble channels. Also, small mounds of sand 0.5–1.0 m high and 1.0–3.0 m wide were observed at the sediment/water surface near DH2 and GH1, further suggesting point sources for sediment discharge into the lake through the ice cover (Nedell *et al.*, in prep.). Therefore, instead of

Table 1. Total dry mass and composition of sediment trap material collected from dive holes (DH) 1, 2, 4 and glacier dive hole (GH) 1 in Lake Hoare, southern Victorian Land, Antarctica (modified from Simmons *et al.*, 1986).[1]

	Total dry mass (g)	Organic matter (g)	Carbonate (g)	Gravel (g)	Sand (g)	Mud (g)	Other[2] (g)
DH 1-A	17.52	0.81	6.61	0.00	0.25	9.85	0
DH 1-B	24.38	0.63	13.79	0.00	0.43	9.53	0
DH 1-C	16.95	1.10	6.04	0.00	0.34	9.47	0
DH 2-A	44.25	3.11	2.46	3.88	32.65	2.15	0
DH 2-B	4.58	0.21	1.52	0.00	0.00	2.85	0
DH 2-C	4.98	0.48	1.41	0.64	0.10	2.35	0
DH 4-A	2.58	0.10	0.22	0.00	2.13	0.13	0
DH 4-B	2.08	0.13	0.18	0.00	1.67	0.10	0
DH 4-C	36.35	0.32	0.89	2.58	32.39	0.17	0
GH 1-A	633.00	39.69	18.38	0.00	572.42	1.51	1.00
GH 1-B	544.00	4.30	15.30	0.00	522.74	2.24	0.02
GH 1-C	856.20	7.53	30.74	0.00	813.97	2.94	1.02

[1] Traps were deployed in Dec/Jan 1982 and retrieved in Nov/Dec 1985; methods for determining composition of trap material in Simmons *et al.* (1986) and Nedell *et al.* (in prep). Traps were composed of an aluminium funnel with a 45 cm diameter opening, 47 cm tall, attached to a plastic container 10 cm in diameter and 20 cm tall.

[2] Other = aluminium oxide material.

receiving most of its sediment from the lake margin and inflowing streams, the majority of sediment at the bottom of Lake Hoare is transported downward through the ice cover. The similarity in grain-size distribution and mineralogy between samples from the ice cover and lake bottom supports this conclusion (Simmons *et al.*, 1986; Nedell *et al.*, 1987b). However, the fine-grained sediment population that appears in the lake bottom samples but not in the samples from the ice cover, is probably brought in via meltstreams. This is corroborated by the observation of an under ice plume of glacial flour 250–300 m from the entrance of the Canada Glacier meltstream into the lake.

It is not clear what controls the total sediment burden on the ice cover. We suggest at least two general possibilities. One possibility is that the ice-sediment system is in steady state; the rate of sediment percolation through the ice cover is sufficient to balance the rate at which sediment is added to the ice cover. In this case, the current sediment burden is the steady state value and therefore its low amount suggests that there must be a fairly efficient mechanism for getting sediment through the ice cover. A second possibility is that the ice-sediment system is cyclic. Sediment builds up on an initially clean ice cover until the level of sediment causes a change in the ice cover sufficient to dump the sediment. Clearly, an extreme case would be when sufficient sediment is loaded on the ice cover which then becomes negatively buoyant. Simmons *et al.* (1986) have discussed a possible ice-sediment interaction cycle. In their cycle,increased sediment loading results in increased surface topography on the ice cover and increased ablation. The increased ablation results in a thinner ice cover until eventually a point is reached at which the sediment, which accumulates at the topographic lows (ponds), can pass through the ice cover. The clean ice cover reseals itself and the cycle begins anew. The changes in ice cover thickness shown in Fig. 5 could be evidence that the ice-sediment system is cycling and the ice cover of Lake Hoare just underwent a transition as part of such a cycle (Simmons *et al.*, 1986).

Another important aspect of the benthic sediment is their involvement with possible groundwater influx. We have obtained the first actual measurements of groundwater flow into a dry valley lake. Seepage meters were placed in two locations in the lake for a pilot study to measure ground-water flow. One location was in approximately 8 m of water beneath the ice near the shore by the meteorological station, and the other was in approximately 27 m of water at the base of the Canada glacier near GH1 (Fig. 1). The seepage meters were constructed of polyvinyl chloride (PVG) and 1 L Nalgene collection bags were used for the collection of the seepage water (see Simmons & Netherton, 1987). Seepage meter water was collected from two areas in the lake during the 1985–86 and 1986–87 austral summers over a six day period. During 1985–86, ~ 625 ml m^{-2} day^{-1} was collected from the site at the glacier's base, and ~ 416 ml m^{-2} day^{-1} from the shallow water site. During the 1986–87 austral summer, ~ 31 ml m^{-2} day^{-1} was collected at both sites also over a six day period. The difference between years may be due to the time period during which the samples were collected. In 1985–86, we collected the samples in early December; whereas, in 1986–87, the samples were collected in mid-January. The importance of this pilot study is that earlier suggestions (Wilson, 1979; Chinn, 1982) of groundwater movement into lakes have been corroborated and deserve further study because groundwater seepage into these lakes could play an important role in nutrient cycling to the benthic community at the sediment water interface.

Conclusions

The environment of the Antarctic lakes could represent a sensitive indicator of short term (< 100 yr) climatic and human-induced changes in the dry valleys. In fact, they may be the only sensitive indicator for changes on these times scales. The other class of physical systems that are in dynamic equilibrium with the climate are the glaciers. Chinn (1985) has discussed this aspect of the structure and equilibrium of the dry valley glaciers and concludes that they respond on many thousand year timescales. Some dry valley glaciers are retreating and some are advancing (Chinn, 1985). The other biological system in the dry valleys is the cryptoendolithic microbial communities described by Friedmann (1982). These communities grow very slowly and again the timescale of response to changes is measured in thousands of years (Friedmann & Weed, 1987).

The time constants in the lake are much shorter. We can estimate the turnover time for the water by dividing the mean depth of Lake Hoare (14.2 m) by the inflow rate (30 cm/yr, assumed to equal the ablation rate). This gives a timescale of ~50 yr. A similar calculation for a 4 m ice cover gives a timescale for lake ice turnover of ~13 years. We have determined the turnover time for atmospheric gases in the lake (O_2 and N_2) by considering the total amount in the lake divided by the sources. This gives a value of ~30 years for O_2 and ~60 years for N_2 (Wharton *et al.*, 1986, 1987). A time constant for biological turnover is more difficult to estimate but is probably also between 10 and 100 years.

We now think that the ice cover has a controlling influence on the lake (*e.g.* light penetration, gas content of water, and sediment deposition). Therefore, small changes in ablation, sand loading, and glacial meltwater (or groundwater) inflow will affect ice cover dynamics and will have a major impact on the lake environment and the biota. These records of changes should be most obvious in the sedimentary record. For example, we know that sediment is deposited through the ice cover, perhaps via cracks which develop when the ice cover thins to ca. 3 m thickness. This sediment buries portions of benthic microbial mat, which will ultimately recolonize the sediment surface after a period of years. The resulting sediment lens in the benthic profile indicates a period of sediment deposition through the ice cover. Thinning of the ice cover (and resultant sediment deposition) probably results from changes in local climate which result in ice cover changes. Consequently, it may be feasible to understand past climate regimes as well as to be able to predict the effect of future modifications of climate (either natural or human-induced changes) on the dry valley lake ecosystem.

There is evidence of a recent change in climatic conditions in the dry valleys. Chinn (1982) has documented a water level rise in virtually all of the dry valley lakes over the period of 1972 to 1982. He suggests that this is related to climate but no causal mechanism is proposed. Our data for ice cover thickness on Lake Hoare also suggests changes in the lake environment, and again we have no causal mechanism to explain it. There is not sufficient data to show whether ice cover thicknesses have changed on the other lakes also.

If the dry valley lakes are to be useful as indicators of climate change in that region of Antarctica, then it is necessary to develop the baseline data that documents the present condition of the lakes. Furthermore, detailed predictive models that relate the biogeochemical cycles and biological processes in the lake to the external environment must be developed.

We feel the following future research objectives are necessary to complete our understanding of the major physical, chemical, and biological interactions that regulate the lake's ecology:

1. More refined models of the climatological controls on the thickness of the ice cover.
2. A re-examination of phytoplankton density changes in relation to changes in lake ice thickness.
3. Development of a carbon cycle model, in particular, quantification of the carbon dioxide sources and sinks and their relationship to carbonate formation particularly in benthic microbial mats.
4. Quantification of sediment loading on the ice cover and dumping.
5. Quantification of benthic microbial biomass accumulation and decomposition rates.
6. Resolution of abiotic/biotic contributions to sediment composition.
7. A sediment profile study of oxygen and oxidation-reduction potentials.
8. Quantification of ground water nutrient and mineral re-cycling processes.

In addition to helping us unravel the climatic and environmental history of the dry valley region, the ice-covered lakes serve as useful models for increasing our understanding of early life on this planet and possibly Mars. The first 2.5 billion years of life on Earth was microbial. These microbes left behind a fossil record of their Precambrian existence in the form of stromatolites. It is a common misconception that stromatolites form *only* in warm and/or saline environments. As discussed previously, microbial mats are forming stromatolites in the Antarctic dry valley lakes. Other studies have shown that several periods of glaciation occurred during the Precambrian (Anderson, 1983; Walter & Bauld, 1983). We suggest that studies of Antarctic lakes may play an important role in the re-interpretation of stromatolite formation during the Precambrian and specifically to their occurrence in Precambrian polar environments.

Another intriguing aspect of research in the dry valley lakes is the connection to extraterrestrial habitats. The Antarctic dry valley lakes have been suggested as analogs of paleolakes on Mars which may have sheltered early Martian life. Geological and climatological studies suggest that conditions on early Mars (> 3 b.y.a.) were very different from what they are today, and were similar to early earth (McKay, 1986). Because life on earth is known to have originated during this early period on earth, the Martian environment could have also been conducive to the origin of life. The record of the origin and early evolution of life on Earth has been obscured by extensive erosional and tectonic activity. However, on Mars much of the ancient heavily cratered terrain, dating back to this early period, remains in pristine condition and may hold a record of events that led up to the origin and early evolution of life.

Recent studies by Nedell *et al.* (1987a), have described an area of ancient (> 3 b.y.a.) lake sediments in the Valles Marineris canyon system on Mars. As the Martian atmosphere thinned and the surface grew cold, these putative Martian paleolakes, like the Antarctic lakes, would have contained liquid water beneath a layer of ice, as opposed to being frozen solid (McKay *et al.*, 1985). In addition to providing a relatively warm, liquid water environment, the process of concentrating atmospheric gases beneath the ice cover could have significantly affected the gas budget in those lakes, possibly enhancing the levels of biologically important gases from the thin Martian atmosphere (Wharton *et al.*, 1987). Also, it is possible that the sediments observed in the Valles Marineris canyons could have been deposited by sediment passing through ice in much the same fashion as observed in Lake Hoare, Antarctica (Nedell *et al.*, 1987a,b; Simmons *et al.*, 1986).

Acknowledgements

This research was supported by the National Science Foundation's Division of Polar Programs (DPP-8416340) and the National Aeronautics and Space Administration's Exobiology Program (NCA2-2 and NCA2-1R675-402). We are grateful for assistance in Antarctica provided by the US Navy and ITT's Antarctic Services. We are especially thankful to the following scientists for their significant input during this multidisciplinary research program: D. Andersen, C. Chin, G. Clow, L. Leary, S. Nedell, D. Schwindler, and Drs. D. Andersen, D. Des Marais, R. Mancinelli, A. Palmisano, and S. Squyres.

References

Adams, W. P. and Lasenby, D. C. 1978. The role of ice and snow in lake heat budgets. Limnol. oceanogr. 23: 1025–1028.

Allnutt, F. T. C., Parker, B. C., Seaburg, K. G., and Simmons, G. M., Jr., 1981. *In situ* nitrogen C_2H_2-fixation in lakes of southern Victoria Land, Antarctica. Hydrobiol. Bull. 15: 99–109.

Anderson, J., 1983. Spatial and temporal distribution of glacial marine sediment, p. 1–99. In B. F. Molnia (ed.), *Global marine sedimentation.* Plenum Press, New York.

Awramik, S. M., Margulis, L., and Barghoorn, E. S., 1976. Evolutionary processes in the formation of stromatolites, p. 149–162. In M. R. Walter (ed.), *Stromatolites*, Elsevier, Amsterdam.

Benoit, R., Hatcher, R., Green, W., 1971. Bacteriological profiles and some chemical characteristics of two permanently frozen Antarctic lakes, p. 281–293. In J. Cairns, Jr.

(ed.) *The structure and function of freshwater microbial communities.* Research Division Monograph no. 3, Virginia Polytechnic Institute and State University, Blacksburg, VA.

Bydder, E. C. and Holdsworth, R., 1977. Lake Vanda (Antarctica) revisited. N.Z. J. Geol. Geophys. 20: 1027-1032.

Cathey, D. D., Simmons, G. M., Jr., Parker, B. C., Yongue, W. H., VanBrunt, M. R., 1982. Protozoan colonization of artificial substrate in two Antarctic lakes. Trans. Amer. Microscop. Soc. 101: 353–367.

Chinn, T. J. H., 1982. Hydrology and climate in the Ross Sea area. J. Roy. Soc. N.Z. 11: 373–386.

—, 1985. Structure and equilibrium of the dry valley glaciers. N.Z. Ant. Rec. vol. 6, Spec. Suppl., pp. 73–88.

Clow, G. D., McKay, C. P., Simmons, G. M., Jr., Wharton, R. A., Jr., 1988. Climatological observations and predicted sublimation rates at Lake Hoare, Antarctica. J. Appl. Met. Clim. In press.

Craig, H. and Hayward, T., 1987. Oxygen supersaturation in the ocean: biological versus physical contributions. Science 235: 199–202.

Friedman, E. I., 1982. Endolithic microorganisms in the Antarctic cold desert. Science 215: 1045–1053.

— and Weed, R., 1987. Microbial trace-fossils formation, biogenous, and abiotic weathering in the Antarctic cold desert. Science 236: 703–705.

Goldman, C. R., Mason, D. T., Hobbie, J. E., 1967. Two Antarctic desert lakes. Limnol. Oceanogr. 12: 295-310.

Hawes, I., 1983a. Nutrients and their effects on phytoplankton in populations in lakes on Signy Island, Antarctica. Polar Biol. 2: 115–126.

—, 1983b. Turbulent mixing and its consequences on phytoplankton development in two ice covered lakes. Bull. British Antarc. Surv. 60: 69–82.

—, 1985. Factors controlling phytoplankton populations in maritime Antarctic lakes, p. 245–252. In W. R. Siegfried, P. R. Condy, R. M. Laws (ed.), *Antarctic nutrient cycles and food webs*, Proc. 4th SCAR Symposium on Antarctic Biology. Springer, Berlin, Heidelberg, New York, Berlin.

Henderson, R. A., Prebble, W. M., Hoare, R. A, Popplewell, K. B., House, D. A., and Wilson, A. T., 1965. An ablation rate for Lake Fryxell, Victoria Land, Antarctica. J. Glaciol. 6: 129–133.

Heywood, R. B., 1984. Antarctic inland waters, p. 279–344. In R. M. Laws (ed.), *Antarctic Ecology*. Academic Press, London.

—, 1972. Antarctic limnology. A review. British Ant. Surv. Bull. 29: 34–65.

Hoare, R. A., 1966. Problems of heat transfer in Lake Vanda, a density stratified Antarctic lake. Nature 210: 787–789.

Hobbie, J. E., 1984. Polar limnology, p. 63–105. In F. B. Taub (ed.), *Lakes and Reservoirs*. Elsevier.

Huxley, L. (Arranger). 1913. Scott's Last Expedition. II. Being the reports of the journeys and the scientific work undertaken by Dr. E. A. Wilson and the surviving members of the expedition. Smith, Elder Co., London, 633 pp.

Light, J. J., Ellis-Evans, J. C., and Priddle, J. 1981. Phytoplankton ecology in an Antarctic lake. Freshwater Biol. 11: 11–26.

Love, F. G., Simmons, G. M., Jr., Wharton, R. A., Jr., Parker, B. C., 1982. Methods for melting dive holes in thick ice and vibracoring beneath ice. J. Sed. Petrol. 52: 644–647.

—, 1983. Modern *Conophyton*-like algal mats discovered in Lake Vanda, Antarctica. Geomicrobiol. J. 3: 33–48.

McKay, C. P., 1986. Exobiology and future Mars missions: The search for Mars' earliest biosphere. Adv. Space Res. 6: 269–285.

—, Clow, G. A., Wharton, R. A., Jr., and Squyres, S. W., 1985. Thickness of ice on perennially frozen lakes. Nature 313: 561–562.

Mikell, A. T., Jr., Parker, B. C., Simmons, G. M., Jr., 1984. Response of an Antarctic lake heterotrophic community to high dissolved oxygen. Appl. Environ. Microbiol. 47: 1062–1066.

Nedell, S. S., Squyres, S. W., and Andersen, D. W., 1987a. Origin and evolution of the layered deposits in the Valles Marineris, Mars. Icarus. 70: 409–441.

—, Andersen, D. W., Squyres, S. W., and Love, F. G., 1987b. Sedimentation in ice covered Lake Hoare, Antarctica. Sedimentology. 34: 1093–1106.

—, —, —, and Wharton, R. A., Jr. Lake Hoare, Antarctica: Sedimentation through a thick perennial ice cover. in prep.

Palmisano, A. C. and Simmons, G. M., Jr., 1987. Spectral downwelling irradiance in an Antarctic lake. Polar Biol. 7: 145–151.

Parker, B. C., Simmons, G. M., Jr., Love, F. G., Wharton, R. A., Jr., Seaburg, K. G., 1981. Modern stromatolites in Antarctic dry valley lakes. BioScience 31: 656–661.

—, —, Seaburg, K. G., Cathey, D. D., Allnutt, F. C. T., 1982a. Comparative ecology of plankton communities in seven Antarctic oasis lakes. J. Plankt. Res. 4: 271–286.

—, —, Wharton, R. A., Seaburg, K. G., Love, F. G., 1982b. Removal of organic and inorganic matter from Antarctic lakes by aerial escape of bluegreen algal mats. J. Phycol. 18: 72–78.

Priddle, J., 1980. The production ecology of benthic plants in some Antarctic lakes. I. *In situ* productivity studies. J. Ecol. 68: 141–153.

Priscu, J. C., Priscu, L. R., Vincent, W. F., Howard-Williams, C., 1987. Photosynthate distribution by microplankton in permanent ice-covered Antarctic lakes. Limnol. Oceanogr. 32: 260–270.

Ragotzkie, R. A. and Likens, G. E., 1964. The heat balance of two Antarctic lakes. Limnol. Oceanogr. 9: 412–425.

Rigler, F. H., 1978. Limnology in the high Arctic: a case history study of Char Lake. Verh. Internat. Verein. Limnol. 20: 127–140.

Rodhe, W., 1956. Can plankton production proceed during winter darkness in subarctic lakes? Verh. Internat. Verein. Limnol. 16: 302–313.

Roulet, N. T. and Adams, W. P., 1984. Illustration of the spatial variability of light entering a lake using an empirical model. Hydrobiol. 109: 67–74.

Scott, R. F., 1905. The Voyage of Discovery. Vol. 2, McMillian and Co., London, pp. 214–215.

Seaburg, K. G.; Kaspar, M.; Parker, B. G., 1983. Photosynthetic quantum efficiencies of phytoplankton from perennially ice covered Antarctic lakes. J. Phycol. 19: 446–452.

—, Wharton, R. A., Jr., Parker, B. C., Simmons, G. M., Jr., 1982. Temperature-growth responses of algal isolates from Antarctic oasis lakes. J. Phycol. 17: 353–360.

Simmons, G. M., Jr., Parker, B. C., Allnutt, F. T. C., Brown, D., Cathey, D., Seaburg, K. G., 1979. Ecological comparisons of oasis lakes and soils. Antarct. J. US 14: 181–183.

—, Wharton, R. A., Jr., Parker, B. C., Anderson, D., 1983. Chlorophyll *a* and adenosine triphosphate levels in Antarctic and temperate lake sediments. Microb. Ecol. 9: 123–135.

—, —, McKay, C. P., Nedell, S., Clow, G., 1986. Sand/ice interactions and sediment deposition in perennially ice-covered Antarctic lakes. Antarct. J. US. 21: 217–220.

—, and Netherton, J., 1987. Groundwater discharge in a deep coral reef habitat-evidence for a new biogeochemical cycle? Proc. Amer. Acad. Underwater Sci. *In press.*

Solopov, A. V., 1967. Oases in Antarctica (Russian Trans.). National Science Foundation, Washington. 46 pp.

Thompson, D. G., Craig, R. M. F., Bromley, A. M., 1971. Climate and surface heat balance in an Antarctic Dry Valley. N.Z. J. Geophys. 14: 245–251.

Tominaga, H., 1977. Photosynthetic nature and primary productivity of Antarctic freshwater phytoplankton. Japan J. Limnol. 38: 122–130.

Vincent, W. F., 1981. Production strategies in Antarctic inland waters: Phytoplankton ecophysiology in a permanently ice covered lake. Ecol. 62: 1215–1224.

—, Vincent, C. L., 1982. Factors controlling phytoplankton production in Lake Vanda (77° S). Can. J. Fish. Aquat. Sci. 39: 1602–1609.

— and Howard-Williams, C., 1985. Ecosystem properties of dry valley lakes. N.Z. Ant. Rec., Vol. 6, Special Suppl. pp. 11–20.

Walter, M. R. and Bauld, J., 1983. The association of sulphate evaporites, stromatolitic carbonates and glacial sediments: examples from the proterozoic of Australia and the Cainozoic of Antarctica. Precambrian Res. 21: 129–148.

Wharton, Jr., R. A., Parker, B. C., Simmons, G. M., Jr., Seaburg, K. G., Love, F. G., 1982. Biogenic calcite structures forming in Lake Fryxell, Antarctica. Nature 295: 403–405.

—, —, —, 1983. Distribution, species composition, and morphology of algal mats in Antarctic dry valley lakes. Phycologia 22: 355–365.

—, McKay, C. P., Simmons, G. M., Jr., and Parker, B. C., 1986. Oxygen budget of a perennially ice covered Antarctic lake. Limnol. Oceanogr. 31: 437–443.

—, —, Mancinelli, R. L., and Simmons, G. M., Jr., 1987. Perennial N_2, supersaturation in an Antarctic lake. Nature. 325: 343–345.

Wilson, A. T., 1979. Geochemical problems of the Antarctic dry area. Nature 280: 205–208.

Geographical Index

Map Index

www.ingramcontent.com/pod-product-compliance
Lightning Source LLC
Chambersburg PA
CBHW081213180926
51613CB00003B/226
* 9 7 8 9 0 6 1 9 3 6 6 0 2 *